Biopesticides in Organic Farming

Biopesticides in Organic Farming

Recent Advances

Edited by
L.P. Awasthi

CRC Press
Taylor & Francis Group
Boca Raton London New York

CRC Press is an imprint of the
Taylor & Francis Group, an **informa** business

First edition published 2021
by CRC Press
6000 Broken Sound Parkway NW, Suite 300, Boca Raton, FL 33487-2742

and by CRC Press
2 Park Square, Milton Park, Abingdon, Oxon, OX14 4RN

CRC Press is an imprint of Taylor & Francis Group, LLC

Library of Congress Cataloging-in-Publication Data

Names: Awasthi, L.P., editor.
Title: Biopesticides in organic farming : recent advances / L.P. Awasthi.
Description: First edition. | Boca Raton, FL : CRC Press, 2021. | Includes bibliographical references and index.
Identifiers: LCCN 2020041032 | ISBN 9780367460174 (hardback) | ISBN 9781003027690 (ebook)
Subjects: LCSH: Natural pesticides. | Organic farming. | Crops--Diseases and pests--Biological control.
Classification: LCC SB951.145.N37 B567 2021 | DDC 632/.95--dc23
LC record available at https://lccn.loc.gov/2020041032

ISBN: 9780367460174 (hbk)
ISBN: 9781003027690 (ebk)

Typeset in Times
by Deanta Global Publishing Services, Chennai, India

This book is dedicated to my most respected Guru

Professor H.N. Verma

Vice Chancellor

Jaipur National University, Jaipur, Rajasthan, India

Through your smiles and stern looks and subtle remarks to do better, I have sailed through life. When I am tempted to give up, I remember your words, urging me to give it another try. Thank you for making me a stronger person.

Professor L.P. Awasthi

Contents

Section VII Role of Nutrients in the Management of Crop Diseases through Biopesticides

Section VIII Management of Crop Diseases through Bacterial Biopesticides

Section IX Management of Crop Diseases through Fungal Biopesticides

Section X Management of Crop Diseases through Viral Biopesticides

Section XI Management of Crop Diseases through Algal Biopesticides

Section XII Management of the Diseases and Pests of Crops through Protozoan Biopesticides

Section XX Biotechnological Trends in Insect Pest Control Strategy

Section XXI Challenges in the Popularization of Biopesticides in Organic Farming

Section XXII Certification Process and Standards of Organic Farming

Section XXIII Marketing and Export Potential of Organic Products

Foreword

Dr. Richard Alan Humber
USDA-ARS Collection of Entomopathogenic Fungal Cultures (ARSEF)
Emerging Pests and Pathogens Research Unit
Robert W. Holley Center for Agriculture and Health

Virtually all agriculture could have been seen as "organic" until the dawn of using artificial chemical agents to control pests and the rise of industrial-scale agriculture to provide the food and other products required by the exploding human population. As the human population exceeded the planet's presumed carrying capacity, agriculture's capacity to produce an adequate food supply has become deeply stressed. As pesticide and herbicide use burgeoned, the target pests' resistance to chemical control agents became an urgent problem, and the cost of such chemical inputs to agriculture was higher than many growers could easily bear. The current trend to insert pesticidal or herbicidal capacities into the genomes of crop plants has bound those adopting such technologies to buy seed from their corporate providers and to implement complex strategies to delay or to prevent the pests' resistance to these introduced genes. Such issues as the costs of using genetically modified crops (corn, cotton, etc.) met strong resistance from many regulatory agencies and by much of the world's people who neither like nor willingly consume such plants.

The idea of using microbes (at first mostly fungal entomopathogens) as biological control agents emerged in the late 1800s but suffered from many setbacks in the earliest uses against target pests. In addition to the vast number of predatory and parasitic arthropods and parasitoids used to control pests, the range of biological control agents used against insects, mites and other invertebrate pests, plant diseases, and weeds has expanded to include viruses, bacteria, fungi (and organisms once treated as fungi), microsporidia, protozoans, and entomopathogenic nematodes. It is equally important to recognize the contributions of plant extracts for pest control as well as the benefits of co-planting crops with other plants that protect a crop against nematodes, other pests, and soil-borne diseases.

The science supporting the uses of so many organisms is complex and relies on a myriad of technologies from molecular biology and genomics through every imaginable aspect of organismal biology, population biology, and ecology. The screening of global biodiversity for useful microbial biocontrol agents is reasonably well documented even though much of the planet has never been surveyed adequately for such agents. While many entomologists prefer to restrict the term "biological control" to using predatory or parasitic insects and parasitoids against other arthropods, the strategies and principles to use insects against insects apply equally to the uses of microbes, nematodes, and other invertebrate agents.

The rigors of selecting the most virulent microbial agents to bring through efficacy and safety testing are well documented. After an initial focus on basic biological aspects of biopesticide development and efficacy, the emphasis usually switches to the technical problems of scaling up their production and how to formulate and apply these agents, and on the legal hurdles of their regulatory registrations. It is not surprising that, after all of this is done, very few agents become available for practical use—fungi such as *Beauveria bassiana* and *Metarhizium anisopliae*, bacteria such as *Bacillus thuringiensis* and the nematode-borne species of *Photorhabdus* and *Xenorhabdus*, a very small number of viruses, and even fewer microsporidia such as species of *Nosema* and *Vairimorpha*. As the scale of applications of biopesticides and plant-derived pest control agents grows, the full range of skills enabling their adoption needs to be better appreciated; this book surveys on the current state of the art for all of these topics.

Today we have come to the point when the oldest, most traditional approach to agriculture returns to an honored place in many countries. Organic agriculture now presses forward with a new level of biological and agronomic knowledge, and a previously unimaginable array of techniques and technologies. The accelerating adoption of organic agriculture is responding to urgent concerns about the environmental and health costs of continuing use of chemical pesticides. This volume addresses many daunting challenges posed by this push to reduce chemical pest control approaches while embracing integrated pest management (combining biopesticides with limited use of chemical pesticides) or strictly organic practices.

A very diverse set of reviews of the development, application, and environmental and regulatory consequences of implementing biopesticides appear here. Its 24 sections comprise 54 chapters by authorities from around the world. Entomopathogenic viruses, bacteria, fungi, and nematodes are covered in detail, while algae and protozoans that are usually little recognized to have any role in pest management are also covered, and there are three chapters about insect parasitoids as key natural enemies. Other chapters treat the diverse spectrum of plant-derived compounds and insect pheromones that offer positive benefits for managing plant pests and diseases. This compendium will be a key reference for scientific professionals, students, legislators and administrators, the leaders and followers in businesses and in agriculture, farmers, and entrepreneurs wanting to provide safe and healthy food and plant products. Readers should be able to gather new ideas, fresh perspectives, and inspiration to advance their knowledge with the ultimate goal of improving the safety and sustainability of global agriculture.

Biopesticides in Organic Agriculture: Recent Advances, edited by Professor L.P. Awasthi, provides a kaleidoscopic perspective on the adoption, adaption, and expansion of biological and biorational approaches to maintain the agricultural and silvicultural world. If seen from a wholly different perspective, this volume contributes significantly to our increasingly urgent need and responsibility to preserve and, indeed, to cherish the world of plants on which all other forms of life on this planet are so totally dependent.

Richard A. Humber

Preface

Organic agriculture has emerged as an important priority area globally in view of the growing demand for safe and healthy food and concerns about environmental pollution associated with the indiscriminate use of agrochemicals. Organic farming is the pathway that leads to life in harmony with nature. It is key to sound development and sustainable environment, makes use of non-conventional natural resources, and conserves soil fertility through implementation of appropriate conservation practices.

Biopesticides play a significant role in sustainable organic farming, which is a present-day need for better human health. Biopesticides are developed from naturally occurring living organisms such as animals, plants, and microorganisms (e.g., bacteria, fungi, and viruses) that can control serious plant-damaging insect pests by their non-toxic ecofriendly mode of actions, therefore globally demanding and paying attention. Biopesticides and their by-products are mainly utilized for the management of pests that are injurious to plants. As costs of using synthetic chemicals became apparent, there has been a resurgence in academic and industrial research of biopesticides. Biopesticides, including entomopathogenic viruses, bacteria, fungi, nematodes, and plant secondary metabolites, are gaining stature as they are alternatives to chemical pesticides and are a major component of many pest control programs. The virulence of various biopesticides such as nuclear polyhydrosis virus (NPV), bacteria, and plant product were tested very successfully and evaluated under field conditions with major success. Biopesticides are effective, ecofriendly, biodegradable, and do not leave any harmful residue in the environment. Various types of pesticides like bacterial pesticides, viral pesticides, botanicals, pheromones, predators, and parasitoids of biological origin are widely used as biopesticides.

Biopesticides in Organic Farming: Recent Advances describes critically reviewed, key aspects of organic agriculture and provides a unique and timely science-based resource for researchers, teachers, extension workers, students, primary producers, and others around the world. There are different sections in this book. The first section provides an overview of organic farming with special reference to biopesticides, followed by the principles of the applications of biopesticides in organic farming, impact of environmental factors on biopesticides in organic farming, opportunities and challenges in the application of biopesticides in organic farming, strategies for crop protection with biopesticides in organic farming, pesticide exposure impacts on health and need of biopesticides in organic farming, and the role of nutrients in the management of crop diseases through biopesticides. The next section deals with the management of various crop diseases through bacterial, fungal, viral, algal, protozoan, and botanical biopesticides, insect sex hormones, natural enemies, parasitoids biopesticides, microbial biopesticides, integrated pest management, biopesticides weapons against agricultural mite pests, biotechnological trends in insect pests control strategy, challenges in the popularization of biopesticides in organic farming, the certification process, and standards of organic farming and marketing and export potential of organic products. The book gathers together a range of specialists with direct experience over many years of biopesticides in organic farming. This book is intended to be a unique and indispensable resource that offers a diverse range of valuable information and perspectives on biopesticides in organic agriculture at a time when the world community is increasingly aware of the problems of our current agricultural practices with importance of creating sustainable agricultural systems for the long-term health of the biosphere as a whole. This book is designed, considering the requirements of undergraduates, postgraduates, researchers, and university professors, as per the organic agriculture course curriculum of different universities. It has chapters, on each and every aspect related to biopesticides in organic agriculture, compiled by researchers and eminent professors at various universities across the globe. I wish the students and various readers who relish working in the field of organic farming will find the format of the book, its level, and the quantity of information contained in the book to be appropriate for easier learning and understanding. The wide spectrum of information in various chapters with the addition of the terms related to organic agriculture and concept statements is presented in a very concise manner.

I am confident that this publication will be useful to university students, professors, researchers, development department officials, extension workers, policymakers, and all those interested in organic farming.

Professor L.P. Awasthi

Acknowledgments

I would like to express my heartfelt thanks to Prof. H.N. Verma, Vice Chancellor, India, Prof. Ahmad Hadidi, USDA, USA, Prof. Rakesh Yamdagni, Ex Vice chancellor, Prof. S M Ilyas, Ex Vice chancellor, Prof A.N. Mukhopadhyay, Ex Vice chancellor and Prof. Narayan Rishi, Director, Amity University, India, Dr. S.K. Dara and Dr. Matthew Russell Gates, USA for their valuable and fruitful suggestions during the preparation of this book.

I wish to acknowledge and express my sincere thanks to my family members for their untiring support, cooperation and encouragement during the course of the compilation of this book.

Grateful thanks are due to all learned contributors for their cooperation in compiling useful information on different aspects of biopesticides in organic farming. Each of them have endeavored to present an update of their specialized aspect; Drs. M. Prakruthi and M.S. Mahesh, India, Surendra K. Dara, USA, Enespa, Prem Chandra, Rishabh Chitranshi, Sushree Suparna Mahapatra, Sudeepta Pattanayak, Siddhartha Das, L. P. Awasthi, C. R. Patil, and Shekarappa, India, Muhammad Haroon Sarwar, Muhammad Farhan Sarwar, Muhammad Sarwar and Muhammad Taimoor Khalid, Pakistan, Vittal Navi, Santosh G., India, Gabriela Cristina Alles, Diouneia Lisiane Berlitz, Maximiano Cassal and Lidia Mariana Fiuza, Federal Institute of Education, Science and Technology, RS, Brazil, Vilmar Machado, Spain, Subramaniam Gopalakrishnan, Vadlamudi Srinivas, Pratyusha Sambangi, and Sravani Ankati, India, Younes Rezaee Danesh, Iran, Semra Demir, Çağlar Sagun, Solmaz Najafi, Turkey, Jéssica Batista Torres Araújo Oliveira, Cassia Renata Pinheiro and Glacy Jaqueline Da Silva, Brazil, Pradipta Banerjee, Pratibha Sharma, Raja Manokaran and Prashant Prakash Jambhulkar, India, Roohi Aslam, Pakistan, Meenakshi Devi, Suman Devi, Neha Upadhyay, Babli, Upasana Mohapatra, Gayatri Biswal, Deeksha Joshi, Monika Upadhyay, Raghvendra Tiwari and B. Meena, India, ASOGWA Evarestus Uche and Theophilus Chinyere Nkasiobi Ndubuaku, Nigeria, D.K. Kulkarni, R.B. Bhagat, P.V. Patil and S P Taware, India, Naveed Akhter Shad, Riffat Batool and Sidra Sarwar, Pakistan, Desam Nagarjuna Reddy, India, Abdul Jabbar Al-Rajab, Canada, Rajeev Kumar, India, Mara Tabakovic-Tosic, Serbia, Sathish Kota, Jaba Jagdish, Ramya Sree, Reddy, Vinod Kukanur, D. K. Rana and Naveen Arora, India, Simone Mundstock Jahnke and Gisele De Souza Da Silva, Brazil, Hany Mohamad Galal EL-Kawas, Mohamad Mohamad Ahamad Khedr, Egypt, Aparna Shree Singh, India, Fernando Belezini Vinha, Alexandre De Sene Pinto, Kássia Cristina Freire Zilch, Gisele De Souza Da Silva, Cleder Pezzini, Roberta Agostini Rohr, Brazil, Rafaela Cristina dos Santos SP; Brazil, María Alejandra Correa, Massiel Pinto, Oscar Valbuena, Marcelo Molinatti, Domenico Pavone, Venezuela, Jatin. K, and Suraj Prasad Mishra, India., Katherine Girón, Brazil, Matthew Russell Gates, USA, R. Elaini, and R. Bouharroud, Morocco, Noreen Akram and Hafiz Sanaullah Babar, Pakistan, Bugiani Riccardo and Bariselli Massimo, Italy, R. Shekhara Naik, India and Meena Thakur, New Zealand.

I am sure, detailed accounts on different aspects of biopesticides in organic farming will be a great help to students, teachers, researchers and extension workers.

Last but not the least; I would like to extend my thanks to Taylor and Francis Publishers, U.S.A, their publishing team Dr. Renu Upadhyay, Commissioning Editor (Chemical and Life Sciences), CRC Press, Taylor & Francis Group, Dr. Jyotsna Jangra, Editorial Assistant, CRC Press, Taylor and Francis Group, Marsha Hecht, Project Editor, Taylor & Francis Group, LLC and Rennie Alphonsa, Project Manager at Deanta Global along with their team for getting this book published in a presentable form.

Professor L.P. Awasthi

About the Editor

Prof. (Dr.) L.P. Awasthi, Dean School of Agriculture, R N B Global University, Bikaner, India, formerly Research Scientist/Professor, Amity University and Professor and Head, Department of Plant Pathology, N.D. University of Agriculture & Technology, Faizabad (U.P.) India, is a distinguished agriculture scientist with a strong background in molecular plant virology, resistance breeding, IDM, IPM, organic agriculture and mushroom production with proven ability to undertake independent or collaborative scientific studies as demonstrated by successfully completing 28 mega research projects, funded by various International funding organizations. His professional experience includes more than 47 years of teaching and conducting research. He has published more than 350 research papers in foreign and Indian journals of repute, edited/ authored six books, published a number of popular articles, laboratory manuals and contributory chapters for different books, guided 77 M. Sc. (Ag.) and 35 Ph.D. students and has been working as an Editor/ Referee in the editorial boards of many Indian and foreign journals of repute.

Professor Awasthi visited Karl-Marx University, Leipzig, University of Berlin, Institute of Plant Pathology, Ascherslavan, and Haale University, Germany as a visiting professor for higher studies.

He has been member of the Phytosanitary Certificate Issuing Authorities, Ministry of Agriculture, Department of Agriculture and cooperation, Directorate of Plant Protection, Quarantine and Storage, Government of India and is presently a special invited member in the State-level Pest and Disease Surveillance Advisory Unit, Government of Uttar Pradesh. India.

Prof. Awasthi has long been associated with many professional societies, and was conferred Plant Pathology Leadership Award 2012, 2016 and 2019 by Indian Phytopathological Society, India, Outstanding Virologist award (2007) by International Virological Association for outstanding contribution, scientific excellence and distinguished services for the cause of Plant Pathological Research, Education and Technology dissemination, which has impacted science of plant pathology in the country. He has been recognized for disseminating mushroom cultivation technologies to improve the socioeconomic conditions of farmers below poverty line and woman empowerment and also for the popularization of organic farming.

Contributors

Rohr Roberta Agostini
Faculty of Agronomy
Federal Univ of Rio Grande do Sul (UFRGS)
Porto Alegre, Rio Grande do Sul, Brasil

Shad Naveed Akhter
National Institute for Biotechnology & Genetic Engineering (NIBGE)
Faisalabad, Pakistan

Noreen Akram
National Institute for Biotechnology and Genetic Engineering (NIBGE)
Faisalabad, Pakistan

Correa Maria Alejandra
Centro de Biotecnología Aplicada, Departamento de Biología, Facultad de Ciencias y Tecnología
Universidad de Carabobo
Venezuela

Gabriela Cristina Alles
Biotech Biological Defense
RS, Brazil

Abdul Jabbar Al-Rajab
Pesticides & Pest Management
Environmental Pollution & Control
Ottawa, Ontario, Canada

Jéssica Batista Torres Araújo Oliveira
Biological Control/*Bacillus thuringiensis*
Sete Lagoas
Minas Gerais, Brazil

Roohi Aslam
NUTECH School of Applied Sciences and Humanities (NUSASH)
National University of Technology (NUTECH)
Islamabad, Pakistan

L.P. Awasthi
School of Agriculture
RNB Global University, Bikaner
Rajasthan, India

Li Bab
Department of Agronomy
CCS Haryana Agricultural University
Hisar, India

Pradipta Banerjee
Department of Biochemistry & Plant Physiology
Centurion University of Technology and Management
MS Swaminathan School of Agriculture
Odisha, India

Massimo Bariselli
Servizio Fitosanitario – Plant Protection Service
Crop Protection Regione Emilia-Romagna
Bologna, Italy

Diouneia Lisiane Berlitz
Science and Technology
Federal Institute of Education
RS, Brazil

R.B. Bhagat
Anantrao Pawar College
Pirangut, Tal. Mulshi
Pune, Bhor

Gayatri Biswal
Department of Plant Pathology
College of Agriculture
Orissa University of Agriculture and Technology
Bhubaneswar, India

Sagun Çağlar
Soil, Fertilizer and Water Resources
Central Research Institute
Ankara-Turkey

Maximiano Correa Cassal
Universidade do Vale do Rio dos Sinos
RS, Brazil

Prem Chandra
School for Environmental Science
B.B.A. University
Lucknow, India

Rishabh Chitranshi
School for Environmental Science
B.B.A. University
Lucknow, India

Pezzini Cleder
Faculty of Agronomy
Federal Univ of Rio Grande do Sul (UFRGS)
Porto Alegre, Rio Grande do Sul, Brasil

Younes Rezaee Danesh
Urmia University, Iran
Parlakhemundi, India

Surendra K. Dara
Entomology and Biologicals Advisor at University of California Cooperative Extension
San Luis Obispo, California

Siddhartha Das
Department of Plant Pathology
Centurion University of Technology and Management
MS Swaminathan School of Agriculture
Parlakhemundi, India

Nagarjuna Reddy Desam
Anantha Lakshmi Institute of Technology & Sciences, Itikalapalli
Ananthapuramu, Andhra Pradesh, India

Pinto Alexandre De Sene
Department ofAgronomy
Moura Lacerda University
Ribeirão Preto, São Paulo, Brazil

Da Silva Gisele De Souza
Faculty of Agronomy
Federal Univ of Rio Grande do Sul (UFRGS)
Porto Alegre, Rio Grande do Sul, Brasil

Meenakshi Devi
Entomology
PDM University
Bahadurgarh, India

Suman Devi
Department of Agronomy
CCS Haryana Agricultural University
Hisar, India

Rafaela Cristina dos Santos
Entomology, Chemical Ecology and Insect Behavior Laboratory
Entomology and Acarology Department – ESALQ/USP
Piracicaba, Brazil

Hany Mohamad Galal El-Kawas
Institute of Plant Protection Research
Agricultural Research Center
Egypt

Lidia Mariana Fiuza
Department of Microbiology and Agrotoxicology
Riograndense Institute of Rice
RS, Brazil

Zilch Cristina Kassia Freire
Faculty of Agronomy
Federal Univ of Rio Grande do Sul (UFRGS)
Porto Alegre, Rio Grande do Sul, Brasil

Matthew Russell Gates
Cannabis Horticultural Association
San Diego, California

Subramaniam Gopalakrishnan
International Crops Research Institute for the Semi-Arid Tropics (ICRISAT)
Patancheru, India

Jagdish Jaba
Department of Entomology
Integrated crop managment
International Crops Research Institute for the Semi-Arid Tropics (ICRISAT)
Patancheru, Hyderabad, India

Simone Mundstock Jahnke
Depto. de Fitossanidade
Universidade Federal do Rio Grande do Sul (UFRGS)
Brazil

Da Silva Glacy Jaqueline
Paranaense University, Umuarama
Paraná, Brazil

Deeksha Joshi
Indian Institute of Sugarcane Research
Lucknow, India

D.K. Kulkarni
BAIF Development Research Foundation
Pune, India

Rajeev Kumar
DPPQS, RCIPMC
Lucknow, India

Vilmar Machado
Department of Zoologia y Antropologia Fisica
Universidade de Murcia
Spain

Sushree Suparna Mahapatra
Department of Plant Pathology
Odisha University of Agriculture and Technology
Bhubaneswar, India

M.S. Mahesh
Department of Food Science and Nutrition
Yuvaraja's College (Autonomous)
University of Mysore
Mysuru, India

Raja Manokaran
Department of Plant Pathology
Rani Lakshmi Bai Central Agricultural University
Jhansi, India

Molinatti Marcelo
Centro de Biotecnología Aplicada, Departamento de Biología, Facultad de Ciencias y Tecnología
Universidad de Carabobo
Venezuela

Pinto Massiel
Centro de Biotecnología Aplicada, Departamento de Biología, Facultad de Ciencias y Tecnología
Universidad de Carabobo
Venezuela

B. Meena
Plant Pathology
Regional Research Station
Vriddhachalam, India

Khedr Mohamad Mohamad Ahamad
Plant Protection Research Institute
Agricultural Research Center
Dokki, Giza, Egypt

Upasana Mohapatra
Department of Biotechnology
University of Agricultural Sciences
GKVK
Bengalru, India

Zahid Mukhtar
National Institute for Biotechnology & Genetic Engineering (NIBGE)
Faisalabad, Pakistan

Sarwar Muhammad
National Institute for Biotechnology & Genetic Engineering (NIBGE)
Faisalabad, Pakistan

Shekhara Naik R.
Department of Food Science and Nutrition
Yuvaraja's College
University of Mysore
Mysuru, India

Arora Naveen
Punjab Agricultural University
Ludhiana, Punjab

Vithal Navi
Department of Agricultural Microbiology
University of Agricultural Sciences
Dharwad, India

Theophilus Chinyere Nkasiobi Ndubuaku
Entomology Section
Cocoa Research Institute of Nigeria (CRIN)
Ibadan, Nigeria

Upadhyay Neha
Department of Plant Pathology
G. B. Pant University of Agriculture and Technology
Pantnagar, Uttarakhand, India

Valbuena Oscar
Centro de Biotecnología Aplicada, Departamento de Biología, Facultad de Ciencias y Tecnología
Universidad de Carabobo
Venezuela

C.R. Patil
Department of Agricultural Microbiology
Institute of Organic Farming
University of Agricultural Sciences
Dharwad, India

P.V. Patil
Department of Horticulture
Anantrao Thopte College
Bhor, Maharashtra, India

Sudeepta Pattanayak
Department of Plant Pathology
Centurion University of Technology and Management
MS Swaminathan School of Agriculture
Parlakhemundi, India

Domenico Pavone
Tecnovita
Valencia, Venezuela

Katherine Girón Pérez
Entomóloga Manejo de pragas agrícolas
Resistência de insetos
Milho, Cana e Soja
Paulínia, Brazil

Cassia Renata Pinheiro
São Paulo University, Piracicaba
São Paulo, Brazil

Jambhulkar Prashant Prakash
Department of Plant Pathology
Rani Lakshmi Bai Central Agricultural University
Jhansi, Uttar Pradesh

M. Prakruthi
Central Food Technological Research Institute
Mysuru, India

Mishra Suraj Prasad
International Crops Research Institute for the Semi-Arid Tropics-Patancheru
Telangana, India

Sambangi Pratyusha
ICRISAT
Patancheru, Telangana, India

Elaini Rachid
Omnium Agricole du Souss
Agadir Ida-Outanane Prefecture, Morocco

D.K. Rana
Indira Gandhi Krishi Vishwavidyalaya
Raipur, India

Ramya Sree Reddy
International Crops Research Institute for the Semi-Arid Tropics (ICRISAT)
Patancheru, Telangana, India

Bugiani Riccardo
Plant Protection Service
Emilia-Romagna Region
Via A. da Formigine 3, Italy

and

Servizio Fitosanitario – Plant Protection Service
Crop Protection Regione Emilia-Romagna
Bologna, Italy

Batool Riffat
National Institute for Biotechnology & Genetic Engineering (NIBGE)
Faisalabad, Pakistan

Babar Hafiz Sanaullah
Drug Regulatory Authority of Pakistan
Ministry of National Health Services, Regulations and Coordination
Islamabad, Pakistan

G.P. Santosh
Department of Agricultural Microbiology
University of Agricultural Sciences
Dharwad

Muhammad Farhan Sarwar
Allied Health Sciences
University of Sargodha
Sargodha, Pakistan

Muhammad Haroon Sarwar
King Edward Medical University
Lahore, Punjab, Pakistan

Muhammad Sarwar
National Institute for Biotechnology & Genetic Engineerings (NIBGE)
Faisalabad, Pakistan

Sidra Sarwar
Department of Botany, Government Post Graduate College
Faisalabad, Pakistan

Kota Sathish
International Crops Research Institute for the Semi-Arid Tropics (ICRISAT)
Patancheru, Telangana, India

Demir Semra
Department of Plant Protection, Faculty of Agriculture
Van Yuzuncu Yil University
Van-Turkey

Pratibha Sharma
Division of Plant Pathology
SKN Agricultural University
Jobner, Jaipur, India

Shekarappa
Department of Agricultural Entomology
Institute of Organic Farming
University of Agricultural Sciences
Dharwad, India

Aparna Shree Singh
Lakshya Society for Social and Environmental Development
New Delhi, India

Enespa Singh
School of Agriculture, SMPDC
University of Lucknow
Lucknow, India

Ranjan Singh
School for Environmental Science
B.B.A. University
Lucknow, India

Najafi Solmaz
Department of Field Crops, Faculty of Agriculture
Van Yuzuncu Yil University
Van-Turkey

Ankati Sravani
ICRISAT
Patancheru, Telangana, India

Vadlamudi Srinivas
International Crops Research Institute for the Semi-Arid Tropics (ICRISAT)
Patancheru, India

Mara Tabakovic-Tosic
Department of Forest Protection
Institute of Forestry
Belgrade, Serbia

Khalid Muhammad Taimoor
Allied Health Sciences
University of Sargodha
Sargodha, Pakistan

S.P. Taware
Plant Science Division
Agharkar Research Institute
Pune

Meena Thakur
Biologist R&D
Renovo Technologies Ltd.
New Plymouth, New Zealand

Raghvendra Tiwari
Indian Institute of Sugarcane Research
Lucknow, India

ASOGWA Evarestus Uche
Entomology Section
Cocoa Research Institute of Nigeria (CRIN)
Ibadan, Nigeria

Monika Upadhyay
Indian Institute of Sugarcane Research
Lucknow, India

Fernando Belezini Vinha
Engenheiro Agronômo
Piracicaba, Brazil

Kukanur Vinod
International Crops Research Institute for the Semi-Arid Tropics (ICRISAT)
Patancheru, Telangana, India

Mukhtar Zahid
National Institute for Biotechnology & Genetic Engineering (NIBGE)
Faisalabad, Pakistan

Section I

Overview of Organic Farming with Special Reference to Biopesticides

1

Overview of Organic Farming with Special Reference to Biopesticides

M. Prakruthi and M.S. Mahesh

CONTENTS

1.1 Introduction

Human lifestyle and well-being widely depend on one's health, which in turn depends upon the quality of the food consumed. Food should be free from chemical fertilizers and pesticides. Organic farming is one of the best ways to produce safe and suitable food for consumption without harming the environment.

Organic farming is a holistic production management system that promotes and enhances agroecosystem health, including biodiversity, biological cycles, and soil biological activity. It uses the management practices considering the regional conditions that require locally adapted systems in preference to the use of off-farm inputs, which is accomplished by employing the biological, agronomic, and mechanical methods wherever possible, instead of synthetic materials, to fulfill any specific function within the system (FAO 1999).

The term *organic farming* was first used by Lord Walter James Northbourne in describing the concept of "the farm as organism" in his book *Look to the Land* (1940). Influenced by Sir Albert Howard's work, Lady Eve Balfour first did a scientific side-by-side comparison of organic and conventional farming (Pratap and Vaidya 2009). The concept of organic farming originated with an establishment of the International Federation of Organic Agriculture Movement (IFOAM) on November 6, 1972, in France. Today, organic food is a growing reality all over the world.

According to the FiBL-IFOAM 2017 survey on certified organic agriculture, nearly 70 million hectares of agricultural land are managed organically. The regions of Oceania (35.9 million hectares or 51 percent of the global organic farmland), Europe (14.6 million hectares or 21 percent), and Latin America (8.0 million hectares or 11 percent) are the largest areas of organically managed agricultural land.

The countries with the largest area of organic agricultural land are Australia (35.7 million hectares), Argentina (3.4 million hectares), and China (3.0 million hectares). In the world arena, the organic agricultural land area increased by 11.7 million hectares, or 20 percent, compared with 2016 (Willer and Lernoud) (Figure 1.1).

Around 2.9 million organic producers were reported in the year 2017, which is 5 percent more than in 2016. Furthermore, India continues to be the country with the highest number of producers (835,200), followed by Uganda (210,352), and Mexico (210,000) (Willer and Lernoud 2009) (Figure 1.2).

1.2 Scope and Objectives

Since the middle of the 19th century to the present time, synthetic pesticides have been an agent for controlling pests. There is no doubt that they have been promising agents for pest control but within more than seven decades of their use, the synthetic

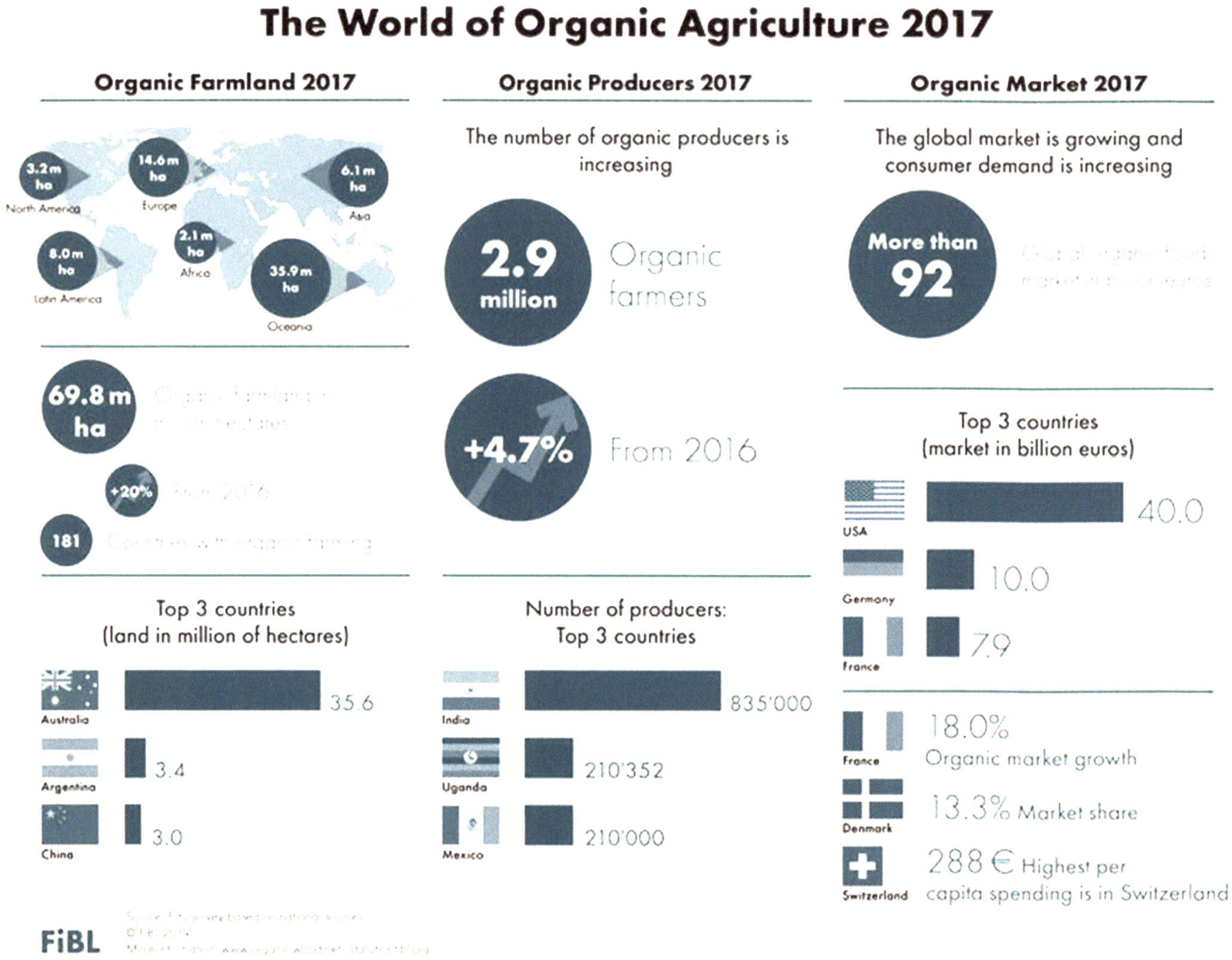

FIGURE 1.1 Organic agriculture worldwide – key indicators. Source: FiBL survey 2019, based on information from private sectors, certifiers, and government.

pesticides have so thoroughly been distributed throughout the animate and inanimate world that they occur virtually everywhere (Dhakal and Singh 2019). The land that used to be productive 50 years ago now shows a declining yield. According to the latest revision of the UN population prospects, world population is projected to grow by 34 percent from 6.8 billion in 2020 to 9.1 billion in 2050. Feeding this increasing population is a great challenge, especially when the productivity of land is declining day by day. Environmental pollution by agrochemical residues is increasing and eroding the natural resource base. Sustainability must be maintained in the production system to feed the burgeoning population of the world. Organic agriculture systems are those that are economically viable and meet society's need for safe and nutritious food while maintaining or enhancing natural resources and the quality of the environment for future generations. It is in harmony with the environment while maintaining buoyancy and dynamism in agricultural growth for meeting basic human needs and conserving natural resources. It aims at producing food that is both nutritious and safe for human health. Because all of the materials are of natural or biological origin, it is very safe to use biopesticides as a potential source of pest control in organic agriculture.

Organic farming promotes the use of crop rotations and encourages balanced host/prey relationships. Nutrients and the organic residues produced on the farm are recycled back to the soil. The goodness of this system helps in ensuring the demerits of modern day agriculture are avoided (Marumegh 2016). Organic farming also has many objectives:

1. To ensure safe and good quality food
2. Ensuring long-term, ecological, system-based organic management
3. Substantiating long-term, biological-based soil fertility
4. Avoiding synthetic inputs at all stages of the organic product chain and exposure of people and the environment to persistent, potentially harmful chemicals
5. Minimizing pollution and degradation of the production/processing unit and surrounding environment from production/processing activities
6. Excluding harmful technologies from the system
7. Avoiding pollution from the surrounding environment
8. Maintaining organic integrity throughout the supply chain
9. Providing organic identity in the supply chain

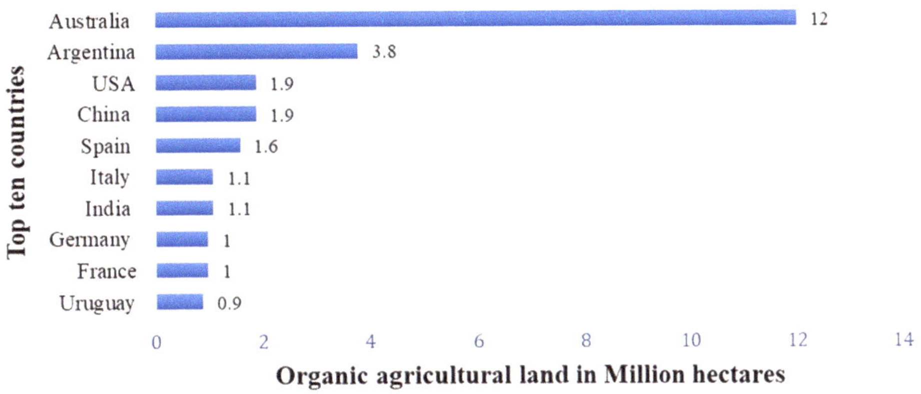

FIGURE 1.2 The ten countries with the largest organic agricultural land.

TABLE 1.1

Comparison between Organic and Inorganic Farming

Organic Farming	Inorganic Farming
Decentralization	Centralization
Independence	Dependence
Community	Competition
Harmony with nature	Domination of nature
Diversity	Specialization
Restraint	Exploitation

1.3 Comparison between Organic and Inorganic Farming

Organic and inorganic agriculture belong to two different paradigms. There is scientific evidence that organic farming fundamentally differs from inorganic farming. Organic farming improves the soil fertility by using biological fertilizers and pesticides to an extent, which cannot be achieved by inorganic farming even if the latter consistently respects some ecological principles (Sharma 2006). The fundamental differences between the two competing agricultural scenarios can be seen in Table 1.1.

1.4 Basic Steps in Organic Farming

There are certain mandatory steps that are need to be followed by a farmer to practice organic farming and also in order to get the certification for the end product as "organic"

1. Conversion of land from conventional management to organic management
2. Management of the entire surrounding system to ensure biodiversity and sustainability of the system
3. Utilization of crop rotation, residue management, organic manures, and biological inputs as alternate sources of nutrients for crop production
4. Employment of management practices, physical and cultural means and by biological control system in order to manage the weeds and pests
5. Maintenance of livestock in tandem with organic concepts to make them an integral part of the entire system

1.5 Components of Organic Farming

Crop rotation, maintenance and enhancement of soil fertility through biological nitrogen fixation, the addition of organic manure and use of soil microorganisms, crop residues, biopesticides, biogas slurry, waste, etc. are the major components of organic farming. Vermiculture has become a major component in biological farming, which is found to be effective in enhancing soil fertility and productivity in a sustainable manner (Figure 1.3).

1.6 Types of Organic Farming

There are various forms and incarnations of organic farming.

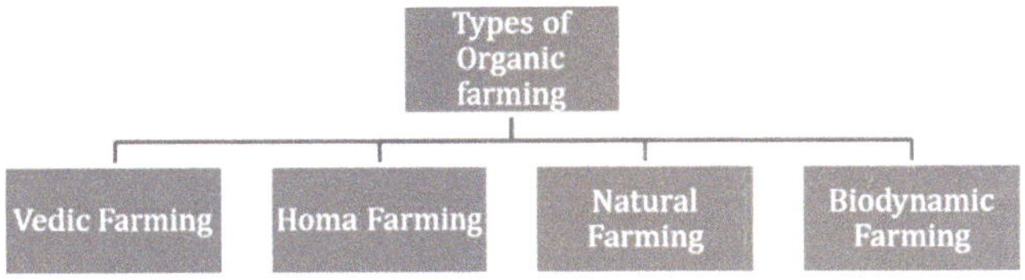

1.6.1 Vedic Farming

Vedic farming is a crop production system found in Indian Vedic literature. Vedic farming clearly depicts the use of natural ingredients on farm organic inputs and also natural/biological and mechanical pest control; it refrains from the use of chemical fertilizers, pesticides, chemical growth regulators, etc. However, natural minerals can be used judiciously as a measure of plant protection, thus resulting in healthy soils, healthy food, healthy environment, and healthy humans and livestock.

1.6.2 Homa Farming

Homa farming is a system of agriculture that may be added to any good organic farming practice. This variant of organic farming involves agnihotra, i.e., burning of whole rice grain mixed with cow ghee (clarified butter) in a copper pyramid of specified

FIGURE 1.3 Components of organic farming. Source: http://agritech.tnau.ac.in/org_farm/orgfarm_introduction.html please quote this reference here.

size, along with chanting of mantras at sunrise and sunset. This practice is said to gather a tremendous amount of energy from the cosmos around the copper pyramid, which goes to the atmosphere along with holy smoke, and exerts dramatic benefits to the crop growth and productivity by way of injecting nutrients into the environment and mitigating the ill-effects of pollution. The ash left after performing agnihotra is touted as complete plant food, although the fact remains that it contains mainly potash.

1.6.3 Natural Farming

Natural farming emphasizes the efficient use of on-farm biological resources and enrichment of soil to ensure high soil biological activity with the usage of jivamruta, which is found to be rich in various beneficial micro-organisms. The use of bijamruta for seed/planting material treatment and jivamruta for soil treatment and foliar spray are important components. Jivamruta has been found to be rich in various beneficial micro organisms.

1.6.4 Biodynamic Farming

Biodynamic farming is a method of farming that aims to treat the farm as a living system that interacts with the environment to build healthy, living soil and to produce food that nourishes and vitalizes and help to develop mankind (Diver 1999). The underlying principle of biodynamic farming is making life-giving compost out of dead material (Proctor and Cole 2012). In this type of organic farming, the chemical fertilizers are totally replaced by microbial (biological) nutrient contributors such as bacteria, algae, fungi, mycorrhiza, and actinomycetes. Furthermore, predators, parasites, and a plethora of natural enemies of pests are employed for biological pest management of crops, which also helps to avoid resorting to chemical pesticides. These agents are introduced into farms through the activities that catalyzed their particular action—composting, green manuring, crop rotations, intercropping, mixed cropping, etc.—as well as through bird perches and trap crops to promote such biological activities (Yadav 2010).

1.7 Merits and Demerits of Organic Farming

Merits	Demerits
Nutritious, poison-free, and tasty food	Lower productivity
Lower growing cost	Requires skill
Enhances soil nourishment	Time-consuming
More energy efficiency	More labor-intensive
Carbon sequestration	Methods are not well established
Less water pollution	
Environment-friendly practices	
Sources of productive labor	

1.8 The Principles of Organic Farming

It is important to understand the guiding principles of organic agriculture in order to understand the motivations of organic farmers, the practices they use, and what they want to achieve (Figure 1.4)

1.8.1 Principle of Health

Health is vital for all living systems. It is not simply about having control over illness; it is vital to develop physical, mental, social, and ecological well-being. Immunity, resilience, and regeneration are key characteristics of health (Meena and Meena 2013).

Organic farming sustains and enhances the health of soil, plant, animal, human, and planet as one and indivisible. The health of individuals and communities from the health of ecosystems are interrelated—healthy soils produce healthy crops that maintain the health of animals and people.

FIGURE 1.4 Principles of organic farming. Source: http://agritech.tnau.ac.in/org_farm/orgfarm_principles.html. Please qoute this here.

The functions of organic farming are to sustain and enhance the health of ecosystems and organisms. In particular, organic farming is meant to produce high-quality, nutritious food that contributes to health care and prosperity. In view of this, it avoids the use of fertilizers, pesticides, animal drugs, and food additives that may have adverse health effects.

1.8.2 Principle of Ecology

Organic farming is based on living ecological systems and cycles, work with them, emulate them, and help sustain them. It is based on ecological processes and recycling. Nourishment and well-being are achieved through the ecology of the specific production environment.

Organic farming helps in balancing ecological cycles. These cycles are universal but their operation is site-specific. This ecological balance is achieved by organic farming through the design of farming systems, establishment of habitats, and maintenance of genetic and agricultural diversity. Organic management is adapted to local conditions, ecology, culture, and scale. Certain methods such as reuse, recycling, and efficient management of materials and energy are employed to reduce the inputs, in order to maintain and improve environmental quality and conserve resources.

The practitioner of organic agriculture should protect as well as benefit the common environment including landscapes, climate, habitats, biodiversity, air, and water.

1.8.3 Principle of Fairness

Organic farming should build on relationships that ensure fairness with regard to the common environment and livelihood opportunities (Nazim Uddin 2015). Fairness is characterized by equity, respect, justice, and stewardship of the shared world, both among people and in their relationships with other living beings.

This principle emphasizes that those involved in organic farming should conduct human relationships in a manner that ensures fairness at all levels and to all parties—farmers, workers, processors, distributors, traders, and consumers (Meena and Meena 2013). Organic farming provides everyone involved with a good quality of life and contributes to food sovereignty and reduction of poverty. It aims to produce good quality food and sufficient supply.

The natural and environmental resources that are used for the production of organic produce should be managed so that it can be held in trust for future generations both socially and ecologically. Fairness demands the systems of production, distribution, and trade that are open, equitable, and that account for real environmental and social costs.

1.8.4 Principle of Care

Organic farming is managed in a precautionary and responsible manner to protect the health and well-being of current and future generations and the environment (IFOAM). Organic farming is a living and dynamic system that responds to internal and external demands and conditions. Practitioners of organic farming can enhance efficiency and increase productivity, but this should not be at the risk of the health and well-being of the agricultural land as well as that of the consumer. Consequently, new technologies need to be imposed and existing methods should be reviewed.

This principle states that prudence and responsibility are the key concerns in management, development, and technology choices in organic farming. Science is necessary to ensure that organic farming is healthy, safe, and ecologically sound. However, scientific knowledge alone is not sufficient. Practical experience, collective wisdom, and traditional and existing knowledge offer valid solutions, tested by time. Organic farming should prevent particular risks by adopting appropriate technologies and rejecting unpredictable ones, such as genetic engineering. Decisions should reflect the values and needs of all who might be affected through transparent and participatory processes (Reed 2010).

1.9 Conclusion and Future Prospects

Statistics on global organic agriculture land shows that 0.9 percent of the land is certified agricultural land, out of which 31 million hectares is farmed with wild organic collections (berries in Finland, honey in Zambia, and medicinal plants in India). There are nearly 1.9 million organic producers in 164 countries (e.g., India 600,000, Uganda 190,000, and Mexico 170,000). Organic agriculture is considered to be safe and organic produce is suitable for consumption, Organic farming also protects and preserves the land and environment. To produce a crop that is said to be organic, there are certain steps and principles that are to be followed. Because of health conditions and lifestyles, consumers are switching to organic products. Thus, organic farming will become more important in the future.

REFERENCES

Dhakal, R. and Singh, D.N. 2019. Bio pesticides: A key to sustainable agriculture. *International Journal of Pure Applied Bioscience*. 7(3): 391–396.

Diver, S. 1999. *Biodynamic Farming & Compost Preparation*. ATTRA.

FAO (Food and Agriculture Organization). 1999. *Guidelines for the Production, Processing, Labelling and Marketing of Organically Produced Foods* (Gl 32 – 1999, Rev. 1 – 2001).

IFOAM (International Federation of Organic Agriculture Movements). 2017. n.d. Available at: https://www.ifoam.bio/en/principles-organic-agriculture/principle-care [Accessed on January 30, 2020].

Marumegh. 2016. Organic farming: Its scope and importance in India. http://www.marumegh.com/index.php/blog/116 [Accessed on January 29, 2020].

Meena, R.P. and Meena, H.P. 2013. Organic Farming: concept and components. *Popular Kheti*. 1(4): 5–14.

Nazim, U.M. 2015. *Status and Future Prospect of Organic Agriculture for Safe Food Security in Bangladesh*. SAARC Agriculture Centre, Dhaka, Bangladesh.

Organic Farming and Sustainable Agriculture. 2007. *Agriculture Today*. 10(3): 48.

Pratap, T. and Vaidya. C.S. 2009. *Organic Farmers Speaks on Economics and Beyond*. Westville Publishing House, New Delhi. p. 5.

Proctor, P. and Cole, G. 2012. *Grasp the Nettle: Making Biodynamic Farming & Gardening Work*. Random House, New York.

Reed, M. 2010. *Rebels for the Soil: The Rise of the Global Organic Food and Farming Movement*. Earthscan, Routledge.

Sharma, A.K. 2006. "A Hand Book of Organic Farming. Agrobios, Agro House, Jodhpur. pp. 33–34.

Willer, H. and Lernoud, J. 2019. *The World of Organic Agriculture - Statistics and Emerging Trends*. International Federation of Organic Agriculture Movements (IFOAM), Bonn, Germany. Available at: http://www.organic-world.net.

Yadav, A.K. 2010. *Organic Agriculture-Concept, Scenario, Principals and Practices*. National Centre of Organic Farming, Ghaziabad. pp. 27–38.

Section II

The Principles of Biopepticide Applications in Organic Farming

2

The Principles of the Application of Biopesticides in Organic Farming

Surendra K. Dara

CONTENTS

2.1 Introduction

Biopesticides are pesticidal materials of biological origin based on botanical extracts, microorganisms, microbial metabolites, or biochemicals that are used for controlling plant pathogens, parasitic nematodes, weeds, or vertebrate or invertebrate pests. Since the use of biopesticides is rapidly increasing in many cropping systems around the world and organic farming depends exclusively on biopesticides, it is important to establish some principles for their application. Regardless of their origin, biopesticides are still pesticides, and they require similar regulatory and safety guidelines and application strategies as chemical pesticides, in order to ensure environmental and human safety and pest control efficacy. This chapter will outline some of the key principles for the application of biopesticides in organic farming.

2.2 Biopesticides in Integrated Pest Management

Integrated pest management (IPM) is a strategy used for managing pests and preventing crop losses in crop production in an economically viable, socially acceptable, and environmentally sustainable manner (Dara 2019). IPM encourages the use of cultural and other management options before using pesticides for environmental and human safety, reducing the risk of resistance development, and preventing secondary pest outbreaks. The use of biopesticides must also follow the same guidelines to avoid the associated risks. Similar to chemical pesticides, biopesticides can also impart selection pressure on pest populations leading to resistance (Dara 2017). While pest resistance to infectious organisms such as entomopathogenic fungi could be relatively slower, resistance to biochemical molecules or infections where toxic microbial metabolites are involved can be faster. Since organic farming has limited biopesticide choices in comparison to conventional farming, which has several groups of pesticides to choose from, there are generally higher chances for the excessive use of biopesticides or repeated use of the same kind of biopesticide in organic farming. To maintain the efficacy of available biopesticides and minimize resistance issues, they should be judiciously used. When necessary, biopesticides should be used in combination or rotation with other biopesticides or pest management tactics.

2.3 Understanding Mode of Action

Understanding the modes of action of biopesticides and using the appropriate one for the appropriate pest, its life stage, and the crop is very important for ensuring the biopesticide's success. For example, azadirachtin, a limonoid compound from the neem tree (*Azadirachta indica*), which acts as an insecticide, insect growth regulator, insect repellent, antifeedant, and a fungicide among others (Mordue 2004; Mondali et al. 2009). Several essential oils contain monoterpene compounds, which are cytotoxic or neurotoxic (Tripathi et al. 2009). Insecticidal oils, including petroleum, plant, and fish oils, disrupt gas exchange, and cause mortality by suffocation (Bográn et al. 2006). Some horticultural oils are also used as fungicides. Neem oil, which has a small amount of azadirachtin, is an insecticide, acaricide, and fungicide, which derives its activity more from the oil than the azadirachtin. Some confuse azadirachtin and neem oil because they are derived from the same tree, but their modes of action are different. In the case of entomopathogens, bacteria, fungi, nematodes, and viruses have unique modes of action (Lacey 2017). Bacteria and viruses need to be ingested by the target pest, fungi infect by contact, and nematodes actively search and infect their hosts. Viruses are very specific to some insect species, while bacteria are specific to certain groups of arthropods, but both are typically used for insects with chewing mouthparts because they need to be ingested. On the other hand, entomopathogenic fungi can be effective against a variety of arthropod pests and are ideal particularly for insects with piercing and sucking mouthparts because they infect by contact. Nematodes thrive under protected environments and are more effective against various soil-dwelling pests. The knowledge of a pest, its biology, vulnerable life stages, and their habitats and choosing the right control strategy with one or more products against one or more life stages is an important part of IPM (Dara 2019).

It is also important to note that biopesticides are perceived to be less effective than chemical pesticides. Although they are a critical part of pest management in organic farming, the stigma of their inefficacy still persists in many parts of the world, so having a good understanding of their effective use is necessary to build confidence in biopesticides.

2.4 Compatibility of Biopesticides

As crop production involves using several agricultural inputs such as fertilizers, plant growth hormones, biostimulants, and other materials along with pesticides, understanding their compatibility and avoiding incompatible combinations or timings are critical to ensure the success of biopesticide applications. To save on the cost of application, growers routinely apply pesticides with other agricultural inputs. Similarly, certain fungicides and insecticides might be applied at once or in succession. If an entomopathogenic fungus is used for insect control and a biofungicide is applied at the same time or soon after the fungal application, the fungicide could have a negative impact on the entomopathogenic fungus. It is necessary to check biopesticide labels or contact the manufacturer for information on their compatibility with other agricultural inputs to avoid issues.

2.5 Storage, Handling, and Application of Biopesticides

Since some biopesticides contain live organisms or biological compounds that have a shorter shelf life than a chemical compound would have, they may require special storage and handling. Manufacturers are trying to improve biopesticide formulations to extend the shelf life and eliminate the need for refrigeration of some biopesticides. However, growers must follow the storage and handling guidelines and avoid storing or transporting them at high temperatures and exposing them to sunlight or other harsh conditions. Applying entomopathogens in the evenings or with materials that can extend their life will help improve their environmental longevity. For example, applying *Steinernema carpocapsae* with a hydrogel and UV-protectants (titanium dioxide and octyl methoxycinnamate) improved the survival of entomopathogenic nematodes and the resulting mortality of the larvae of the greater wax moth, *Galleria mellonella* (Dit et al. 2016). Conidial production, field persistence, and pathogenicity improved when microsclerotia granules of *Metarhizium brunneum* were applied with a hydromulch containing water, wheat straw, and a tackifier against the adults of the Asian longhorned beetle, *Anoplophora glabripennis* (Goble et al. 2016). Similarly, applying certain biopesticide combinations can be more efficacious than individual formulations. Metabolic acids, such as oxalic and citric acids, improved the pathogenicity of *Beauveria bassiana* and *Metarhizium anisopliae* against the first instar larvae of the potato tuber moth, *Phthorimaea operculella* (Sabbour 2002). The combination of *B. bassiana* with azadirachtin resulted in improved efficacy against the two-spotted spider mite, *Tetranychus urticae*, in laboratory studies (Hernández et al. 2012), and against the rice root aphid, *Rhopalosiphum rufiabdominale*, and the honeysuckle aphid, *Hyadaphis foeniculi*, in organic celery in the field (Dara 2015). Additionally, an additive effect between spinosad and *B. bassiana* at sublethal doses was also seen against the black cutworm, *Agrotis ipsilon*, in laboratory studies (Gosselin et al. 2009).

2.6 Avoiding the Negative Impact on the Environment and Non-Target Organisms

Care must be taken to avoid the potential short-term or long-term negative impacts of certain biopesticides on the environment. Some ecologists are concerned about the impact of microbial biostimulants, soil amendments, and biopesticides on the microbial diversity in the crop environment. While there is no evidence that beneficial microbial applications have a long-term impact on the soil microbiome, such concerns should be taken into consideration to avoid potential negative impacts of biopesticides. While several biopesticides are compatible with biological agents (Lyons et al. 2012; Wu et al. 2016) and are reported to be safer for pollinators (Al Mazra'awi et al. 2006; Kapongo et al. 2008a and 2008b), there are also some examples of a negative impact of biopesticides on natural enemies (Cottrell and Shapiro-Ilan 2003; Roy et al. 2007). Growers should exercise caution and verify labels and compatibility information when there is a potential risk. Although biopesticides are derived from natural sources such as plants or microorganisms, their active ingredients are used at very high concentrations that are not usually present in nature. Repeated applications of biopesticides and generally shorter postharvest and restricted entry intervals compared to chemical pesticides can increase biopesticide residues in soil, water, harvested fruits and vegetables, and higher human exposure. Consequently, care must be taken to ensure the appropriate use of biopesticides and to reduce environmental and human risk.

2.7 Following Regulatory Guidelines

In general, biopesticide development and use are subjected to strict regulatory guidelines. However, there are certain examples where the same strain or different strains of a microorganism are available as both biopesticides and biological soil amendments, which are less regulated than pesticides. In some countries, botanical extracts or local microbial strains can be formulated on a small scale with minimal or no regulation. While this may not pose a big risk in many situations, the misidentification of an organism, contamination of the culture with a pathogen, or use of botanical extracts or other materials whose short- or long-term health risks have not been thoroughly investigated can lead to major problems. It is critical to obtain the technical advice of local agricultural departments or qualified researchers when homemade biopesticides are used. In many cultures, biopesticides have been used in agriculture for centuries, and as science and technology advanced, the knowledge of their effective and safe use has also increased. Modern communication tools make scientific information readily accessible and outreach more effective. Following regulatory and scientific guidelines will help with the appropriate use of biopesticides for effective pest management and further promote their use in conventional agriculture.

2.8 Conclusions and Future Prospects

Biopesticides are effective tools for pest management and an integral part of IPM in organic farming. Understanding their modes of action and use strategies and following guidelines for their effective use will enhance their efficacy, reduce the risk of resistance development, and promote sustainable crop production practices.

REFERENCES

Al Mazra'awi, M. S., Shipp, J. L., Broadbent, A. B., and Kevan, P. G. 2006. Dissemination of *Beauveria bassiana* by honey bees (Hymenoptera: Apidae) for control of tarnished plant bug (Hemiptera: Miridae) on canola. *Environmental Entomology.* 35: 1569–1577. https://doi.org/10.1093/ee/35.6.1569.

Bográn, C. E., Ludwig, S., and Metz, B. 2006. Using oils as pesticides. *Texas A&M AgriLife Extension Publication E-419.* https://cdn-ext.agnet.tamu.edu/wp-content/uploads/2018/11/E-419_-Using-Oils-As-Pesticides.pdf (Accessed 1 June 2020).

Cottrell, T. E., and Shapiro-Ilan, D. 2003. Susceptibility of a native and an exotic lady beetle (Coleoptera: Coccinellidae) to *Beauveria bassiana. Journal of Invertebrate Pathology.* 84: 137–144. https://doi.org/10.1016/j.jip.2003.09.003.

Dara, S. K. 2015. Root aphids and their management in organic celery. *CAPCA Adviser.* 18 (5): 65–70.

Dara, S. K. 2017. Insect resistance to biopesticides. *UCANR E-Journal of Entomology and Biologicals.* http://www.ucanr.org/blogs/blogcore/postdetail.cfm?postnum=25819 (Accessed 1 June 2020).

Dara, S. K. 2019. The new integrated pest management paradigm for the modern age. *Journal of Integrated Pest Management.* 10: 12. https://doi.org/10.1093/jipm/pmz010.

Dito, D. F., Shapiro-Ilan, D. I., Dunlap, C. A., Behle, R. W., and Lewis, E. E. 2016. Enhanced biological control potential of the entomopathogenic nematode, *Steinernemacarpocapsae*, applied with a protective gel formulation. *Biocontrol Science and Technology.* 26: 835–848. https://doi.org/10.1080/09583157.2016.1159659.

Goble, T. A., Gardescu, S., Fisher, J. J., Jackson, M. A., and Hajek, A. E. 2016. Conidial production, persistence and pathogenicity of hydromulch formulations of *Metarhizium brunneum* F52 microsclerotia under forest conditions. *Biological Control.* 95: 83–93. https://doi.org/10.1016/j.biocontrol.2016.01.003.

Gosselin, M.-E., Bélair, G., and Brodeur, J. 2009. Toxicity of spinosad and *Beauveria bassiana* to the black cutworm, and the additivity of sublethal doses. *Biocontrol Science and Technology.* 19: 201–217. https://doi.org/10.1080/09583150802663285.

Hernández, M. M., Martínez-Villar, E., Peace, C., Pérez-Moreno, I., and Marco, V. 2012. Compatibility of the entomopathogenic fungus *Beauveria bassiana* with flufenoxuron and azadirachtin against *Tetranychus urticae. Experimental and Applied Acarology.* 58: 395–405. https://doi.org/10.1007/s10493-012-9594-1.

Kapongo, J. P., Shipp, L., Kevan, P., and Broadbent, B. 2008a. Optimal concentration of *Beauveria bassiana* vectored by bumble bees in relation to pest and bee mortality in greenhouse tomato and sweet pepper. *BioControl.* 53: 797. https://doi.org/10.1007/s10526-007-9142-9.

Kapongo, J. P., Shipp, L., Kevan, P., and Sutton, J. C. 2008b. Co-vectoring of *Beauveria bassiana* and *Clonostachysrosea* by bumble bees (*Bombus impatiens*) for control of insect pests and suppression of grey mould in greenhouse tomato and sweet pepper. *Biological Control.* 46: 508–514. https://doi.org/10.1016/j.biocontrol.2008.05.008.

Lacey, L. A. 2017. *Microbial Control of Insect and Mite Pests: From Theory to Practice.* New York: Academic Press.

Lyons, D. B., Helson, B. V., Bourchier, R. S., and Jones, G. C. 2012. Effects of azadirachtin-based insecticides on the egg parasitoid *Trichogrammaminutum* (Hymenoptera: Trichogrammatidae). *The Canadian Entomologist.* 135: 685–695. https://doi.org/10.4039/n02-113.

Mondali, N. K., Mojumdar, A., Chatterje, S. K., Banerjee, A., Datta, J. K., and Gupta, S. 2009. Antifungal activities and chemical characterization of neem leaf extracts on the growth of some selected fungal species in vitro culture medium. *Journal of Applied Sciences and Environmental Management.* 13: 49–53.

Mordue, A. J. 2004. Present concepts of the mode of action of azadirachtin from neem. In *Neem: Today and in the New Millennium*, eds. O. Koul, and S. Wahab, 229–242. Dordrecht: Springer. https://doi.org/10.1007/1-4020-2596-3_11.

Roy, H. E., Brwon, P. M. J., Rothery, P., Ware, R. L., and Majerus, M. E. N. 2007. Interactions between the fungal pathogen *Beauveria bassiana* and three species of coccinellid: *Harmonia axyridis, Coccinellaseptempunctata* and *Adaliabipunctata.* In *From Biological Control to Invasion: The Ladybird Harmonia axyridis as a Model Species*, eds. H. E. Roy, and E. Wajnberg, 265–276. Dordrecht: Springer. https://doi.org/10.1007/978-1-4020-6939-0_18.

Sabbour, M. M. 2002. The role of chemical additives in enhancing the efficacy of *Beauveria bassiana* and *Metarhizium anisopliae* against the potato tuber moth *Phthorimaeaoperculella*(Zeller) (Lepidoptera: Gelechiidae). *Pakistan Journal of Biological Sciences.* 5: 1155–1159. http://dx.doi.org/10.3923/pjbs.2002.1155.1159.

Tripathi, A. K., Upadhyay, S., Bhuiyan, M., and Bhattacharya, P. R. 2009. A review on prospects of essential oils as biopesticide in insect-pest management. *Journal of Pharmacognosy and Phytotherapy* 1: 52–63.

Wu, S., Xie, H., Li, M., Xu, X., and Lei, Z. 2016. Highly virulent *Beauveria bassiana* strains against the two-spotted spider mite, *Tetranychusurticae*, show no pathogenicity against five phytoseiid mite species. *Experimental and Applied Acarology.* 70: 421–435.

Section III

Impact of Environmental Factors on Biopesticides in Organic Farming

3

Impact of Environmental Factors on Biopesticides in Organic Farming

Enespa and Prem Chandra

CONTENTS

3.1 Introduction

Codex Alimentarius Guidelines (established by the United Nations' Food and Agricultural Organization (FAO) and the World Health Organization) and the International Federation of Organic Agriculture Movements' (IFOAM) Basic Standards regulated organic farming internationally (Aldanondo-Ochoa & Almansa-Sáez, 2009). IFOAM (2008) based the standards on four principles such as (i) ecosystem: without exploiting it organic agriculture should fit the cycles and balances in nature by spending local resources, reprocessing, recycle and well-organized management of ingredients and energy, (ii) health: without expending inorganic fertilizers, artificial pesticides, animal medicines and foodstuff additives organic agriculture is projected to yield high-quality food which may have adversarial health effects, (iii) objectivity: provided decent life, food sovereignty contribution, poverty reduction, improve animal welfare and take future groups into justification, (iv) maintenance: deterrent and accountability have to be applied before accepting new machineries for organic farming, and substantial hazards should be prohibited by declining erratic technologies, such as hereditary manufacturing. Codex Alimentarius Guidelines and IFOAM Basic Standards afford a minimum standard for nationwide and provincial criteria international (Zagata, 2010; Lund & Algers, 2003). Enhancement of soil fertility using efficient local resources to a farming system refers to organic agriculture. Although the earlier use of inherently modified microorganisms, agrochemicals, and several artificial chemical compounds used as additives in food (Gavrilescu & Chisti, 2005). High-quality organic food and its additional value rely on agricultural reforms recognized on environmental rotations, and objectives aimed at lessening the environmental impact of food production, preserving the long-term sustainability of soil, and reducing to a minimum the use of non-renewable resources (Shah & Wu, 2019). To compare the environmental impacts of organic and conventional farming systems, several approaches have been used.

3.2 What Is a Biopesticide?

Generally, natural compounds or genetic modification agents such as pheromones, hormones, plant regulators, insect growth regulators (biochemical pesticide), fungi, bacteria, or genetically modified microorganisms (microbial pesticides) are defined as biopesticides in accordance with the United Nations Food and Agriculture Organization Standards (World Health Organization, 2018). Microbial, botanical, and animal pesticides are a division of biopesticides according to their sources and composition (Oguh et al., 2019). Biopesticides are crop protection products defined by the U.S. Environmental Protection Agency (EPA). Canola oil and baking soda with pesticidal applications are examples of biopesticides (Surgan et al., 2010).

3.3 Limitations and Benefits of Biopesticides

For the environment, the biopesticides have profits and restrictions affect human life or agricultural products. The benefits of biopesticides include:

1. The capability to make another mode of action available for old-fashioned products in most IPM programs makes them a critical component.

2. These types of foodstuffs should be practical before infections to be effective are thought to turn on plant defense responses.
3. Most have host-specific targets, have short residual time, truncated toxicity, and reduce risk to nontarget organisms or the atmosphere.
4. They tend to persist at levels ineffective for control and are incompetent with natural soil microflora.
5. Biopesticides display negligible effects in the atmosphere and on humans due to low persistent, or are decomposable and self-perpetuating; they are less destructive on beneficial flora and fauna because of less time registration than conventional chemical products.
6. Providing the resistance management assistances in agricultural programs, they can extend the life of conventional chemicals.
7. Like decreased preharvest restrictions and application in ecologically sensitive areas, tolerance exemptions documents biopesticides with no Maximum Residue Levels (MRLs) to be used on crops proposed for export.
8. Highly safety of workers and the through reentry intervals allowed by law.
9. Enhanced plant health, revenues, and quality like value-added profits.

3.4 Soil Environment

Soil plays a key role in agriculture is one of the most important natural resources. Micro and macronutrient-enriched soil is necessary for the growth and development of healthy plants. In addition to the manufacture function, filtering, buffering, and transformation are the other properties of the soil, and it is the environment for microorganisms. It also develops a socioeconomic function in soil habitat (Kumar & Gopal, 2015).

Within organic farming, the following positive changes are taking place:

- Soil organic matter (up to 30% higher organic carbon content)
- Biological activity of soil increased (by 30–100 %), biomass decomposition indicator
- Total edaphic biomass increased (by 50–80%)
- Root colonization by mycorrhiza increased and richness in higher saprophytic fungi
- Use of acceptable resources by soil microorganisms more efficient
- Physical and chemical soil properties improved soil structure
- Erosion threat reduction and hydroscopicity improved

3.5 Impact on the Environment

On both fronts, biopesticides provide valuable tools to growers by transporting clarifications that are effective highly in treatment pests and diseases without producing negative effects on the environment. T active and inert elements are generally recognized as safe (GRAS) (Abdel-Shafy & Mansour, 2018). Biopesticides have partial toxicity to wildlife, birds, fish, bees, and other valuable soil microorganisms. They maintain valuable insect populations easily biodegradable in the environment and harshly reduced conventional pesticide applications in resistance management programs through their effective use (Ghorbani et al., 2009). Biopesticides could have a positive impact on the environment due to the microbial antagonist(s) (Laengle & Cass, 2011). To control various plant pathogens, several microbial antagonists have been used. The genus *Trichoderma* and *Pseudomonas* were determined and tested forcefully among the microbial antagonists (Howell, 2003). No Maximum Residue Levels (MRLs) have been seen in biopesticides so that no residue in soil remains, which will make sustainable to agriculture. In the growth and development environment similar to other soil microorganisms, microbial antagonists need favorable conditions28O'Callaghan, 2016). The best temperature for growing some *Trichoderma* isolates was at a range of 25–30°C and is similar for other microorganisms growing temperature (Kough & Edelstein, 2012). The temperature and soil pH for growing *Streptomyces* spp. were in a range of 20–37°C and 4.5–9.0, respectively. Carrier media for conveying biopesticides were involved in several organic materials, such as animal broth, organic materials, or organic waste products, in addition to microbial content (Alvarez et al., 2012; Chandra and Singh, 2016).

3.6 Biopesticides and Environmental Issues

Through very strict regulatory procedures, biopesticides are industrialized to function with reasonable inevitability and negligible effect on human health and the atmosphere; public concerns amplify the adverse effects of artificial agrochemicals in the environment (Acquavella et al., 2004). In the global market, fewer chemical pesticides are available, and some pesticide withdrawal from the market, either regulatory or commercial reasons due to pesticide resistance problems. Therefore, the ICM method needs to be arranged to counter the degradation of the agro-ecosystem due to the ongoing rigorous agriculture (Knight &Norton, 1989). The use of biofertilizers, biopesticides, soil and water conservation practices, IPM, conservation of biological diversity is included. Crop plants have allowable isolating genes operational against particular pests and a better understanding of genes from microorganisms. They are being organized to insect pests and disease control of crop plants (McSweeney & Mackie, 2012). The use of biopesticides benefits in farming and public health platforms is significant. The residual effects are a matter of significant concern for consumers, not lies in the environment in fruits and vegetables. The value of biopesticides can be identical to the conventional pesticides when used as a constituent of IPM. Its benefits are linked with the safety of the environment, target-specificity and efficiency in less capacity, decomposition naturally and suitability in IPM biopesticide interest. So, to manage the environmental pollutions, biopesticides are one of the favorable substitutes (Usta, 2013). They replaced several lethal chemical pesticides. However, the biopesticide potential for supporting ecological agriculture has been known for years. An assessment of the growing demands has expanded interest for healthy and safe organic food (Kumar, 2012). Although the use of agrochemicals is essential to meet the ever-growing demands of food, feed and fodder, prospects do exist in particular crops and niche areas where biopesticides can be used as a constituent of IPM (Thakore, 2006). The biopesticide market's key drivers are the growing organic food market and easier registration than chemical pesticides, and demands increased for residue-free crop production. Some of the endangered species there are raising alarms for conservation of biodiversity and threats established against the requirement to increase agricultural production without extreme dependence on chemical pesticides (Marrone, 2014).

3.7 Biopesticides for Organic Farming

Organic produce to be pesticide-free or chemical-free perhaps does not mean. Under several states' regulations, there are various chemical sprays and powders available for organic farmers to use (Gazaresen et al., 2020). It must be derived from natural sources and must not be an artificial chemical if a pesticide is used in biological farming. In the past three years, it is also essential that the biopesticides must be practical using apparatus that has not been used to apply any artificial chemical compounds. The land used must not have any treatment with man-made chemicals (Damalas & Koutroubas, 2020). Crop rotation, cover crops, disease-resistant variations, and good seedbed preparation are chiefly several agricultural practices applied to control the pests and diseases in organic farming. However, these practices are not always adequate to defend crops from losses (Moura et al., 2020). So, to ensure and enhance the quality of their organic products, organic farmers turn toward biopesticides.

3.8 Environmental Effects of Organic Farming

Mainly in Europe and North America in current years, several observations have compared the environmental impacts of organic and conventional farming systems (Barbieri et al., 2017). Several researchers have reviewed the previous observations scientifically to determine the environmental impacts of organic and conventional farming on energy use and greenhouse gas emissions (GHGE) (Colnenne-David et al., 2017). Regarding the environmental performance, the current observations have produced several conclusions due to the transformations in the approaches engaged, including alterations in data sources, size of sample, statistical studies, and measurement (Drovandi et al., 2017). Comparing the environmental impacts of organic and conventional farming found that organic farming was additional likely to result in less energy use and lower GHGE production per unit of land but higher energy use and emissions per unit of output (Lee & Choe, 2019). However, due to modifications in farm features, data sources, determination methods, and types of studies, the outcomes varied. The result requires variability of observations in the identification of the essential variables that were linked with the contradictory outcomes found in these observations.

3.9 Conclusion and Future Prospects

The microbial antagonist needs favorable conditions for growth and development, similar to other soil microorganisms in the environment. For conveying biopesticide, carrier media involved several organic materials, like animal broth, organic materials, or organic waste products besides the microbial content. To stop hastening agricultural-mediated global climate change, the negative impacts of agricultural systems have to be reduced and replaced by alternatives such as reducing synthetic nitrogen fertilizers. As a substitute, organic fertilizers in guanos, slurries, or other organic residues and the leguminous crop cultivation should be stimulated. In addition to reducing energy, challenging procedures, and harmful environmental effects such as phosphorus and nitrogen leaching and enriching the soil properties with SOC, this might ease depletion SOC.

REFERENCES

Abdel-Shafy, HI; Mansour, MS; Solid waste issue: Sources, composition, disposal, recycling, and valorization. *Egypt J Pet*, 2018, *27*(4), 1275–1290.

Acquavella, JF; Alexander, BH; Mandel, JS; Gustin, C; Baker, B; Chapman, P; Bleeke, M; Glyphosate bio monitoring for farmers and their families: Results from the farm family exposure study. *Environ Health Perspect*, 2004, *112*(3), 321–326.

Aldanondo-Ochoa, AM; Almansa-Sáez, C; The private provision of public environment: Consumer preferences for organic production systems. *Land Use Policy*, 2009, *26*(3), 669–682.

Alvarez, A; Benimeli, CS; Saez, JM; Fuentes, MS; Cuozzo, SA; Polti, MA; Amoroso, MJ; Bacterial bio-resources for remediation of hexachlorocyclohexane. *Int J Mol Sci*, 2012, *13*(11), 15086–15106.

Barbieri, P; Pellerin, S; Nesme, T; Comparing crop rotations between organic and conventional farming. *Sci Rep*, 2017, *7*(1), 1–10.

Chandra, P; Singh, E; Applications and mechanisms of plant growth-stimulating rhizobacteria. In *Plant-Microbe Interaction: An Approach to Sustainable Agriculture*, 2016 (pp. 37–62). Springer, Singapore.

Colnenne-David, C; Grandeau, G; Jeuffroy, MH; Doré, T; Ambitious environmental and economic goals for the future of agriculture are unequally achieved by innovative cropping systems. *Field Crops Res*, 2017, *210*, 114–128.

Damalas, CA; Koutroubas, SD; Botanical pesticides for eco-friendly pest management: Drawbacks and limitations. *Pestic Crop Prod: Physiol BiochemAction*, 2020, 181–193.

Damalas, CA; Koutroubas, SD; Botanical pesticides for eco-friendly pest management: Drawbacks and limitations. *Pestic Crop Prod: Physiol BiochemAction*, 2020 (pp 181–193). John Wiley & Sons, USA

Drovandi, CC; Holmes, C; McGree, JM; Mengersen, K; Richardson, S; Ryan, EG; Principles of experimental design for big data analysis. *Stat Sci Rev J Inst Math Stat*, 2017, *32*(3), 385.

Gavrilescu, M; Chisti, Y; Biotechnology: A sustainable alternative for chemical industry. *Biotechnol Adv*, 2005, 23(7–8), 471–499.

Gazaresen, RK; Tyagi, P; Pandey, KN; Organic farming and naturally grown products: Need of training intervention for its growth and healthy society. *Our Heritage*, 2020, *68*(30), 5200–5215.

Ghorbani, R; Wilcockson, S; Koocheki, A; Leifert, C; Soil management for sustainable crop disease control: A review. In *Organic Farming, Pest Control and Remediation of Soil Pollutants*, 2009 (pp. 177–201), Springer, Dordrecht.

Howell, CR; Mechanisms employed by *Trichoderma* species in the biological control of plant diseases: The history and evolution of current concepts. *Plant Dis*, 2003, *87*(1), 4–10.

Knight, AL; Norton, GW; Economics of agricultural pesticide resistance in arthropods. *Annu Rev Entomol*, 1989, *34*(1), 293–313.

Kough, JL; Edelstein, R; EPA regulatory requirements for plant-incorporated protectants. In *Regulation of Agricultural Biotechnology: The United States and Canada*, 2012, (pp. 163–174). Springer, Dordrecht.

Kumar, BL; Gopal, DS; Effective role of indigenous microorganisms for sustainable environment. 3 Biotech, 2015, *5*(6), 867–876.

Kumar, S; Biopesticides: A need for food and environmental safety. *J Biofertil Biopestic*, 2012, *3*(4), 1–3.

Laengle, T; Cass, L; *Categories of Biopesticides and Related Products*, 2011. http://www4.agr.gc.ca/AAFC-AAC/display-afficher.do?id=1288727518132&lang=eng

Lee, KS; Choe, YC; Environmental performance of organic farming: Evidence from Korean small-holder soybean production. *J Cleaner Prod*, 2019, *211*, 742–748.

Lund, V; Algers, B; Research on animal health and welfare in organic farming: A literature review. *Livest Prod Sci*, 2003, *80*(1–2), 55–68.

Marrone, PG; The market and potential for biopesticides. In *Biopesticides: State of the Art and Future Opportunities*, 2014, (pp. 245–258). American Chemical Society, Washington, DC.

McSweeney, C; Mackie, R; Commission on genetic resources for food and agriculture. In *Micro-organisms and Ruminant Digestion: State of Knowledge, Trends and Future Prospects.* Background Study Paper No. 61, 2012, (pp. 1–62). Food and Agriculture Organization of the United Nations, Rome, Italy.

Moura, DR; Yamada, JK; de Albuquerque, LC; de Carvalho Pontes, N; Crop rotation reduces the density of volunteer plants in processing tomato fields and the inoculum of bacterial spot. *Eur J Plant Pathol*, 2020, *156*(1), 299–304.

O'Callaghan, M; Microbial inoculation of seed for improved crop performance: Issues and opportunities. *Appl Microbiol Biotechnol*, 2016, *100*(13), 5729–5746.

Oguh, CE; Okpaka, CO; Ubani, CS; Okekeaji, U; Joseph, PS; Amadi, EU; Natural pesticides (biopesticides) and uses in pest management: A critical review. *Asian J Biotechnol Genet Eng*, 2019, 1–18.

Omokolo, ND; Impact of effective and indigenous microorganisms manures on *Colocassiaesculenta* and enzymes activities. *Afr J Agric Res*, 2013, *8*(12), 1086–1092.

Rundlöf, M; Nilsson, H; Smith, HG; Interacting effects of farming practice and landscape context on bumble bees. *Biol Conserv*, 2008, 141(2), 417–426.

Shah, F; Wu, W; Soil and crop management strategies to ensure higher crop productivity within sustainable environments. *Sustainability*, 2019, *11*(5), 1485.

Surgan, M; Condon, M; Cox, C; Pesticide risk indicators: Unidentified inert ingredients compromise their integrity and utility. *Environ Manage*, 2010, *45*(4), 834–841.

Thakore, Y; The biopesticide market for global agricultural use. *Ind Biotechnol*, 2006, *2*(3), 194–208.

Usta, C; Microorganisms in biological pest control—a review (bacterial toxin application and effect of environmental factors), current progress in biological research, Dr. Marina Silva-Opps. 2013, InTech. doi, 10, 55786.

World Health Organization; *International Code of Conduct on Pesticide Management: Guidelines for the Registration of Microbial, Botanical and Semiochemical Pest Control Agents for Plant Protection and Public Health Uses*, 2018. Food and Agriculture Organization of the United Nations, Rome, Italy.

Zagata, L; How organic farmers view their own practice: results from the Czech Republic. *Agric Hum Values*, 2010, *27*(3), 277–290.

Section IV

Opportunities and Challenges in the Application of Biopesticides in Organic Farming

4

Opportunities and Challenges in the Application of Biopesticides in Organic Farming

Rishabh Chitranshi, Enespa, Ranjan Singh, and Prem Chandra

CONTENTS

4.1 Introduction

The production and supply of food has altered greatly due to the massive increasing population in the present scenario. High yielding varieties, chemical fertilizers, and pesticides developed by researchers resulted in remarkable growth in agricultural productivity (Timsina, 2018). The use of synthetic chemical pesticides has severely affected both the abiotic and biotic components of the environment. While the earlier is represented by pesticide deposits in water, soil, air, food etc., the latter includes toxicity in plants, physiological defects, illnesses, mortality, changes in population, genetic complaints, gene erosion, etc. in vegetation, animals, insects and another organisms. The present pest management strategy depends heavily on synthetic chemical pesticides that adversely affect the beneficial organisms, pesticide residues in food, feed, and fodder, and ecological pollution. Inoculation of harmful chemicals and pesticide residues into the food chain and their bioaccumulation may initiate several unexpected consequences (Özkara et al., 2016). While prospective and opportunity of biopesticides and biofertilizers for promoting sustainable agriculture has been known for years, organic farming has developed because of the growing demands for safe and healthy food and environmental pollution (Thakore, 2006). The industrialization of agriculture organic agriculture arose as a reaction and linked environmental and social complications. Over conventional agriculture, whether organic agriculture delivers overall advantages is, however, debatable. The main criticism of organic agriculture is that it is not as productive when food production has to rise significantly to feed a growing, more prosperous worldwide populace (Röös et al., 2018). They point out that the current agricultural farm will not avail, and further agricultural land expansion into virgin ecosystems will result from organic agriculture expansion with the rising worldwide demand for food (Crowder & Reganold, 2015).

4.2 Biopesticides

Biopesticides are made from the microbes, substances found naturally in the soil. They are derived from nematodes, plants, or microorganisms in the form of their products such as phytochemicals, microbial products, or byproducts used to control insect or fungal infections before or after sowing the crop (Senthil-Nathan, 2015). It also observed that biological treatments of disease management and crop protection have a healthy future for sustainable agriculture with lower investment and less harmful effects (Enespa & Chandra, 2017).

4.2.1 Types of Biopesticides

Biopesticides are classified into three categories: (i) plant-incorporated protectants (PIPs), (ii) microbial pesticides, and (iii) nanopesticides.

4.2.1.1 Plant-Incorporated Protectants (PIPS)

Plant-incorporated protectants (PIPs) purify the plant unsuitable for pest insects. This does not require insect lethality and could be achieved through the expression of a suitable repellent. However, from a practical point of view, protectants typically act by compromising insect viability by disrupting normal physiology (Matten et al., 2012). Insecticidal interactions have attained this mostly by targeting the insect nervous system, which can be accomplished through oral delivery and absorption following surface contact (Braverman et al., 2010). PIPs, however, must be ingested, overcome both digestive and physical barriers, and then reach the target site where they act. Systems for overcoming similar challenges in achieving activity in target hosts by oral delivery exist in nature and can be thought of as models for developing PIP machineries. For example, bacteria and other microorganisms that utilize foreign hosts (such as insects) to promote self-propagation have evolved methods for attacking, restraining, and often killing the host when the main route of admission is through the digestive system (Tangtrakulwanich & Reddy, 2014). The most commonly used biopesticides are living organisms, which are pathogenic for the pest of interest. These include bio fungicides such as *Trichoderma* sp., bioherbicides like *Phytophthora*, and bioinsecticides like *Bacillus thuringiensis* and *B. sphaericus*. The latent benefits to agriculture and public health programs through biopesticides are considerable (Koskela et al., 2010).

4.2.1.2 Microbial Pesticides

Microbial biopesticides are naturally occurring or genetically altered bacteria, fungi, algae, viruses, or protozoans pest-specific. There are at least 1500 naturally occurring insect-specific microorganisms, 100 of which are insecticidal. These are also known as microbial biocontrol agents, which are proved as an effective alternative to chemical insecticides (Usta, 2013). The Pathogenicity of these microorganisms may be mostly as a target based on specific pests or species. Various observations have confirmed that the toxins produced by microbial pathogens are peptides and are very significant in assembly, noxiousness, and specificity (Salyers et al., 1994). These microbial pesticides offer an alternative for ecological safety as well. Microbial pesticides can work individually or in combination with other pests (21Raivio, 2005).

In the latest organic period, bacteria like *Pseudomonas*, *Rhizobium*, *Bacillus*, *Azotobacter Azospirillum*, *Streptomyces*, *Thiobacillus*, and fungi *Trichoderma* and *Coniothyrium* are very effective for agriculture farming and used and proved their capability in separate fields (Din et al., 2019). Bio-formulations of microbes can be an alternative solution to harmful chemicals. Additionally, they are economically cheap, environmentally friendly, and non-toxic natural products (Beneduzi et al., 2012).

4.2.1.3 Nanobiopesticides

The use of safe and ecofriendly biocontrol agents as a substitute for chemical pesticides is extensive global nowadays. Researchers are working to determine the probability of improving the efficiency of biocontrol agents by renovating them into nanoparticles for cost-effective and sustainable agriculture (Macfadyen et al., 2009). In this kind of technique, secretion of a bacterium, *Photorhabdus luminescens* was transformed into nanoparticles and found that its efficiency was boosted significantly. The nano form of the biocontrol agent has been tested against two sucking pests of cotton, *Tetranychus macfarlanei*, a species of mite, and *Aphis gossypii*, a species of aphid (Koul, 2019).

4.3 Benefits of Biopesticides

Individual products differ in significant ways, but the following list of beneficial characteristics applies to microbial insecticides overall. The toxic action of microbial insecticides is often specific to a single group or species of insects. This specificity means that most microbial insecticides do not directly affect beneficial insects (including predators or parasites of pests) in treated areas.

- Most microbial insecticides can be used in conjunction with synthetic chemical insecticides because, in most cases, the microbial product is not deactivated or damaged by residues of conventional insecticides. Because their residues present no hazards to humans or other animals, microbial insecticides can be applied even when a crop is almost ready for harvest.

4.4 Organic Farming

Organic farming is a method of agriculture based on the use of organic fertilizers and biosynthetic pesticides with crop rotation to maintain the fertility of the land to produce healthy and nutritious food (Reeve et al. 2016). Organic farming is one of the best techniques in recent agroeconomic trends compared to any other traditional farming methods, as it not only has maximum production at low cost but also has no adverse or harmful effects on the environment and the surrounding climatic conditions (Mahapatra et al., 2009). Currently, we are moving toward the second green revolution, and we will have to emphasize increasing agricultural production as well as reducing environmental pollution, water management, maintaining soil quality, and protecting crops from disease (Pingali, 2012). It would not be wrong to say that all these problems can be solved effectively by organic farming. Therefore, the need of the hour is to encourage organic farming, discouraging the use of chemical fertilizers. That's why most consumers today consider organic food as safer and healthier than conventionally produced food. The popularity of organic food is growing rapidly in most developed countries in the world.

4.5 Principles of Organic Farming

Its main objective is to sustain the psychological, ecological, and social welfare of human beings and the health and well-being of soil, plants, animals, and the earth through organic agriculture (Vaarst & Alrøe, 2012). This provides pollution and chemical-free, nutritious foods for humans. Organic farming provides a good quality of life not only among humans but other living beings and also helps in improving the economic condition, so through this, natural resources should be judiciously used and conserved for future generations. Organic agriculture should be based on living ecosystems. Organic farming methods should fit the ecological balance and cycle in nature (Gomiero et al., 2011).

4.6 Benefits of Organic Farming

Food grown from organic farming is full of nutrients such as vitamins, enzymes, minerals, and other micronutrients compared to traditional agriculture as farms are managed and nourished by using crop rotation in organic farming (Badgley et al., 2007). Organic fertilizers, biopesticides, and fungicides made by the microbes found in nature are used in organic farming which is environment friendly and does not have any adverse effect on the land.

4.7 Risks and Opportunities Associated with Increasing Yields in Organic Production

4.7.1 Biodiversity

Compared with conventional farming, organic farming normally increases the crop and landscape heterogeneity, which augments biodiversity. According to one meta-analysis, overall species productivity on organic farms is on average 34% (95% CI: 26–43%) more than on conventional farms (Shennan et al., 2017). Though, on organic farms between the organism groups, the magnitude of the positive effects fluctuates widely such as for pollinators and predators species richness is 50% (95% CI 27-77%) and 12% (95% CI 1–24%) higher, correspondingly. The positive effects also show large variation across landscapes shows a positive effect, e.g., lower effects in more diverse landscapes. Though, the benefits of organic production for biodiversity have been shown to be greatest at field level in some cases, while increases at farm or landscape level may be lesser (Chandra & Enespa, 2019).

4.7.2 Emissions of Greenhouse Gases

In Northern Europe, the climate effect from agriculture arises mainly from emissions of nitrous oxide (N_2O) from soils, by use of nitrogenous fertilizer, carbon dioxide (CO_2) emissions from organic soils (12%), methane (CH_4) from enteric fermentation in ruminants (26%) and emissions of N_2O and CH_4 from manure management (5%) driven largely (Shen et al., 2020). The production of mineral fertilizers is also a significant source of GHG in conventional agriculture. The transportation and production of some organically suitable fertilizers require non-negligible quantities of fossil energy input. Therefore, organic farming does not use energy-demanding mineral fertilizers (Hu et al., 2020).

4.7.3 Nutrient Losses

Loss of N and phosphorus (P) from agricultural systems to waterways is a serious problem causing Eutrophication is a serious problem caused by loss of nitrogen (N) and phosphorus (P) from agricultural systems to water streams mainly in coastal areas. Mainly from manure management agriculture is also the main contributor to airborne NH_3 emissions (Naher et al., 2020). In organic farming N has great potential to increase yields in increased input of nutrients.

4.7.4 Soil Fertility

Loss of soil organic carbon (SOC), nutrient reduction, soil compaction, and heavy metal deposition are several anthropogenic pressures affected agricultural soils. A key characteristic of soil fertility is a high SOC level that encourages soil structure, plant root development, aeration, water-holding capacity, chemical buffering capacity, soil microbial activity, and nutrients through mineralization continuously (Haddix et al., 2020). With high nutrient availability, the increased use of fertilizers, i.e., biogas digestate, or forthcoming introduction of renewable mineral fertilizers in organic farming could provide the potential to increase yields through increased precision in fertilizer application (Chandra & Singh, 2016).

4.8 Conclusion and Future Prospects

In various areas such as agriculture and food production, international relations, direct marketing, and process auditing the organic movement has a range of strengths. However, balancing organic principles with viable pressures and flexibility to maintain important issues needs to be addressed while also following international management. To challenge its critics as increasing numbers of successful initiatives are established in various countries, organic agriculture will continue. Biopesticides bring more income for the farmers and support permanence and sustainability of agroecosystem through lower cost and higher price of the product. However, there are still many limitations to be reduced.

REFERENCES

Badgley, C.; Perfecto, I. Can organic agriculture feed the world? *Renew Agric Food Syst*, 2007, *22*(2), 80–86.

Beneduzi, A.; Ambrosini, A.; Passaglia, L.M. Plant growth-promoting rhizobacteria (PGPR): Their potential as antagonists and biocontrol agents. *Genet Mol Biol*, 2012, *35*(4) (Supplement, 1044–1051.

Braverman, M.P.; Baron, J.J.; Kunkel, D.L. An overview and future trends of US biopesticide regulations. *Outlooks Pest Manage*, 2010, *21*(3), 132–134.

Chandra, P.; Enespa. Soil–microbes–plants: Interactions and ecological diversity. In *Plant Microbe Interface*, 2019 (pp. 145–176). Springer, Cham.

Chandra, P.; Singh, E. Applications and mechanisms of plant growth-stimulating rhizobacteria. In *Plant-Microbe Interaction: An Approach to Sustainable Agriculture*, 2016 (pp. 37–62). Springer, Singapore.

Crowder, D.W.; Reganold, J.P. Financial competitiveness of organic agriculture on a global scale.*Proc Natl Acad Sci U S A*, 2015, *112*(24), 7611–7616.

Din, M.; Nelofer, R.; Salman, M.; Khan, F.H.; Khan, A.; Ahmad, M.; Khan, M. Production of nitrogen fixing *Azotobacter* (SR-4) and phosphorus solubilizing *Aspergillus niger* and their evaluation on *Lagenaria siceraria* and *Abelmoschus esculentus*. *Biotechnol Rep*, 2019, *22*, e00323.

Enespa, Chandra P. Microbial volatiles as chemical weapons against pathogenic fungi. In *Volatiles and Food Security*, 2017, (pp. 227–254). Springer, Singapore.

Gomiero, T.; Pimentel, D.; Paoletti, M.G. Environmental impact of different agricultural management practices: Conventional vs. organic agriculture. *Crit Rev Plant Sci*, 2011, *30*(1–2), 95–124.

Haddix, M.L.; Gregorich, E.G.; Helgason, B.L.; Janzen, H.; Ellert, B.H.; Cotrufo, M.F. Climate, carbon content, and soil texture control the independent formation and persistence of particulate and mineral-associated organic matter in soil. *Geoderma*, 2020, *363*, 114160.

Hu, X.; Xia, B.; Chong, H.Y.; Skitmore, M.; Buys, L. Improving the sustainable retirement village framework: From theory to practice. *J Clean Prod*, 2020, *248*, 119290.

Koskela, J.; Vinceti, B.; Dvorak, W.; Bush, D.; Dawson, I.; Loo, J.; Jamnadass, R. The use and exchange of forest genetic resources for food and agriculture. In *FAO Commission on Genetic Resources, Background Stud* Paper No. 44, 2010 (1–67). UNEP, Colombia.

Koul, O. Nanobiopesticides: An introduction. In *Nano-Biopesticides Today and Future Perspectives*, 2019 (pp. 1–15). Academic Press. NX Amsterdam, Netherlands.

Lau, H.; Shum, P.K.; Nakandala, D.; Fan, Y.; Lee, C. A game theoretic decision model for organic food supplier evaluation in the global supply chains. *J Cleaner Prod*, 2020, *242*, 118536.

Macfadyen, S.; Gibson, R.; Polaszek, A.; Morris, R.J.; Craze, P.G.; Planqué, R.; Memmott, J. Do differences in food web structure between organic and conventional farms affect the ecosystem service of pest control? *Ecol Lett*, 2009, *12*(3), 229–238.

Mahapatra, B.S.; Ramasubramanian, T.; Chowdhury, H. Organic farming for sustainable agriculture: Global and Indian perspective. *Indian J Agron*, 2009, *54*(2), 178–185.

Matten, S.R.; Frederick, R.J.; Reynolds, A.H. United States Environmental Protection Agency insect resistance management programs for plant-incorporated protectants and use of simulation modeling. In *Regulation of Agricultural Biotechnology: The United States and Canada*, 2012 (pp. 175–267). Springer, Dordrecht.

Naher, U.A.; Choudhury, A.T.; Biswas, J.C.; Panhwar, Q.A.; Kennedy, I.R. Prospects of using leguminous green manuring crop *Sesbania rostrata* for supplementing fertilizer nitrogen in rice production and control of environmental pollution. *J Plant Nutr*, 2020, *43*(2), 285–296.

Özkara, A.; Akyıl, D.; Konuk, M. Pesticides, environmental pollution, and health. In *Environmental Health Risk-Hazardous Factors to Living Species*, 2016 (pp. 1–27). Intech Open, London, U.K.

Pingali, P.L. Green revolution: Impacts, limits, and the path ahead. *Proc Natl Acad Sci U S A*, 2012, *109*(31), 12302–12308.

Raivio, T.L. Micro review: Envelope stress responses and Gram-negative bacterial pathogenesis. *Mol Microbiol*, 2005, *56*(5), 1119–1128.

Reeve, J.R.; Hoagland, L.A.; Villalba, J.J.; Carr, P.M.; Atucha, A.; Cambardella, C.; Delate, K. Organic farming, soil health, and food quality: Considering possible links. In *Advances in Agronomy*, 2016 (Vol. 137, pp. 319–367). Academic Press, Chennai.

Röös, E.; Mie, A.; Wivstad, M.; Salomon, E.; Johansson, B.; Gunnarsson, S.; Watson, C.A. Risks and opportunities of increasing yields in organic farming. A review. *Agron Sustainable Dev*, 2018, *38*(2), 14.

Salyers, A.A.; Whitt, D.D.; Whitt, D.D. *Bacterial Pathogenesis: A Molecular Approach*, 1994 (No. QR201. B34 S34 2002). ASM Press, Washington, DC.

Senthil-Nathan, S. A review of biopesticides and their mode of action against insect pests. In *Environmental Sustainability*, 2015 (pp. 49–63). Springer, New Delhi.

Shen, W.; Xue, H.; Gao, N.; Shiratori, Y.; Kamiya, T.; Fujiwara, T.; Senoo, K. Effects of copper on nitrous oxide (N_2O) reduction in denitrifiers and N_2O emissions from agricultural soils. *Biol Fertil Soils*, 2020, *56*(1), 39–51.

Shennan, C.; Krupnik, T.J.; Baird, G.; Cohen, H.; Forbush, K.; Lovell, R.J.; Olimpi, E.M. Organic and conventional agriculture: A useful framing? *Annu Rev Environ Resour*, 2017, *42*, 317–346.

Tangtrakulwanich, K.; Reddy, G.V. Development of insect resistance to plant biopesticides: An overview. In *Advances in Plant Biopesticides*, 2014 (pp. 47–62). Springer, New Delhi.

Thakore, Y. The biopesticide market for global agricultural use. *Ind Biotechnol*, 2006, *2*(3), 194–208.

Timsina, J. Can organic sources of nutrients increase crop yields to meet global food demand?*Agronomy*,2018, *8*(10), 214.

Usta, C. Microorganisms in biological pest control: A review (bacterial toxin application and effect of environmental factors). *Current Prog Biol Res*, 2013, 287–317. Intech Open, DOI: 10.5772/55786.

Vaarst, M.; Alrøe, H.F. Concepts of animal health and welfare in organic livestock systems. *J Agric Environ Eth*, 2012, *25*(3), 333–347.

Section V

Strategies for Crop Protection with Biopesticides in Organic Farming

5

Impact of Crop Protection Strategies with Biopesticides on Long, Sustainable Organic Agriculture

Sushree Suparna Mahapatra, Sudeepta Pattanayak, Siddhartha Das, and L. P. Awasthi

CONTENTS

5.1 Introduction

Production of different crops is governed by various factors, which may be both biotic and abiotic. To achieve maximum yield, there is a requirement of optimum climatic conditions, irrigation, nutrition, and protection against adverse conditions. Several biotic and abiotic agents directly or indirectly cause crop loss. Biotic factors such as weeds, insect pests, and disease, resulting in microorganisms such as nematodes, fungi, bacteria, viruses, protozoa, etc., play a major role in increasing economic loss in crop production. These factors, on a global scale, cause around 68% of crop loss. The loss caused by the weeds accounts to be the highest among all the biotic agents, which is around 34%. The loss caused by insect pests and microbial disease is 18% and 16%, respectively. To safeguard crops against these adverse factors, several crop protection strategies are enforced according to the need. But these strategies may incorporate the use of both physical and chemical entities. The use of chemical pesticides, fungicides, fertilizers, and antibiotics may worsen the situation. So, a need-based use of chemicals and, more importantly, ecofriendly protection, have to be manifested.

The ecofriendly methods like the use of resistant cultivars, biological, botanicals, cultural, physical controls, and application of biotechnology have to be employed for sustainable agriculture. Therefore, small-scale farmers require crop protection measures that are cheap (so that farmers can afford them), simple (so that they can be applied under particular circumstances), cost-effective (so that they can enable farmers to make profits and prosper), and sustainable (so that the production is also sustained).

5.2 Need for Crop Protection

Plant protection is said to be a collection of proper pest management and weather parameters, which ensures high yield along with minimal wastage. To maintain sustainability, there is a need to make proper use of available resources, which does not hamper future needs. So sustainable agriculture aims at increasing the productivity of the land by proper management practices, which does not have any ill effect on the ecological system. High crop yields can be obtained and sustainability in agriculture if plants are protected from diseases and pests and other adverse conditions (Cook, 1986). This will result in better growth of the plants by getting favorable growth conditions and withstand the foreign attack.

5.2.1 Chemical Control

Chemical control is based on substances that are toxic to the harmful pests attacking beneficial crops. When these chemical pesticides are applied to protect plants from pests, diseases or overgrowth by weeds, they are considered plant protection products. Pesticides are grouped into five main categories depending on the target organisms: fungicides, herbicides, insecticides, acaricides, nematicides against fungi, weeds, insects, mites, and nematodes, respectively. According to Walker (1982), decisions regarding preventive measures should be made considering the pest, disease, or weed density level, the action threshold, or the economic threshold, based on the benefit/cost ratio.

DeDatta, S.K. (1981) suggested alternatives to synthetic herbicides by incorporating quarantine measures, eradication, and reduction of the weed population. Excessive use of chemicals often leads to the initiation of resistance against them. There are few active ingredient groups of fungicides available for chemical disease control of the predominant diseases (triazoles, strobilurins, morpholines, SDHIs, chloronitriles), which makes the implementation of anti-resistance strategies difficult. Effective use of pesticides can be achieved by:

1. Reducing the number of pesticides per season, which can limit the exposure of the pest population to the fungicide.
2. Avoiding the use of fungicides in situations where the target pest is already well established in the crop.
3. Composite application- Mixing or alternating pesticides with different modes of action.
4. By manipulating the dose of application, i.e., avoiding multiple low doses and promoting high doses.

There are several advantages and disadvantages associated with the use of chemical pesticides. Certain advantages associated with the use of chemical pesticides are:

- Increase in productivity of a land
- Control of storage loss
- Fast action and easy availability
- Highly effective

On the other side, chemicals are disadvantageous as they can be detrimental to the environment, create pest resurgence, toxic to humans, animals, and beneficial insects and produce considerable damage to the ecosystem by eliminating flora and fauna.

5.2.2 Cultural Control

Cultural control as a strategy for crop protection is of utmost importance as Froud Williams (1984) stated, "Before the widespread use of selective herbicides, in addition to the use of uncontaminated crop seed, weed control was facilitated by a combination of crop rotation and cultivation, the tenets of good husbandry." It involves the manipulation and management of the environmental conditions so that it becomes unsuitable for the pests and weeds to grow and sustain. It includes farm sanitation, tillage, proper irrigation, spacing, aeration of the soil for proper crop growth, which limits the pest population. These methods, if properly undertaken, create unfavorable conditions for weeds, pests, and pathogens to grow and propagate. During field preparation, the land should be weed-free and devoid of any infected debris of the previous crop. This can be achieved by proper tillage and soil solarization. Irrigation channels should be cleaned as they are an important source of spreading weeds and diseases. Proper and clean planting method needs to be adopted. Uninfected seeds should be used, sowing operation with drills reduce the germinating weeds. The crop variety is one of the most important factors that is needed to be taken care of. Short stature, erect leaved varieties should be sown for weed control since they permit more light than tall and leafy traditional varieties. The use of resistant varieties is the best way to avoid the incidence of diseases or pest attack. This in turn reduces the use of chemicals. Timely and optimum irrigation should be applied with proper drainage. A proper crop rotation is the basis for every more sustainable farming system. It lays the basis for other methods such as crop protection (Hans et al., 2005).

5.2.3 Physical Control

Physical weed control methods include hand weeding, hoeing, mechanical cultivation, mulching, burning, mowing, and power. It can employ both pre- and postsowing and harvesting. Several oils and botanicals are used for banding the trees to protect from insects. Mechanical control is the oldest form of protection against harmful organisms. They don't have any adverse effects on the environment, making it an important component of sustainable agriculture. They are labor-intensive as a result of which chemical control methods are used over mechanical ones.

5.2.4 Integrated Crop Protection Method

Novel modified farming systems in crop production are required since the demand for high-quality ecofriendly products is rising to reduce production risks and increase the income of the farmers. Integrated crop protection can be defined as the prevention or minimization of economic damage to crops caused by harmful species with a minimum of negative effects on the environment. It majorly focuses on the sustainability of agricultural production, production of quality harvests by exerting the least impact on the environment. It considers the reduction of harmful byproducts, emission, and destruction of the beneficial biota. The use of natural resources, various regulatory mechanisms, and non-chemical methods are the backbone of integrated crop protection. According to Han et al. (2005), for designing integrated crop protection strategies, a three-step approach is followed by:

1. optimizing prevention
2. establishing the need for control
3. choosing the actual control measures

Marliac et al. (2015) identified crop protection strategies that are classified as (1) ecologically intensive strategy which promotes ecofriendly methods of management, (2) the substitution strategy mainly based on the use of chemical pesticides, (3) the technologically intensive strategy which makes use of innovative technological methods. They concluded that organic farming is the way to achieve sustainable practices and low impact on the environment by exploiting the strategies mentioned above.

Preventive measures are more preferred than curative ones. Prevention of colonization, sanitation, clean seeds or vegetative propagating materials, crop rotation, adjustment of planting time, removal of weeds, fencing against vertebrates, sealing, or repelling against storage pests are employed. A synchronization is a basis for the prevention of pest attack. Susceptible plants can be grown at a specific time when the invaders are less pervasive.

Biopesticides are living organisms (bacteria, viruses, fungi, etc.) or their products (phytochemicals, microbial products) or by-products (semi-chemicals), which can be used for the management of pests that are injurious to crops.

The advantages of using biopesticides are:

1. They are non-toxic and non-pathogenic to non-target organism
2. high specificity
3. mostly compatible with other chemicals
4. in some cases, the pathogenic microorganisms can become established in a pest population or its habitat and provide control during subsequent pest generations or seasons
5. enhance the root and plant growth

Biopesticides include biofungicides, bioherbicides, bioinsecticides, and baculovirus. Bacterial biopesticides are the most widely used out of all the biopesticides. They are cheaper than the other modes of pest bioregulation. One of the *Bacillus* species, *Bacillus thuringiensis,* has developed many molecular mechanisms to produce insecticidal toxins, and most of the toxins are coded for by several Cry genes (Schnepf et al., 1998).

5.3 Registration of Biopesticides

Biopesticides can be registered under the Insecticides Act, 1968, as per the government of India. Registration of pesticides is the registration committee (RC) constituted under Section 5 of the Insecticides Act, 1968. The registration committee considers the registration of pesticides after scrutinizing their formulae and verifying claims regarding efficacy and safety to human beings and animals. It specifies dosage, manner of use, application time and technique, and precautionary measures printed on the labels and leaflets. For registration of biopesticide, there needs to be a provisional registration under Section 9(3B) for the first-time introduction. The application for registration has to be submitted after inclusion in the schedule to the act, sample import permit, and data generation under Indian condition. Registration under Section 9(3) and 9(4) of the Act ME-TOO is made for permanent registration.

The following is a list of registered biopesticides under the Insecticides Act, 1968 (Gupta and Dikshit, 2010):

- NPV of *Helicoverpa armigera*
- NPV of *Spodoptera litura*
- *Bacillus thuringiensis* var. *galleriae*
- *Bacillus sphaericus*
- *Trichoderma harzianum*
- *Beauveria bassiana*
- Need-based pesticides

5.4 Relationship between Integrated Pest Management (IPM) and Organic Crop Protection (OCP)

Both IPM and OCP are holistic approaches that are adopted with the sole objective of increasing crop protection by reducing the losses caused by harmful entities by incorporating environmentally friendly protection measures. IPM is one of the components of organic farming that considers methods of organic origin and negligible chemical use. Both approaches amins at protection of crops by keeping the pest population below the economic threshold limit by avoiding the use of chemical pesticides. IPM is operated following the principles of organic farming and crop protection. The major difference between the concepts lies in the extent of the use of chemicals in both cases. OCP completely restricts the use of chemicals, whereas, in IPM strategies, need-based chemical pesticides are allowed. IPM and OCP are used interchangeably many times. But both of them are rather continuous approaches that enrich each other.

5.5 Analysis of Decision Making in Crop Protection

The decisions for crop protection are taken by farmers basing on certain factors: the farmer's objective, the perception of the pest attack and the extent of damage caused, control measures available to him, and the decision rules by which it operates (Norton, 1976). The main objective of crop protection is to safeguard the crop against harmful species. Different weeds and pests have different effects on the crop (Letourneau et al., 2006). Weed affects the normal developmental mechanism of the plants by interfering with the available solar radiation, nutrients, and water. Pathogens and pests interfere with the metabolism of the plant by reduction of chlorophyll content and quality of the crop strand, which results in the reduction of the nutrient status. The crop protection methodology should be based on the extent and nature of damage caused to the crop. The research outcome, when combined with the farmer's need on-field, brings out the best crop protection strategy. The critical period and stage of attack should be studied for adequate measures.

5.6 Conclusion and Future Prospects

In today's world of deteriorating environmental conditions and increasing global warming, there is an urgent need to change the present way of exploitation of natural resources and step into a sustainable one. There is a requirement of research advances in the interaction of weeds, pests, and diseases with crop production and chalk out proper strategies to protect the crop. Attention to enhancement of nutrient efficiency of soil, inherent resistance, water holding capacity, botanicals, composts is needed. Scientific knowledge of various microbial formulations for pest and disease control has to be improved.

Since sustainable agriculture aims at present use of natural resources without hampering the future and the environment as well, use must adopt organic and biointensive strategies for crop protection. As a result of which we can get rid of the harmful organisms along with maintaining the population of the beneficial organisms. Due to excessive chemical pesticide threats in today's world, there is an urgent need to shift to ecofriendly cropping methods. The application of biopesticides is an important part of the integrated management system in agricultural production. By combining the safety and performance of these natural products, we can make optimum use of them. For this, proper decision-making strategy must be taken care of for need base application and optimum outcome. In the current situation of degrading environmental conditions, it is the responsibility of both researchers and farmers to make correct decisions to enhance agricultural production along with environmental safety.

REFERENCES

Cook, R. J. (1986). Plant health and the sustainability of agriculture, with special reference to disease control by beneficial microorganisms. Biological Agriculture & Horticulture, 3(2–3), 211–232.

De Haan, J. J.; Wijnands, F. G.; Sukkel, W. Integrated crop protection as a system approach. In *Working Group "Integrated Protection in Field Vegetable Crops": Proceedings of the Meeting, Deinze, Belgium*, 2005, *28*, 61–72.

DeDatta, S. K. Weed control in rice in South and Southeast Asia. In *Weeds and Weed Control in Asia: Proceedings of a Seminar on Weeds and Their Control*, 1981,1–24.

Froud-Williams, R. J.; Chancellor, R. J.; Drennan, D. S. H. The effects of seed burial and soil disturbance on emergence and survival of arable weeds in relation to minimal cultivation. *Journal of Applied Ecology*, 1984, *21* 629–641.

Gupta, S.; Dikshit, A. K. Biopesticides: An ecofriendly approach for pest control. *Journal of Biopesticides*, 2010, 3(Special Issue), 186.

Letourneau, D.; van Bruggen, A. *Crop Protection in Organic Agriculture*, 2006, 93–121. Australia: CSIRO Press.

Marliac, G.; Penvern, S.; Barbier, J. M.; Lescourret, F.; Capowiez, Y. Impact of crop protection strategies on natural enemies in organic apple production. *Agronomy for Sustainable Development*, 2015, 35(2), 803–813.

Norton, G. A. Analysis of decision making in crop protection. *Agro-Ecosystems*, 1976, 3, 27–44.

Schnepf, E.; Crickmore, N. V.; Van Rie, J.; Lereclus, D.; Baum, J.; Feitelson, J.; Dean, D. H. *Bacillus thuringiensis* and its pesticidal crystal proteins. *Microbiology and Molecular Biology Reviews*, 1998, 62(3), 775–806.

Walker, P. T. Crop losses: The need to quantify the effects of pests, diseases and weeds on agricultural production. *Agriculture, Ecosystems & Environment*, 1983, 9(2), 119–158.

Walker, R. H. Crop manipulation in integrated weed management systems. *Weed science*, 1982, 30, 17–24.

6

Strategies for the Application of Biopesticides in Crop Protection

C. R. Patil and Shekharappa

CONTENTS

6.1 Introduction

The post-green revolution period in India although helped initially to increase the food grain production but eventually has raised a question mark on the sustainability in production. The losses caused due to insect pests are 23.3% in the post-independence period as compared to 7.2% during the pre-independence period in India (Dhaliwal et al., 2013). On looking back and carefully analyzing the events and strategies adopted, two reasons that emerged and appeared to have threatened sustainability are indiscriminate and excessive use of agrochemical which possibly resulted in soil, air, and water pollution. Ultimate concerns emerging from this era are current problems with soil, human health, and threatened ecosystem. To restore these, there is an urgent need to address ecological, social, and economical issues in a concerted manner. In view of this there is growing emphasis on ecofriendly and sustainable approaches in agriculture production. Consequently, scientists, policymakers and all stakeholders worldwide are keen to embrace the principles of organic farming for its eco-friendly and low-cost production systems, which are key for sustainability.

Chemical pesticides have caused ecological imbalance, pesticide resistance, destruction of natural enemies, and increased level of pesticide residues on this earth (Lingappa et al., 2003). Both domestic and international markets are discarding products having pesticide residues. This has provided greater impetus to develop alternative non-chemical approaches to pest management.

6.2 Strategies for Insect Pest Management

Crop protection is a key aspect of crop production. Considering the fact that crop protection influences, quality, economics, and safety of produce, strategies adopted in protecting crops from pests become relevant. Presently, available technologies aim to provide an ecologically sound pest management program with sustainable use of renewable natural resources. These technologies are discussed below.

6.2.1 Physical Methods

These methods aim to reduce pest populations by using devices that affect them physically or alter their physical environment. Use of certain physical forces for eradication of insects including desiccation, hot or cold treatment, light traps picking and destroying, separating and protecting produce from attack by insect pests by using storage techniques and physical separation are some general strategies.

6.2.2 Mechanical Methods

These methods are aimed at the direct destruction of the pests as well as their removal from cultivated plants by means of manual devices such as handpicking, exclusion by screens and barriers, trapping and suction devices, sticky traps, fishmeal trap, clipping, pruning, and crushing.

6.2.3 Cultural Practices

Cultural control of pests and diseases involves changes or modifications in the cultural practices in crop husbandry in such a way as to create conditions unfavorable for the development of insect pest populations and favorable for development of its defenders. The various cultural practices are discussed in the following sections.

6.2.3.1 Time of Sowing

Most insect pests or diseases attack plants only in a certain life stage. Therefore, it is crucial that this vulnerable life stage doesn't correspond with the period of high pest density and thus that the optimal planting time is chosen. Manipulation of sowing dates in order to create a balance in the agroecosystem that would favor growth, development, and production of a crop and be detrimental to insect pests is an important practice to avoid a pest problem. Alteration in planting time helps to minimize pest damage by causing asynchrony between the host plant and the pest or synchronizing insect pests with their natural enemies or crop production with available alternate host plants of the pest.

Early sowing is found to reduce shoot fly, head bug, and midge damage in sorghum and millets, white grub damage in groundnut and mustard aphid damage in *Brassica*, and gall midge and leaf folder in rice crops. Timely sowing reduces bollworm damage in cotton. Early sowing of safflower by the end of September or in early October followed by groundnut in December–January reduces aphid and thrips damage. For the management of sunflower necrosis disease, sowing after August 15 is recommended.

6.2.3.2 Seed Rate and Plant Spacing

Use of healthy seeds, i.e., weed/disease-free seeds, is very important and imperative. In addition, sufficient distance between the plants reduces the spread of a disease. Good aeration of the plants allows leaves to dry off faster, which hinders pathogen development and infection. Thus, seed rate and planting geometry influence pest occurrence. Higher seed rate is found most useful in some of the crops such as maize and sorghum to compensate the lost seedlings killed by stem borers and shoot fly, respectively, as their occurrence is specific to the age of the crop. A higher seed rate helps to reduce peanut bud necrosis disease (PBND) in groundnut. This management involves removal of infested plants and maintaining normal plant population to obtain higher yields.

Plant spacing can influence crop pest interrelations and reduce pest damage. Plant spacing has a direct effect on a host plant, as it influences plant health, growth and development of the plant, and the micro-environment of the pest species. Close spacing tends to produce favorable micro-environments for some pests but may reduce losses due to weeds.

Closer spacing in groundnut lowers the incidence of thrips, leafhoppers, and leaf miner and also increases parasitism in the latter. Close spacing of groundnut controls rosette disease since the aphid vector; *Aphis craccivora* alights less frequently compared with the situation in widely spaced groundnut. There is also a positive correlation between the incidence of *Chilo partellus* (stem borer) and plant population in sorghum. On the other hand, closer spacing in cotton increases the damage caused by leafhopper, whitefly, and bollworms.

6.2.3.3 Soil Tillage

Fall plowing is often helpful in reducing the overwintering population of *Helicoverpa* spp., Red headed hairy caterpillar (RHHC), root grub, and several species of cutworms that undergo diapause in the soil during winter. It also helps in exposing the resting pupae to their natural enemies like birds and also to the action of sun and wind. Similarly, the pathogens present in soil are killed since they are exposed to high temperature after plowing.

Deep plowing minimizes the incidence of armyworm in cereal crops, white grubs in groundnut and chili, pupae of hairy caterpillars attacking groundnut, green gram, etc. Tilling soils infested with grasshopper helps to expose eggs to desiccation and predation by insects and birds. It has been observed that higher egg pod density in the area being tilled especially between February and April, just before the rains begin, was found to be effective in controlling different species of grasshoppers. Racking up and hoeing of the soil around melon plants and fruit trees helps to destroy fruit fly pupae.

6.2.3.4 Plant Diversity

Intrafield diversity helps in the deterrence of insect pest colonization. Crop diversity can be enhanced by intercropping, trap cropping, border cropping, banker cropping, sacrificing cropping ecofeast, crop rotation, water management, crop rotation water management, sanitation, closed season, harvesting practices, and by the presence of weeds or by crops grown in adjacent fields. If interplanted crops are also suitable host plants for a particular pest, they may reduce feeding damage on the main crop by diverting the pest.

6.2.3.5 Crop Rotation

Rotation of host by non-host plants often provides an effective and economical means of reducing pest populations to subeconomic levels. Crop rotation is most effective against pests that have a narrow host range and dispersal capacity.

Cotton should be rotated with non-preferred hosts such as ragi, maize, groundnut, cowpea, or soybean to minimize the incidence of insect pests. Similarly, rotation of groundnut with non-leguminous crops is recommended as a method for minimizing the damage by leaf miner. This plays a very important role in disease management as well.

6.2.3.6 Water Management

Managing water effectively has important implications in insect control and disease spread. Soil moisture, in the form of irrigation or rainfall, can influence both the host and the soil microflora and fauna.

Small insect pests are easily washed off from plants by overhead irrigation or heavy rainfall and soil insects may be killed by the pressure of swelling soil particles in saturating soils, but these practices can help pathogenic fungal spores to germinate and even spread. Therefore, caution must be exercised depending on situation. In arid and semi-arid regions, where irrigation is essential for crop growth, irrigated crops may act as a magnet for certain actively moving pests, while it may prevent damage caused by termites. All of these clearly indicate the importance of irrigation in pest management, and adequate knowledge about the impact on each crop is essential before applying this strategy.

6.2.3.7 Sanitation

This practice involves the removal and/or destruction of crop residues/weeds, volunteer plants, and nearby wild host plants to eliminate a pest by destroying its food and shelter. Clean cultivation as a method for minimizing the incidence of insect pests was advocated more than a century ago by Balfour, in 1887. These measures are useful and widely applied in production of several major crops such as cotton, wheat, sorghum, groundnut, tobacco, etc.

Clean cultivation may be achieved by plowing directly or shredding and chopping, burning residues, or raking and scooping them into piles for burning. Removal of cotton sticks by August 1 every year was enforced under the Cotton Pests Act 1911, which helped in reducing the incidence of pink bollworm in Madras state.

Destruction of alternate host plants helped to reduce incidence of shoot fly, stem borer and head bug in sorghum. Similarly, in cotton, the incidence of spotted bollworms could be minimized by removal of alternate hosts. Avoiding ratooning of pigeonpea during the off season helped to reduce the carryover of pod fly and eriophid mite (*Aceria cajani*), an important vector of pigeonpea sterility mosaic virus. Selective pruning of pigeonpea can be useful in reducing the incidence of leaf folder and spotted caterpillar.

6.2.3.8 Closed Season

The enforcement of a closed season especially helps to break the food cycle of monophagous insect pests and disrupt the disease cycle. Crop holiday breaks the pest succession most effectively when combined with a sanitation method of destruction of old plants and crop debris.

6.3 Host Plant Resistance

Host plant resistance (HPR) is the result of interaction between plant and insect. Plant resistance is described as the relative amount of heritable qualities that influence the ultimate degree of damage done by the insect. In practical agriculture, resistance represents the ability of a certain variety to produce a larger crop of good quality than ordinary varieties at the same level of insect population.

Considerable progress has been made in India in identification and utilization of resistance for crop pests. Insect resistant varieties have been developed for rice (gall midge, stem borers, brown plant hopper and green leafhopper) maize (maize stem borer and pink stem borer), sorghum (shoot fly and midge), groundnut (*Spodoptera*, leaf miner, leafhoppers, and thrips), red gram (*Helicoverpa*) and cotton (bollworms and leafhoppers), etc.

Varieties possessing multiple resistance to a number of insect pests and diseases are ideal in organic production systems. Cultivars with multiple resistance to insects, nematodes, and pathogens and tolerance to abiotic stresses (drought, soil mineral toxicity, etc,) have been developed in various crops, viz. sorghum, wheat, cowpea, alfalfa, maize, soybean, and rice.

6.4 Biological Control

The term biological control has been applied to include virtually all pest control measures except the application of chemical pesticides. However, it is generally agreed that the term should be defined as the use of natural enemies as pest control agents," and macroorganisms that include parasitoids and predators. As this is a relatively permanent and sustainable approach, it suits organic production systems. When the natural enemies consist of micro-organisms such as bacteria, fungi, protozoa, nematodes, or viruses the term microbial control is often used.

Biological control practices involve mainly three major techniques, viz. conservation, introduction, and augmentation.

Several species of natural enemies including spiders, insects, mites, nematodes, birds, fungi, have been recorded as bioregulators of the population of injurious insects. Among them, only a few are efficient in controlling the pest population. Egg parasitoid, viz. *Trichogramma*, is one bioagent that has been used extensively and regularly to suppress lepidopterous insects in many crops, viz., sugarcane, cotton, rice, tomato, etc. *Bracon kirkpatriki*, *Chelonus blackburni*, *Apanteles angleti*, *Rogos aligerensis*, *Campoletis chloridaea*, etc., are some of the larval parasitoids that have greater potentiality in biocontrol programs on various crops. Standardization of mass multiplication techniques is required for establishing viable commercial production units.

Among different predators, ladybird beetles, *Menochilus sexmaculatus*, *Coccinella* sp., *Scymnus* sp., several species of spiders, mirid bugs and Lace wing bug, and *Chrysoperla carnea* were found to regulate the pest population.

Birds (both carnivorous and omnivorous) are predators in various crop ecosystems. The capacity of birds to locate prey from certain distances because of their better eyesight and modification of their beaks have made them excellent predators. Black drango, Indian mynas, cattle egrets, and sparrows are efficient predatory birds. Attraction of birds to any crop ecosystem is possible by providing places (perches) for alighting/nesting and water source. Bird perches can be inanimate, e.g., branched twigs, long poles with a cross, etc., or animate, e.g., stray planting of tall growing cereal crops such as maize, sorghum, and stiff fiber plants like pundi (mestha) plants, etc.

6.5 Botanicals

Plants are the richest source of renewable active chemicals that check insect populations. These are secondary metabolites that disrupt the fundamental, physiological, or biochemical process of the insect. As many as 2,121 plant species have been reported to possess pest control properties, of which 1,005 are insecticidal, 384 are antifeedants, 297 are repellants, 27 are attractants and 31 have growth-inhibiting properties (Table 6.5) (Lingappa et al, 2003). These native plants can be utilized very effectively under organic plant protection. One among them is neem, and India has an estimated 18 million neem trees with a potential production to the tune of 0.7 million tons of fruit. It is suggested that on average, if planted on the border of one hectare field, four grown-up neem trees should yield neem seeds required to protect the crop. About 67 commercial neem-based formulations are currently available.

All parts of the neem tree possess insecticidal activity, but seed kernel is the most active and effective part. Neem bark, leaf, fruit, and oil as well as extracts with water (both hot and cold) have exhibited activity against insect pests. To date, 450 to 500 species of insects have been tested with neem products and 413 are reported to be susceptible at varied concentrations. The neem products exhibit repellent, antifeedant, ovicidal and growth inhibitory effects, and reduction in insect vigor, longevity, and fecundity. Azadirachtin, the most important biologically active component of neem, shows phagorepellent and toxic effects at 0.1 to 1000 parts per million (ppm) when incorporated into the diets of different insect species.

6.6 Biorational Approaches

Utilization of chemicals that affect insect behavior growth or reproduction for suppression of insect population is often referred to as biorational control. The utilization of pheromones has become important in organic pest management.

6.6.1 Pheromones

Pheromones are substances that are secreted to the outside by an individual and received by a second individual of the same species in which they evoke a specific response, i.e., definite behavior pattern or developmental process. Pheromones have currently emerged as viable alternative methods for monitoring, mass trapping, mating disruption, attraction, and destruction.

Traps baited with sex pheromones are widely used in practice for the detection of certain species in specific areas. If five moths are trapped per trap per day for three consecutive days, the control measures are suggested. It also helps in interventions based on threshold levels.

Mass trapping with pheromones as a direct control method is to lure away the large proportion of individuals from certain populations. This technique is successful in low pest population densities where moths have a low dispersal tendency or where populations have been physically isolated. A modification of the mass trapping techniques has been employed for the control of tephritid fruit flies for several decades.

The greatest success with the use of pheromones for the control of insects has been obtained with mating disruption technique. Pink bollworm (*Pectinophora gossypiella*), a key pest of cotton, has been successfully managed in a number of countries by using controlled release formulation of its female sex pheromone.

6.7 Conclusion and Future Prospects

It can be known that a number of practices originating from traditional knowledge to the adoption of scientific principles can provide handful of opportunities for managing pests especially the insect pests in organic production system. Furthermore, these practices should become routine as most of them can be practiced along with other ecofriendly approaches and fits into traditional practices. However, in organic production systems, creation of biodiversity, environmental protection including flora and fauna takes the topmost priority, and when practiced these principles of pest management presented in this chapter ensure sustainable agricultural production free of chemical residues. The rule is healthy food from a healthy environment that contributes to sustenance of life.

REFERENCES

Balfour S. G. E. *The Agricultural Pests of India and of Eastern and South Asia.* 1887, Agro Biological Publications, New Delhi, India.

Dhaliwal G. S., Singh R., and Jindal V. *A Text Book of Integrated Pest Management.* 2013, Kalyani Publishers, New Delhi, pp. 1–617.

Lingappa S., Shekharappa, and Patil R. K. *Integrated Pest Management.* 2003, University of Agricultural Sciences, Dharwad, pp. 1–100.

Section VI

Pesticide Exposure Impacts on Human Health and Need of Biopesticides in Organic Farming

7

Pesticides Exposure Impacts on Human Health and Need for Biopesticides in Organic Farming

Muhammad Haroon Sarwar, Muhammad Farhan Sarwar, Muhammad Taimoor Khalid, and Muhammad Sarwar

CONTENTS

7.1 Introduction

In a broadest sense, pesticides are used to reduce damage to crops, fruits and vegetables from insects, rodents, weeds and microorganisms to increase plants yield. Exposure of general population to synthetic pesticides occurs primarily through eating of food and drinking water contaminated with pesticide. residues. Synthetic pesticides have harmful health effects on humans and the environment at higher doses and may cause serious illness and even death. Chronic poisoning may produce an acute toxic effect which will necessitate toxicological emergency action and a chronic management of the patient until the body load is reduced or poison eliminated from the body (Akbar et al., 2005; Sarwar, 2004; Sarwar & Salman, 2015a).

Exposure of the general population to pesticides occurs mainly through eating of food and drinking water contaminated with pesticides. In contrast, substantial exposure to pesticides can also occur when living close to a workplace that uses pesticides or even when workers bring home contaminated articles (Jaga & Dharmani, 2003). Non-occupational exposure originating from pesticide residues in food, air and drinking water generally involves low doses and is chronic (or semi-chronic) (Harris & Gaston, 2004; Sarwar et al., 2011; 2007).

As a result of pesticide uses on crops (Figure 7.1) otherwise in or around the home (Figure 7.2), individuals can be exposed during the preparation and application of pesticides or even after applications are completed. In contrast, delayed exposure can occur through inhalation of residual air concentrations or exposure to residues found on surfaces, clothing, bedding, food, dust, discarded pesticide containers, or application equipment (Davis et al., 1992). Pesticide mishandling such as transferring the products from their original packages into household containers and also lack of compliance with instructions of the label can also be sources of exposure (Figure 7.3).

For instance, separate pictograms and or signal words are provided on labels as warning signs of potential pesticide exposure.

An enormous volume of evidence exists on the relation between exposure to pesticides and elevated rate of chronic diseases, such as cancers, diabetes, and neurodegenerative disorders (Parkinson's, Alzheimer's, amyotrophic lateral sclerosis, reproductive disorders) besides respiratory and cardiovascular diseases, chronic nephropathies, autoimmune, rheumatoid arthritis, chronic fatigue syndrome and aging (Sarwar, 2016b; Sarwar & Salman, 2015b).

These diseases, disorders, and chronic health problems tax corporate, government, community, and individual resources by trillions of dollars annually, but with pressure to feed a growing world population with alternatives available (Mostafalou & Abdollahi, 2013).

These pesticides can play a role in the cancer process by either non-genotoxic mechanisms such as promotion, peroxisome proliferation, and hormone imbalance (Hodgson & Levi, 1996; Sarwar et al., 2014), or by affecting the carcinogenic process in a variety of ways, both by altering the genome and by providing a growth advantage for neoplastic cells (Williams et al., 1992; Sarwar et al., 2019). Also, the pesticide used by appliers and exposure of the general population of crop-growing regions are associated with increased birth anomalies (Garry et al., 1996).

FIGURE 7.1 Foliar spraying of crops.

FIGURE 7.2 Spraying of home.

Because of the basic similarities between mammalian and insect nervous systems, insecticides (organochlorines, organochlorines, and carbamates) have been designed to attack the insect nervous system and produce acute and chronic neurotoxic effects on mammals (Tanner & Longston, 1990). Aplastic anemia has been described as an idiosyncratic immunologic response to exposure to organochlorine pesticides (Hayes & Laws, 1991). Also, allergic responses, especially allergic dermatitis, can be seen with many classes of pesticides (Hogan, 1990). An exposure tochlorpyrifos during a development period is known to produce immune competence (Navarro et al., 2001).

Some of the other effects that are known to be induced by pesticides in humans include cancer (various types), infertility, liver damage, kidney damage, premature death, hyperactivity in children, bronchitis, defective sight and blindness, birth defects, blood diseases, nervous system damage, heart defects, and sudden death. These illnesses are very devastating with high morbidity and mortality rates, but as they are preventable, they can be best prevented (Sarwar & Salman. 2015c; Sarwar et al., 2015).

7.2 Synthetic Pesticides

Synthetic pesticides are designed to be stable, have a good shelf life, and be easy to distribute. They are also designed to target pests and have low toxicity to non-target animals and the environment (Gilden et al., 2010). Because there are many kinds of potential pests, so there are several kinds of pesticides. The following are some examples:

- **Insecticides:** Reduce the destruction and contamination of growing and harvested crops by insects and their larvae or nymphs.
- **Herbicides:** Also known as weedkillers, can destroy or inactivate weeds to improve crop yield.

FIGURE 7.3 Warning signs of potential pesticide exposure.

- **Rodenticides:** Important for controlling destruction and contamination of crops by vermin as well as mouse and rodent-borne injuries.
- **Fungicides:** Especially important for the protection of harvested crops and seeds from fungal rot.
- **Acaricides:** Acaricide or miticide is a pesticide that kills members of the arachnid subclass Acari, which includes ticks and mites.
- **Nematicide:** Nematicide is a type of chemical pesticide used to kill plant-parasitic nematodes.
- **Bactericide:** Bactericide is a chemical agent that helps to prevent the formation of bacteria.
- **Virucide:** A virucide is any physical or chemical agent that deactivates or destroys viruses. This differs from an antiviral drug, which inhibits the proliferation of the virus.
- **Molluscicides:** Also known as snail or slug pellets or baits, molluscicides are pesticides used against mollusks, which are usually used in agriculture or gardening, to control gastropod pests, specifically slugs and snails which damage crops or other valued plants by feeding on them.
- **Slimicide:** Slimicide or slime agent is a broad-spectrum antimicrobial pesticide used to kill slime-producing microorganisms such as algae, bacteria, fungi, and slime molds.
- **Pediculicide:** An agent that destroys lice parasites from human hair.
- **Avicide:** An avicide is any substance that can be used to kill birds.
- **Algaecide:** An algaecide or algicide is a biocide used for killing and preventing the growth of algae.
- **Fumigants:** Fumigants are used in gaseous form as pesticides to reach inaccessible areas. Their examples include phosphine, cynide, carbon tetrachloride, ethylene oxide, and methyl bromide.

Insecticides are structured into classes that share common chemical structures and modes of action or the way they kill an insect. Carbamates, neonicotinoids, and pyrethroids are examples of insecticide classes.

- **Organochlorines:** Including dichlorodiphenyltrichloroethane (DDT), these have been largely banned or restricted due to negative effects on the environment.
- **Organophosphates:** Insecticides that target the nervous system. Several of them have been banned or restricted due to toxic accidental exposures.

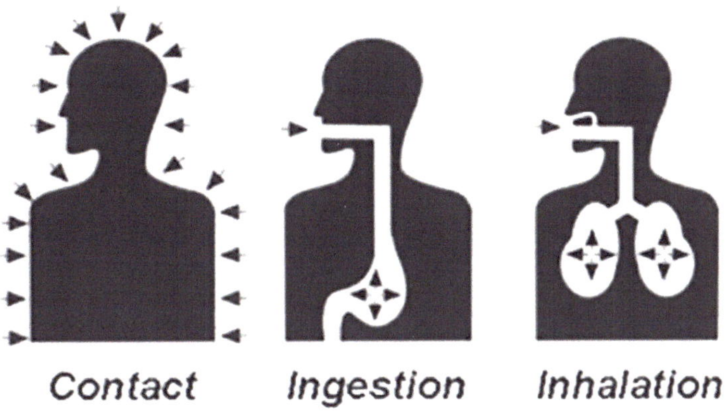

FIGURE 7.4 Main routes of pesticide entry.

- **Carbamates:** Insecticides affect the nervous system similarly to organophosphates, but they are less toxic as their effects wear off more quickly.
- **Pyrethroids:** Also affect the nervous system, and they are laboratory-produced version of a natural pesticide that is found in chrysanthemums.
- **Neonicotinoids:** Insecticides are used on leaves and trees, and they are currently under scrutiny of unintended harm to bees.

Rodenticides have been used to kill small animals like rats and mice. Fluoroacetate is too toxic for use in the home. The most prominent effects of acute poisoning in man from ingestion or inhalation of fluoroacetate are vomiting and convulsions. It induces irregular heartbeat, exhaustion, and coma, and death is usually due to respiratory failure. Pindone is a coumarin anticoagulant with actions similar to warfarin, which is also used as a rodenticide. In large doses, they can cause vascular collapse (Osman, 2011).

Herbicides are now used extensively to destroy noxious weeds. Most of the herbicide exposures result in eye or skin irritations. Chlorophenoxy compounds are used to control broad-leaf weeds. Dinitrophenolsare is also used extensively in weed control, and their toxic effect is due to the uncoupling of oxidative phosphorylation. Thus, the metabolic rate is increased with a subsequent rise in body temperature. Other symptoms of acute poisoning include nausea, restlessness, sweating, tachycardia, rapid respiration, fever, and cyanosis.

Most fungicides that are used to control fungal diseases on seeds and plants are not particularly toxic except the mercurials. Both and ethylenebis dithiocarbamate have low acute toxicity; however, they may have teratogenic (substance capable of interfering with the development of a fetus causing birth defects) and carcinogenic (substance that causes cancer) effects and can produce disulfiram-like effects when alcohol is ingested. In humans, in the environment and during cooking of contaminated food, the ethylenebis dithiocarbamate is broken down to form ethylenethiourea, which is carcinogenic and teratogenic.

Fumigant is used to fumigate grains for insect pest control. Fumigant phosphine is slowly released from aluminum phosphine tablets in the presence of atmospheric moisture. Poisoning by inhalation may cause a fall in blood pressure, pulmonary edema, collapse, convulsion, and coma. The first sign of chronic poisoning is toothache, followed by swelling of the jaw and necrosis of the mandible (phossy jaw). There may be anemia and spontaneous fractures (Alewu & Nosiri, 2011).

FIGURE 7.5 Personal protective equipment.

Farmers' exposure mainly occurs during the preparation and application of the pesticide spray solutions and the cleaning-up of spraying equipment. Farmers who perform manual labor in areas treated with pesticides can face major exposure from direct spray, drift from neighboring fields, or by contact with pesticide residues on the crop or soil. The dermal and inhalation routes of entry are typically the most common routes of farmers' exposure to pesticides than ingestion (Figure 7.4). Dermal exposure during usual pesticide handling takes place in body areas that remain uncovered by protective clothing, such as the face and the hands. Farmers' exposure to pesticides can be reduced through less use of pesticides and the correct use of the appropriate type of personal protective equipment (Figure 7.5) in all pesticide handling (Damalas et al., 2016; Sarwar, 2016c).

7.3 Forensic Toxicology

Forensic toxicology concerns the medical and legal aspects of the harmful effects of chemicals on animals, including man. In forensic toxicology, an attempt is made to identify the chemical in the tissue by chemical analysis and in case of death, to establish the cause or circumstances of death. The methods for the management and treatment of acute and chronic poisoning, including antidotes, have been developed (Alavanja, 2009). Thus, forensic toxicology deals with medical and legal aspects of the harmful effects of chemicals on the human body (Sarwar, 2020).

Crops are treated with biopesticides, producing of foods that are better for people's health and the planet. Many farmers use these as part of their Integrated Pest Management (IPM) programs (Sarwar, 2013a; 2013b; 2014; 2017; Sarwar & Hamza, 2013), so they can rely less on higher-risk pesticides and effectively produce higher crop yields and quality with lower impact on the environment and human health.

7.4 Organic Farming

Organic farming is a technique, which involves the cultivation of plants and rearing of animals in natural ways. This process involves using of biological materials, avoiding synthetic substances to maintain soil fertility, pest control, and ecological balance, and minimizing pollution and wastage. It involves much more than choosing not to use pesticides, fertilizers, genetically modified organisms, antibiotics, and growth hormones (Paull, 2019). Consumers purchase organic foods for many different reasons. Many want to buy food products free of chemical pesticides or grown without conventional fertilizers, as shown in Figure 7.6.

The organic program has created rules to keep organic products free from extremely harmful pesticides, such as synthetic pesticides. Some naturally occurring ingredients, such as plant extracts, insect pathogens, and fungal derivatives, may be used so long as they are not indicated as prohibited. Some pesticides can be used as long as they do not contaminate crops, soil, or water. Among these are alcohols, chlorine materials, copper sulfate, hydrogen peroxide, ozone gas, peracetic acid, soap-based algicide or demossers, and sodium carbonate peroxyhydrate. Among substances strictly prohibited is ash from manure burning, arsenic, calcium chloride, lead salts, potassium chloride, sodium fluoaluminate (mined), sodium nitrate, strychnine, and tobacco dust (nicotine sulfate) (Sarwar, 2016d; 2016e).

7.5 Need for Organic or Biopesticides

Organic pesticides or biopesticides are naturally occurring chemicals, but they may be artificially reproduced in labs for use in organic farming. At the same time, chemical pesticides are neurotoxic to pests and often to people, so biopesticides use other modes of action, including mating disruption, anti-feeding, suffocation, and desiccation.

Organic farming makes use of biopesticides, or naturally occurring pesticide chemicals that have evolved in plants. First, "organic" does not mean "pesticide-free"; rather, it refers to specialized pesticides that occur in nature and are used instead of synthetic pesticides. Secondly, "natural" does not mean "non-toxic"; but organic pesticides can also be harmful to health and the environment. Biopesticides are certain types of pesticides derived from such natural materials as animals, plants, bacteria, and certain minerals. For example, canola oil and baking soda have pesticidal applications and are considered biopesticides. As a direct result, start-ups in the biopesticide space have proliferated and created a competitive, cutting-edge arena for innovation (Olson, 2015).

7.5.1 Classes of Biopesticides

Biopesticides fall into three major classes.

7.5.1.1 Biochemical Pesticides

Biochemical pesticides are naturally occurring substances that control pests by non-toxic mechanisms. Conventional pesticides, by contrast, are generally synthetic materials that directly kill or inactivate the pest. Biochemical pesticides include substances that interfere with mating, such as insect sex pheromones and various scented plant extracts that attract insect pests to traps (Sarwar, 2015a).

7.5.1.2 Microbial Pesticides

Microbial pesticides consist of a microorganism (bacterium, fungus, virus, or protozoan) as the active ingredient. Microbial pesticides can control many different kinds of pests, although each separate active ingredient is relatively specific for its target pest or pests. For example, some fungi control certain weeds, and other fungi kill specific insects. The most widely used microbial pesticides are subspecies and strains of *Bacillus thuringiensis* or Bt. Each strain of this bacterium produces a different mix of proteins and kills one or a few related species of insect larvae. While some Bt ingredients control moth larvae found on plants, other Bt ingredients are specific for larvae of flies and mosquitoes. The target insect species are determined by whether the particular Bt produces a protein that can bind to a larval gut receptor, thereby causing the insect larvae to starve (Sarwar, 2015b).

FIGURE 7.6 Crops, fruits, and vegetables from organic or ecological farming.

7.5.1.3 Plant Incorporated Protectants

Plant incorporated protectants are pesticidal substances, which plants produce from genetic material that has been added to the plant. For example, scientists can take a gene for the Bt pesticidal protein and to introduce genes into the plant's genetic material. The plant, instead of Bt bacterium, manufactures the substance that destroys the pest instead of the Bt bacterium. The protein and its genetic material, but not the plant itself, which are pesticidal substances that plants produce from genetic material, are added to the plant. For example, scientists can take the Btpesticidal protein gene and introduce it into the plant's genetic material. Then the plant manufactures the substance that destroys the pest (Sarwar, 2015c).

There are too many types to outline here, but a few examples of important organic pesticides are given, although this list is not comprehensive, but it illustrates a few important concepts.

- **Rotenone:** This is an insecticide used in combination with other organic pesticides. It is naturally produced as a beetle deterrent by several tropical plants and notoriously toxic to fish.
- **Copper sulfate:** It destroys fungi and some weeds. Although classified as a biopesticide, it is industrially produced and toxic to humans and the environment at high levels.
- **Horticultural oils:** This refers to oil extracts from various plants with anti-insect effects. These differ in their ingredients and potential side effects, and some can harm beneficial insects like bees.
- **Bt toxin:** It is produced by bacteria and is effective against several kinds of insects; Bt toxin has been introduced into some types of genetically modified organisms (GMO).

7.5.2 Advantages of Using Biopesticides

Research on organic products shows structurally fewer pesticide residues than on conventional products; this is one of the reasons why manufacturers of baby food usually opt for organic products, because babies are particularly sensitive to the effects of pesticide residue. In addition to employing natural enemies (such as parasitic wasps) and preventive measures (soil life, biodiversity, crop rotation), organic farming also uses natural resources such as plant extracts (Sarwar, 2015d).

The advantages of using biopesticides include; biopesticides are usually inherently less toxic than conventional pesticides; generally affect only the target pest and closely related organisms in contrast to broad-spectrum, conventional pesticides, which may affect organisms as different such as birds, insects, and mammals; often are effective in very small quantities, and a lot decompose quickly, resulting in lower exposures and largely avoiding pollution problems caused by conventional pesticides; and when used as a component of IPM programs biopesticides can greatly reduce the use of conventional pesticides while crop yields remain high. To use biopesticides effectively and safely; however, users need to know a great deal about managing of pests and must carefully follow all label directions(Sarwar, 2016f).

In addition to pesticides, some non-organic products, if not commercially available in organic form, can be used as ingredients in products labeled "organic." The reality is that some organic farmers do use pesticides. Still, such products are primarily derived from natural substances and must go through a strict regulatory approval process to ensure they are not harmful to the environment and human health. Perhaps the most important point in the discussion about the use of pesticides in organic agriculture is that the organic rules require that approved synthetic pesticides be used only as a last resort. For many consumers, pesticide residues on food are a big concern and organic foods have been shown to contain far less pesticide residues than conventional foods (Sarwar & Salman, 2015d).

Neem *Azadirachta indica* A. Juss has a pesticide or insecticide formulation based botanical product that contains azadirachtin as an active ingredient. Azadirachtin is found to be very effective for over 600 species of insects. This is the best among insecticides due to properties that it is a natural product non-toxic;100% biodegradable and environment-friendly; completely safe for humans and beneficial insects like honeybee, parasites, and predators; can be mixed with other synthetic pesticide and also enhances their action; and a broad-spectrum insecticide. It exhibits various behavioral responses such as antifeedant, feeding deterrent, repellent, oviposition deterrent, and physiological responses such as insect growth regulator, molting inhibitor, reproduction inhibitor, antifertility, etc.

As a result, for improved aids of biopesticides in controlling of annoyances, apply these as soon as pests are observed, apply during immature pest stages; under high pest population, use it every 10 to 15 days depending upon the requirement; after spraying, if the rain comes, then again apply the formulation; good mixing is required before application; mix a fresh formulation for every application, and spraying during the evening hours is preferable (Rajput et al., 2003). Use of this botanical product on a large scale might contribute to the decline in mortalities of public health, wildlife and aquatic species, and protect the landscape (Sarwar, 2015e; 2015f; 2015g; 2016g; 2016h).

7.6 Biotechnology and Biopesticides

Biological pesticides are produced through biotechnology by harnessing the pest-fighting abilities of existing plants and microbes. Biotech-based pesticides, therefore, are becoming viable alternatives to chemical pesticides. It is concerned with the targeted modification of the genetic material of bacteria or plants, for example, to stimulate them to biosynthesize desired products.

Then there is the consideration over organic-approved Bt insecticide versus genetically engineered Bt or *B. thuringiensis* bacterium. The organic industry is hypocritical by allowing a foliar Bt spray to kill insects while opposing GM or genetically modified crops contain genes from the Bt. It is a completely different situation; Bt toxins in GM sweet corn are inside the kernel and remain there when the corn is cooked and eaten. People who eat that sweet corn will get a high dose of Bt toxins. By contrast, the Bt foliar spray breaks down quickly and does not remain on the plant. There is zero human exposure to foliar Bt spray, implying that GMO Bt is the same as foliar spray is an untruth."

7.7 Conclusions and Future Prospects

New tools or techniques with greater reliability than those already existing are needed to predict the potential hazards of pesticides, and reduce adverse effects on human health and the environment. The use of appropriate and well-maintained spraying equipment, along with taking all precautions that are required in all stages of pesticide handling, could minimize human exposure to pesticides and their potential adverse effects on the environment. Farmers need to be encouraged to reduce or eliminate pesticides, with the introduction of incentives to the farmers to help them shift from synthetic pesticides to biopesticides and organic farming. For general people, to reduce the risk from pesticides, they have to eat organically and ecologically grown food, wash and peel vegetables and fruits, and grow their own food. Organic farming can be a viable alternative production method for farmers, but one key to success is being open to alternative organic approaches to solve production problems. Any pesticides used during organic production must be registered and pay attention to pesticide product labels in their entirety because every ingredient, active or inert, noted in the ingredient list must comply with the rule. A biopesticide cannot be considered as non-toxic because it is natural in origin, but may generally be considered only as less toxic to humans.

REFERENCES

Akbar, A.; Sarwar, M.; Ahmad, N.; Tofique, M. Evaluation of different granular insecticides for the suppression of rice stem borers. In *Proceeding of 25th Pakistan Congres of Zoology*, March 1–3, Sindh Agriculture University, Tandojam. 2005, 25: 49–55.

Alavanja, M. C. R. Pesticide use and exposure extensive worldwide. *Reviews on Environmental Health*. 2009, 24 (4): 303–309.

Alewu, B.; Nosiri, C. Pesticides and human health. In: *Pesticides in the Modern World—Effects of Pesticides Exposure*, Stoytcheva, M. (Ed.). InTech, London. 2011, 231–250.

Damalas, C. A.; Koutroubas, S. D. Farmers' exposure to pesticides: Toxicity types and ways of prevention. *Toxics*. 2016, 4 (1): 1.

Davis, J. R.; Brownson, R. C.; Garcia, R. Family pesticide use in the home, garden, orchard, and yard. *Archives of Environmental Contamination and Toxicology*. 1992, 22: 260–266.

Garry, V. F.; Schreiemachers, D.; Harkins, M. E.; Griffith, J. Pesticide appliers, biocides, and birth defects in rural Minnesota. *Environmental Health Perspectives*. 1996, 104: 394–399.

Gilden, R. C.; Huffling, K.; Sattler, B. Pesticides and health risks. *Journal of Obstetric, Gynecologic, and Neonatal Nursing*. 2010, 39 (1): 103–110.

Harris, C.; Gaston, C. P. Effects of refining predicted chronic dietary intakes of pesticide residues: A case study using glyphosate. *Food Additives and Contaminants*. 2004, 21: 857–864.

Hayes, W. J.; Laws, E. R. *Handbook of Pesticide Toxicology*. Academic Press, San Diego.1991, p. 1576.

Hodgson, E.; Levi, P. E. Pesticides: An Important but underused model for the environmental health sciences. *Environmental Health Perspectives*. 1996, 104 (1): 97–106.

Hogan, D. J. Pesticides and other agricultural chemicals. In: *Occupational Skin Disease*, Adams, R. M. (Ed.). W. B. Saunders Co, Philadelphia. 1990.

Jaga, K.; Dharmani, C. Sources of exposure to and public health implications of organophosphate pesticides. *Pan American Journal of Public Health*. 2003, 14: 171–185.

Mostafalou, S.; Abdollahi, M. Pesticides and human chronic diseases: Evidences, mechanisms, and perspectives. *Toxicology and Applied Pharmacology*. 2013, 268 (2): 157–177.

Navarro, H. A.; Basta, P. V., Seidler, F. J.; Slokin, T. A. Neonatal chlorpyrifos administration elicits in immune function in adulthood: A neural effect? *Developmental Brain Research*. 2001, 130 (2): 249–252.

Olson, S. An Analysis of the biopesticide market now and where it is going. *Outlooks on Pest Management*. 2015, 5: 203–206.

Osman K. A. Pesticides and human health. In: *Pesticides in the Modern World—Effects of Pesticides Exposure*, Stoytcheva, M. (Ed.). InTech, London. 2011, 205–230.

Paull, J. Organic agriculture in Australia: Attaining the global majority (51%). *Journal of Environment Protection and Sustainable Development*. 2019, 5 (2): 70–74.

Rajput, A. A.; Sarwar, M.; Bux, M.; Tofique, M. Evaluation of synthetic and some plant origin insecticides against *Helicoverpa armigera* (Hubner) on chickpea. *Pakistan Journal of Biological Sciences*. 2003, 6 (5): 496–499.

Sarwar, M. Biopesticides: An effective and environmental friendly insect-pests inhibitor line of action. *International Journal of Engineering and Advanced Research Technology*. 2015a, 1 (2): 10–15.

Sarwar, M. Commonly available commercial insecticide formulations and their applications in the field. *International Journal of Materials Chemistry and Physics*. 2015b, 1 (2): 116–123.

Sarwar, M. Concept of integrated insect pests management. *Pakistan & Gulf Economists*. 2004, 23 (46–47): 39–41.

Sarwar, M. Development and boosting of integrated insect pests management in stored grains. *Journal of Agriculture and Allied Sciences*. 2013a, 2 (4): 16–20.

Sarwar, M. Implementation of integrated pest management tactics in rice (*Oryza sativa* L.) for controlling of rice stem borers (Insecta: Lepidoptera). *Rice Plus Magazine*. 2014, 6 (1): 4–5.

Sarwar, M. Indoor risks of pesticide uses are significantly linked to hazards of the family members. *Cogent Medicine*. 2016a, 3: 1155373.

Sarwar, M. Information on activities regarding biochemical pesticides: An ecological friendly plant protection against insects. *International Journal of Engineering and Advanced Research Technology*. 2015c, 1 (2): 27–31.

Sarwar, M. Inorganic insecticides used in landscape settings and insect pests. *Chemistry Research Journal*. 2016b, 1 (1): 50–57.

Sarwar, M. Insecticide risk exposes threat to aquatic life in surface water bodies and its remedying. *International Journal for Research in Applied Chemistry*. 2016c, 1 (6): 1–9.

Sarwar, M. Integrated control of insect pests on canola and other brassica oilseed crops in Pakistan. In: *Integrated Management of Insect Pests on Canola and Other Brassica Oilseed Crops*, Reddy, G. V. P. (Ed.). CABI, UK. 2017, pp. 193–221 (p. 408).

Sarwar, M. Integrated pest management (IPM)—A constructive utensil to manage plant fatalities. *Journal of Agriculture and Allied Sciences*. 2013b, 2 (3): 1–4.

Sarwar, M. Outdoors agricultural insecticides pose worth global risks and espousal of safety practices among farmers. *International Journal for Research in Agricultural Research*. 2016d, 1 (7): 1–9.

Sarwar, M. Pesticide's dangers to wildlife and safeguard dealings in reducing of the risks. *International Journal for Research in Health Sciences and Nursing*. 2016e, 1 (7): 15–25.

Sarwar, M. Potential uses of synergists in insecticides resistance management accompanied by their contributions as control agents and research tools. *Chemistry Research Journal.* 2016f,1 (3): 21–26.

Sarwar, M. The dangers of pesticides associated with public health and preventing of the risks. *International Journal of Bioinformatics and Biomedical Engineering.* 2015d, 1 (2): 130–136.

Sarwar, M. The killer chemicals as controller of agriculture insect pests: The conventional insecticides. *International Journal of Chemical and Biomolecular Science.* 2015e, 1 (3): 141–147.

Sarwar, M. The killer chemicals for control of agriculture insect pests: The botanical insecticides. *International Journal of Chemical and Biomolecular Science.* 2015f, 1 (3): 123–128.

Sarwar M. Typical flies: Natural history, lifestyle and diversity of Diptera. In: *Life Cycle and Development of Diptera*, Sarwar, M. (Ed.). InTech, London. 2020, p. 50.

Sarwar, M. Usage of biorational pesticides with novel modes of action, mechanism and application in crop protection. *International Journal of Materials Chemistry and Physics.* 2015g, 1 (2): 156–162.

Sarwar, M. Usage spots of biological insecticides in consort with target insect pests or vectors and application in habitat. *International Journal of Entomology and Nematology.* 2016g, 3 (1): 14–20.

Sarwar, M. A. Glance at pesticides usage: Remunerations and complications associated with insecticides putting in practice. *International Journal for Research in Agricultural Research.* 2016h, 1 (7): 10–19.

Sarwar, M.; Hamza, A. Adoption of integrated pest management strategy in rice (*Oryza sativa* L.). *Rice Plus Magazine.* 2013, 5 (9): 6–7.

Sarwar, M.; Salman, M. Insecticides resistance in insect pests or vectors and development of novel strategies to combat its evolution. *International Journal of Bioinformatics and Biomedical Engineering.* 2015a, 1 (3): 344–351.

Sarwar, M.; Salman, M. Overall notable health challenges about the toxicity of pesticides concerning to end users. *International Journal of Bioinformatics and Biomedical Engineering.* 2015b, 1 (3): 323–330.

Sarwar, M.; Salman, M. The paramount benefits of using insecticides and their worldwide importance in food production. *International Journal of Bioinformatics and Biomedical Engineering.* 2015c, 1 (3): 359–365.

Sarwar, M.; Salman, M. Success stories of eco-friendly organically acceptable insecticides as natural products discovery. *International Journal of Materials Chemistry and Physics.* 2015d, 1 (3): 392–398.

Sarwar, M.; Ahmad, N.; Bux, M.; Nasrullah; Tofique, M. Comparative field evaluation of some newer versus conventional insecticides for the control of aphids (Homoptera: Aphididae) on oilseed rape (*Brassica napus* L.). *The Nucleus.* 2011, 48 (2): 163–167.

Sarwar, M.; Akbar, A.; Ahmad, N.; Khan, G. Z.; Bux, M.; Tofique, M. Field performance of systemic foliar and granular insecticides against rice stem borers (*Scirpophaga* spp.) in rice crop. *Proceedings of the 26th Pakistan Congress of Zoology*, March 1–3, 2007, Multan. 2007, 27: 89–94.

Sarwar, M. H.; Mughal, A. R.; Mughal, S.; Sarwar, M. Concerns of heart diseases and mediations to encourage healthful actions for their deterrence. *International Journal of Bioinformatics and Biomedical Engineering.* 2015, 1 (2): 70–76.

Sarwar, M. H.; Raees, A. R.; Rauf, A.; Sarwar, M. Great ways to encourage better health habits and their preferment in the community. *Specialty Journal of Medical Research and Health Science.* 2019, 4 (2): 60–69.

Sarwar, M. H.; Sarwar, M. F.; Sarwar, M. Understanding the significance of medical education for health care of community around the globe. *International Journal of Innovation and Research in Educational Sciences.* 2014, 1 (2): 149–152.

Tanner, R. W.; Langston, J. W. Do environmental toxins cause Parkinson's disease? A Critical Review. *Neurology.* 1990, 40: 17–30.

Williams, G. M.; Verna, L. K.; Whysner, J. Mechanisms of chemical carcinogensis: Application to safety assessment of pesticides. *Brighton Crop Protection Conference, Pests and Diseases*, November 23–26, 1992, Brighton. 1992.

Section VII

Role of Nutrients in the Management of Crop Diseases through Biopesticides

8

Microbial Agents as Biopesticides for Nutrients: Management in Organic Agriculture

C. R. Patil, Vithal Navi, G. P. Santosh, and Shekharappa

CONTENTS

8.1 Introduction

Microorganisms pioneer all environments that challenge life. Wherever microorganisms exist they perform a huge number of diverse functions, most of which have made this planet a sustainable habitat. Many of their natural functions have found application as low cost technologies to solve environmental problems emerging from anthropogenic activities. Therefore the microbial sources are inevitable to ensure sustainability. Use of microbial agents for mobilizing one or more plant nutrients is an eco-friendly sustainable practice in agriculture.

Every civilization has its own traditional knowledge of employing microorganisms for deriving benefits (Grenier, 1998; IIRR, 1966). Many cultivation practices in ancient India and until the beginning of the green revolution were heavily relying on the preparations and uses of several indigenous knowledge–based practices such as effective microorganisms (EM) as integral parts of farming. Liquid organic manures, such as Panchagavya, Beejamruth, Jeevamruth, etc., are obtained as the products of microbial fermentation and used in crop production. The native flora in them play a significant role as known in Panchagavvya (Balasaheb & Patil, 2015). Many farming systems, especially the organically managed farming systems, rely on the native EM sources. Together with biofertilizers, organic manures and the liquid organic manures (*Jeevamruth*, *Beejamruth*, *Panchagavya*, etc.,) form the most important part of organic nutrient management strategies.

8.2 Effective Microorganisms (EM)

Mixed cultures of effective microorganisms (EM), referred to as EM technology, was developed in the 1980s by Teruo Higa, a professor at the University of the Ryukyus, Okinawa, Japan (Higa, 1991). EM formulations contain beneficial microorganisms that can integrate the soil-plant microbiological equilibrium, which include lactic acid bacteria, photosynthetic bacteria, actinomycetes, yeast, and mycorrhizal fungi. Apart from EM, other beneficial microorganisms, especially those that mobilize one or more nutrients to plants (called biofertilizers), are important sources of nutrients in organic farming.

8.3 Biofertilizers

Biofertilizers are the preparations containing primarily active strains of living microorganisms which colonize the rhizosphere or interior of the plant and promote growth by increasing the supply or availability of primary nutrients and growth regulators to the target crop when applied to seed, plant surface, or soil (Vessey, 2003). The beneficial effects of biofertilizers are; absorption and fixation of atmospheric nitrogen in root-zone, providing 30 to 50 percent of crop nitrogen requirements, production of plant growth hormones such as auxins and cytokinins, enhancing the seed germination efficiency, plant immunity, and the yield by about 25 percent (Yadav et al., 2006), assisting in the uptake of mineral nutrients from the soil, excretion of antibiotics that protect against minor root pathogens. Biofertilizers can mobilize one or more plant nutrients and hence can replace chemical fertilizers partly or completely. Based on the nutrient that they mobilize to plants biofertilizers can broadly be grouped as nitrogen fixers, P-solubilizers and P- mobilizers, Potash solubilizers, Zinc mobilizers, and NPK consortium.

8.3.1 Nitrogen Fixing Biofertilizers

Soil bacteria are capable of converting unavailable dinitrogen (N_2) from the atmosphere to a utilizable form, ammonia (NH_3), catalyzed by an enzyme called *nitrogenase* that is unique to these bacteria. These bacteria are called diazotrophs and their process is called biological nitrogen fixation (BNF). Some common BNF systems are discussed in the following section.

8.3.1.1 Rhizobium *in Association with Legumes*

The nitrogen-fixing ability of legumes is being used in agriculture for more than a century. This is an important approach in organic production system especially for the management of nitrogen through N_2 fixing green manure crops and biofertilizers. This ability of legumes comes because of their symbiotic association with a bacterium in soil that forms root nodules. The leguminous species are specific for a certain kind of specific nodule bacteria. There are groups of legumes that can be modulated by different strains of *Rhizobium*. Each group of these legumes and their specific *Rhizobium* constitute a cross-inoculation group. An estimated 200 to 300 kg ha^{-1} year^{-1} of atmospheric nitrogen is fixed by legume–*Rhizobium* symbiosis (Peoples et al., 1995a and b).

Rhizobium–legume association has the potential to fix on average about 20 to 60 kg atmospheric nitrogen per hectare per season, although *Rhizobium* with the potential to fix between 80 to 150 kg N ha^{-1} in 90 days has been reported.

8.3.1.2 Frankia *in Association with Non-Legume Trees*

Frankia spp., associated with certain non-legume plants such as those belonging to *Causurina*, *Alnus*, *Allocasuarina*, *Myrica*, *Alder*, *Allocasuarina*, *Hippophaee*, etc., referred to as actinorrhizal plants, have potential applications in forestry. *Frankia* is an actinomycete that forms nodules in the roots of host plants and fixes atmospheric nitrogen. *Frankia* obtained from one host is able to form nodules in several others. Hence, nodules collected are crushed and used to inoculate fresh seedlings.

8.3.1.3 Azotobacter

Azotobacter fixes on average 20 to 40 kg N ha^{-1} year^{-1}. Inoculation of seeds with *Azotobacter* has shown to improve growth and yield in a variety of cereals, oilseeds, vegetables, cash crops, plantation crops, seedlings, and forest trees. *Azotobacter* also serves as a broad-spectrum inoculant used for various crops like tea, coffee, and all forest, fruit, and flower plants. It improves the percentage of seed germination, helps in root proliferation, accelerates plant growth, and increases crop yield up to 10 to 15 percent. *Azotobacter* fixes atmospheric nitrogen in soil and thus reduces the use of chemical N-fertilizers to the extent of 20 to 40 kg ha^{-1}. *Azotobacter* synthesizes antibiotic substances that control or suppress various fungal, bacterial, and viral plant diseases.

8.3.1.4 Azospirillum

Azospirillum is found to be associated with forage grasses, cereal crops and occurs in roots and rhizosphere of many members of gramineae, particularly in the tropics. Dual inoculation of *Azospirillum* and phosphate-solubilizing bacteria has synergistic effect on plants. Apart from fixing nitrogen, this organism is also known to produce growth-promoting substances. *Azospirillum* inoculation helps to save 20 to 30 kg N ha^{-1} and also in increasing yield.

8.3.1.5 Blue Green Algae

Blue green algae (BGA), also referred to as cyanobacteria, are a diverse group of photosynthetic prokaryotes frequently occurring in rice fields and are a very useful source of organic nitrogen for low land paddy. In addition to fixing nitrogen, BGA convert insoluble P-sources into soluble forms (Irisarri et al., 2001). They promote the growth of paddy crop by supplying fixed nitrogen through exudation and microbial degradation of dead algal cells. BGA applications at 10 kg ha^{-1} after one week of transplantation contribute 25 to 30 kg N ha^{-1} per season and increased grain yield of 10 to 15 percent.

8.3.1.6 Azolla

Azolla is an aquatic fern and is popular as a nitrogen fixer and green manure in wetland cultivation, particularly in rice crop under submerged soil conditions. The leaf of azolla contains blue green algal symbiont, i.e., *Anabaena azollae*, which is responsible for nitrogen fixation from the atmosphere. Azolla doubles its biomass in three to five days and fixes 40 to 60 kg N ha^{-1}. It can be applied as mono or dual crop with rice and has the potentiality to supply all nutrients required by rice under organic production system. Azolla, along with other organic sources of nutrients, contributes to sustainable organic rice production.

8.3.2 Phosphorus Mobilizing Microorganisms

8.3.2.1 Phosphorus Solubilizing Microorganisms

Phosphate-solubilizing biofertilizers containing species of *Pseudomonas*, *Bacillus*, *Penicillium*, *Aspergillus*, etc., have the ability to solubilize or mobilize insoluble phosphorus such as rock phosphate, tricalcium phosphate, iron, and aluminum phosphate by production of organic acids and by protonation of the environment. Depending on the soil type and crop about 33 to 50 percent of P requirement can be met from Phosphorus solubilizing bacteria (PSB). Combination of rock phosphate equivalent to P requirement of crops and PSB can meet the entire P requirement which is economical and sustainable P-management strategy.

8.3.2.2 Mycorrhiza

The role of Arbuscular mycorrhizal fungi (AMF) in enhancing plant growth is well known. AMF shows synergestic interaction with other Plant Growth Promoting Rhizobacteria (PGPR). Inoculation of AMF to plants is known to improve uptake of phosphorus and other diffusion-limited nutrients such as K, Cu, Mn, and Zn (Manjunath & Habte, 1988). The contribution of AMF to plant growth and nutrition especially in soils with low available P is highly significant.

8.3.3 Potassium Solubilizers

Potassium solubilizing biofertilizers include certain bacteria such as *Bacillus edaphicus*, *B. circulans*, *Psuedomonas*, *Berkholderia*, etc., and fungi such as *Asperigillus terreus* and *Asperigillus niger*. The main mechanism of K-solubilization is by production of acids. Whenever used K-solubilizing biofertilizers help to increase K availability for plant uptake by about 13 to 15 percent.

8.3.4 Zinc Solubilizing Bacteria

A new group of biofertilizers recently gaining importance and hence have been covered under Fertilizer Control Order (FCO). Microorganisms especially bacteria which have the ability to act on zinc oxides and solubilize them; thus, enhancing available zinc to crop plants is being developed. Application of zinc solubilizers is known to bring sustainability in agricultural production.

8.3.5 NPK Consortia

It contains a mixture of biofertilizer strains capable of fixing atmospheric nitrogen and also solubilizing insoluble phosphorus and potassium in soil thus single inoculum capable of mobilizing N, P, and K (Siti et al., 2016). This is becoming very popular as one inoculum helps in uptake of a minimum of three plant nutrients N, P, and K. However, the strains should be compatible (not competing) and efficient in mobilizing the nutrient. It contains a nitrogen fixer, a P-solubilizer, and a K-solubilizer listed in the previous section.

Different biofertilizer products and their prescribed standards are covered under FCO 1985 (amended in 2006 and further amended in 2009).

8.3.6 Plant Growth Promoting Microorganisms

Consists of diverse groups of microorganisms capable of promoting plant growth by elaborating plant hormones and growth-promoting metabolites that influence plants harboring them. These include PGPR strains capable of elaborating IAA, GA, cytokinins, and similar growth-promoting substances produced by them in the milieu of plant roots or foliage wherever they are applied. They have a significant impact on rooting, flowering, flower retention, stay green trait, and enhanced shelf life. These are useful as amendments for propagating material especially using grafting, layering, cuttings, etc. Different biofertilizer products covered under FCO 1985 (amended in 2006 and further amended in 2009) include *Rhizobium*, *Azospirillum*, and *Azotobacter* as nitrogen fixers, P-solubilizing bacteria and P-solubilizing fungi, K-solubilizers, zinc solubilizers, and NPK consortia. Other biofertilizer products not covered under FCO 1985 but popularly used in organic farming, especially for residue management and plant growth promotion, are important but currently do not have prescribed standards. These include waste decomposer, compost culture, etc.

8.3.7 Waste Decomposer

A product containing inoculums of PGPR and can be multiplied at field by farmers and useful for the conversion of biowaste into organic manure and also for seed treatment and crop protection. Developed at the National Centre of Organic Farming, Ghaziabad, Utter Pradesh, this product is widely used by farmers with simple protocols for mass production and application.

8.3.8 Composting Culture

A consortium of four lingo-cellulolytic fungi named compost culture for rapid and uniform decomposition of agroresidues contains four lingo-cellulolytic fungi, namely: *Pleurotus* spp., *Phaenerochaete chrysosporium* (Accession No. MF196240), *Aspergillums awamori*, and *Trichoderma viridae* (Accession No. MF196239). This consortium developed at the Department of Agricultural Microbiology, University of Agricultural Sciences, Dharwad, Karnataka State, when used at the rate of 1 kg per ton of residues is known to bring about rapid and uniform degradation of agroresidues (Patil et al., 2019). The technology is useful for recycling agroresidues both under *ex-situ* (compost pits) and *in-situ* (field) by farmers to effectively and rapidly convert them to useful manure.

8.4 Methods of Inoculating Biofertilizers

8.4.1 Seed Inoculation

Seed treatment is a cheap, effective, and less laborious method of inoculating biofertilizers. The requirements of liquid formulation for seed treatment may vary from 2 to 4 ml per kg of seeds while about 12 to 15 g of powder formulations per kg seeds may be adequate.

8.4.2 Seedling Inoculation

This method is useful to inoculate seedlings just before transplanting. The seedlings are dipped in a suspension prepared by mixing biofertilizers and water in the ratio of 1:10. During treatment, the roots of seedlings must be immersed in suspension and kept for about 20 minutes. The treated seedlings are transplanted immediately. About 4 to 5 kg of carrier biofertilizer or 2 to 3 liters of liquid formulation may be required for treating seedlings required over 1 acre of land.

8.4.3 Inoculating Setts/Tuber

Sugarcane setts, tubers of potato, rhizomes of vegetable, fruits, and flower crops are inoculated by spraying sieved suspension prepared by mixing 1 kg of biofertilizers in 15 to 20 liters of water. Mixing up of the materials gently during spraying may help for better coverage of the biofertilizer. The treated setts/tubers have to be sown without any delay.

8.4.4 Soil Application

To treat 1 acre of land by this method the quantity of biofertilizer required is around 8 kg of carrier-based or 3 liters of liquid formulation. The biofertilizer is mixed thoroughly with about 200 to 250 kg of fine-sieved Farm yard manure (FYM) or vermicompost and applied in furrows or planting spots for effective inoculation. The method is also useful for inoculating arbuscular mycorrhizal (AM) fungi under field conditions. In this method, more than one biofertilizer can be mixed at the same ratio and applied to soil just before sowing or planting. In addition to these commonly used methods, drenching root-zone with biofertilizers and fertigation are being practiced by farmers especially with modern water irrigation systems in place.

8.4.5 Inoculation Methods for Arbuscular Mycorrhizal Fungi

Seedlings are raised in sterilized or unsterilized soil supplied with selected arbuscular mycorrhizal fungi in small nursery beds or containers and planted out when mycorrhizal colonization is well established. At nursery stage, inoculating about 2 to 2.5 kg of AM biofertilizer for a nursery bed of 3 m^2 is ideal for most crops.

8.5 Biofertilizers Current Status

Biofertilizers production is always demand-driven, and creation of demand among farmers is one of the most important steps required towards promotion of biofertilizers (Parr et al., 1994).

From the contents of a brochure from a global summit on setting up a biofertilizers and biopesticides unit (Biotechnology Government of Gujarat, Vibrant Gujarat, January 10–13, 2017), the current growth rate of the biofertilizer industry is observed to be around 14 percent and the total production is around 52,000 to 58,000 metric tons, including unorganized sectors. This is expected to reach 80,000 metric tons in the next five years (Department of Fertilizers, Ministry of Chemicals and Fertilizers). In contrast to this, the total requirement of biofertilizers in India is around 5,28,000 metric tons as estimated on crops covered. This clearly indicates that the demand is far more than the actual production and there is immense scope for entrepreneurship in this industry. Biofertilizer production methods and also biofertilizers that receive organic certification are considered as inputs in organic agriculture.

REFERENCES

Balasaheb G.B and Patil C. R. Bacterial succession during panchagavya making as revealed by DGGE analysis, 2015. *Journal of Pure and Applied Microbiology.* 2015, 9(4): 3005–3013.

Grenier, L. *Working with Indigenous Knowledge: A Guide for Researchers.* 1998, IDRC, Ottawa, Canada.

Higa, T. Effective microorganisms: A biotechnology for mankind. p. 8–14. In J. F. Parr, S. B. Hornick, and C. E. Whitman (eds.), Proceedings of the First International Conference on Kyusei Nature Farming. 1991, U.S. Department of Agriculture, Washington, DC.

IIPR (International Institute of Rural Reconstructions). *Recording and Using Indigenous Knowledge: A Manual.* 1996, International Institute of Rural Reconstructions, Philippines.

Irisarri, P., Gonnet, S. and Monza, J. Cyanobacteria in Uruguayan rice fields: Diversity, nitrogen fixing ability and tolerance to herbicides and combined nitrogen. *Journal of Biotechnology.* 2001, 91: 95–103.

Manjunath, A. and Habte, M. Development of vesicular-arbuscular mycorrhizal infection and uptake of immobile nutrients in *leucaena leucocephala. Plant and Soil.* 1988, 106: 97–103.

Patil, C. R., Jagadeesh, K. S., Srinivasalu, G. and Dasar G. V. Farm waste recycling using consortium of lignocellulolytic fungi—A farmers participatory approach. *Current Journal of Applied Science and Technology.* 2019, 37(6): 1–6.

Parr, J. F., Hornick, S. B., Kaufman, D. D. *Use of Microbial Inoculants and Organic Fertilizers in Agricultural Production.* 1994, ASPAC Food & Fertilizer Technology Center.

Siti Ramlah Ahmad Ali, Mohd Najib Ahmal, Mohd Fahmi Keni, Syrifah azura Syed Ibrahim, Norman Hj Kamaruddin and Ahmad Kushairi, EcoB8: Microbial consortia based Biofertilizers. Technical Report. MPOB Information Series: 2016, ISSN 1511–7871.

Vessey, J. K. Plant growth promoting rhizobacteria as biofertilizers. *Plant and Soil.* 2003, 255(2): 571–586.

Yadav, K., Kumar, S., Murtuza, M. and Varshney, S. K., Effect of carrier based and liquid inoculants on the nodulation and grain yield of pigeonpea. *International Chickpea and Pigeonpea Newsletter.* 2006, 13: 38–39.

Section VIII

Management of Crop Diseases through Bacterial Biopesticides

9

Bacillus Thuringiensis: Characterization of New Strains and Use as Bioinsecticide to Spodoptera frugiperda (Lepidoptera: Noctuidae) in Southern Brazil

Gabriela Cristina Alles, Diouneia Lisiane Berlitz, Vilmar Machado, Maximiano Correa Cassal, and Lidia Mariana Fiuza

CONTENTS

9.1 Introduction

It is estimated that in the world 15% of crop losses are due to insect attacks (James, 2015). The residual effects produced by insecticides, such as environmental pollution, toxicity, and induced resistance in insects have motivated the use of microorganisms as an alternative and an efficient means of control (Crickmore, 2006).

Bacillus thuringiensis strains have shown a wide range of specificity for insects of different orders such as Lepidoptera, Coleoptera, Hymenoptera, Homoptera, Mallophaga, and Diptera (Van Frankenhuyzen, 2000; Cavados et al., 2001; Pinto et al., 2003; De Maagd et al., 2003). Intensive screening programs have isolated and characterized *B. thuringiensis* from samples of soil, plant surfaces, dead insects, stored grains, and aquatic habitats (Bravo et al., 1998; Reyes-Ramirez & Ibarra, 2008). Commercial insecticides derived from this bacterium have a long history of successful use in the biocontrol of insect pests (Bravo et al., 2011; Federici, 2005), in agriculture, forestry, and disease vectors (Van Frankenhuyzen, 2000; Thiéry et al., 1996).

The entomopathogenic activity of this bacterium is principally due to the presence of insecticidal crystal proteins (Cry proteins). Genes that express these proteins are designated as *cry* genes due to the crystalline phenotype and are usually located on plasmids, with large molecular weight (Schnepf et al., 1998; Höfte & Whiteley, 1989). Currently, more than 800 *cry* genes have been described, which are divided into 78 classes (Crickmore, 2020: www.lifesci.sussex.ac.uk/home/Neil_Crickmore/).

The *cry* genes are located in chromosome and plasmids, but some strains present these genes only into plasmids. The *cry*1Ac gene of the strain *B. thuringiensis* subsp. *kustaki* HD37, for example, is located on a plasmid of 50MDa (Lereclus et al., 1993) while other strains can show a different pattern, such as the subspecies *B. thuringiensis aizawai*, which contains five genes: four located in the chromosome or the megaplasmid and a fifth one of 45MDa in the plasmid (Sanchis et al., 1998).

In Brazil, *B. thuringiensis* has been used as a bioinsecticide for the control of many pests, especially *Spodoptera frugiperda* (Lepidoptera: Noctuidae). Attacks of this caterpillar reach various crops, mainly corn and rice, but also peanuts, cotton, soybeans, alfalfa, and forage grasses (Virla et al., 2008). Most products of *B. thuringiensis* used to control lepidopteran pests are based on spore-crystal mixture produced by the strain HD-1, *B. thuringiensis* subsp. *kurstaki,* which has a wide spectrum of larvicidal activity within the order Lepidoptera (Fiuza & Berlitz, 2009; Navon, 1993). However, insects belonging to the family Noctuidae, which includes various *Spodoptera* species, are insensitive to these products. In addition, there are reports of evolution on resistance of this pest to HD-1 and also to the chemical insecticide, showing that its control presents serious practical limitations (Yu, 1992; Diez-Rodrigues & Omoto, 2001).

The need to develop safe pesticides has accelerated the search for new strains of *B. thuringiensis* with different modes of action (Bernhard & Utz, 1993). Using data on *B. thuringiensis* encompassed in the Brazilian state of Rio Grande do Sul, the aim of

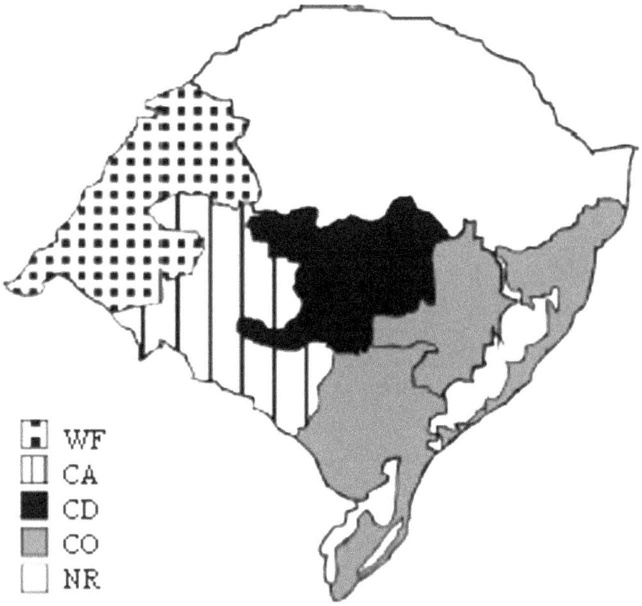

FIGURE 9.1 Representative map of Rio Grande do Sul regions; WF: West Frontier; CA: Campaign; CD: Central Depression; CO: Coast; NR: No rice fields area.

this study was the characterization of strains of *B. thuringiensis*, isolated from rice-producing regions, with high toxicity to *S. frugiperda*, biologic activity determined by qualitative bioassays and protein profiles of the protoxin crystals by SDS-PAGE.

9.2 Materials and Methods

9.2.1 Bacterial Strains and Their Origins

The isolates of *B. thuringiensis* used in this study were collected from soil samples in irrigated rice areas of Rio Grande do Sul (RS) (Figure 9.1). *B. thuringiensis thuringiensis* 4412 was used as positive control.

9.2.2 Phenotypic Tests

Strains of *B. thuringiensis* were evaluated by biochemical, physiologic, and cytomorphologic specific tests for taxonomic studies of bacteria belonging to the genus *Bacillus*, according to the protocols by Gordon et al. (1973), Cowan and Steel (1974), and Claus and Berkeley (1986), respectively, and were observed in differential interference contrast microscope Zeiss and using the software AxioVision SF64.

9.2.3 Antibiograms of Susceptibility

The antibiogram profiles of resistance and susceptibility were performed according to the technique of antibiotic diffusion from discs impregnated with known amounts, according to the Committee for Clinical Laboratory Standards. The antimicrobials used and the interpretation of results of the measurements of the halos of resistance were compared with standard measures for each antimicrobial tested (Table 9.1).

9.2.4 Bioassays against *S. frugiperda*

Second-instar larvae of *S. frugiperda* were collected in fields of rice and maize and maintained at 25 ± 2°C, 80% Relative Humidity (RH), and for a 12-hour photoperiod. The bioassays were conducted in Biological Oxygen Demand (B.O.D.) chambers, at 25 ± 2°C, 80% RH, and for a 12-hour photoperiod. The bacterial isolates were grown in Usual Glucosed Medium (De Barjac and Lecadet, 1976) at 28°C and 180 rpm for 48 h. Cultures were centrifuged at 4500 rpm for 15 min, the supernatant was discarded and the pellet was washed in phosphate buffer. The bacterial concentration was determined with a Neubauer chamber and an optical microscope (400×). All bioassay data were corrected by Abbott's formula (Abbott, 1925). The culture corresponding to 1×10^{10} cells/mL was added to Poitout diet (Poitout and Bues, 1974), previously organized in mini-plates (30 mm of diameter), where 20 larvae of 2nd instar were individualized for each isolate. In the control group, the culture was substituted by sterile distilled water. Mortality was observed until the seventh day after treatment application.

9.2.5 Protein Profiles of the Protoxin Crystals

The protein profiles of the crystal components of *B. thuringiensis* strains were determined through sodium dodecyl sulfate–polyacrylamide gel electrophoresis (SDS-PAGE) (Laemmli, 1970) with 10% gels. The isolates were grown in Usual Glucosed Medium at 33°C ± 1°C under 175 rpm agitation for 48 hours until cellular lysis with release of spores and endotoxin crystals. Bacterial suspensions were centrifuged (10,000 rpm for 20 min) and the pellet was suspended in 1 mL of distilled water. Next, 75 mL of these suspensions was mixed with 25 mL of sample buffer (Tris–HCl 0.5 M pH 6.8; SDS 10%; b-mercaptoethanol 4%; glycerol 8%; bromophenol blue 0.1%) and boiled at 100°C for 10 min as previously described by Lecadet et al. (1992). Samples were immediately evaluated with electrophoresis.

9.3 Results

The phenotypic and genotypic evaluation presents specific parameters for *B. thuringiensis* (Table 9.2).

From the antibiotic resistance profiles, all strains of *B. thuringiensis* were evaluated to be 100% resistant to Penicillin G (PN) and Vancomycin (VA) (Table 9.3). Additionally, in all strains, the sensitivity profile showed 100% inhibition to the antibiotic Chloramphenicol (CO). For the rest, the strains presented different resistance profiles and sensitivity among the evaluated antibiotics, reporting a significant heterogeneity according to the source region.

Data regarding mortality of *S. frugiperda* larvae are shown in Figure 9.2. It shows that, among the 20 strains of *B. thuringiensis* subjected to bioassays, only MTOX 3420-11 had corrected mortality greater than 88% compared with the reference strain *B. thuringiensis thuringiensis* 4412.

The strain MTOX 3420-11 was evaluated in differential interference contrast microscopy using a Zeiss microscope and the software AxioVision SF64. It was possible to visualize the bacterial spore and crystal within the cell (Figure 9.3). Protein

TABLE 9.1

Parameters for Interpretation of the Diameters of Inhibition Zones in Antibiograms

Antibiotics*	Abbreviation	Concentration	Resistance (mm)	Sensitivity (mm)
Cefoxitin	CT	30µl	≤ 17	≥ 18
Chloramphenicol	CO	10µl	≤ 17	≥ 18
Streptomycin	ET	10µl	≤ 14	≥ 15
Erythromycin	EL	15µl	≤ 17	≥ 18
Penicillin G	PN	10µl	≤ 21	≥ 22
Tetracycline	TT	30µl	≤ 18	≥ 19
Vancomycin	VA	5µl	≤ 17	≥ 18

TABLE 9.2

Sources and Region of Isolation for *Bacillus thuringiensis* Strains Used in This Study

		Colony Morphology			
Strains	**Region of Origin**[1]	**Form**[2]	**Edge**	**Size**[3]	**Structure**
MTox 2835-4	CO	C	Smooth	L	dense
MTox 1547-1	CO	C	Smooth	L	dense
MTox 3467-4	WF	C	Serrated	L	dense
MTox 3280-1	CA	C	Smooth	A	dense
MTox 3420-5	CO	C	Smooth	L	dense
MTox 3419-1	CA	C	Smooth	L	dense
MTox 3420-6	CA	C	Smooth	L	dense
MTox 3420-10	CA	C	Smooth	L	dense
MTox 3420-11	CA	C	Smooth	L	dense
MTox 1893-7P	CO	C	Wavy	S	granulated
MTox 1893-13	CO	C	Wavy	A	dense
MTox 1893-14	CO	C	Smooth	S	dense
MTox 1893-16	CO	C	Wavy	S	granulated
MTox 1893-24	CO	C	Wavy	S	granulated
MTox 1458-1	WF	C	Wavy	S	granulated
MTox 1458-2	WF	C	Wavy	S	granulated
MTox 1458-4	WF	C	Wavy	L	granulated
MTox 1490-5	WF	C	Wavy	L	granulated
MTox 3133-5	CD	C	Smooth	A	granulated
MTox 1829-2	CD	C	Smooth	L	dense

[1] WF: West Frontier; CA: Campaign; CD: Central Depression; CO: Coast; [2]C: circular; [3]S: small, A: average, L: large

analysis results by SDS-PAGE showed the proteins of the strain MTOX 3420-11: 97kDa, 80kDa, 55kDa, 36kDa.

9.4 Discussion

The methods for phenotypic and biochemical characters used for classifying *B. thuringiensis* have both contributed to the establishment of a useful classification (Iriarte et al., 2000). Here we report 20 characterized *B. thuringiensis* strains, isolated from soil samples collected in irrigated rice areas of Rio Grande do Sul (RS). The genotypic methods used in the present study demonstrated the possibility of distinguishing the strains since those methods are based on bacterial genome analysis (Iriarte et al., 2000).

Resistance and sensitivity characteristics can be associated with plasmids, which can be transferred between bacteria of the same species or related species of different taxonomic groups. *B. thuringiensis* and *B. cereus* are genetically related species. It is known that they transfer plasmids containing genes encoding the crystal proteins (Gonzalez et al., 1982). It is also noteworthy that the heterogeneity and profile of multi-resistance strains of *B. thuringiensis* have already been demonstrated by Silva (2002). The use of antibiotics in combination with other techniques such as PCR analysis and optical microscopy are used to characterize and identify new isolates of *B. thuringiensis* and differentiate them from other bacteria (Monnerat et al., 2001).

The results obtained in pathogenic strains also revealed that one showed high toxicity (larval mortality >85%) against *S. frugiperda* being the most promising. Different researchers searched for new strains of *B. thuringiensis* with insecticidal activity to *S. frugiperda*, such as Polanczyk et al. (2003), testing 53 isolates, Monnerat et al. (2007) testing a collection of 1400

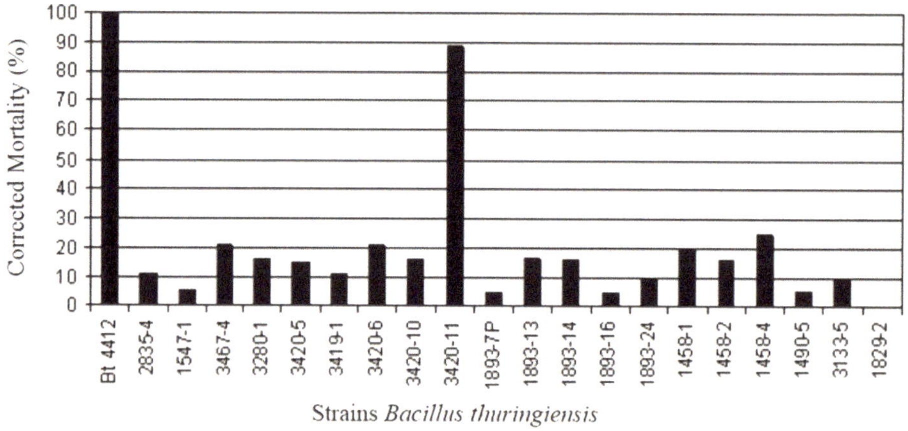

FIGURE 9.2 Corrected mortality of *Bacillus thuringiensis* strains submitted to bioassays with *Spodoptera frugiperda*.

TABLE 9.3

Profile of *Bacillus thuringiensis* Strains Submitted to Growth with Different Antibiotics

	Antibiotics						
Strains	**CT**	**CO**	**ET**	**EL**	**PN**	**TE**	**VA**
Mtox 2835-4	R	S	R	R	R	S	R
Mtox 1547-1	R	S	R	R	R	S	R
Mtox 3467-4	S	S	S	S	R	S	R
Mtox 3280-1	S	S	R	R	R	R	R
Mtox 3420-5	R	S	R	R	R	S	R
Mtox 3419-1	S	S	R	R	R	S	R
Mtox 3420-6	S	S	R	R	R	S	R
Mtox 3420-10	S	S	R	R	R	S	R
Mtox 3420-11	S	S	R	R	R	S	R
Mtox 1893-7P	R	S	R	R	R	R	R
Mtox 1893-13	R	S	R	R	R	R	R
Mtox 1893-14	R	S	R	R	R	R	R
Mtox 1893-16	R	S	R	R	R	R	R
Mtox 1893-24	R	S	R	R	R	R	R
Mtox 1458-1	S	S	S	S	R	S	R
Mtox 1458-2	S	S	S	S	R	S	R
Mtox 1458-4	S	S	S	S	R	S	R
Mtox 1490-5	S	S	S	S	R	S	R
Mtox 3133-5	R	S	S	R	R	R	R
Mtox 1829-2	R	S	S	R	R	R	R
% R²	50	0	65	75	100	40	10
% S²	50	100	35	25	0	60	0

Legend: Cefoxitin (CT); Chloramphenicol, (CO); Streptomycin, (ET); Erythromycin, (EL); Penicillin G (PN); Tetracycline (TE) and Vancomycin (VA). (R) Resistance, (S) sensitivity. % R: response in the percentage of resistance for each of the antibiograms. % S: response in the percentage of sensitivity for each of the antibiograms.

isolates, and other researchers from Embrapa. It is very difficult to find a strain with insecticidal potential, especially against *S. frugiperda*.

Some authors studied the interaction of VIP and Cry proteins for the control of *S. frugiperda*. However, in the insect midgut, Cry1Ia10 and Vip3Aa might compete for the same receptor and showed an antagonistic effect when the proteins were combined (Bergamasco et al., 2013).

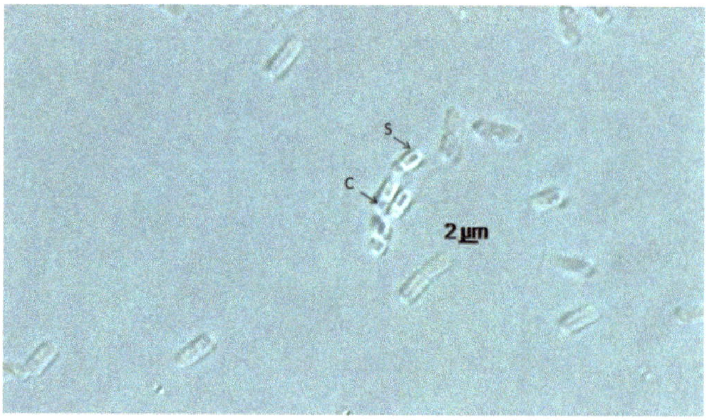

FIGURE 9.3 Cytological characteristics of the strain *Bacillus thuringiensis* MTOX 3420-11 in differential interference contrast microscopy. S-Spore and C-protein crystal.

The protein profile in SDS-PAGE of the strain MTox 3420-11 showed majoritarian bands around 200, 116, and 97 kDa, characteristic of the presence of Cry proteins (data not shown). According to Fazion et al. (2011), Cry proteins of 130 kDa have specific action mainly on insects of the order Lepidoptera. Pavani (2013), showed proteins that the Cry1 class, when activated, have molecular weight of 70 kDa.

The molecular techniques have useful applications such as the determination of persistence of bacteria in the environment, which can be evidenced by the presence of *cry* genes (Bourque et al., 1993). Praça et al. (2004), worked with *B. thuringiensis* strains for controlling *S. frugiperda* and *Anticarsia gemmatalis* and showed that the mortality obtained in these species is probably due to the gene *cry1*.

9.5 Conclusion and Future Prospects

The combination of classical and molecular methods helps us to group isolates, gives more details than a single classification method and, therefore, makes an important contribution to our knowledge of intra-specific differences for this important and useful bacterial species. The characterization of native strains with activity against pests gives new tools to be introduced in management programs. This will decrease the disturbance of ecosystem balance by reducing the release of chemical pesticides or other exogenous biological products.

REFERENCES

Abbott WS (1925). A method of computing the effectiveness insecticide. *J. Econ. Entomol.* 18:265–267.

Bergamasco VB, Mendes DRP, Fernandes OA, Desidério JA, Lemos MVF (2013). *Bacillus thuringiensis* Cry1Ia10 and Vip3Aa protein interactions and their toxicity in *Spodoptera* spp. (Lepidoptera). *J. Invertebr. Pathol.* 112:152–158.

Bernhard K, Utz R (1993). Production of *Bacillus thuringiensis* insecticides for experimental and commercial uses. In Entwistle, PF, Cory, JS. (Eds.) *Bacillus thuringiensis*, pp. 255–267. Wiley, Chichester UK.

Bourque SN, Valéro JR, Mercier J, Lavoie MC, Levesque RC (1993). Multiplex polymerase chain reaction for detection and differentiation of the microbial insecticide *Bacillus thuringiensis*. *Appl. Environ. Microbiol.* 59:523–527.

Bravo A, Likitvivatanavong S, Gill SS, Soberón M (2011). *Bacillus thuringiensis*: A story of a successful bioinsecticide. *Insect Biochem. Mol. Biol.* 41:423–431.

Bravo A, Sarabia S, López L, Ontiveros H, Abarca C, Ortiz A, Ortiz M, Lina L, Villalobos F, Peña Nuñez-Valdez GM, Soberon M, Quintero R (1998). Characterization of *cry* genes in Mexican *B. thuringiensis* strain collection. *Appl. Environ. Microbiol.* 64:4965–4972.

Cavados CFG, Fonseca RN, Chaves JQ, Rabinovitch L, Araújo-Coutinho CJPC (2001). Identification of entomopathogenic *Bacillus* isolated from Simulium (diptera, simuliidae) larvae and adults. *Mem. Inst. Oswaldo Cruz.* 96:1017–1021.

Claus D, Berkeley RCW (1986). Genus *Bacillus* Cohn 1872. In: Sneath PHA, Mair NS, Sharpe ME, Holt JG (Eds.) *Bergey's Manual of Systematic Bacteriology*, 2nd ed. Williams and Wilkins, Baltimore, pp. 1105–1141.

Cowan ST, Steel KJ (1974). *Cowan and Steel's Manual for Identification of Medical Bacteria*. Cambridge University Press, London.

Crickmore, N. (2006). Beyond the spore-past and future developments of *Bacillus thuringiensis* as a biopesticide. *J. Appl. Microbiol.* 101:616–619.

Crickmore, N (2020). Available in: http://www.lifesci.sussex.ac.uk/home/Neil_Crickmore/Bt/.

De Barjac H, Lecadet MM (1976). Dosage biochimique d'exotoxine thermostable *Bacillus thuringiensis* d'après l'inhibition d'RNA-polymerases bacteriennes. *C.R. Acad. Sci.* 282:2119–2122.

De Maagd RA, Bravo A, Berry C, Crickmore N, Schnepf E (2003). Structure, diversity and evolution of protein toxins from sporeforming entomopathogenic bacteria. *Annu. Rev. Genet.* 37:409–433.

Diez-Rodrigues GI, Omoto C (2001). Herança da resistência de *Spodoptera frugiperda* (J.E.Smith) (Lepidoptera: Noctuidae) a lambdacialotrina. *Neotrop. Entomol.* 30:311–316.

Fazion FAP, Scarpassa JA, Ricieto APS, Helene LCF, Vilas-Bôas GT (2011). Previsão da atividade inseticida de linhagens de *Bacillus thuringiensis* através da análise do perfil de proteínas Cry. In *Anais Eletrônico VII EPCC—Encontro Internacional de Produção Científica Cesumar CESUMAR*. Centro Universitário de Maringá, Maringá, Brazil, 4p.

Federici BA (2005). Insecticidal bacteria: an overwhelming success for invertebrate pathology. *J. Invertebr. Pathol.* 89:30–38.

Fiuza LM, Berlitz DL (2009). Produtos de *Bacillus thuringiensis*: Registro e comercialização. *Biotecnol. Ciênc. Desenvolv.* 38:58–60.

Gonzalez JM, Brown BJ, Carlton BC (1982). Transfer of *Bacillus thuringiensis* plasmids coding for delta-endotoxin among strains of *B. thuringiensis* and *B. cereus*. *Proc. Natl. Acad. Sci. USA*. 79:6951–6955.

Gordon RE, Haynes WC, Pang CN (1973). *The Genus Bacillus (Agricultural Handbook: No. 427)*. US Department of Agriculture, Washington, DC.

Höfte H, Whiteley HR (1989). Insecticidal crystal proteins of *Bacillus thuringiensis*. *Microbiol. Rev.* 53:242–255.

Iriarte J, Dumanoir VC, Bel YM, Porcar MD, Ferrandis MM, Lecadet J, Ferre Caballero P (2000). Characterization of *Bacillus thuringiensis* ser. *balearica* (Serotype H48) and ser. navarrensis (Serotype H50): Two novel serovars isolated in Spain. *Curr. Microbiol.* 40:17–22.

James C (2015). Global status of commercialized biotech/GM crops. In *International Service for the Acquisition of Agri Biotech Applications*: *No. 51*. ISAAA, Ithaca, NY.

Lecadet MM, Chaufaux J, Ribier J, Lereclus D (1992). Construction of novel Bacillus thuringiensis strains with different insecticidal activities by transduction and transformation. *Appl. Environ. Microbiol.* 58:840–849.

Lereclus D, Delécluse L, Lecadet MM (1993). Diversity *of Bacillus thuringiensis* toxins and genes. In Entwistle P, Cory JS, Baley MJ, Higgs S (Eds.) *Bacillus thuringiensis, An Environmental Biopesticide: Theory and Practice*. Wiley, Chichester, UK, pp. 37–69.

Monnerat RG, Batista AC, Medeiros PTS, Martins E, Melatti V, Praça L, Dumas V, Demo C, Gomes ACM, Falcão R, Berry C (2007). Characterization of Brazilian *Bacillus thuringiensis* strains active against *Spodoptera frugiperda*, *Plutella xylostella* and *Anticarsia gemmatalis*. *Biol. Control.* 41:291–295.

Monnerat RG, Silva SF, Wernwck JOS, Dias JMCS (2001) Métodos de coleta, isolamento, caracterização e armazenamento de estirpes de *B. sphaericus*. *Embrapa Recursos Genéticos e Biotecnologia* (Circular Técnica, 9).

Navon A (1993). Control of lepidopteran pests with *Bacillus thuringiensis*. In Entwistle PF et al. (Eds). *Bacillus thuringiensis, An Environmental Biopesticide*: Theory and Practice. Wiley, Chichester, pp. 125–146.

Pavani CD (2013). Efeitos da interação e toxicidade das proteínas Cry1 e Vip3A de *Bacillus thuringiensis*, Berliner em *Spodoptera frugiperda* (J. E. Smith) (Lepidoptera: Noctuidae). Dissertação- Universidade Estadual Paulista, Faculdade de Ciências Agrárias e Veterinárias, 44p.

Pinto LMN, Azambuja AO, Dihel E, Fiuza LM (2003). Pathogenicity of *Bacillus thuringiensis* isolates from two species of *Acromirmex* (Hymenoptera: Formicidae). *Braz. J. Biol.* 63(2): 301–306.

Poitout S, Bues R (1974). Élevage des chenilles de vint huit espèces de lépidoptères Noctuidae et deux espèces d'Arctiidae sur milieu artificiel simple. Particularités de l'élevage selon les espèces. *Ann. Zool. Ecol. Anim.* 6:431–441.

Polanczyk R, da Silva RFP, Fiuza LM (2003). Screening of *Bacillus thuringiensis* isolates pathogenic to *Spodoptera frugiperda* (JE Smith) (Lepidoptera Noctuidae). *Arq. Inst. Biol.* 70(1):69–72.

Praça LB, Batista ACE, Martins ÉS (2004). Estirpes de *Bacillus thuringiensis* efetivas contra insetos das ordens Lepidoptera, Coleoptera e Diptera. *Pesqui. Agropecu. Bras.* 39:11–16.

Reyes-Ramírez A, Ibarra JE (2008). Plasmid patterns of *B. thuringiensis* type strains. *Appl. Environ. Microbiol.* 74:125–129.

Sanchis V, Lereclus D, Menou G, Chaufaux J, Lecadet MM (1998). Multiplicity of δ endotoxinas genes with different insecticidal specificities in *Bacillus thuringiensis aizawai* 7.29. *Mol. Microbiol.* 2:393–404.

Schnepf E, Crickmore N, Van Rie J, Lereclus D, Baum J, Feitelson J, ZeiGler DR, Dean DH (1998). *Bacillus thuringiensis* and its pesticidal crystal proteins. *Microbiol. Mol. Biol. Rev.* 62:775–806.

Silva CCA (2002). Aspectos do sistema imunológico dos insetos. *Biotecnol. Cienc. Desenv.* 24:68–72.

Thiéry I, Back C, Barbazan P, Sinègre G (1996). Applications de *Bacillus thuringiensis* et de *B. sphaericus* dans la démoustication et la lutte contre les vecteurs de maladies tropicales. *Ann. Inst. Pasteur Actual.* 7:247–260.

Van Frankenhuyzen K (2000). Application of *Bacillus thuringiensis* in forestry. In Charles JF, Delécluse A, Nielsen-LeRoux C (Eds.) *Entomopathogenic Bacteria: From Laboratory to Field Application*. Kluwer Associate Publishing, Norwell, MA, pp. 371–376.

Virla E, Alvarez A, Loto F et al. (2008). Fall Armyworm strains (Lepidoptera: Noctuidae) in Argentina, their associate host plants and response to different mortality factors in laboratory. *Fla. Entomol.* 91:63–69.

Yu SJ (1992). Detection and biochemical characterization of insecticide resistance in fall armyworm (Lepidoptera: Noctuidae). *J. Econ. Entomol.* 85:675–682.

10

*Management of Fungal Diseases of Chickpea (*Cicer arietinum* L.) through Plant Growth Promoting Actinobacteria and Their Secondary Metabolites*

Subramaniam Gopalakrishnan, Vadlamudi Srinivas, Pratyusha Sambangi, and Sravani Ankati

CONTENTS

10.1 Introduction

Chickpea (*Cicer arietinum* L.) is an important pulse crop and ranks third in overall production after bean and pea on a world basis. It is grown in 33 countries over an area of about 11.5 million hectares (Bidyarani et al. 2016). Chickpea is mainly used as food because of its high protein (12–31%) and carbohydrate (52–71%) contents (Mergaand Haji 2019). Global yields of chickpea (968 kg ha^{-1}) have been stagnant for the past five decades in spite of using various conventional and molecular breeding approaches and extensive use of synthetic fertilizers and pesticides (FAOSTAT 2014). Productivity of chickpea may be considerably improved if the adverse effects of biotic stresses (such as *Ascochyta* blight, dry root rot, *Fusarium* wilt, collar rot, and *Botrytis* gray mold) are addressed. Management of fungal diseases of chickpea is difficult, as no single control measure is fully effective. Some of the control measures such as advanced sowing date, solarization of soil, use of pathogen-free seed and fungicide-treated seed are usually employed to control the diseases, but with limited success. The use of resistant cultivar is the most efficient control measure but the effectiveness of disease resistance is restricted by the occurrence of several races of the pathogen. Hence, there is a need to use biological options to manage plant pathogens.

Biocontrol is a natural method to manage plant pathogens and it usually occurs by one or more distinct mechanisms such as parasitism, antibiotics production, induced systemic resistance, and competition for nutrients (Bach et al. 2016). Biocontrol agents are widely reported to manage plant pathogens of important crops including chickpea. Of which, *Bacillus*, *Pseudomonas*, and *Trichoderma* were widely reported. In recent years, actinobacteria and their secondary metabolites are exploited for controlling plant pathogens as they are the most prolific source for the production of bioactive metabolites (Bach et al. 2016; Gopalakrishnan and Srinivas 2019). This study explores how actinobacteria and their metabolites are exploited for the management of the fungal diseases of chickpea.

10.2 Actinobacteria and Their Metabolites

Actinobacteria are Gram-positive, aerobic, spore-forming, filamentous bacteria belong to the order Actinomycetales. All members of the order Actinomycetales consist of high G+C content (> 50%) in their genomic DNA (Ventura et al. 2007). Actinobacteria produce secondary metabolites of agricultural interest. There is a growing interest in the use of secondary metabolites, such as toxins, proteins, hormones, amino acids, and antibiotics from actinobacteria for the control of plant pathogens as these are readily degradable, highly specific, and less toxic to nature (Doumbou et al. 2001). Actinobacteria are commonly found in soil, compost, and fresh and marine water ecosystems.

Among the actinobacteria, *Streptomyces* is the predominant genus and principal contributors of secondary metabolites followed by *Micromonospora*, *Nocardia*, *Frankia*, *Nonomurea*, *Actinomadura*, *Microbispora*, *Mycobacterium*, *Actinoplanes*, *Saccharopolyspora*, and *Verrucosispora* (Martinez-Hidalgo et al. 2014). Such metabolites hold fungicidal, bactericidal, insecticidal, and PGP traits and can fill the need for biological agents.

10.3 Mechanisms Involved in Antagonistic Traits of Actinobacteria

Actinobacteria facilitate plant growth either by direct stimulation (such as fixed nitrogen, soluble phosphate, chelate iron and produce phytohormones and cell wall degrading enzymes) or by indirect stimulation (by inhibiting phytopathogens). Some of the important PGP and biocontrol traits of importance are as described in the following sections.

10.3.1 Siderophore Production

Siderophores are produced by many actinobacteria, which bind Fe^{3+} from the environment and make it available for plant growth. Actinobacteria compete with other rhizosphere plant pathogens for iron, hence competition for iron is also a possible mechanism to control the phytopathogens. Siderophore production was estimated as per the protocol of Schwyn and Neilands (1987). Numerous species of actinobacteria including *Streptomyces* and *Amycolatopsis* have been reported to produce siderophores and inhibit phytopathogens (Wang et al. 2014; Alekhya and Gopalakrishnan 2016).

10.3.2 Cellulase Production

Cellulase degrades cellulose on the plant cell wall, thus destroying the adhesion of the pathogen to the root surface of the plant. Cellulase-producing actinobacteria are reported to have nutrient mineralization, organic matter decomposition, and biocontrol traits on cellulose cell wall-bearing pathogens such as *Pythium* and *Phytophthora* (Lima et al. 1998). Actinobacteria including *Streptomyces* are known to produce cellulase (Gopalakrishnan et al. 2011; 2014) and inhibit pathogens of chickpea. The protocol of Hendricks et al. (1995) was used to determine the cellulase production.

10.3.3 Protease and Lipase Production

Protease and lipase production confers antibiosis to antagonistic microbes by degrading the proteins and lipids, respectively. Protease and lipase producing bacteria act as biocontrol agents on protein/lipid cell wall-bearing plant pathogens (Lima et al. 1998). Actinobacteria are widely demonstrated to produce protease and lipase and observed to inhibit both fungal pathogens and insect pests (Yandigeri et al. 2015). The protease and lipase production was measured as per Bhattachaya et al. (2009).

10.3.4 Hydrocyanic Acid (HCN) Production

HCN is reported as an anti-fungal secondary metabolite produced by many PGP microorganisms including actinobacteria (*Streptomyces* in particular) and play an important role in disease suppression (Bhosale and Kadam 2015). HCN was estimated qualitatively by the sulfocyanate colorimetric method (Gopalakrishnan et al. 2011).

10.3.5 Chitinase Production

Chitin, a linear polymer of beta-1, 4-linked N-acetyl glucosamine, is synthesized in all major groups of organisms including animals, plants, and fungi. The pathogenic fungal cell wall is composed of chitin. The actinobacteria capable of producing chitinase can degrade pathogenic cell wall and thus inhibit those (Yandigeri et al. 2015). Several chitinolytic enzymes have been identified in various *Streptomyces* species and a correlation between chitinolysis and production of bioactive compounds has also been reported (Yandigeri et al. 2015). Chitinase production by actinobacteria was estimated as per the protocols of Hirano and Nagao (1988).

10.3.6 β-1,3-glucanase Production

β-1,3-glucan is present in the cell walls of fungal pathogens. The lysis of β-1,3-glucan by β-1,3-glucanase producing bacteria leads to leakage of cell contents and thus inhibiting the pathogen invasion (Singh et al. 1999; Gopalakrishnan et al. 2014). β-1,3-glucanase production in actinobacteria was done as per the protocols of Singh et al. (1999) and Gopalakrishnan et al. (2014). Actinobacteria including *Streptomyces* and *Amycolatopsis* are reported to produce β-1,3-glucanase and play a role in disease suppression in chickpea (Alekhya and Gopalakrishnan 2016).

10.3.7 Indole Acetic Acid (IAA) Production

IAA is the common product of L-tryptophan metabolism produced by several bacteria including actinobacteria. IAA producing bacteria are known to promote root elongation and plant growth. It was measured as per the protocols of Patten and Glick (1996). Actinomycetes including *Streptomyces*, *Amycolatopsis*, and *Nocardiopsis*are widely reported to produce IAA and promote plant growth (Shutsrirung et al. 2013; Alekhya and Gopalakrishnan et al. 2016).

10.3.8 Other Cell Wall Degrading Enzymes

The other important cell wall degrading enzymes that play a role in the mechanism of pathogen inhibition include nucleases, xylanases, amylases, peptidases, peroxidases, ligninases, pectinase, hemicellulase, and keratinase. These enzymes also contribute to inhibiting phytopathogens because the cell walls of most fungal and bacterial pathogens consist of polymers like chitin, glucan, cellulose, proteins, and lipids (Yandigeri et al. 2015).

10.4 Actinobacteria and Their Role in Biocontrol of Fungal Diseases of Chickpea

Streptomyces is the most prominent and largest genus of actinobacteria. *Streptomyces* protect plants from plant pathogens by various mechanisms including antibiosis (Soltanzadeh et al. 2016), competition for nutrients and space (Li et al. 2010), production of siderophores (Sadeghi et al. 2012), induction of systemic resistance (Singh and Gaur 2017), production of extracellular enzymes (Gopalakrishnan et al. 2011; Vijayabharathi et al. 2018).

Five strains of *Streptomyces* (CAI-24, CAI-121, CAI-127, KAI-32, and KAI-90) isolated from vermicompost, were reported to have antagonistic potential against *Fusarium* wilt in chickpea under wilt sick field conditions (Gopalakrishnan et al. 2011). Amini et al. (2016) reported efficient control of *Fusarium* wilt in chickpea by five strains of *Streptomyces* in Iran. Soltanzadeh et al. (2016) reported *Streptomyces antibioticus* S3, *Streptomyces antibioticus* S12, and *Streptomyces peruviensis* S40, having antagonistic potential against black root rot of chickpea caused by *Fusarium solani* f. sp. *pisi*. Singh and Gaur (2017) demonstrated six endophytic *Streptomyces* viz., *S. diasataticus*, *S. fradiae*, *S.olivochromogenes*, *S. collinus*, *S. ossamyceticus*, and *S. griseus* to control collar rot of chickpea which led to an induced resistance against *S. rolfsii*. Misk and Franco (2011) reported a total of 11 *Streptomyces* having antagonistic potential against *Botrytis* gray mold disease in chickpea. The *Streptomyces* strains were also found to have antagonistic potential against *Phytophthora* root rot disease and PGP traits in chickpea when co-inoculated with *Mesorhizobium ciceri* WSM1666 (Misk and Franco 2011).

Streptomyces were reported to have broad-spectrum biocontrol and PGP potentials. *Streptomyces* with broad-spectrum properties offer effective novel strategies not only for controlling plant pathogens and insect pests that attack crops but also for crop growth and yield. Alekhya and Gopalakrishnan (2014) reported a total of eight strains of *Streptomyces* (BCA-657, BCA-671, BCA-679, BCA-687, BCA-690, CAI-67, CAI-70, and CAI-98) having broad-spectrum antagonistic potential against important fungal diseases of chickpea such as *Fusarium* wilt, dry root rot, and *Botrytis* gray mold. Nine strains of *Streptomyces* (SAI-13, SAI-29, VAI-7, VAI-30, AC-5, AC-6, AC-10, AC-18, and AC-19) were reported to have a broad spectrum of antagonistic potential against fungal diseases of chickpea including *Fusarium* wilt, dry root rot, and collar rot (Sreevidya et al. 2016; Anusha et al. 2019).

PGP potential of *Streptomyces* was reported extensively in chickpea (Gopalakrishnan et al. 2015). However, PGP by other members of the actinomycete family is rarely reported. PGP potentials of *Amycolatopsis* was demonstrated in chickpea under field conditions (Alekhya and Gopalakrishnan 2016).

Secondary metabolites from *Streptomyces* are reported to control *Helicoverpa armigera*, one of the important polyphagous insect pests of many legume crops including chickpea, causing yield losses of up to 100% (Vijayabharathi et al. 2014). An insecticidal compound, diketopiperazine, cyclo (Trp-Phe), was purified from the extracellular extract of *Streptomyces griseoplanus* SAI-25 against *H. armigera* of chickpea (Sathya et al. 2016). One more metabolite was also purified from the culture filtrate of *Streptomyces* sp. CAI-155 as *N*-(1-(2,2-dimethyl-5-undecyl-1, 3-dioxolan-4-yl)-2-hydroxyethyl)stearamide (Gopalakrishnan et al. 2016) against *H. armigera* of chickpea. Among all the actinobacteria which produce secondary metabolites of commercial interest. Streptomycetes have the ability to produce enormous varieties of metabolites.

10.5 Conclusion

The agricultural sector is in need of newer biocontrol agents and their significant secondary metabolites. Biocontrol often appears promising in specialized environments, especially where disappointing results were frequently observed in the field as several factors determine the survival and delivery of the antagonist. Therefore, the strategy to combat the disease should be to integrate different methods of pest control including disease-resistant cultivars, biocontrol using actinobacteria in specific and safe cultural practices. There is a need to also evaluate these PGP actinobacteria under different field conditions (multi-location trials) for commercialization and to improve sustainability in agricultural production, especially in chickpea.

REFERENCES

Alekhya, G., and Gopalakrishnan, S. 2014. Characterization of antagonistic *Streptomyces* as potential biocontrol agent against fungal pathogens of chickpea and sorghum. *Philippine Agricultural Scientist*. 97:191–198.

Alekhya, G., and Gopalakrishnan, S. 2016. Exploiting plant growth-promoting *Amycolatopsis* spp. in chickpea and sorghum for improving growth and yield. *Journal of Food Legumes*. 29:225–231.

Amini, J., Agapoor, Z., and Ashengroph. 2016. Evaluation of *Streptomyces* spp. against *Fusarium oxysporum* f. sp. *ciceri* for the management of chickpea wilt. *Journal of Plant Protection Research*. 56:257–264.

Anusha, B.G., Gopalakrishnan, S., Naik, M.K., and Sharma, M. 2019. Evaluation of *Streptomyces* spp. and *Bacillus* spp. for biocontrol of Fusarium wilt in chickpea (*Cicer arietinum* L.). *Archives of Phytopathology and Plant Protection*. 52:1–26.

Bach, E., Seger, G.D.D.S., Fernandes, G.D.C., Lisboa, B.B., and Passaglia, L.M.P. 2016. Evaluation of biological control and rhizosphere competence of plant growth-promoting bacteria. *Applied Soil Ecology*. 99:141–149.

Bhattacharya, A., Chandra, S., and Barik, S. 2009. Lipase and protease producing microbes from the environment of sugar beet field. *Indian Journal of Agricultural Biochemistry*. 22:26–30.

Bhosale, H.J, and Kadam, T.A. 2015. Diversity and a comparative account on plant growth-promoting characteristics of actinomycetes in roots and rhizosphere of *Saccharum offcinarum*. *International Journal of Current Microbiology and Applied Sciences*. 4:230–244.

Bidyarani, N., Prasanna, R., Babu, S., Hossain, F., and Saxena, A.K. 2016. Enhancement of plant growth and yields in chickpea (*Cicer arietinum* L.) through novel cyanobacterial and biofilmed inoculants. *Microbiological Research*. 188:97–105.

Doumbou, C.L., Hamby Salove, M.K., Crawford, D.L., and Beaulieu, C. 2001. Actinomycetes, promising tools to control plant diseases and to promote plant growth. *Phytoprotection* 82:85–102.

FAOSTAT. 2014. Statistical database. http://faostat.fao.org/site/339 /default. aspx.

Gopalakrishnan, S., Pande, S., Sharma, M., Humayun, P., Kiran, B.K., Sandeep, D., Vidya, M.S., Deepthi, K., and Rupela, O. 2011. Evaluation of actinomycete isolates obtained from herbal vermicompost for biological control of *Fusarium* wilt of chickpea. *Crop Protection*. 30:1070–1078.

Gopalakrishnan, S., and Srinivas, S. 2019. Management of soil-borne diseases of grain legumes through broad-spectrum actinomycetes having plant growth-promoting and biocontrol traits. *Plant Microbe Interface*. 5:129–144.

Gopalakrishnan, S., Srinivas, V., Alekhya, G., Prakash, B., Kudapa, H., and Varshney, R.K. 2015. Evaluation of *Streptomyces* sp. obtained from herbal vermicompost for broad spectrum of plant growth-promoting activities in chickpea. *Organic Agriculture*. 5:123–133.

Gopalakrishnan, S., Vadlamudi, S., Bandikinda, P., Sathya, A., Vijayabharathi, R., Rupela, O., Kudapa, B., Katta, K., and Varshney, R.K. 2014. Evaluation of *Streptomyces* strains isolated from herbal vermicompost for their plant growth-promotion traits in rice. *Microbiological Research*. 169:40–48.

Gopalakrishnan, S., Vijayabharathi, R., Sathya, A., Sharma, H.C., Srinivas, V., Bhimineni, R.K., Gonzalez, S.V., Melø, T.M., and Simic, N. 2016. Insecticidal activity of a novel fatty acid amide derivative from *Streptomyces* species against *Helicoverpa armigera*. *Natural Product Research*. 30:2760–2769.

Hendricks, C.W., Doyle, J.D., and Hugley, B. 1995. A new solid medium for enumerating cellulose-utilizing bacteria in soil. *Applied and Environmental Microbiology*. 61:2016–2019.

Hirano, S., and Nagao, N. 1988. An improved method for the preparation of colloidal chitin by using methane sulfonic acid. *Agricultural and Biological Chemistry*. 52:2111–2112.

Li, Q., Ning, P., Zheng, I., Huang, J., Li, G., and Hsiang, T. 2010. Fumigant activity of volatiles of *Streptomyces globisporus* JK-1 against *Penicillium italicum* on *Citrus microcarpa*. *Postharvest Biology and Technology*. 58:157–165.

Lima, L.H.C., Marco, J.L., and Felix, J.R. 1998. Enzimas hidroliticas envolvidas no controle biologico por miciparasitisma. In: Melo IS, Azevedo JL, (Eds). *Controle Biologico*. Jaguraiuna: EMBRAPA-Meio Ambiente, Vol. 11, pp. 263–304.

Martinez-Hidalgo, P., Olivares, J., Delgado, A., Bedmar, E., and Martínez-Molina, E. 2014. Endophytic *Micromonospora* from *Medicago sativa* are apparently not able to fix atmospheric nitrogen. *Soil Biology and Biochemistry*. 74:201–203.

Merga, B., and Haji, J. 2019. Economic importance of chickpea: Production, value, and world trade. *Cogent Food and Agriculture*. 5:1.

Misk, A., and Franco, C. 2011. Biocontrol of chickpea root rot using endophytic actinobacteria. *Biocontrol*. 56:811–822.

Patten, C., and Glick, B.R. 1996. Bacterial biosynthesis of indole-3-acetic acid. *Canadian Journal of Microbiology*. 42:207–220.

Sadeghi, A., Karimi, E., Dahaji, P.A., Javid, M.G., Dalvand, Y., and Askari, H. 2012. Plant growth-promoting activity of an auxin and siderophore producing isolate of *Streptomyces* under saline soil conditions. *World Journal of Microbiology and Biotechnology*. 28:1503–1509.

Sathya, A., Vijayabharathi, R., Vadlamudi, S., Sharma, H.C., and Gopalakrishnan, S. 2016. Assessment of tryptophan based diketopiperazine, cyclo (L-Trp-L-Phe) from *Streptomyces griseoplanus* SAI-25 against *Helicoverpa armigera* (Hübner). *Journal of Applied Entomology and Zoology*. 51:11–20.

Schwyn, B., and Neilands, J.B. 1987. Universal chemical assay for the detection and determination of siderophores. *Analytical Biochemistry*. 160:47–56.

Shutsrirung, A., Chromkaew, Y., Pathom-Aree, W., Choonluchanon, S., and Boonkerd, N. 2013. Diversity of endophytic actinomycetes in mandarin grown in northern Thailand, their phytohormone production potential and plant growth promoting activity. *Soil Science and Plant Nutrition*. 59:322–330.

Singh, S.P., and Gaur, R. 2017. Endophytic *Streptomyces* spp. underscore induction of defence regulatory genes and confers resistance against *Sclerotium rolfsii* in chickpea. *Biological Control*. 104:44–56.

Singh, P.P., Shin, Y.C., Park, C.S., and Chung, Y.R. 1999. Biological control of *Fusarium* wilt of cucumber by chitinolytic bacteria. *Phytopathology*. 89:92–99.

Soltanzadeh, M., Nejad, M.S., and Bonjar, G.H.S. 2016. Application of soil-borne actinomycetes for biological control against *Fusarium* wilt of chickpea (*Cicer arietinum*) caused by *Fusarium solani* f. sp. *pisi*. *Journal of Phytopathology*. 164:967–978.

Sreevidya, M., Gopalakrishnan, S., Kudapa, H., and Varshney, R.K. 2016. Exploring PGP actinomycetes from vermicompost and rhizosphere soil for yield enhancement in chickpea. *Brazilian Journal of Microbiology*. 47:85–95.

Ventura, M., Canchaya, C., Fitzgerald, G.F., Gupta, R.S., and Van Sinderen, D. 2007. Genomics as a means to understand bacterial phylogeny and ecological adaptation: The case of Bifidobacteria. *Antonie van Leeuwenhoek*. 91:351–372.

Vijayabharathi, R., Gopalakrishnan, S., Sathya, A., Kumar, M.V., Srinivas, V., and Mamta, S. 2018. *Streptomyces* sp. as plant growth- promoters and host-plant resistance inducers against *Botrytis cinerea* in chickpea. *Biocontrol Science and Technology*. 28:1–24.

Vijayabharathi, R., Ratna Kumari, B., Satya, A., Srinivas, V., Rathore, A., Sharma, H.C., and Gopalakrishnan, S. 2014. Biological activity of entomopathogenic actinomycetes against lepidopteran insects (Noctuidae: Lepidoptera). *Canadian Journal of Plant Science*. 94:759–769.

Wang, W., Qiu, Z., Tan, H., and Cao, L. 2014. Siderophore production by actinobacteria. *Biometals*. 27:623–631.

Yandigeri, M.S., Malviya, N., Solanki, M.K., Shrivastava, P., and Sivakumar, G. 2015. Chitinolytic *Streptomyces vinaceusdrappus* S5MW2 isolated from Chilikalake, India enhances plant growth and biocontrol efficacy through chitin supplementation against *Rhizoctonia solani*. *World Journal of Microbiology and Biotechnology*. 31:1217–1225.

11

The Role of Plant-Growth Promoting Bacteria in Sustainable Plant Disease Management

Younes Rezaee Danesh, Semra Demir, Solmaz Najafi, and Çağlar Sagun

CONTENTS

11.1 Introduction

The world population is continuously increasing and there is a real challenge to feed this huge population, an endeavor that requires agricultural productivity to be increased. Thus, more than ever, obtaining high yields is the main challenge for agriculture. In addition, in recent years, both producers and consumers are increasingly focusing on health and quality of food and on their organoleptic and nutritional properties. The decrease in crop productivity is the incidence of various biotic and abiotic stresses. Biotic factors include stresses due to phytopathogens and pests such as fungi, nematodes, viruses, and insects. Phytopathogenic organisms have co-existed with plants since the inception of agriculture evolution. Despite being a natural phenomenon, their existence has adversely affected agriculture. Microbial attacks can cause a minor infection resulting in solely a reduction of plant-growth capacities or can even be the origin of much more severe damage leading to plant death in the worst case. Plant diseases are responsible for the loss of at least 10% of global food production, representing a threat to food security. To prevent and manage these diseases, producers have become increasingly dependent on agrochemicals, especially over the past few decades, as agricultural production has intensified. However, despite the great effectiveness and ease of utilization of these products, their use or misuse has in fact led to problems associated with significant pollution of soils and ground water reservoirs, the accumulation of undesirable chemical residues in the food chain, the emergence of fungicide-resistant strains of pathogens, not to mention health concerns for growers. Consequently, nowadays there is an increasing demand from consumers and authorities for safer, rational, sustainable, and ecofriendly strategies. Sustainable agricultural development is best defined as the management and conservation of agricultural resource base in an appropriate, economically viable, and socially acceptable way. Thus, sustainable agriculture should ensure availability, access, utilization and stability of resources to present and future generations. Exploitation of microbe-based management will be an alternative approach to control this disease. In nature, soil harbors numerous beneficial microorganisms with potential genes for governing resistance and promoting plant growth which can be well exploited for managing the plant diseases. Plant-growth-promoting rhizobacteria (PGPRs) commonly called rhizobacteria, have received prominent attention because of their multifarious activities to improve plant growth and disease control ensuring crop yields in a sustainable way. The application of PGPR to crops offers an environment-friendly approach by reducing chemical inputs (Prasad et al., 2015). An incredible demand for the use of PGPR bio-fertilizers is observed day by day with an increase in the importance of organic agriculture with minimum input of chemicals. These bio-fertilizer inoculants could be a particular isolate or group of organisms that confer plant health.

11.2 PGPRs: Definition and Functions

Soil bacterial species growing in the plant rhizosphere that grow in, on, or around plant tissues and stimulate plant growth by numerous mechanisms are jointly referred to as PGPR (plant growth-promoting rhizobacteria). The term PGPR was coined by Kloepper and Schroth to explain the helpful microorganism population that colonizes the roots of plants within the rhizosphere, once attached to the root surface, to support plant growth and plant growth promotion activities. PGPR can be classified into completely different subgroups on the basis of their application: (1) bio-fertilizers (increasing the availability of nutrients to plants), (2) phytostimulators (plant growth promotion, usually through phytohormones such as indole-3-acetic acid), (3) rhizoremediators (degrading organic pollutants), and (4) biopesticides (controlling diseases, chiefly by the production of antibiotics and antifungal metabolites) (Bhattacharya & Jha, 2012). In the last few years, the number of PGPRs has been found to increase mainly due to their role in the rhizosphere. Various species of bacteria such as *Pseudomonas*, *Azospirillum*, *Azotobacter*, *Klebsiella*, *Enterobacter*, *Alcaligenes*, *Arthrobacter*, *Burkholderia*, *Bacillus*, and *Serratia* have been reported to enhance the plant growth and suppress phytopathogens. PGPRs are divided into two classes: biocontrol plant growth-promoting bacteria (biocontrol PGPB) and plant growth-promoting bacteria (PGPB) (Bashan & Holguin, 1998). This classification may include beneficial bacteria that are not rhizosphere bacteria, but this classification system has not been widely accepted. PGPR are known to provide essential nutrients for plant growth or to enhance their availability, to play an important role in pathogen suppression, and to improve the health of soil in the long term, and therefore are potentially important in reducing the use of chemical fertilizers and pesticides. The PGPRs enhance plant growth and development by direct and/or indirect mechanisms. Direct mechanisms elicit growth promotion by biological nitrogen fixation (BNF), production of hormones such as indole-3-acetic acid (IAA), Gibberellic acid (GA3), Cytokinin and phosphate solubilization or mobilization, production of siderophores for sequestering of iron (Fe) from the soil and supply it to the plants and synthesis of hydrogen cyanide, etc. (Keel & Defago, 1997). Some strains improve the innate ability to tolerate the stresses like acidity, salinity, drought, etc., in addition to production of 1-aminocyclopropane-1-carboxylate (ACC) deaminase enzyme to lower the ethylene synthesis and synthesis of fungal cell wall lytic enzymes. Secondly, indirect mechanisms include suppression of harmful/deleterious rhizosphere microbes through a variety of mechanisms such as rhizosphere competition, rhizospheric engineering, quorum sensing, production of volatile organic compounds, enzyme production, induced systematic resistance (ISR), reduction or prevention of deleterious effects of phytopathogens on plant growth by biosynthesis of stress-related phytohormones such as jasmonic acid (JA) or ethylene, and biosynthesis of antimicrobial molecules (Jain et al., 2014). These mechanisms can work autonomously or simultaneously with one another. Their application as crop inoculants for bio-fertilization would be an attractive option to reduce the use of chemical fertilizers. In addition, PGPR can adapt well to harsh environments, including drought stress, salt stress, high temperatures, dryness, or heavy rainfall in tropical countries and contaminated environments, indicating that they could contribute to ameliorating plant crops in areas with poor agricultural potential. The successful establishment of an introduced PGPR depends on its compatibility/establishment with the crop and also on its interaction with indigenous microflora. An ideal PGPR should be rhizosphere competent, enhance plant growth, be easy to mass multiply, possess a broad spectrum of action, have consistent biological control activity, be safe to the environment, and be compatible with other rhizobacteria (Barea, 2015).

11.3 PGPRs Functions in Sustainable Agriculture and Biocontrol

Sustainable agriculture needs IPM, and part of the IPM approaches is the use of non-chemical control methods. The non-chemical methods include use of biocontrol microbes, predators and parasites, and plants. Dissemination and adoption of ecofriendly approaches of pest management reached only a smaller section of farmers, and hence, future extension efforts should be concentrated to reach a larger range of farmers. Plant pest and disease management includes the use of beneficial microorganisms for the effective and sustained production of agricultural and horticultural crops. Numerous species of soil bacteria and fungi which flourish in the rhizosphere of plants stimulate plant growth both directly and indirectly. Plant-growth-promoting rhizobacteria (PGPR), viz., *Pseudomonas*, *Bacillus*, and fungal antagonist *Trichoderma*, have been well exploited for the management of plant diseases in economically important agricultural and horticultural crops. Definition of biocontrol differs depending on the target, number, type and source of biological agents, and the level of human intervention. Most simply, biological control is the suppression of damaging activities of one organism by other organisms, usually the natural enemies. In case of plant diseases, the term applies to the use of microbial antagonists to suppress diseases as well as the use of host-specific pathogens to control weed populations. Antagonistic PGPR offers a highly effective, economical, and environment-friendly alternative to the use of synthetic pesticides. The modes of action of the antagonistic PGPRs against various soil-borne plant pathogens are different, including biosynthesis of antibiotics, production of hydrolytic enzymes, production of siderophore, competition for substrates and the sites on the roots, and induced systemic resistance (ISR) (Ramamoorthy et al., 2001; Kim et al., 2003). Successful bacterial antagonists often show a synergistic combination of mechanisms responsible for a successful interaction.

11.4 Bioformulation of PGPRs

Bioformulations may be defined as biologically active products in which one or more effective and beneficial microbial strains are present in an easy to use and economical carrier material. Although biocontrol with PGPR is an acceptable green approach, the registration of biocontrol agents for commercial availability is very low. Handling these products requires care, but these are not properly handled by the people who use it. This is the reason that there are so many PGPRs that perform very well in controlled conditions but not in field conditions, and hence the commercialization of these products is not easy. Because bioformulation is a living product, utmost care is needed at all steps, from production to end-use, to maintain the microbial load and vigor. Most of the applications of bioformulations are meant under field applications for which the use of suitable carrier materials is very important in order to maintain cell viability under adverse environmental conditions. A good quality and efficient formulation promotes survival of bacteria, maintaining available population sufficient to exude growth-promoting effects on plants. Therefore, plant-growth-promoting rhizobacterial bioformulations refer to microbial preparations that may be considered as partial or complete substitutes for chemical fertilizers and pesticides and most importantly offer an environmentally sustainable approach to increase crop production and health (Singh et al., 2014). The development of formulations with increased shelf life and a broad spectrum of action with consistent performance under field conditions can pave the way for commercialization of the technology at a faster rate. The development of a successful inoculant involves several critical elements such as strain selection, selection of a carrier, mass multiplication, formulation of the inoculant, and packaging and marketing. The functions of formulations include stabilizing the microorganism, aiding in its delivery, protecting the microbe at the target zone, and enhancing the functionality of the microbe against the target (if a BCA). The isolation of antagonistic strains, screening, fermentation methods, mass production, formulation viability, toxicology, industrial linkages, quality control, and field efficacy are the different stages in the process of commercialization under which abundance of work has been performed. Moreover, factors which affect the commercial success of PGPR strains involve economical and viable market demand, consistent and broad-spectrum market, longer shelf life, safety and stability, low capital costs, and easy availability of career materials (Nandakumar et al., 2001). The availability of the commercial formulation of biocontrol agents (BCAs) in the market for control of plant diseases is still in its infancy due to different degrees of limitations. However, the development of commercial formulation of biocontrol agent (BCA) is an important area of microbiological research, concerning mainly on the development of low cost, most effective, and suitable for all types of farming systems and preservation of microbial activity for a period long enough to enable delivery of an effective product for field application (Ramamoorthy et al., 2001). The most successful and effective product would be the one that can be used for disease management with existing machinery or methods. The biological control agents, therefore, are generally formulated as wettable powders, dusts, granules, and aqueous or oil-based liquid products, with various additives to attain all the desirable attributes along with their biocontrol activity. Today, a number of biocontrol agents are available for the control of fungal plant diseases (Ardakani et al., 2009). Even though many small, privately owned companies with limited product lines are developing and are in the process of registration, there still are large companies with diverse product lines that include a variety of agrochemicals, and biotechnological products playing an important role in the development and marketing of products. These products are standalone or formulated as mixtures of microorganisms. The biopesticide industry alliance is establishing a certification process to ensure standards for efficacy, quality, and consistency for lifting the global marketing perception. In this process, it is also important to concentrate on the training of growers, formulation of biocontrol microorganisms, and the role of environmental factors (Ardakani et al., 2009). To develop a successful PGPR formulation, rhizobacteria should possess certain characteristics such as high rhizosphere competence; high competitive saprophytic ability; enhanced plant growth; ease for mass multiplication; broad spectrum of action; excellent and reliable control; safety to environment, humans, and animals; compatible with other rhizobacteria; and tolerance to desiccation, heat, oxidizing agents, and UV radiations (Nakkeeran et al., 2005). To be an ideal formulation, it should have increased shelf life. It should not be phytotoxic to plants. It should be dissolved well in water and should release the bacteria. It should be cost-effective and should give reliable control of plant diseases. It should be compatible with other agrochemicals, and the carriers (for carrying the microbe) must be cheap and readily available for formulation development (Nakkeeran et al., 2005). The formulations are in two forms, dry powder (solid) and liquid suspensions, which can be prepared in several ways. Formulation design also involves the type of formulation in which the choice of carrier is an important task. Many workers have evaluated the superiority of liquid formulations over the solid carrier-based formulations.

11.5 Obstacles and Challenges in Formulation Development

Production technology of bioformulations requires proper care and aid of sophisticated equipment to ensure availability of quality products in the market (Arora et al., 2010). During transport and distribution maintaining the temperature of the formulation is an important task. This, in turn, increases the cost of the product. Shelf life of formulation is another obstacle because the storage of bioformulations requires special facilities and skills, which most producers, shopkeepers, and farmers do not possess. Shelf life is a culmination of several factors such as production technology, carrier and packaging material used, and mode and distance of transport. All these are desired to sustain the shelf life (Arora et al., 2010). Shelf life of formulation also depends on the performance of strain in field and laboratory conditions. To avoid this problem of reduced survival rate of strain, the preparation should be air-dried and lyophilized. Inconsistent field performance is the major obstacle associated with marketing of formulations. Many PGPRs work well in laboratory conditions, but they are unable to survive in fields, and therefore, this is an obstacle in the industrialization of formulation. These failures have raised concerns about the perspective of the great practical potential offered by microbial releases into soils. A key factor involved in the failure is the rapid decline of the size of the population of active cells. The active cells decline to a level at which the formulation becomes ineffective when introduced into the soil and the desired objective is not achieved. As soil is a heterogeneous system with a mixed biota, the formulation added in soil is subjected to rhizosphere competence, interaction with already adopted soil microflora, and survival in a new habitat. A number of factors such as soil composition, temperature, relative humidity, composition of root exudates, presence of recombinant plasmids, as well as the interactions with other soil biota can affect the persistence of a PGPR on the root system making it difficult to predict the behavior of the bacterial strain under natural conditions. Therefore, PGPRs that are effective in the laboratory may not show any significant impact on plants under field conditions (Glick et al., 1999). The degree of success of biocontrol-mediated disease management depends upon the biotic, abiotic, and procedural factors. There should be rigorous testing of biocontrol agents (BCAs) for target specificity to ensure that the biocontrol organism attacks only the target organism. To monitor the biological control, the following points need to be clarified: select the biocontrol agents with the potential to control the target pathogen; release only the safe and approved agents; strict legislation should be made to regulate the BCAs; appropriate protocols should be used for release and documentation; monitor the impact on target pathogen; avoid further application of biocontrol agent when effective or control is achieved;

and proper assessment of changes in plant and animal communities along with the complete ecosystem should be done.

11.6 Conclusion and Future Perspectives

The continuous increase in the world total population (9 billion by 2050) and the imminent global climate change are starting to challenge the sustainability of our current food supplies and energy resources. Furthermore, plant pests, currently responsible for up to 40% of the losses in the main crops worldwide, will continue hampering plant growth and agricultural production. A significant body of evidence shows that synthetic pesticides and fertilizers represent important risks to the environment and human health. In consequence, governments are starting to restrict the use of some synthetic agrochemicals in order to promote more sustainable agricultural practices (Glare et al., 2012), and ecologically friendly alternative management strategies to chemical fertilizers and pesticides are urgently needed. The bacterial control agent will completely replace chemicals in the foreseeable future and is very doubtful. But we can at least expect some reduction in the use of chemicals. Thus far, most approaches include the single antagonist concept; however, a biological system approach that involves integration of various factors for disease suppression might provide a better alternative. The biological control agents could be used as one component of the integrated management program to achieve the best possible results. Biological control really developed as an academic discipline during the 1970s and is now a mature science supported both in the public and private sectors. Research related to biological control is published in scientific journals. Additionally, some academic journals are specifically devoted to this discipline (Heydari & Pessarakli, 2010). Several research schemes, institutions, awareness programs, monitoring techniques, funds for boosting agriculture, and small business innovation are taken in hand. Such ventures are intended to be conduits for academic research that can be used to develop new products of commercial importance (Spadaro & Gullino, 2005). Much has been learned from the biological control research conducted over the past 40 years. But, in addition to learning the lessons from the past, biocontrol researchers need to look forward to define new and different questions and the answers to which will facilitate new biocontrol strategies, technologies, and applications. In recent times, advances in computing, molecular biology, analytical chemistry, and statistics have added new dimensions to research that aimed at characterizing the structure and functions of biological control agents, pathogens, and host plants at molecular, cellular, physiological, and ecological levels. Ecological factors play very important roles in the performance and potentiality of biological control agents (BCAs). To develop efficient and effective biological control system, the following factors need to be clarified and studied, viz., (1) distribution of various pathogens and their natural antagonists in the environment; (2) optimum field conditions which influence suppressive capacities of control agents; (3) response of native and introduced populations to different management practices; (4) studies of factors responsible for successful colonization and expression of biocontrol traits in fields like signal molecules of plant and microbial origin that regulate expression of biocontrol traits, mechanism of defense induction in plants; (5) search for more effective strain variants for broad base field applications; (6) use of rDNA technology for manipulation of plants and biological control agents; (7) development of proper formulations to enhance field activities of known biocontrol agents; (8) role of other gene products involved in pathogen suppression; (9) efficacy of using consortium of novel strains in place of individual agents; and (10) development of effective biocontrol–cultivar combinations by plant breeders.

REFERENCES

Ardakani, S.; Heydari, A.; Khorassani, N.; Arjmandi, R.; Ehteshami, M. Preparation of new biofungicides using antagonistic bacteria and mineral compounds for controlling cotton seedling damping-off disease. *J. Plant. Prot. Res.* 2009, *49*, 49–55.

Arora, N. K.; Khare, E.; Maheshwari, D. K. Plant growth promoting rhizobacteria: Constraints in bioformulation, commercialization, and future strategies. In: *Plant Growth and Health Promoting Bacteria, Microbiology Monographs*; Maheshwari, D. K. Ed.; Springer, Berlin/Heidelberg, 2010; pp. 97–115.

Barea, J. M. Future challenges and perspectives for applying microbial biotechnology in sustainable agriculture based on a better understanding of plant-microbiome interactions. *J. Soil Sci. Plant. Nutr.* 2015, *15*, 261–282.

Bashan, Y.; Holguin, G. Proposal for the division of plant growth-promoting rhizobacteria into two classifications: Biocontrol-PGPB (plant growth-promoting bacteria) and PGPB. *Soil Biol. Biochem.* 1998, *30*, 1225–1228.

Bhattacharya, P. N.; Jha, D. K. Plant growth-promoting rhizobacteria (PGPR): Emergence in agriculture. *World J. Microbiol. Biotechnol.* 2012, *28*(4), 1327–1350.

Glare, T.; Caradus, J.; Gelernter, W.; Jackson, T.; Keyhani, N.; Köhl, J.; Marrone, P.; Morin, L.; Stewart, A. Have biopesticides come of age? *Trends Biotechnol.* 2012, *30*, 250–258.

Glick, B. R.; Patten, C. L.; Holguin, G.; Penrose, D. M. *Biochemical and Genetic Mechanisms Used by Plant Growth Promoting Bacteria*. Imperial College Press, London, 1999.

Heydari, A.; Pessarakli, M. A review on biological control of fungal plant pathogens using microbial antagonists. *J. Biol. Sci.* 2010, *10*(4), 273–290.

Jain, S.; Vaishnav, A.; Kasotia, A.; Kumari, S.; Choudhary, D. K. Plant growth-promoting bacteria elicited induced systemic resistance and tolerance in plants. In: *Emerging Technologies and Management of Crop Stress Tolerance*, 2014; p. 109.

Keel, C.; Defago, G. Interactions between beneficial soil bacteria and root pathogens: mechanisms and ecological impact. In: *Multitrophic Interactions in Terrestrial System*; Gange, A. C., Brown, V. K. Eds.; Blackwell Science, Oxford, 1997; pp 27–47.

Kim, K. J.; Yang, Y. J.; Kim, J. G. Purification and characterization of chitinase from *Streptomyces* sp. M-20. *J. Biochem. Mol. Biol.* 2003, *36*(2), 185–189.

Nakkeeran, S.; Fernando, D. W. G.; Siddiqui, Z. A. Plant growth promoting rhizobacteria formulations and its scope in commercialization for the management of pests and diseases. In: *Biocontrol and Biofertilization*; Siddiqui, Z. A. Ed.; Springer, Dordrecht, 2005; pp. 257–296.

Nandakumar, R.; Viswanathan, R.; Babu, S.; Sheela, J.; Raghuchander, T.; Samiyappan, R. A new bio-formulation containing plant growth promoting rhizobacterial mixture for the management of sheath blight and enhanced grain yield in rice. *Biocontrol.* 2001, *46*, 493–510.

Prasad, R.; Kumar, M.; Varma, A. Role of PGPR in soil fertility and plant health. In: *Plant Growth-Promoting Rhizobacteria and Medicinal Plants*; Egamberdieva, D., Shrivastava, S., Varma, A. Eds.; Springer, Switzerland, 2015; pp. 247–260.

Ramamoorthy, V.; Viswanathan, R.; Raghuchander, T.; Prakasam, V.; Samiyappan, R. Induction of systemic resistance by plant growth promoting rhizobacteria in crop plants against pests and diseases. *Crop Prot.* 2001, *20*, 1–11.

Singh, S.; Gupta, G.; Khare, E.; Behal, K. K.; Arora, N. K. Effect of enrichment material on the shelf life and field efficiency of bioformulation of *Rhizobium* sp. and P-solubilizing *Pseudomonas fluorescence. Sci. Res. Rep.* 2014, *4*, 44–50.

Spadaro, D.; Gullino, M. L. Improving the efficiency of biocontrol agents soilborne pathogens. *Crop Prot.* 2005, *24*, 601–613.

12

Killing Bacteria as Agents of Insect Pest Control

Muhammad Sarwar

CONTENTS

12.1 Introduction

To reduce the environmental footprint of pests control, bacteria are the ideal tools to develop smart precision bioinsecticides and prevent disease transmission by insects. The bacterial cell envelope encases the cytoplasm and all its components, as bacteria do not have a membrane-enclosed nucleus (Sarwar, 2015a; Sarwar, 2015b; Sarwar, 2015c). This trend is also the result of modern legislative frameworks fostering bioinsecticides in Integrated Pest Management (IPM) programs (Sarwar, 2004; 2013a; 2017).

The entomopathogens belonging to the Gammaproteo bacteria class include the entomopathogenic nematode symbionts *Photorhabdus* species and *Xenorhabdus* species, *Serratia* species, *Yersinia entomophagy* Hurst and *Pseudomonas entomophila* Pen are also important (Ruiu, 2015).

A *Coccobacillus* is a very short rod type of bacterium with a shape intermediate between cocci (spherical bacteria) and bacilli (rod-shaped bacteria). The bacterium *Coccobacillus acridiorum* D'Herelle is considered a pathogen of locusts and grasshoppers. It is an etiological agent of a highly virulent disease, responsible for the widespread destruction of locusts. The initial discovery is a *Coccobacillus* present in cricket and locust intestines, provoking the insects' rapid death (Morange, 2017).

A more recent discovery is that of Betaproteo bacteria species, which shows broad-spectrum insecticidal properties. This group includes specific strains of *Burkholderia* species and *Chromobacterium* species. Certain Actinobacteria species have gained high scientific and commercial interest in the production of various metabolites acting as potent insecticides. This is the case for *Streptomyces* and *Saccharopolyspora* species (Glare et al., 2012).

As a result of modern studies, microbial pest management dominated by *Bacillus thuringiensis*, which is novel bacterial species with innovative modes of action, have been discovered and formulated as new biopesticidal products (Ruiu, 2013).

A highly dense protein crystal, the Bt toxin kills victims by first paralyzing their midgut and their entire bodies. Its short residual period also makes it an ideal candidate for pest management in fruits and vegetables (Lacey et al., 2001).

Bt bacterium synthesizes parasporal crystalline inclusions containing Cry and Cyt proteins, Some of are toxic against a wide range of insect orders, nematodes and human-cancer cells. Bt also synthesizes insecticidal proteins during the vegetative growth phase, subsequently secreted into the growth medium. These proteins are commonly known as vegetative insecticidal proteins (Vips) and hold insecticidal activity against lepidopteran, coleopteran and some homopteran pests. A less well characterized secretory protein with no amino acid similarity to Vip proteins has shown insecticidal activity against coleopteran pests and is termed Sip (secreted insecticidal protein) (Palma et al., 2014). The purpose of the present chapter is to give an overall and concise picture of insect pathogenic bacteria's knowledge advancements and their use as bioinsecticidal products for integrated pest management (IPM).

12.2 Pathogenic Bacteria of Insect Pests

12.2.1 Bacillaceae

The family Bacillaceae comprises rod-shaped bacteria that form endospores, wherein principal subdivisions are anaerobic spore-forming bacilli of the genus *Clostridium* and aerobic, facultative anaerobic endospore formers constituting genus *Bacillus*.

12.2.1.1 Bacillus thuringiensis Berliner

Bacterium *Bacillus thuringiensis* or *Bt*, is a gram-positive, soil-dwelling bacterium commonly used as a biological pesticide (Figures 12.1 & 12.2). *B. thuringiensis* also occurs naturally in the gut of caterpillars of various types of moths and butterflies, as well on leaf surfaces, aquatic environments, animal feces, insect-rich environments, and flour mills and grain storage facilities. Bt bacteria make proteins that are toxic to some insects when eaten but not toxic to humans because, like all mammals, these vertebrates cannot activate them. During sporulation, many Bt strains produce crystal proteins called δ-endotoxins that have insecticidal action. Many crystal-producing Bt strains do not have insecticidal properties (Roh et al., 2007).

There are several subspecies of *B. thuringiensis (Bacillus thuringiensis* subspecies *kurstaki* (Btk), species *israelensis* (Bti) and subspecies *aizawa* (Bta). Commonly used as insecticides, the Bt subspecies *israelensisis*, commonly used to control mosquitoes and gnats, while subspecies *tenebrionis* proved effective against some coleopterans (Madigan & Martinko, 2005).

The types of responses below are based on the susceptibility of caterpillars to Bt toxins (Ruiu et al. 2013).

Type I response: Midgut paralysis in the host insect occurs within a few minutes after delta-endotoxin is ingested. Symptoms include cessation of feeding, an increase in hemolymph pH, vomiting, diarrhea and sluggishness. General paralysis and septicemia occur in 24-48 hours resulting in the death of the insect. Examples of insects that show Type I response include silkworm, tomato hornworm and tobacco hornworm.

Type II response: In a host insect, midgut paralysis occurs within a few minutes after the ingestion of delta-endotoxin, but there will be no general paralysis. Septicemia occurs within 24 to 72 hours. Examples of host insects include inchworms, alfalfa caterpillar and cabbage butterfly.

Type III response: In an insect host, midgut paralysis occurs after delta-endotoxin is ingested followed by cessation of feeding. Insect may move actively as there will be no general paralysis. Mortality occurs in 48 to 96 hours and higher mortality occurs if spores are ingested. Insect examples include Mediterranean flour moth, corn earworm, gypsy moth, spruce budworm.

Type IV response: In this case, insects are naturally resistant to infection and older instars are less susceptible than the younger ones. Midgut paralysis occurs after delta-endotoxin is ingested followed by cessation of feeding. The insect may move actively as there will be no general paralysis. Mortality occurs in 72 to 96 or more hours, and higher mortality occurs if spores are ingested. Cutworms and armyworms are examples for this category of response.

A small number of entomopathogenic bacteria have been commercially developed to control insect pests (Table 12.1). These

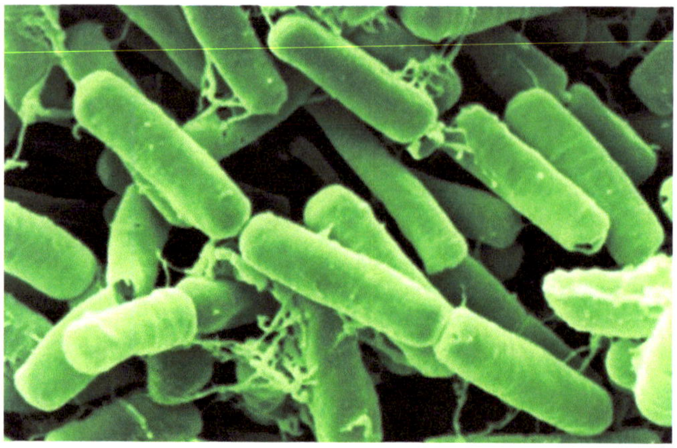

FIGURE 12.1 *Bacillus thuringiensis.*

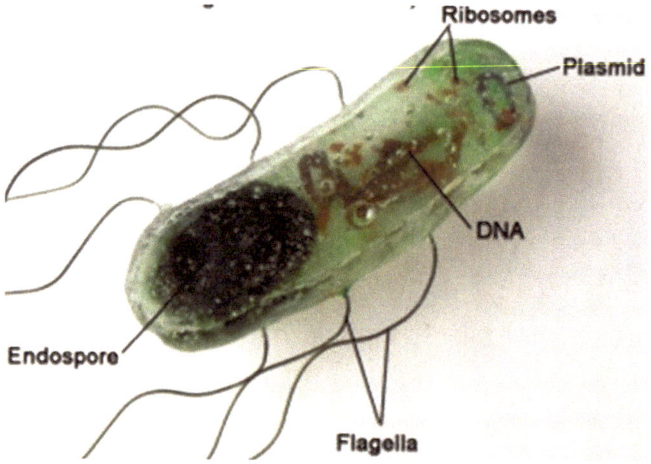

FIGURE 12.2 Bottom view of *B. thuringiensis* showing internal structures.

TABLE 12.1

Examples of successful genes usage originated from *Bacillus thuringiensis* for integration against insects resistance

Pathogenic Bacteria	Host Range	Uses
Bacillus thuringiensis var. *kurstaki*	Caterpillars (larvae of moths and butterflies)	Effective for foliage-feeding caterpillars and Indian meal moth in stored grain. Deactivates rapidly in sunlight; apply in the evening or on overcast days and direct some spray to lower surfaces or leaves. Does not cycle extensively in the environment.
Bacillus thuringiensis var. *israelensis*	Larvae of *Aedes* and *Psorophora* mosquitoes, black flies and fungus gnats	Effective against larvae only and active only if ingested. Mosquitoes *Culex* and *Anopheles* are not controlled at normal application rates. Does not cycle extensively in the environment.
Bacillus thuringiensis var. *tenebrinos*	Larvae of Colorado potato beetle, elm leaf beetle adults	Effective against Colorado potato beetle larvae and the elm leaf beetle. Like other Bts, it must be ingested for action. It is subjected to breakdown in ultraviolet light and does not cycle extensively in the environment.
Bacillus thuringiensis var. *aizawai*	Wax moth caterpillars	Used only for control of wax moth infestations in honeybee hives.
Bacillus popilliae and *Bacillus lentimorbus*	Larvae (grubs) of Japanese beetle	The main illinois lawn grub (annual white grub *Cyclocephala* sp.), that is not susceptible to milky spore disease.
Bacillus sphaericus	Larvae of *Culex, Psorophora* and *Culiseta* mosquitos, larvae of some *Aedes* sp.	Active only if ingested, for use against *Culex, Psorophora* and *Culiseta* species; also effective against *Aedes vexans* (Meigen). Remains effective in stagnant or turbid water

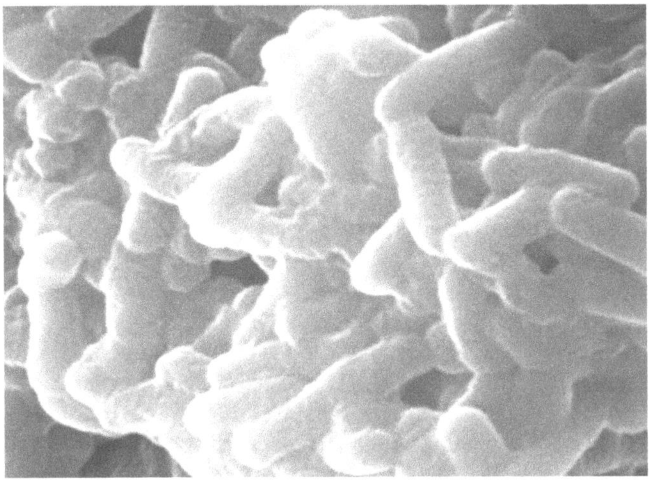

FIGURE 12.3 *Lysinibacillus sphaericus.*

pathogens combine the advantages of chemical pesticides and Mast Cell Activation Syndrome (MCAS): they are fast-acting, easy to produce at a relatively low cost, easy to formulate, have a long shelf life, and allow delivery using conventional application equipment and systemics (in transgenic plants). Unlike broad-spectrum chemical pesticides, *B. thuringiensis* toxins are selective and negative environmental impact is very limited. The Bt genes used in insect-resistant transgenic crops belong to the Cry and vegetative insecticidal protein families of toxins (Lacey et al., 2015).

12.2.1.2 *Lysinibacillus sphaericus*

Entomopathogenic strains belonging to the *L. sphaericus* (Figure 12.3) species group are featured by the production of spherical endospores closely associated with parasporal crystals containing an equimolar ratio of binary protein toxins (BinA and BinB) (Baumann et al., 1991). The insecticidal mode of action includes damages to the microvillar epithelial cells in the midgut comparable to the ones known for *B. thuringiensis* (Charles et al., 2000).

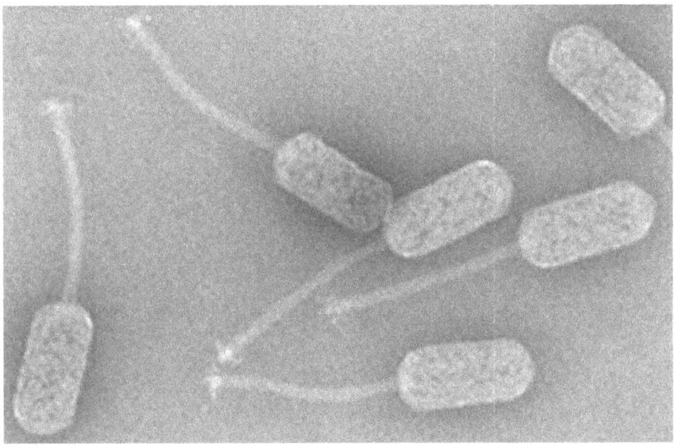

FIGURE 12.4 *Paenibacillus* larvae.

12.2.1.3 *Paenibacillus* Species

The genus *Paenibacillus* includes different species showing pathogenicity against insects like the causative agent of the honeybee disease American Foulbrood, *P. larvae* subsp. *larvae* (Sarwar, 2016a; 2016b; 2016c; 2016d; 2016e) (Figure 12.4). On the other hand, the spore-formers *P. popilliae* (Dutky) Pettersson et al. (Figure 12.5) and *P. lentimorbus* (Dutky) Pettersson et al. (Figure 12.6) are the causal agents of milky disease in phytophagous coleopteran larvae. The production of parasporal inclusions within the sporangial cells has been observed in *P. popilliae.* However, homology between a 80 kDa parasporal protein this species produces and *Bt* Cry toxins has been demonstrated. After spores are ingested by the host, they germinate in the midgut (Zhang et al., 1997).

12.2.1.4 *Brevibacillus laterosporus*

Bacterium *Brevibacillus laterosporus* (Figure 12.7) is a pathogen of invertebrates and a broad-spectrum antimicrobial species (Ruiu et al., 2007).

The whole genome of *B. laterosporus* has been published (Djukic et al., 2011), which reveals the potential to produce different toxins. Certain strains showing toxicity against the

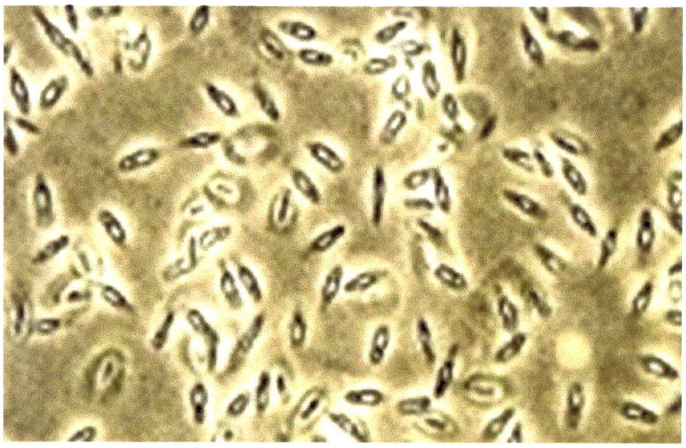

FIGURE 12.5 *Paenibacillus popilliae.*

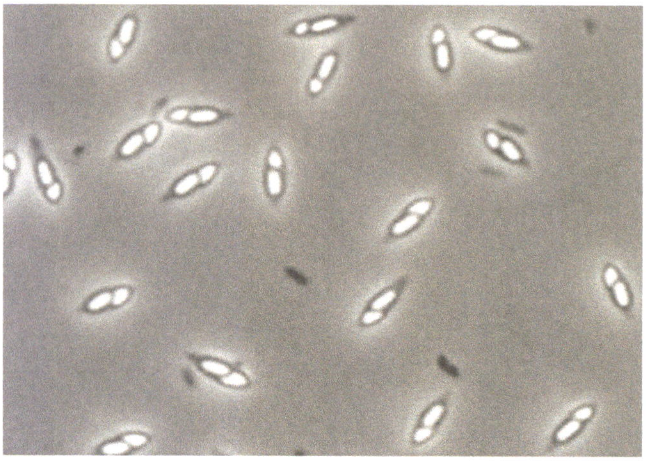

FIGURE 12.6 *Paenibacillus lentimorbus.*

corn rootworms (*Diabrotrica* species) and other coleopteran larvae, produce insecticidal secreted proteins (ISPs) that act as binary toxins in the insect midgut and have high homology with *B. thuringiensis* vegetative insecticidal proteins (Warren, 1997).

Specific strains toxic to mosquitoes produce parasporal inclusion bodies reminiscent of those produced by *B. thuringiensis* (Zubasheva et al., 2010). Spores of a strain lacking parasporal crystals are highly toxic to the house fly *Musca domestica* L., and the mode of action implies histopathological changes in the midgut with disruption of the microvillar epithelium (Ruiu et al., 2012).

12.2.2 Impacts on Nontarget Organisms

It has been determined that nontarget organisms are not exposed to high enough amounts of the Bt toxins to have an adverse effect on the population. Soil-dwelling organisms (Saxena & Stotzky, 2000). A new phenomenon called colony collapse disorder began affecting beehives (Sarwar, 2016f; 2016g; 2016h; 2016i). Initial speculation on possible causes included new parasites, pesticide use and the use of Bt transgenic crops (Rose et al., 2007).

Some isolates of *B. thuringiensis* produce a class of insecticidal small molecules called beta-exotoxin, the common name

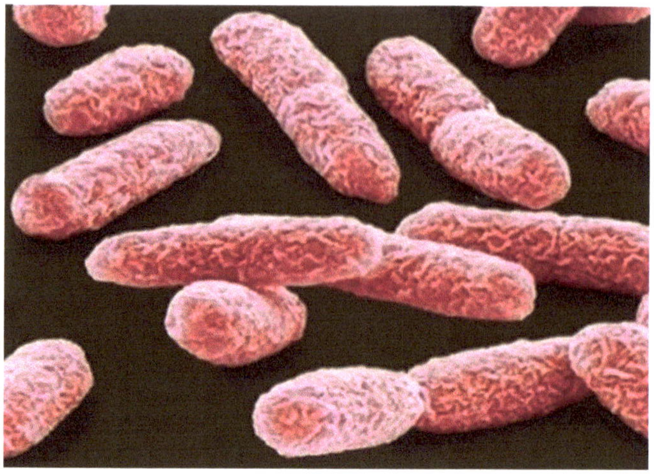

FIGURE 12.7 *Brevibacillus laterosporus.*

for which is thuringiensin. Beta-exotoxins are known to be toxic to humans and almost all other forms of life (Ruiheng, 2016).

12.2.3 Plant-Incorporated Protectants

Genetically modified plants possess a gene or genes that have been transferred from a different species. Successful examples to show *B. thuringiensis* genes (originated from *B. thuringiensis*) integration for insect resistance in rice include; Cry 1A(b) for striped stem borer, leaf folder and striped stem borer; Cry 1A(b)/Cry 1A(c) against leaf folder and yellow stem borer; Cry 1A(c) for yellow stem borer and striped stem borer; Cry 2A (c) against leaf folder and yellow stem borer; Cry 2A/ Cry 1A(c) used for leaf folder and yellow stem borer; and Cry 1Ie in case of corn borer (Gomezet al., 2002).

Bacterium *B. thuringiensis* and its products have been formulated into various forms for application as biological control agents. An alternative and highly successful method for delivering the toxins to the target insect has been to express the toxin-encoding genes in transgenic plants (Aliet al., 2010).

12.2.4 Clostridiaceae

12.2.4.1 Clostridium bifermentans

A strain of *Clostridium bifermentans* (C.b.m.) (Figure 12.8), isolated, shows high toxicity against mosquitoes and black flies. During sporulation, this bacterium produces three major proteins involved in the insecticidal action. These include the mosquitocidal protein Cbm71 showing homology to *B. thuringiensis* delta endotoxins (Barloy et al., 1996).

12.2.5 Gamma Proteobacteria

12.2.5.1 Photorhabdus Species and *Xenorhabdus* Species

Photorhabdus and *Xenorhabdus* species producing an insecticidal toxins complex (Tc) have high potential for pest management. Generally, Tcs are high-molecular weight and multisubunit proteins that include three components, A, B and C, orally active against different insects (Waterfield et al., 2005).

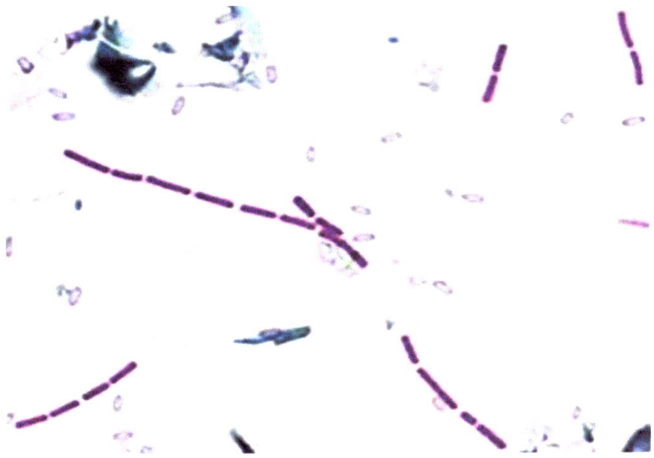

FIGURE 12.8 *Clostridium bifermentans.*

FIGURE 12.9 *Photorhabdus luminescens.*

Photorhabdus insect-related (Pir) proteins, produced by *P. luminescens* (Thomas and Poinar) (Figure 12.9), show similarity to *B. thuringiensis* delta-endotoxins and have been proposed to be mimics of the juvenile hormone esterases interfering with insect development regulation (Ffrench-Constant et al., 2007). It has been shown that certain *Photorhabdus* species exploit lipopolysaccharide modifications to resist the action of insect antimicrobial peptides, while *X. nematophila* (Poinar and Thomas) interferes with the expression mechanisms of host AMPs (Ji et al., 2004). New species are continuously being identified and characterized (Tailliez et al., 2006).

12.2.5.2 Serratia *Species*

The most common is the production of toxin complexes analogous to those produced by *Xenorhabdus* species and *Photorhabdus* species, *S. entomophila* Grimont et al. (Figure 12.10), a pathogen of the grass grub *Costelytra zealandica* (White) (Coleoptera: Scarabaeidae), produces Sep proteins (SepA, SepB, SepC), a group of insecticidal toxins showing similarities to the insecticidal toxins of *P. luminescens* (Hurst et al., 2000).

Pathogenicity of *S. marcescens* Bizio is increased by the action of a serralysin metalloprotease it secretes. This bacterium can press cellular immunity by decreasing the adhesive properties of immune surveillance cells of the insect host (Ishii et al., 2014).

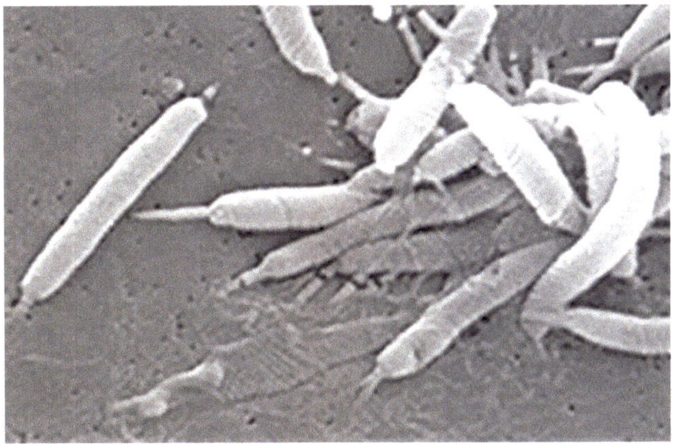

FIGURE 12.10 *Serratia entomophila.*

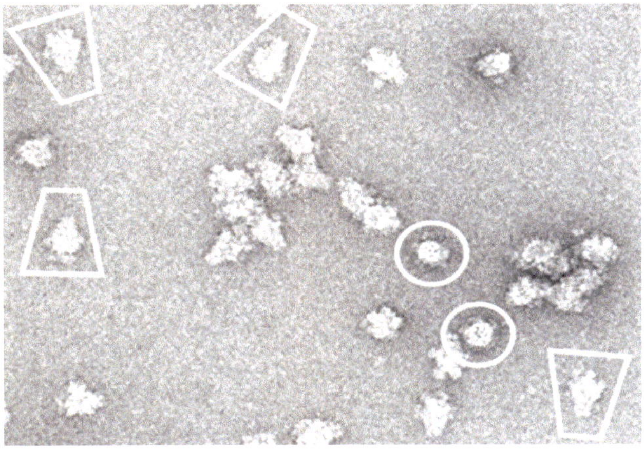

FIGURE 12.11 *Yersinia entomophaga.*

12.2.5.3 Yersinia entomophaga

Sequestered from grass grub *C. zealandica*, *Y. entomophaga* (Figure 12.11) is a nonspore-forming entomopathogenic bacterium characterized by the production of an insecticidal toxin complex (Yen-Tc) (Hurst et al., 2004). Promising studies conducted in field conditions with insecticidal formulations containing *Y. Entomophaga* against the pasture pest porina (*Wiseana* species larvae) (Ferguson et al., 2012).

12.2.5.4 Pseudomonas entomophila

Host-pathogen interactions have been studied in experiments with *Drosophila melanogaster* Meigen (Diptera: Drosophilidae), which highlights a specific postingestion immune response. Recent complete genome sequencing of *P. Entomophila* revealed a specific secretion system and the associated toxins probably responsible for its entompathogenic properties (Vodovar et al., 2005) (Figure 12.12).

12.2.6 Betaproteobacteria

12.2.6.1 Burkholderia Species

The potential of *Burkholderia* species as biocontrol agents against different plant pathogens has also been reported (Burkhead et al.,

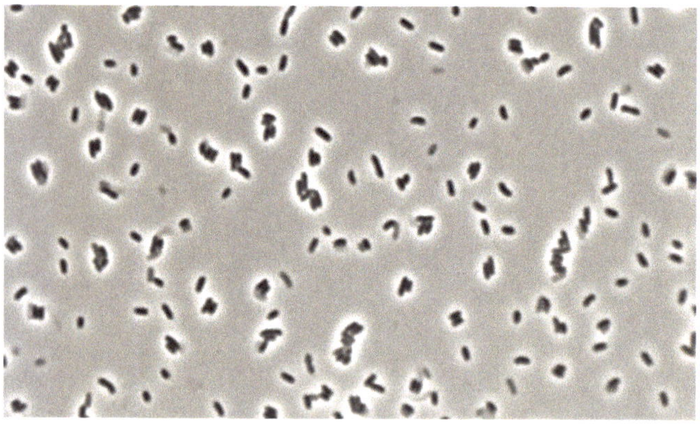

FIGURE 12.12 *Pseudomonas entomophila.*

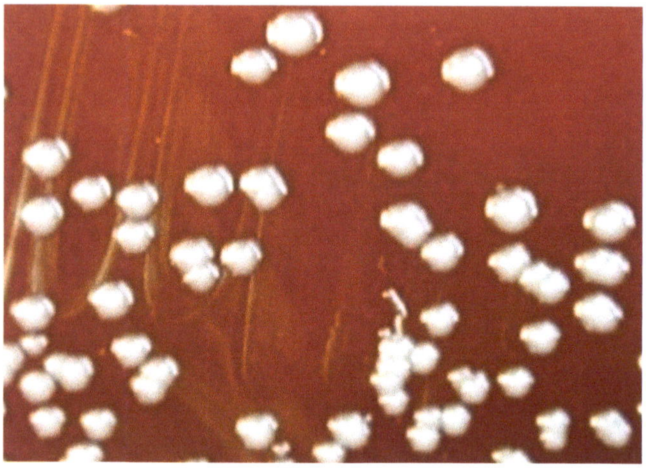

FIGURE 12.13 *Burkholderia* species.

1994). More recently, insecticidal properties of a new strain isolated in soil and identified as *B. rinojensis* sp. nov., have been discovered. Insecticidal and miticidal properties are maintained after heat-treatment, hence commercial formulations against a variety of chewing and sucking insects, and mites are based on heat-killed cells and spent fermentation media (Sarwar, 2013b; 2019). Interestingly, another study highlighted the biochemical properties of natural compounds produced by fermentation of a *Burkholderia* species strain showing high homology to *B. rinojensis* (He et al., 2014) (Figure 12.13).

12.2.6.2 Chromobacterium Species

A strain of *Chromobacterium subtsugae* (Figure 12.14) isolated from a soil sample and named PRAA4-1T, has been discovered to show high insecticidal activity against insect species in different orders. Among the bacterial metabolites, *C. subtsugae* synthesizes the tryptophan derivative violacein, which confers a typical violet color to its colonies (Martin et al., 2007).

More recently, a new strain identified as *Chromobacterium* (Csp_P), isolated from the midgut of *Aedes aegypti* L. (Diptera: Culicidae) has shown to be able to colonize the insect midgut and display entomopathogenic and antipathogen properties (Ramirez et al., 2014).

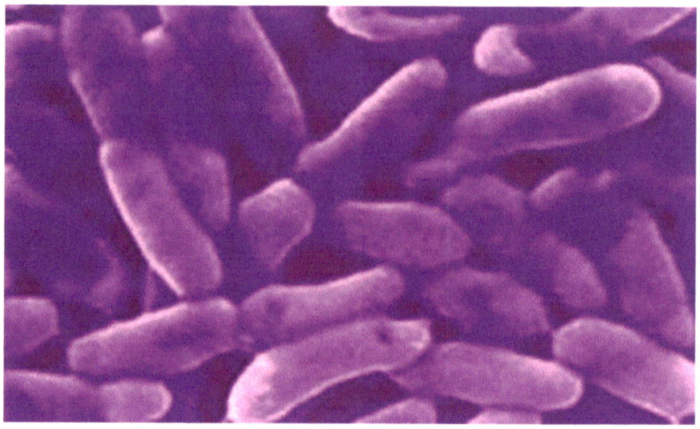

FIGURE 12.14 *Chromobacterium subtsugae.*

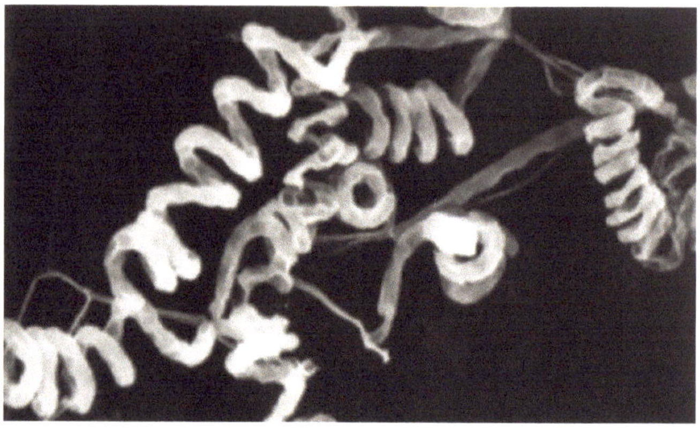

FIGURE 12.15 *Streptomyces avermitilis.*

12.2.7 Actinobacteria

12.2.7.1 Streptomyces Species

Streptomyces species are associated with herbivorous insects that take advantage of their cellulolytic properties. Other species and strains in this genus produce a variety of metabolites acting as potent toxins against either phytopathogenic microbials or insect pests (Copping & Menn, 2000). Insecticidal and anthelmintic activities of avermectins produced by the soil Actinomycete *Streptomyces avermitilis* MS & Dwas (Figure 12.15), have been discovered. The enhancement of GABA binding generates a cascade of events resulting in the inhibition of neurotransmission and paralysis of the neuromuscular systems (Bloomquist, 1996).

Insecticides based on avermectins include a mixture of avermectin B1a and avermectin B1b, known as abamectin, which act by contact and ingestion and have a limited plant translaminar activity.

12.2.7.2 Saccharopolyspora spinosa

The *Saccharopolyspora spinosa* Mertz and Yao (Figure 12.16) has been discovered during a screening program where the insecticidal activity of the isolate A83543 emerged (Mertz & Yao, 1990). The major component is spinosyn A, whose structure comprises a tetracyclic polyketideaglycone to which a neutral saccharide substituent is attached (Waldron et al., 2000).

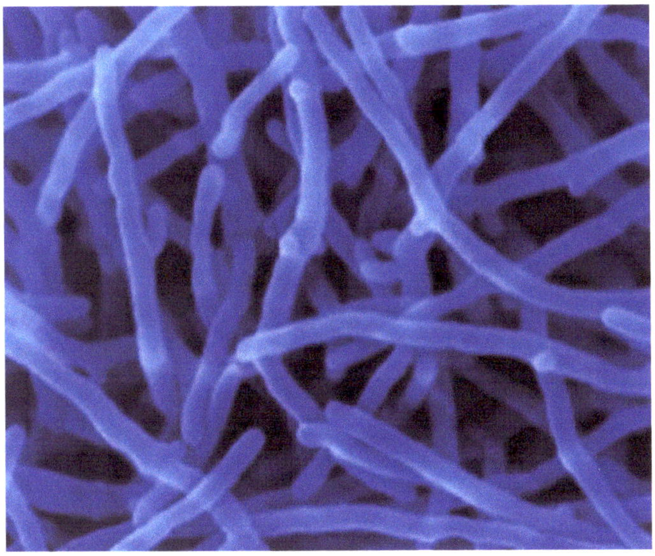

FIGURE 12.16 *Saccharopoly sporaspinosa.*

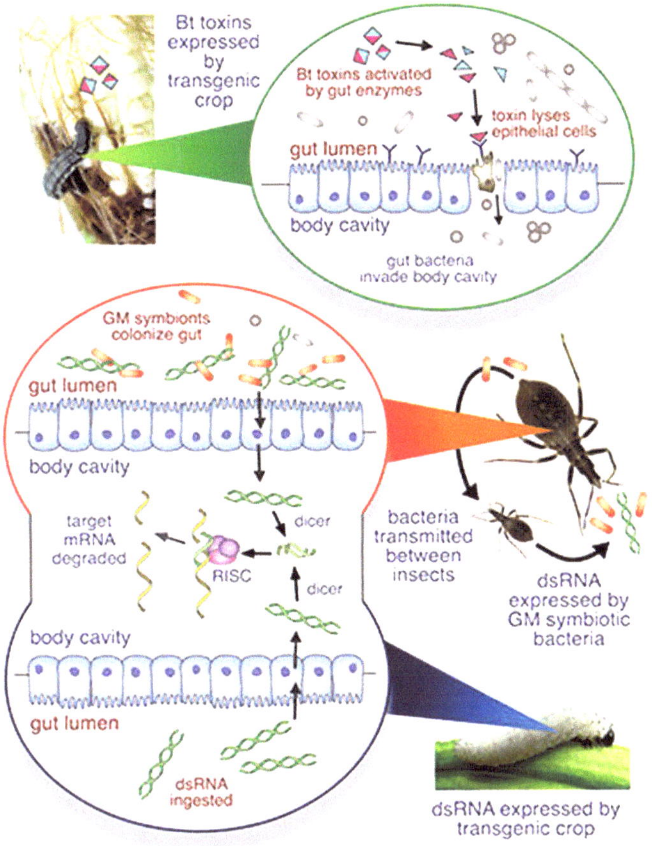

FIGURE 12.17 Bio-insecticide strategies using genetic modification approaches.

The insecticidal mode of action of spinosyns has been demonstrated that their interaction with g-aminobutyric acid receptors and nicotinic acetylcholine receptors, eventually leads to disruption of neuronal activity and consequent insect paralysis and death (Perry et al., 2007).

TABLE 12.2

Development of Some *B. thuringiensis* Transgenic Crops for Resistance to Insects

Crop target	Gene	Target Pest
Corn	*Cry 1A(b)*	European corn borer
Soybean	*Cry 1A(c)*	Bollworm and Bud worm
Tobacco	*Cry 2aa2*	Cotton bollworm
Sugar cane	*Cry 1A(b)*	Stem borer
Potato	*Cry 5*	Potato tuber moth
Alfalfa	*Cry 1C*	Leaf worm
Tomato	*B. thuringiensis* (k)	Tobacco hornworm, tomato pinworm and tomato fruit worm
Brassica	*Cry 1A(c)*	Pod borer
Cotton	*Cry 1A(b)/(c)*	Lepidoptera
Cotton	*Cry 2A*	Pink bollworm

12.3 Pathogenic Bacteria in Integrated Insect Pest Management

Insect pathogenic bacteria and their derived products represent the active substances of various biopesticides. Development of some *B. thuringiensis* transgenic crops for insect resistance is given in Table 12.2 (Lambert et al., 1992).

The effectiveness of entomopathogenic bacteria is often associated with a proper application in the field. For instance, in the case of products that act by ingestion (Bt-based products), timing is critical to ensure that the bacterial toxins remain stable in the environment until they are ingested by their target insect stage (Vega & Kaya, 2012; Sarwar, 2016j). These include dispersants, surfactants, wetters, spreaders, drift control agents, pH buffers, antifoam agents, carriers, phagostimulants and attractants (Satinder et al., 2006).

12.4 Bacteria-Mediated RNA Interference to Control Insect Pests

Major bioinsecticide strategies using genetic modification approaches have been presented in Figure 12.17.

12.5 Conclusion and Future Prospects

Bacterial biopesticides are the most common and cheaper form of microbial pesticides. The variety of entomopathogenic bacteria-based active substances authorized for commercialization and of their relative commercial products is significantly increasing worldwide. Consequently, their market is expanding into new segments, thus broadening their pest management use, traditionally relegated to niche contexts (forest, public health, protected crops). Transgenic Bt and dsRNA crops work best against insects that feed exclusively on one crop species. But many pests are polyphagous and they can feed on a variety of plants. In this situation, plant-based biocides are unlikely to offer much protection to other crops in the vicinity and it is hardly feasible to engineer every species of vulnerable crop with a pest-specific biocide. These innovative technologies are likely to increase our arsenal of toxins still further and with increased efficiency.

REFERENCES

Ali, S.; Zafar, Y.; Ali, G. M.; Nazir, F. *Bacillus thuringiensis* and its application in agriculture. *Afr. J. Biotechnol.* 2010, 9(14):2022–2031.

Barloy, F.; Delècluse, A.; Nicolas, L.; Lecadet, M. M. Cloning and expression of the first anaerobic toxin gene from *Clostridium bifermentans* subsp. *Malaysia* encoding a new mosquitocidal protein with homologies to *Bacillus thuringiensis* delta-endotoxins. *J. Bacteriol.* 1996, 178:3099–3105.

Baumann, P.; Clark, M.A.; Baumann, L.; Broadwell, A.H. *Bacillus sphaericus* as a mosquito pathogen: Properties of the organism and its toxins. *Microbiol. Rev.* 1991, 55:425–436.

Bloomquist, J. R. Ion Channels as Targets for Insecticides. *Annu. Rev. Entomol.* 1996, 41:163–190.

Burkhead K. D.; Schisler D. A.; Slininger P. J. Pyrrolnitrin production by biological-control agent *Pseudomonas cepacia* B37W in culture and in colonized wounds of potatoes. *Appl. Environ. Microbiol.*1994, 60:2031–2039.

Charles, J. F.; Silva-Filha, M. H.; Nielsen-LeRoux, C. Mode of action of *Bacillus sphaericus* on mosquito larvae: Incidence on resistance. In: *Entomopathogenic Bacteria*: From Laboratory to Field Application; Charles, J. F., Delecluse, A., Nielsen-LeRoux, C. (Eds.). Kluwer Academic: London, UK, 2000; pp. 237–252.

Copping, G. L.; Menn, J. J. Biopesticides: A review of their action, applications and efficacy. *Pest Manage. Sci.* 2000, 56:651–676.

Djukic, M.; Poehlein, A.; Thurmer, A.; Daniel, R. Genome sequence of *Brevibacillus laterosporus* LMG 15441, a pathogen of invertebrates. *J. Bacteriol.* 2011, 193:5535–5536.

Ferguson, C. M.; Barton, D. M.; Harpe, L. A. Swaminathan J., Van Koten, C.; Hurst, M. R. H. Survival of *Yersinia entomophaga* MH96 in a pasture ecosystem and effects on pest and non-target invertebrate populations. *New Zealand Plant Prot.* 2012, 65:166–173.

Ffrench-Constant, R. H.; Dowling, A.; Waterfield, N. R. Insecticidal toxins from *Photorhabdus* bacteria and their potential use in agriculture. *Toxicon.* 2007, 49:436–451.

Glare, T.; Caradus, J.; Gelernter, W.; Jackson, T.; Keyhani, N. Kohl, J.; Marrone, P.; Morin, L.; Stewart, A. Have biopesticides come of age? *Trends Biotechnol.* 2012, 30:250–258.

Gomez, I.; Sanchez, J.; Miranda, R.; Bravo, A.; Soberon, M. Cadherin-like receptor binding facilitates proteolytic cleavage of helix α-1 in domain I and oligomer pre-pore formation of *Bacillus thuringiensis* Cry1Ab toxin. FEBS Lett. 2002, 513(2–3):242–246.

He, H.; Ratnayake, A. S.; Janso, J. E.; He, M.; Yang, H. Y.; Loganzo, F.; Shor, B.; O'Donnell, C. J.; Koehn, F. E. Cytotoxic spliceostatins from *Burkholderia* sp. and their semisynthetic analogues. *J. Nat. Prod.* 2014, 77:1864–1870.

Hurst, M. R.; Glare, T. R.; Jackson, T. A.; Ronson, C. W. Plasmid-located pathogenicity determinants of *Serratia entomophila*, the causal agent of amber disease of grass grub, show similarity to the insecticidal toxins of *Photorhabdus luminescens. J. Bacteriol.* 2000, 182:5127–5138.

Hurst, M. R. H.; Glare, T. R.; Jackson, T. A. Cloning Serratia entomophila anti-feeding genes – a putative defective prophageactive against the grass grub Costelytra zealandica. J. Bacteriol. 2004,186: 5116–5128.

Ishii, K.; Adachi, T.; Hamamoto, H.; Sekimizu, K. *Serratia marcescens* suppresses host cellular immunity via the production of an adhesion-inhibitory factor against immune surveillance cells. *J. Biol. Chem.* 2014, 289:5876–5888.

Ji, D.; Kim, Y. An entomopathogenic bacterium, *Xenorhabdus nematophila*, inhibits the expression of an antimicrobial peptide, cecropin, of the beet armyworm, *Spodoptera exigua. J. Insect Physiol.* 2004, 50:489–496.

Lacey, L. A.; Frutos, R.; Kaya, H. K.; Vail, P. Insect pathogens as biological control agents: Do they have a future? *Biol. Control.* 2001, 21:230–248.

Lacey, L. A.; Grzywacz, D.; Shapiro-Ilan, D. I.; Frutos, R.; Brownbridge, M.; Goettel, M. S. Insect pathogens as biological control agents: Back to the future. *J. Invertebr. Pathol.* 2015, 132:1–41.

Lambert, B.; Hofte, H.; Annys, K.; Jansens, S.; Soetaert, P.; Peferoen, M. Novel *Bacillus thuringiensis* insecticidal crystal protein with a silent activity against coleopteran larvae. *Appl. Environ. Microbiol.* 1992, 58(8):2536–2542.

Madigan, M. T.; Martinko, J. M. *Brock Biology of Microorganisms* (11th ed.). Pearson Prentice Hall, England. 2005; p. 992.

Martin, P. A. W.; Hirose, E.; Aldrich, J. R. Toxicity of *Chromobacterium subtsugae* to Southern stink bug (Heteroptera: Pentatomidae) and corn rootworm (Coleoptera: Chrysomelidae). *J. Econ. Entomol.* 2007, 100:680–684.

Mertz, F. P.; Yao, R. C. *Saccharopolyspora spinosa* sp. nov. isolated from soil collected in a sugar mill rum still. *Int. J. Syst. Bacteriol.* 1990, 40:34–39.

Morange, M. What history tells us XLIII Bacteriophage: The contexts in which it was discovered. *J. Biosci.* 2017, 42(3):359–362.

Palma, L.; Munoz, D.; Berry, C.; Murillo, J.; Caballero, P. *Bacillus thuringiensis* toxins: An overview of their biocidal activity. *Toxins.* 2014, 6(12):3296–3325.

Perry, T.; McKenzie, J. A.; Batterham P. A Da6 knockout strain of *Drosophila melanogaster* confers a high level of resistance to spinosad. *Insect Biochem. Mol. Biol.* 2007, 37:184–188.

Ramirez, J. L.; Short, S. M.; Bahia, A. C.; Saraiva, R. G.; Dong, Y.; Kang, S.; Tripathi, A.; Mlambo, G.; Dimopoulos, G. *Chromobacterium* Csp_P reduces malaria and dengue infection in vector mosquitoes and has entomopathogenic and in vitroanti-pathogen activities. *PLOS Pathog.* 2014, 10:e1004398.

Roh, J. Y.; Choi, J. Y.; Li, M. S.; Jin, B. R.; Je, Y. H. *Bacillus thuringiensis* as a specific, safe, and effective tool for insect pest control. *J. Microbiol. Biotechnol.* 2007, 17(4):547–559.

Rose, R.; Dively, G. P.; Pettis, J. Effects of Bt corn pollen on honey bees: Emphasis on protocol development. *Apidologie.* 2007, 38(4):368–377.

Ruiheng, Y. Structural basis of transcription inhibition by the nucleoside-analog inhibitor thuringiensin. Rutgers University–Graduate School, New Brunswick. 2016; p. 36.

Ruiu, L. *Brevibacillus laterosporus*, a pathogen of invertebrates and a broad-spectrum antimicrobial species. *Insects.* 2013, 4:476–492.

Ruiu, L.; Floris, I.; Satta, A.; Ellar, D. J. Toxicity of a *Brevibacillus laterosporus* strain lacking parasporal crystals against *Musca domestica* and *Aedes aegypti. Biol. Control.* 2007, 43:136–143.

Ruiu, L. Insect pathogenic bacteria in integrated pest management. *Insects*. 2015, 6:352–367.

Ruiu, L.; Satta, A.; Floris, I. Emerging entomopathogenic bacteria for insect pest management. *Bull. Insectol.* 2013, 66:181–186.

Ruiu, L.; Satta, A.; Floris, I. Observations on house fly larvae midgut ultrastructure after *Brevibacillus laterosporus* ingestion. *J. Invertebr. Pathol.* 2012, 111:211–216.

Sarwar, M. Biology and ecology of some predaceous and herbivorous mites important from the agricultural perception. In: *Pests Control and Acarology*; Haouas, D., Hufnagel, L. (Eds.). IntechOpen Ltd., London, UK. 2019; p. 29.

Sarwar, M. Biopesticides: An effective and environmental friendly insect-pests inhibitor line of action. *Int. J. Eng. Adv. Res. Technol.* 2015a, 1(2):10–15.

Sarwar, M. Challenges due to bacterial infections of the honey bees and contributions to manage pest problems. *Int. J. Entomol. Res.* 2016a, 1(1):4–10.

Sarwar, M. Concept of integrated insect pests management. *Pak. Gulf Economists.* 2004, 23(46 & 47):39–41.

Sarwar, M. Comparing abundance of predacious and phytophagous mites (Acarina) in conjunction with resistance identification between Bt and non-Bt cotton cultivars. *Afr. Entomol.* 2013a, 21(1):108–118.

Sarwar, M. Fungal diseases of honey bees (Hymenoptera: Apidae) that induce considerable losses to colonies and protocol for treatment. *Int. J. Zool. Stud.* 2016b, 1(1):8–13.

Sarwar, M. A guide to the honey bee's protozoan nosema disease and treatment recommendations for pest control. *Int. J. Res. Appl. Phys.* 2016c, 1(7):1–11.

Sarwar, M. Information on activities regarding biochemical pesticides: An ecological friendly plant protection against insects. *Int. J. Eng. Adv. Res. Technol.* 2015b, 1(2):27–31.

Sarwar, M. Insect pests of honey bees and choosing of the right management strategic plan. *Int. J. Entomol. Res.* 2016d, 1(2):16–22.

Sarwar, M. Integrated control of insect pests on canola and other brassica oilseed crops in pakistan. In: *Integrated Management of Insect Pests on Canola and Other Brassica Oilseed Crops*; Reddy, G. V. P. (Ed.). Centre for Agriculture and Bioscience International, Oxfordshire, UK. 2017; p. 408.

Sarwar, M. Integrated pest management (IPM): A constructive utensil to manage plant fatalities. *J. Agric. Allied Sci.*2013b, 2(3):1–4.

Sarwar, M. Microbial insecticides: An ecofriendly effective line of attack for insect pests management. *Int. J. Eng. Adv. Res. Technol.* 2015c, 1(2):4–9.

Sarwar, M. Mite culprits for causing mortality and reduction in population of honey bee colonies and measures for pests control. *Int. J. Res. Appl. Chem.* 2016e, 1(7):10–22.

Sarwar, M. Mites: The tiny killers to push honeybee colonies into collapse and integrated pest management. *Int. J. Res. Appl. Phys.* 2016f, 1(7):12–21.

Sarwar, M. Pervasiveness of non-infectious and non-pest-related disorders accompanying with honey bees colony and control of downfalls. *Int. J. Entomol. Res.* 2016g, 1(2):35–40.

Sarwar, M. Predations on honey bees (Arthropoda) by vertebrate pests (Chordata) and control of nuisance. *Int. J. Zool. Stud.* 2016h, 1(2):12–17.

Sarwar, M. Prevalence of multiple viral diseases associated with honey bees colony collapse and control of disorders. *Int. J. Zool. Stud.* 2016i, 1(2):29–34.

Sarwar, M. Usage spots of biological insecticides in consort with target insect pests or vectors and application in habitat. *Int. J. Entomol. Nematol.* 2016j, 3(1):14–20.

Satinder, K. B.; Verma, M.; Tyagi, R. D.; Valero, J. R. Recent advances in downstream processing and formulations of *Bacillus thuringiensis* based biopesticides. *Process Biochem.* 2006, 41:323–342.

Saxena, D.; Stotzky, G. Insecticidal toxin from *Bacillus thuringiensis* is released from roots of transgenic *Bt* corn in vitro and in situ. *FEMS Microbiol. Ecol.* 2000, 33(1):35–39.

Tailliez P., Pages S., Ginibre N., Boemare N. New insight into diversity in the genus *Xenorhabdus*, including the description of ten novel species. *Int. J. Syst. Evol. Microbiol.* 2006, 56:2805–2818.

Vega, F. E.; Kaya, H. K. *Insect Pathology* (2nd ed.). Elsevier, London, UK. 2012; p. 504.

Vodovar, N.; Vinals, M.; Liehl, P.; Basset, A.; Degrouard, J.; Spellman, P.; Boccard, F.; Lemaitre, B. Drosophila host defense after oral infection by and entomopathogenic *Pseudomonas* species. *Proc. Natl. Acad. Sci. USA.* 2005, 102:11414–11419.

Waldron, C.; Madduri, K.; Crawford, K.; Merlo, D. J.; Treadway, P.; Broughton, M. C.; Baltz, R. H. A cluster of genes for the biosynthesis of spinosyns, novel macrolide insect control agents produced by *Saccharopolyspora spinosa. Antonie van Leeuwenhoek.* 2000, 78:385–390.

Warren, G. W. Vegetative insecticidal proteins: Novel proteins for control of corn pests. In: *Advances in Insect Control: The Role of Transgenic Plants*; Carozzi, N. B., Koziel, M. G. (Eds.). Taylor & Francis, London, UK, 1997; pp. 109–121.

Waterfield, N.; Kamita, S. G.; Hammock, B. D.; Ffrench-Constant, R. The *Photorhabdus* Pir toxins are similar to a developmentally regulated insect protein but show no juvenile hormone esterase activity. *FEMS Microbiol. Lett.* 2005, 245:47–52.

Zhang, J.; Hodgman, T. C.; Krieger, L.; Schnetter, W.; Schairer, H. U. Cloning and analysis of the cry gene from *Bacillus popilliae. J. Bacteriol.* 1997, 179:4336–4341.

Zubasheva, M. V.; Ganushkina, L. A.; Smirnova, T. A.; Azizbekyan, R. R. Larvicidal activity of crystal-forming strains of *Brevibacillus laterosporus. Appl. Biochem. Microbiol.* 2010, 46:755–762.

13

Biological Control of Pests Using Bacillus thuringiensis

Jéssica Batista Torres Araújo Oliveira, Cassia Renata Pinheiro, and Glacy Jaqueline Da Silva

CONTENTS

13.1 Introduction

Biological control is one of the pillars of Integrated Pest Management (IPM), increasingly being implemented in the fields for providing multiple environmental benefits. Among these benefits it can be highlighted the decrease in the application of chemical insecticides, absence of residues, does not promote the emergence of resistant populations of insects—pests, avoids the contamination of soil and water, thus offering a balance of the ecosystem and sustainable and healthy production. This practice consists of using the natural resources found in the environment such as predators, parasitoids, and entomopathogenic microorganisms (fungi, bacteria, viruses, and nematodes) (Alves, 2008).

Currently, the biological control most used in agriculture is microbial, with emphasis on the entomopathogenic bacteria *Bacillus thuringiensis* (Bt) used in the control of several agricultural and forest pests. This bacteria is characterized by producing protein crystals that have an insecticidal effect during the sporulation phase (Tirado Montiel et al., 2001). German Ernst Berliner isolated the same bacteria again, which was found in dead caterpillars of the moth *Ephestia kuehniella* Zell Lepidoptera: Pyralidae, in which it had spores and crystals on its body. The biologist renamed the bacteria *Bacillus thuringiensis* var. sotto (Siegel, 2000).

In 1927, the researcher Mattes was successful in controlling the European corn borer (*Ostrinia nubilalis* (Hübner) (Lepidoptera: Pyralidae)) using Bt in field experiments confirming its insecticidal effect (HUSZ, 1930). Faced with this challenge, companies sought to invest in production processes to improve the efficiency and toxicity of the bacteria to obtain a product with more quality and standardization (Sanahuja et al., 2011). Due to the possibility of improvement, a relentless search for the quality of this technology was started (Perlak et al., 1991).

In 1986 the United States and France began their first experiments in fields with transgenic tobacco plants, with promising results. However, it was only in 1995 that the first Bt-based transgenic commercial crops were approved on the market for corn, cotton, and potatoes (Sanahuja et al., 2011). Bt is among the most applied entomopathogens in agriculture and one of the most harmless microbial products to humans and the environment, with more than 26 countries already adopting these technologies in their crops (Lacey et al., 2015).

13.2 *Bacillus thuringiensis*

Bacillus thuringiensis is a gram-positive, aerobic bacteria and can be found on various substrates such as soil, water, plant surfaces, dead insects, cobwebs, and stored grains (Bravo et al., 1998; Valicente & Barreto, 2003). One of the remarkable features of this bacteria is the ability to produce crystal protein, also known as δ-endotoxins, during the stationary or sporulation phase (Bravo et al., 2017). Figure 13.1 will show morphology of cells from *Bacillus thuringiensis.* The arrows point to a crystal and a spore These crystals have specific insecticidal properties for several species of insects, with emphasis on the orders Lepidoptera, Diptera, and Coleoptera. However, there are subspecies of *B. thuringiensis* that present toxic crystals against other insect orders such as Hymenoptera, Hemiptera, Orthoptera, in addition to other organisms such as some nematodes, protozoa, and mites.

The δ-endotoxins are made up of Cry proteins that are activated in the form of protoxins. Its action depends on activation processes that take place inside the insect's digestive system, which has an alkaline pH (Bravo, 1992; Bravo et al., 2017). Protoxins bind to membrane receptors in the intestinal epithelial cells, causing cell lysis and rupture of the midgut, leading to death of the larva by starvation and septicemia (generalized infection) (Bravo, 1992; Vilas-Boas et al., 2012).

Other types of endotoxins can be produced by several strains of this entomopathogen, such as cytolytic endotoxins, also known as Cyt, which promote insecticidal activities for larvae

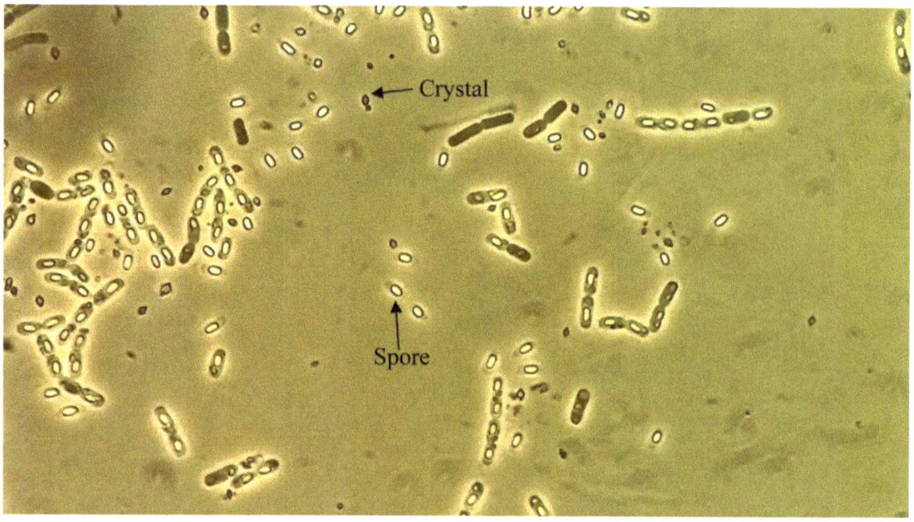

FIGURE 13.1 Cell morphology of the 1641 strain of *Bacillus thuringiensis*. The arrows point to a crystal and a spore (Courtesy of JBTA Oliveira).

of dipterans such as flies and mosquitoes. These endotoxins are cleaved, releasing the active toxin, which binds directly to the lipids of the insect's intestinal cell membrane, promoting the formation of pores. In some organisms they act by disrupting the lipid bilayer of the membranes (Crickmore et al., 2015). Some studies have shown a synergistic effect between the Cry and Cyt proteins, with greater toxicity, where one enhances the effect of the other (Bravo et al., 2007).

Another family of Bt protein used in pest management is the Vip protein (Vegetative Insecticidal Proteins) which is produced during the vegetative phase and has a lethal effect mainly for insects of the order Lepidoptera (Estruch et al., 1996). These proteins are divided into four families, Vip1, Vip2, Vip3, and Vip4, differentiated according to the composition of amino acids. The Vip1 and Vip2 proteins are toxic to some species of coleoptera, while the Vip3 toxins are specific to lepidopterans. Some research is being conducted to define the toxicity of Vip4. Vips do not have the inclusion of protein crystals; however, their mode of action is similar to that of Cry proteins, causing paralysis of the intestine and lysis of the epithelium cells of the insect's midgut (Yu et al., 1997).

Some other proteins have been identified in *B. thuringiensis*, such as the Sip protein (secreted insecticidal protein) found in some Bt strains. It can be found in the culture supernatant during the fermentation period, which was characterized by toxicity to coleopteran larvae (Palma et. al., 2014). There is also β-exotoxin, which is a thermostable toxin efficient in the control of many insects, but this toxin was characterized by having pathogenicity to vertebrates, preventing its use in the manufacture of Bt-based biopesticides (Polanczyk & Alves, 2003).

13.3 Production of Biopesticides Based on *Bacillus thuringiensis*

The biopesticide production process consists of choosing a genetically stable and highly efficient strain to combat insect pests, storage, followed by a fermentative process, formulation, and finally product quality analyzes (Dias, 1992). In the case of Bt biological products, these are usually composed of a mixture of spores, crystals, and few cells (Capalbo et al., 2005. However, the culture medium is a limiting factor in the manufacture of Bt-based products, which when produced on a large scale have a high cost in the fermentation process (Valicente et al., 2010).

Culture media for the production of *Bacillus thuringiensis* generally require a source of carbon, another source of nitrogen and mineral salts (Tiradomotiel et al., 2001). Several studies have been done in search of alternative sources of nutrients to compose the medium of culture, aiming at cost reduction during the fermentation process (Digmant & Sthaly, 1983; Valicente et al., 2010). In addition to the culture medium being of great importance in the fermentation process, the physical parameters of fermentation also play an important role in the quality of the product.

According to Dias (1992), fermentation must be monitored according to temperature, pH, and oxygenation. The ideal temperature for Bt production varies between 28° to 30°C. Temperatures above the recommended can inhibit the production of crystals, whereas lower temperatures can decrease cell division, impairing efficiency (Moraes et al., 2001). The pH can be monitored by adding acids or bases, maintaining the value around seven, or using buffered media that are efficient throughout the fermentation process by increasing crystal production (Içgen et al., 2002).

Fermentation can take 24 to 72 hours depending on the conditions submitted. Care during the entire process is extremely relevant due to the ease of other microorganisms contamination (Dias, 1992).

13.4 Management

Valicente (2008) reports that for greater product efficiency, it is necessary to monitor the crop weekly to detect the level of attack by insects. Injuries caused by defoliating caterpillars can start one week after seed germination, forcing the producer to be alert to the initial signs of these pests attack. Preferably, applications should be carried out in the late afternoon, which has a lower incidence of UV rays. This is the main limiting factor in the efficiency of these products in the field, as the sun's rays can make the efficiency of Bt-based products unviable (Valicente, 2008). Cry proteins can remain in plants for only a few days, due to the climatic conditions of the environment and the presence of proteinases that facilitate the degradation of proteins. Therefore, the application of the product must be reapplied according to the monitoring (Sanahuja et al., 2011).

13.5 Management in Transgenic Plants

Pest management in planting of transgenic plants deserves special attention due to the emergence of resistance genes in insect pest populations. This occurs because there is a great deal of selection pressure and also the use of some inappropriate techniques (Tabashink et al., 2014, Gassmann & Clifton, 2017). High concentration of toxins is characterized by the ability of a plant to express huge amounts of Bt toxins sufficiently to kill 99% of heterozygous insects (Epa, 1998, Gassmann & Clifton, 2017).

The use of cultures with pyramid genes is the application of different Bt toxins with different modes of action, against the same species of pests aiming at the delay of resistance (Roush, 1998, Gassmann & Clifton 2017). In the refuge area, plants that do not have Bt technology are grown. This tactic aims to develop a population of insects susceptible to the toxin, favoring the mating between a resistant species with another non-resistant one, generating a population of susceptible insects (Carrière et al., 2010, Gassmann & Clifton, 2017). The size of the refuge area depends on which crop will be planted. It is important that this area is close to the culture with transgenic technology, to facilitate the reproduction of susceptible individuals (Machado & Fiuza, 2011).

It is important to emphasize that for successful production using Bt technology, it is necessary to adapt these tactics in the field together with rigorous monitoring to prolong the use of transgenics, without the emergence of pests with resistance genes.

13.6 Conclusion and Future Prospects

Currently, the use of biological control in agriculture has grown exponentially due to the great demand for more sustainable, productive, and balanced agriculture. Therefore, to be successful in the use of this technology, it is necessary to have knowledge of the culture, biological cycle of the pest, and climatic conditions to determine the correct time to start with biological control to maintain a population balance of the pests. In this way there is no danger of any economic damage. Often, in certain situations, biological pest control must be associated with other control measures, which do not interfere with the efficiency of the technology. Despite the great demand, research, and range of biological products commercialized, biological control has not yet reached levels when compared to chemical insecticides. Therefore, there is a need to spread more management techniques and usage information for farmers.

REFERENCES

Alves, S.B.; Lopes, R.B.; Pereira, R.M.; Tamai, M.A. O *Controle Microbiano de Pragas na América Latina*. Piracicaba, Brazil: FEALQ, vol. 14, 414p, 2008.

Bravo, A. et al. Characterization of cry genes in a Mexican *Bacillus thuringiensis* strain collection. *Applied and Environmental Microbiology*, 64, 4965–4972, 1998.

Bravo, A., Jansens, S., Peferoen, M. Immunocy to chemical localization of *Bacillus thuringiensis* insecticidal crystal proteins in intoxicated insects. *Journal of Invertebrate Pathology*, 60(3), 237–246, 1992.

Bravo, A., Gill, S. S., Soberó, M. Mode of action of *Bacillus thuringiensis* cry and cyt toxins and their potential for insect control. *Toxicon*, 49(4), 423–435, 2007.

Bravo, A., Pacheco, S., Gómez, I., Garcia-Gómez, B., Onofre, J., Soberón, M. Insecticidal proteins from *Bacillus thuringiensis* and their mechanism of action. Chapter 53 In *Bacillus thuringiensis Polanczyk and Crickmore and Fiuza*. 2017. Springer International Publishing, Cham, Switzerland

Capalbo, D. M. F. et al. *Bacillus thuringiensis*. *Revista Biotecnologia Ciência e Desenvolvimento*, 34, 69–85, 2005.

Carriere, Y., Crowder, D. W., Tabashnik, B. E. Evolutionary ecology of insect adaptation to Bt crops. *Evolutionary Applications*, 3, 561–573, 2010.

Crickmore, N., Baum, J., Bravo, A., Lereclus, D., Narva, K., Sampson, K., Schnepf, E., Sun, M., Zeigler, D. R. *Bacillus thuringiensis* Toxin Nomenclature. 2015. www.btnomenclature.info.

Dingman, D. W., Stahly, D. P. Medium promoting sporulation of Bacillus larvae and metabolism of medium components. *Applied and Environmental Microbiology*, 46(4), 860–869, 1983.

EPA. Final Report of the FIFRA Scientific Advisory Panel Subpanel on *Bacillus thuringiensis* (Bt) Plant-Pesticides and Resistance Management, 1998. Accessed at http://archive.epa.gov/scipoly/sap/meetings/web/pdf/finalfeb.pdf.

Estruch, J. J., Warren, G. W., Mullins, M. A., Nye, G. J., Craig, J. A., Koziel, M. G. Vip3A, a novel *Bacillus thuringiensis* vegetative insecticidal protein with a wide spectrum of activities against lepidopteran insects. *Proceedings of the National Academy of Sciences*, 93(11), 5389–5394, 1996.

Gassmann, A.J.; Clifton, E.H. current and potential applications of biopesticides to manage insect pests of maize. In *Microbial Control of Insect and Mite Pests*. Ames, IA: Iowa State University, 2017.

Husz, B. Field experiments on the application of *Bacillus thuringiensis* against the corn borer. *International Corn Borer Investigations Science Report*, 3, 91–98, 1930.

Içgen, Y., Içgen, B., Özcengiz, G. Regulation of crystal protein biosynthesis by *Bacillus thuringiensis*: I. Effects of mineral elements and pH. *Research in Microbiology*, 153(9), 599–604, 2002.

Lacey L. A., Grzywacz D., Shapiro-Ilan D. I., Frutos R., Brownbridge M., Goettel M. S. Insect pathogens as biological control agents: Back to the future. *Journal of Invertebrate Pathology*, 132, 1–41, 2015.

Machado, V., Fiuza, L. M. Manejo da resistência: Na era das plantas transgênicas. *Oecologia Australis*, 15(2), 291–302, 2011.

Mattes, O. Parasitare Krankheiten der Mehlmottenlarven und Versuche uber ihre Verwendbarkeit als biologisches Bekiampfungsmittel. *Sitzber. Ges. Beforder. Ges. Naturw. Marburg*., 62, 381–417, 1927.

Moraes, I. O., Capalbo, D. M. F., Arruda, R. O. M. Produção de Bioinseticidas. Lima, U. A., Aquarone, E., Borzani, W., Schmidell, W. (eds.). *Processos Fermentativos e Enzimáticos*. São Paulo: Edgard Blücher, vol. 3, p. 249–278, 2001.

Palma, L., Muñoz, D., Berry, C., Murillo, J., Caballero, P. *Bacillus thuringiensis* toxins: An overview of their biocidal activity. Toxins, 6(12), 3296–3325, 2014.

Perlak, F. J., Fuchs, R. L., Dean, D. A., McPherson, S. L., Fischoff, D. A. Modification of the coding sequence enhances plant expression of insect control genes. *Proceedings of the National academy of Sciences of the United States of America*, 88, 3324–3328, 1991.

Polanczyk, R., Alves, S. B. *Bacillus thuringiensis*: *Uma Breve Revisão*. Texcoco: Agrociencia, vol. 7, 2003.

Roush, R. T. Two-toxin strategies for management of insecticidal transgenic crops: can pyramiding succeed where pesticide mixtures have not? *Philosophical Transactions of the Royal Society of London, Series B*, 353, 1777–1786, 1998.

Sanahuja, G., Banakar, R., Twyman, R. M., Capell, T., Christou, P. Bacillus thuringiensis: A century of research, development and commercial applications. Plant Biotechnology Journal, 9(3), 283–300, 2011.

Siegel, J. P. Bacteria. In *Field Manual of Techniques in Invertebrate Pathology* (Lacey, L. L. Kaya, H. K. eds.). Dordrecht, Netherlands: Kluwer Scientific, pp. 209–230, 2000.

Tabashnik, B. E., Mota-Sanchez, D., Whalon, M. E., Hollingworth, R. M., Carrière, Y. Defining terms for proactive management of resistance to Bt crops and pesticides. *Journal of Economic Entomology*. 107, 496–507, 2014.

Tirado-Montiel, M. I.; Tyagi, R. D.; Valero, J. R. Wastewater treatment sludge as a raw material for the production of *Bacillus thuringiensis* based biopesticides. New York: Water Research, vol. 35, p. 38, 2001.

Valicente, F. H., Barreto, M. R. Bacillus thuringiensis Survey in Brazil: Geographical distribution and insecticidal activity against *Spodoptera frugiperda* (J. E. Smith) (Lepidoptera: Noctuidae). *Neotropical Entomology*, 32, 639–644, 2003.

Valicente, F. H., de Toledo Picoli, E. A., de Vasconcelos, M. J. V., Carneiro, N. P., Carneiro, A. A., Guimarães, C. T., Lana, U.G. Molecular characterization and distribution of *Bacillus thuringiensis* cry1 genes from Brazilian strains effective against the fall armyworm, Spodoptera frugirperda. *Biological Control*, 53(3), 360–366, 2010.

Valicente, F. H; Mourão, A. H. C. Use of by-products rich in carbon and nitrogen as a nutrient source to produce *Bacillus thuringiensis* (Berliner)-based biopesticide. *Neotropical Entomology*, 37(6), 702–708, 2008.

Vaughn, T.; Cavato, T.; Brar, G.; Coombe, T.; DeGooyer, T.; Ford, S.; Groth, M.; Howe, A.; Johnson, S.; Kolacz, K.; Pilcher, C.; Purcell, J. A method of controlling corn rootworm feeding using *Bacillus thuringiensis* protein expressed in transgenic maize. *Crop Science*, 45, 931–938, 2005.

Vilas-Bôas, G. T., Alvarez, R. C., dos Santos, C. A., Vilas-Boas, L. A. Fatores de virulência de *Bacillus thuringiensis*: o que existe além das proteínas Cry. *EntomoBrasilis*, 5(1), 1–10, 2012.

Yu, C. G., Mullins, M. A., Warren, G. W., Koziel, M. G., Estruch, JJ. The *Bacillus thuringiensis* vegetative insecticidal protein Vip3A lyses midgut epithelium cells of susceptible insects. *Applied and Environmental Microbiology*, 63(2), 532–536, 1997.

Section IX

Management of Crop Diseases through Fungal Biopesticides

14

Crop Protection through Fungal Biopesticides: An Endeavor of Modern Organic Agriculture

Sudeepta Pattanayak, Siddhartha Das, and Pradipta Banerjee

CONTENTS

14.1 Introduction

The population of India is dramatically increasing at the rate of 1.8 percent annually. Keeping in view of the increasing population, the production of food grain would increase by about 2 million tonnes per year (Paroda, 1999). Although in the contemporary period of time, India has gained self-sufficiency in food grain production, a question mark is still arising in food security because of the tremendous use of pesticides. The enormous use of synthetic chemicals may increase production on a temporary basis while leaving a scratch on the environment by reducing the quality of groundwater through leaching and build-up of chemicals in the environment. All this is only to "increase the productivity." In the name of productivity, we have diverted our path from sustainable agriculture to unsustainability.

Several factors are playing crucial roles in the loss of yield as well as quality in various crops. Weeds, insects, and diseases are mainly responsible for yield loss. It was reported that approximately 26% yield loss is due to insects and 20% yield loss (Bhattacharjee et al., 2014) occurs due to plant diseases. The highest yield loss occurs due to weed infestation, which accounts for 34%. Due to the tremendous use of pesticides, pests become resistant, and then the use of pesticides must increase repeatedly to kill them. This is one of the most important and major pinpointed causes for the increase in suicide cases among farmers.

Keeping the threats of the future in view, the system coming into focus is Organic Farming. For a long time, scientists have been busy developing ecofriendly and toxic-free methods to control plant disease and pests. The best substitutes for synthetic chemicals are biopesticides or biological control of pests and diseases. Biopesticides are very effective in small amounts, non-hazardous to the environment, target-specific, and leave no toxic residual effect in ecology.

14.2 Fungal Biopesticide

Fungi are one of the most important pathogens of several pests mainly associated with a lot of insects, weeds, and pathogens having more than 1000 known species. Among all known species of pathogenic fungi, many genera and species attack against a wide range of pests. The fungal biocontrol agents belong to four major groups, i.e., laboulbeniales, pyrenomycetes, hypomycetes, zygomycetes, etc.

14.3 Why Are Fungal Insecticides Ecofriendly?

Fungal biopesticides:

(a) have a narrow target range and a very specific mode of action

(b) are slow-acting
(c) have relatively critical application times
(d) suppress the pest population
(e) have a short shelf life and limited field persistence
(f) are safer to human society and the environment than synthetic pesticide
(g) have no residual problems

Table 14.1 gives examples of biocontrol fungal agents against different micro-organisms, their targets, trade names, and producing companies. Table 14.2 shows entomopathogenic fungi with their targets, while Table 14.3 show the list of fungal biopesticide against several weeds.

14.4 Fungal Biocontrol Agents

14.4.1 Trichoderma

Trichoderma spp. has been used as an important antagonistic fungus against a wide range of plant pathogens for a long time (Khalili et al., 2012; Gveroska et al., 2012). It can be easily identified by its green-colored spore and its sweet smell due to the presence of 6-pentyl-á-pyrone, a volatile compound (Adnan et al., 2019). It is more efficient in both alkaline and acidic soils and has the ability to control soil and airborne pathogens. The increased popularity of *Trichoderma* spp. as an antagonistic fungus in biocontrol is due to its high aggressiveness towards pathogens, high reproductive capability, tolerance to unfavorable conditions, ability to modify the rhizosphere region, ability to use the plant nutrients, and high efficiency in promoting plant growth and defense mechanisms.

Trichoderma controls the pathogens through several mechanisms, such as competition, where pathogens compete for food and nutrients, parasitism, i.e., Trichoderma as a parasite on other pathogens, fungistasis, and induced systemic resistance (ISR), which means inhibiting the growth and development of the pathogen at different sites of the plant.

14.4.2 Metarhizium

Metarhizium anisopliae, formerly known as *Entompthora anisopliae*, was first recognized from infected larvae of the *Anisopliae austrica*, wheat cockchafer in 1879 from Ukraine followed by *Cleonus punctiventis*, and in 1833 it was renamed *Metarhizium anisopliae* by Sorokin (Tulloch, 1976; Rai et al., 2014). The genus *Metarhizium* is known to attack a vast range of agricultural as well as forest insects (Rai et al., 2014). Due to the characteristic green color of the conidial cells of this genus, they are recognized as the pathogen of "green muscardine," a disease in insects (del Prado et al., 2008). The taxonomy of this genus is very complex. *M. flavoride* and *M. anisopliae* were identified by Tulloch in 1976 by the morphology of their conidia and conidiogenous cell. *M. anisopliae* again subdivides into *majus*, having short conidia up to 9 μm, and *anisopliae*, having long conidia up to 18 μm. *M. anisopliae* comprises four genetic groups (Driver et al., 2000).

TABLE 14.1

List of Fungal Species with Their Targets for Biological Control of Plant Pathogens (Fravel et al., 1998; Burges, 1998)

Fungal Species	Product	Formulation	Target	Producer	Application Method
Tricoderma harzinaum, T. Polysporum	Binab T	Wettable powder and pellets	Take all disease, wilt root rot	Bion-Innovation, Sweden	Spray or drench to fruits after harvesting
Tricoderma spp.	Bio-Fungus	Wettable powder, sticks, and granules	*Pythium* spp., *Phytopthora, Sclerotinia, Rhizoctonia solani, Verticillium, Fusarium*	Zaena Bio	Spray or drench
Tricoderma harzinaum	Trichodex	Liquid and powder	*Botrytis cinerea*	Makhteshim-Agan, DeCeuster, Belgium	Spraying or drenching
Glaiocladiumvirens	SoilGard	Granules	Damping off and several root pathogens	ThermoTrilogy, USA	Soil incorporation
Ampelomyces quisqualis	AQ10 Biofungicide	Water dispersible granules	Powdery mildew	EcogenInc.USA	Spray

TABLE 14.2

List of Fungal Species with Their Targets for Biological Control of Insects (Fravel et al., 1998; Burges, 1998)

Fungal Species	Product	Target	Producer
Beauveria bassiana	Conidia	Coffee berry borer	Live Systems Technology, Colombia
Beauveria bassiana	Boverosil	Colorado beetle	Czechoslovakia
Beauveria bassiana	Ostrinil	Corn borer	Natural plant protection, France
Beauveria bassiana	Mycotrol GH	Locust and grasshopper	Mycotech, USA
Beauveria brongniartii	Schweizer Beauveria	Cockchafers	Eric Schweizer, Switzerland
Lagenidium giganteu	Laginex	Mosquito larvae	AgraQuest, USA
Metarhizium anisopliae	Biogreen	Scarab larvae	Bio-care Technology, Australia

TABLE 14.3

List of Fungal Species with Their Targets for Biological Control of Weeds (Fravel et al., 1998, Templeton &Heiny, 1989)

Fungal Species	Product	Target Weed	Supplier or Country Where Registered
Colletotrichum coccodes	Velgo	*Abutilon theophrasti* in corn and soybeans	USA, Canada
Colletotrichum gloeosporioides f. sp *cuscutae*	Luboa 2	*Cuscuta chinensis* in soybeans	PR China
Colletotrichum gloeosporioides f. sp *aeschynomene*	Collego	*Aeschynomene virginica* in rice	Encore Technologies, USA
Colletotrichum gloeosporioides f. sp *Malvae*	Biomal	*Malva pusilla* in lentils and wheat	Canada

14.4.3 Beauveria

This genus comprises approximately 49 *spp.* while nearly 22 are known to be pathogenic (Kirk, 2003). The most important and commonly used fungus in this genus is *Beauveria bassiana*, previously known as *Tritirachium shiotae*. It was renamed after the Italian scientist and lawyer Agostino Bassi, who identified it as the pathogen of white muscardine disease in silkworms (Zimmermann, 2007). This fungus grows naturally in agriculture and forest soils (Sandhu et al., 2004). It can be isolated from soils or from insect bodies (Beilharz et al., 1982). It can be grown in simple media in vitro (Goettel & Inglis, 1997). This fungal spp. is known for having high host specificity, e.g., *Helicoverpa armigera*, American bollworm, Colorado potato beetle, termites, and codling moth.

14.4.4 Verticillium

This fungus plays an important role by controlling a vast range of insect in warm, humid, tropical, and subtropical regions (Nunez et al., 2008). This fungus was used to control several species of whiteflies, aphids, and green peach aphids in chrysanthemum in greenhouse conditions (Hamlen et al., 1979). It was observed that it is also useful to control *Trialeurodes vaporariorum* in greenhouse conditions in South Korea (Kim et al., 2002). It usually lives on the underside of leaves and attack nymphs and adults (Nunez et al., 2008).

14.4.5 Paecilomyces

Under this genus, the most important fungi, *Paecilomyces fumosoroseus*, is reported to have most potential against whiteflies by causing a disease known as "yellow muscardine" (Nunez et al., 2008). It mainly attacks the nymphs of the whitefly (Kim et al., 2002). The mycelia of the fungus cover the whole body of the nymph and attach to the undersides of leaves. It is observed to grow mainly in a humid environment (Wraight et al., 2000) in both open as well as greenhouse conditions.

14.4.6 Nomuraea

The most effective entomopathogenic fungi under this genus, *Nomuraea rileyi*, is known to attack several insects in the lepidoptera and coleoptera families (Ignoffo, 1981). It can also be used against *Bombyx mori*, *Heliothiszea*, *Trichoplusiani*, etc.

14.5 Limitations

The most important factors for the efficacy of fungal bioagent inoculum are their virulence and viability (Doust and Roberts, 1983). Several isolates of *Metarhizzium* and *Beuveria spp.* show their potential in field conditions (Easwaramoorthy, 2003). A number of fungal genera can't resist the field environment after application in crop plants as they are exposed to various abiotic stress conditions (Rangel et al., 2005a). Some fungal genera are found to be susceptible to solar radiation as UV light is harmful to them (Bjorn, 2006). Soil temperature is another factor that lessens their survival in field conditions, as the DNA may become damaged due to the dry heat and wet heat, resulting in membrane disorganization and protein denaturation. The pathogenicity, mode of virulence, reproduction, and spread are affected directly and indirectly by the environmental factors and abiotic stress (Hasan, 2014). Use of natural strains may disturb the ecosystem balance as these cannot be genetically modified. Some fungal genera often show phytotoxicity towards plants, animals, and mammals. Therefore, it is essential to create awareness among end-users about the biosafety, how to apply, and time of application, so that the environment and ecology should not be endangered at any cost. To maintain the shelf life of the biocontrol agents, they should be stored at the given temperature, which causes problems for farmers due to the unavailability of proper storage facilities.

14.6 Future Trends and Conclusion

14.6.1 Research Issues

- Development of noble strain to increase its efficacy
- Compatibility of different agrochemicals with biocontrol agents
- Increase shelf life issue
- Make the BCA resistant against different abiotic stress

14.6.2 Industrial Issues

- More products with high efficacy, less risk hazard to the ecosystem, and compatibility with different agrochemicals should be developed
- At the time of production, the quality should be checked at every step
- Cost-effective products should be developed
- The industries should work based upon the market and farmers' demand

14.6.3 Regulatory Issues

- Strict punishment must be implemented against the companies producing false and unauthorized products
- Before producing any BCA product, the company must be aware of all policy, guidelines, safety rules, and acts enacted by the government

Fungal biocontrol agents are playing an important role in management of pests in agriculture as well as forests in comparison with several agrochemicals. Continuous evolution of pathogens is observed every year and to counter them, the fungal bioagents are used for their multifaceted interactions. With less risk hazard to the ecosystem, high host specificity, and potentiality, their demand is increasing in the global market. They not only control the pest but also help in root colonization, improvement of plant health, root mass, and increase the yield. They can also be used as seed treatment agents so pathogen attacks will be reduced after sowing in the field. These bioagents can be mixed with other chemicals. The government, non-government, and industrial sectors should focus on different training and awareness programs to disseminate knowledge about the use of (Biocontrol agents) BCAs, safety measures, time of application, and every related detail related. They can't completely eradicate the pest but still lessen the severity of the pest attack in comparison with chemical pesticides. The National Farmer Policy 2007 emphasizes the promotion of biopesticides for increasing production and lessening the hazards to farmers as well as the ecosystem. These fungal bioagents are easy to use, ecofriendly, and cost-effective, but more research is required to develop more potential strains.

REFERENCES

Adnan, M.; Islam, W.; Shabbir, A.; Khan, K.A.; Ghramh, H.A.; Huang, Z.; Chen, H.Y.H.; Lu, G. Plant defense against fungal pathogens by antagonistic fungi with *Trichoderma* in focus. *Microbial Pathogenesis.* 2019, 129, 7–18.

Beilharz, V.C.; Parberry, D.G.; Swart, H.J. Dodine: A selective agent for certain soil fungi. *Transactions of the British Mycological Society.* 1982, 79, 507–511.

Bhattacharjee, R.; Dey, U. An overview of fungal and bacterial biopesticides to control plant pathogens/diseases. *African Journal of Microbiology Research.* 2014, 8(17), 1749–1762.

Bjorn, L.O. Stratospheric ozone, ultraviolet radiation, and cryptogams. *Biological Conservation.* 2006, 135, 326–333.

Burges, H.D. *Formulation of Microbial Pesticides.* Kluwer Academic, Dordrecht, The Netherlands. 1998.

del Prado, E.N.; Iannacone, J.; Gomez, H. Effect of two entomopathogenic fungi in controlling *aleurodicus cocois* (Curtis, 1846) (Hemiptera: *Aleyrodidae*). *Chilean Journal of Agricultural Research* 2008, 68 21–30.

Doust, R.A.; Roberts, D.W. Studies on the prolonged storage of *Metarhizium anisopliae* conidia: Effect of temperature and relative humidity on conidial viability and virulence against mosquitoes. *Journal of Invertebrate Pathology.* 1983, 41, 143–150.

Driver, F.; Milner, R.J.; Trueman, J.W.H. A taxonomic revision of *Metarhizium* based on a phylogenetic analysis of rDNA sequence data. *Mycology Research.* 2000, 104, 134–150.

Easwaramoorthy, S. Entomopathogenic fungi. In: *Biopesticides and Bioagents in Integrated Pest Management of Agricultural Crops* (R.P. Srivastava, ed.). International Book Distributing Co., India. 2003, 341–379.

Fravel, D.R.; Connick, J.W.; Lewis, J.A.; *Formulation of Microbial Biopesticides* (H.D. Burges, ed.). Kluwer Academic, Dordrecht, The Netherlands. 1998.

Goettel, M.S.; Inglis, G.D. Fungi: Hyphomycetes. In: *Manual of Techniques in Insect Pathology* (L.A. Lacey, ed.). Academic Press, London. 1997, 213–250.

Gveroska, B.; Ziberoski, J. *Trichoderma harzianum* as a biocontrol agent against *Alternaria alternata* on tobacco. *Applied Technologies & Innovations.* 2012, 7, 67–76.

Hamlen, R.A. Biological control of insects and mites on European greenhouse crops: Research and commercial implementation. In: Proceedings of the Florida State Horticultural Society, United States. 1979, 92, 367–368.

Hasan, S. Entomopathogenic fungi as potent agents of biological control. *International Journal of Engineering & Technology Research.* 2014, 2, 234–237.

Ignoffo, C.M. The fungus *Nomuraea rileyi* as a microbial insecticide. In: *Microbial Control of Pests and Plant Diseases* (H.D. Burges, ed.). Academic Press, London, UK. 1981, 513–538.

Khalili, E.; Sadravi, M.; Naeimi, S.; Khosravi, V. Biological control of rice brown spot with native isolates of three *Trichoderma* species. *Brazilian Journal Of Microbiology.* 2012, 43, 297–305.

Kim, J.J.; Lee, M.H.; Yoon, C.S.; Kim, H.S.; Yoo, J.K.; Kim, K.C. Control of cotton aphid and greenhouse whitefly with a fungal pathogen. *Journal of National Institute of Agricultural Science and Technology.* 2002, 7–14.

Kirk, P.M. Index fungorum. 2003. Available at http://www.indexfungorum.org (28/10/2009).

Nunez, E.J.; Iannacone.; Omez, H.G. Effect of two entomopathogenic fungi in controlling aleurodicuscocois (Curtis, 1846) (Hemiptera: Aleyrodidae). *Chilean Journal of Agricultural Research* 2008, 68 21–30.

Paroda, R.S.; Mangala, R.; Gautam P.L.; Kochhar, S.; Singh, A.K. *National Action Plan on Agrobiodiversity in India.* National Academy of Agricultural Sciences, Indian Council of Agricultural Research and Indian Society of Plant Genetic Resources, New Delhi, 1999, 56p.

Rai, D.; Updhyay, V.; Mehra, P.; Rana, M. Pandey A.K. Potential of entomopathogenic fungi as biopesticides. *Indian Journal of Science Research and Technology.* 2014, 2(5), 7–13.

Rangel, D.E.N.; Braga, G.U.L.; Anderson, A.J.; Roberts, D.W. Variability in conidial thermo tolerance of *Metarhizium anisopliae* isolates from different geographic origins. *Journal of Invertebrate Pathology.* 2005a, 88, 116–125.

Sandhu, S.S.; Vikrant, P. Myco-insecticides: Control of insect pests. In: *Microbial Diversity: Opportunities & Challenges* (S.P. Gautam, S.S. Sandhu, A. Sharma and A.K. Pandey, eds.). Indica Publishers, New Delhi, India. 2004.

Templeton, G.E.; Heiny, D.K. Improvement of fungi to enhance mycoherbicide potential. In: *Biotechnology of Fungi for Improving Plant Growth* (J.M. Whipps, R.D. Lumsden, eds.). Cambridge University Press, UK. Chapter 6, 1989, 127–151.

Tulloch, M. The genus *Metarhizium.* Transaction of the Britannic Mycology Society. 1976, 66, 407–411.

Wraight, S.P.; Carruthers, R.I.; Jaronski, S.T.; Bradley, C.A.; Garza, C.J.; Wraight, S.G. Evaluation of the entomopathogenic fungi *Beauveria bassiana* and *Paecilomyces fumosoroseus* for microbial control of the silver leaf whitefly, *Bemisiaargentifolii. Biological Control.* 2000, 17, 203–217.

Zimmermann, G. Review on safety of the entomopathogenic fungus *Beauveria bassiana* and *Beauveria brongniartii. Biocontrol Science Technology* 2007, 17, 553–596.

15

Management of the Diseases of Vegetable Crops through Fungal Biopesticides

Pratibha Sharma, Prashant Prakash Jambhulkar, and Raja Manokaran

CONTENTS

15.1 Biological Control of Diseases of Vegetable Crops

Trichoderma spp. is a widely recognized beneficial microbe due to its multifaceted applications in agriculture (Figure 15.1) such as biocontrol agents, plant growth promoter, application in pesticide degradation (Sharma et al., 2016), and opportunistic endophytism (Leon et al., 2018). Some of the prominent species like *Trichoderma harzianum*, *T. asperellum*, *T. virens*, and *T. Koningii* can manage especially the soilborne plant diseases (Sharma et al, 2014). Presently, the consortium of microbes is also being used for diseases of different crops (Jambhulkar et al., 2018).

15.1.1 Brinjal (*Solanum melongena* L.)

Eggplant or brinjal (*Solanum melongena* L.) is one of the most commonly grown vegetable crops in India. Fungi such as *A. solani*, *F. solani*, *C. gloeosporioides*, *B. cinerea*, *Penicillium* sp., Rhizopusniger, *Curvularialunata*, and *Botryodiplodia theobromae* are important pathogens which cause damage to the fruit throughout the year. Eco-friendly management of collar rot caused by *S. rolfsii* was studied by Jadon (2009), who found *T. viride* to be most effective in reducing the percent disease incidence. Likewise, disease management using Trichoderma was carried out by many groups against leaf spot disease of brinjal (Balaji and Ahir, 2011) where the use of plant extracts and biocontrol agent was evaluated under in vitro and in vivo conditions. In another study, the effect of consortium of *T. harzianum* isolates on the growth attributes and management of rot caused by *S. sclerotiorum* were also studied. Jakatimath et al. (2017) examined different isolates of *Trichoderma* viz., *Trichoderma harzianum* strains *T*-21, *T*-28, *T*-29, *T*-72, and *T*-p against *Phomopsis vexans* causing fruit rot of brinjal. Among the isolates, *T. harzianum T*-p exhibits maximum antagonistic ability (70.66%) against the pathogen. Root-knot nematode infection in brinjal is one of the major diseases that cause serious losses in crop yield. Osman et al. (2018) reported that the integrated management of *M. incognita* and root rot disease infecting eggplant can be controlled by application of individual or combined application of *T. harzianum*, Bt, NPK, and chicken manure. Seed treatment with *T. harzianum* reduced bacterial wilt and fruit rot caused by *R. solanacearum* and *P. vexans* by 25–40 and 14–20 percent at nursery and also increased marketable yield in the field (Nahar et al., 2019).

15.1.2 Cauliflower and Cabbage

Cauliflower (*Brassica oleracea* L. var. *botrytis*) and cabbage (*Brassica oleracea* L. var. *capitata*) are the most important winter vegetables of India, but now they are grown throughout the year. Major pathogens which cause crop losses are club root caused by *Plasmodiophora brassicae*, stem and fruit rot caused by *Sclerotinia sclerotiorum*, black rot caused by *Xanthomonas campestris*, wire stem caused by *Rhizoctonia solani*, leaf spot caused by *A. brassicae*, leaf blight caused by *A. brassiciola*, and damping-off caused by *R. solani*, *Pythium*, *Phytophthora*, *Corticium*, and *Fusarium*. The antagonistic potential of *Trichoderma* and *Aspergillus* species were evaluated on *S. sclerotiorum* causing rots in cabbage and cauliflower by Sharma et al. (2001) and, further, the role of Trichoderma spp. in an IDM program on the management was confirmed against *Sclerotinia sclerotiorum* (Sharma et al., 2001; Zewain et al., 2004). Sharma and Sain (2005) reported thc use of bioagents

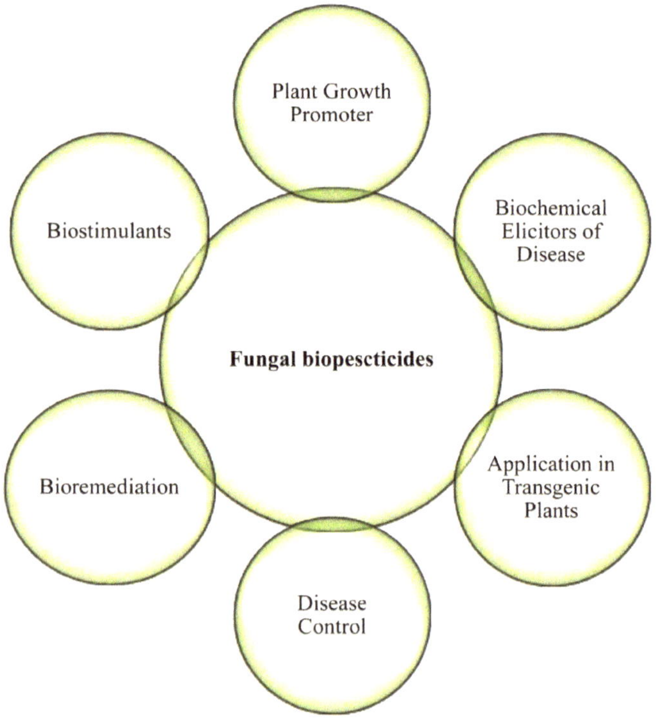

FIGURE 15.1 Multiple applications of fungal biopesticides in organic agriculture.

and abiotic compounds for inducing resistance against damping-off of cauliflower caused by *P. aphanidermatum*. Zewain et al. (2005) evaluated an integrated disease management program on stalk rot of cauliflower. In a study on the effect of soil solarization on sclerotial viability of *S. sclerotiorum* of cauliflower, Sharma et al. (2005) observed the reduction in viability of sclerotial bodies, which also promoted the biocontrol activity. In a study on integrated management of major pests of cauliflower by Ahuja et al. (2012), the compatible strains of *Trichoderma harzianum* TNHU27 and *Pseudomonas aeruginosa* PJHU15 were reported to be effective against *S. sclerotiorum*, causing white rot of cauliflower (Ram et al., 2019).

15.1.3 Chilli (*Capsicum annuum* L.)

Chili (*Capsicum annum* L.) is another important vegetable crop also used as a spice and rich in vitamins C, A, and B. Some of the major fungal diseases like damping-off caused by *S.rolfsii*, *F. oxysporum*, *Pythium* spp., *R. solani*, and *Phytophthora* spp. are widely prevalent in the country. Several groups revealed that isolates of *T. harzianum*, *T. viride* (TVC3), and *T. hamatum* are effective biocontrol agents against anthracnose and damping-off and chili dieback disease. In a study performed by Sharma et al. (2004) on integrated management of chili dieback and anthracnose, the disease reduction was found to be highest in IPM module followed by chemical module. In another study, positive interaction of *T. harzianum* with *R. solani* was observed in the rhizospheres of chili cultivars by Mathur et al. (2006), indicating its potential for biological management of chili diseases. The biocontrol potential of *Trichoderma* spp. against important soilborne diseases of vegetable crops was evaluated by Kapoor (2008). Kaur et al. (2006) demonstrated the in vitro effect of Trichoderma species on *C. capisci* causing fruit rot. Chanchaichaovivat et al. (2007) screened four antagonistic yeast strains, namely *Pichiaguillier mondii*, *Candida musae*, *Issatchenkia orientalis*, and *Candida quercitrusa*, for the management of postharvest chili anthracnose caused by *C. capsici*. *P. guilliermondii* strain R13 showed maximum efficacy in reducing disease incidence at lower temperature (18°C). An in vitro experiment was also conducted to evaluate the effect of eight isolates of *Trichoderma* species by Muthukumar et al. (2011), who found a *Trichoderma* isolate TVC3 that was able to inhibit the mycelial growth of pathogen as well as promote the plant growth. In another study, the concept of using coco-peat enriched with *T. harzianum* was suggested for better plant growth and reduced incidence of tomato wilt and chili root rot while raising disease-free and healthy seedlings, which also reduced wilt incidence in tomato (Sriram et al., 2010). In another study, fungal isolates were tested for biocontrol of anthracnose disease in greenhouse and field conditions by Vasanthakumari et al. (2013), who found that the test antagonistic fungi *F. oxysporum*, *C. globosum* and *T. harzianum* could be used to control anthracnose and promote plant growth, and increase yield of chili in field. Elevation of defense network mechanism against *Colletotrichum capsici* by phyllopsheric *Trichoderma* isolate BHUF4 in chili was studied by Saxena et al. (2016). Consortial application of biocontrol agents can be an important aspect of plant disease management. Chemeltorit et al 2017 studied the combined application of *T. hamatum* THSW13 and *P. aeruginosa* BJ10–86 for the management of damping-off and collar rot caused by *P. capsici*.

15.1.4 Onion (*Allium cepa*)

Onion (*Allium cepa*) is another important vegetable crop of India. This crop is severely infected by major diseases viz., foliar blight of onion caused by *Alternaria alternata*, purple blotch of onion caused by *A. porri*, damping-off and basal rot caused by *Fusarium oxysporum*. f. sp. *cepae* and *Pythium* sp., white rot caused by *Sclerotium rolfsii*, and black mold caused by *Aspergillus niger*. Purple blotch of onion is a very important disease in onion growing areas. Effective isolates of *T. harzianum*, i.e., Th3, Th-30 of *T. harzianum*, and Tv-12, Tv-15 of *T. viride*, were reported to express a high level of disease reduction and growth promotion in susceptible onions (Prakasam and Sharma, 2012). Biological management and IDM of foliar blight have also been reported by Yadav et al. (2014) and Shahnaz et al. (2013). In vitro evaluation of plant extracts, bioagents, and fungicides against purple blotch and Stemphylium blight of onion suggested that *T. viride* is effective in inhibiting the growth of pathogen (Mishra and Gupta, 2012). Elshahawy et al. (2017) reported three mycoparasitic sclerotial isolates. *Chaetomium globosum* (Chg6) Kunze, *Clonostachys rosea* (Cr12) (Link) Schroers, and *Penicillium oxalicum* (Po9) were observed to be antagonists against the most pathogenic isolate (Sc2) of *Stromatinia cepivora* (Berk.) Wetzel, the causal agent of onion white rot. The highly effective strain was of *P. oxalicum* (Po9), which reduced the disease incidence (12.2%), compared to *C. globosum* (Chg6) (13.6%) and *C. rosea* (Cr12) (17.5%). They were also reported to enhance vegetative growth of live onion plants in a greenhouse and increased onion bulb yield in the field compared with the control. In a study conducted by Bunbury-Blanchette et al. (2019) against Fusarium basal rot of onion by *F. oxysporum* f. sp. *Cepae* (FOC), seven species of *Trichoderma* viz., *T. atroviridae*, *T. brevicompactum*, *T. gamsii*, *T. hamatum*, *T. harzianum*, *T. koenengiopsis*, and *T. viridae* were reported to be antagonistic in dual culture. Soil application with the spore suspension 1×10^7 cfu per gram of *T. atroviride* and *T. harzianum* significantly reduced *Fusarium* basal rot symptoms in greenhouse. Similarly, El-Mougy et al. (2019) reported 61.8% FBR reduction in field by the use of *T. viride*. *T. asperellum* (To) have proved to be an effective biocontrol agent against leaf blight disease caused by *Stemphylium vesicarium*. Zapata-Sarmiento (2020) reported 20.8% reduction in disease severity index of the leaf blight.

15.1.5 Potato (*Solanum tuberosum*)

Potato (*Solanum tuberosum*) is one of the most important food crops worldwide. There are many threatening diseases which reduce crop yield, such as early blight by *A. solani*, late blight by *Phythophthora infestans*, common scab (*Streptomyces scabies*), dry rot (*Fusarium* spp.), black scurf, and bacterial wilt (*R. solanacearum*). The antagonistic fungal population may be considered as one of the factors which also plays an important role in germination and yield of potato (Dwivedi, 1988). Methods for friendly potato disease management allow organic tropical Indian production systems to prosper though there are reports of use of Trichoderma bioformulations in management of late blight of potato in Northeastern India (Basu, 2009). Success in mycoparasitism of potato scurf pathogen (*R. solani*) by use of *T. harzianum* by Pandey and Pundhir (2013) can be taken as the future of biological management of this important disease. Larkin and Tavantzis (2013) used a consortium of bacterial and fungal antagonists for management of major soilborne diseases of potato. *Bacillus subtilis* GB03 and *R. solani* hypovirulent isolate Rhs1A1, reduced multiple soilborne diseases, stem and stolon canker by 20–38%, black scurf by 30–58%, and common scab by 10–34%. Late blight is one of the highly destructive diseases of potato caused by *P. infestans*. By using *Trichoderma* isolate HNA14, Yao et al. (2016) reported lowest disease index (16.0) as compared to the control 21 days after treatment. The efficacy of date palm compost and its associated fungi against potato stem canker and black scurf was reported by Khaldi et al. (2016). *Aspergillus terreus* A2, *A. nidulans* A3, *A. fumigatus* A6, and *A. candidus* A13 showed maximum antagonistic effect against pathogen *R. solani*. Maximum reduction in black scurf severity of 96.4–97.1% was observed in cattle manure and date palm compost (CMC). Biological control of potato cyst nematode *Globodera pallida* by using green fluorescent protein (GFP) tagged *T. harzianum* strain ThzID1-M3 was reported by Contina et al. (2017).

15.1.6 Tomato (*Solanum lycopersicum* L.)

Tomato (*Solanum lycopersicum*) is an important vegetable crop grown in almost all parts of India. However, the crop is susceptible to a wide range of diseases such as late blight caused by *Phytophthora infestans*, early blight caused by *Alternaria solani*, wilt caused by *F. oxysporum* f. sp. *lycopersici*, damping-off caused by *P. aphanidermatum*, collar rot caused by *Sclerotiumrolfsii*, Buckeye rot caused by *Phytophthora parasitica*, bacterial leaf spot caused by *X. campestris* pv. *vesicatoria*, and bacterial wilt caused by *R. solanacearum*. In control of collar rot of tomato by *Trichoderma* isolates (Dutta and Das, 2002), *T. harzianum* was the most inhibitory to *S. rolfsii*, showing 61.5% inhibition of the mycelial growth of the pathogen. In vitro studies have shown that *T. harzianum* isolates inhibited the growth of *P. aphanidermatum* up to 50% and was found to be compatible with Bavistin, Captan, and Vegfrugard, respectively (Sharma et al., 2003). The role of *Trichoderma* species in systemic resistance of tomato and cauliflower against stalk rot pathogen *S. sclerotiorum* was documented (Sharma and Sain, 2004). The effect of antagonists on damping-off of tomato was widely studied by many other workers as well. Manoranitham et al. (2000) reported that the application of *T. viride* and *P. fluorescens* effectively checked the preemergence and postemergence damping-off of tomato caused by *P. aphanidermatum* under pot culture experiments. Likewise, Sharma and Sain (2003) evaluated commercial and laboratory formulations of bioagents and plant nutrients against wilt of tomato. Jayaraj et al. (2006) developed a carrier-based formulation of *T. harzianum* strain M1 for control of damping-off of tomato caused by *P. aphanidermatum*. The biocontrol potential of *Trichoderma* spp. against important soilborne diseases of vegetable crops has been reviewed (Kapoor, 2008). Baranwal et al. (2011) reported the management of tomato wilt caused by *Fusarium oxysporum* f. sp. *lycopersici* by using bioagents and organic amendments. Induction of systemic resistance by *T. harzianum* strain Th3 against tomato wilt pathogen *F. oxysporum* f. sp. *lycopersici* (FoL) shows increased production of total phenol, polyphenol oxidase, peroxidase was reported by Raja et al. (2016). De Medeiros et al. (2017) reported the efficacy *of Trichoderma atroviridae* an effective strain Ta against *Meloidogyne javanica*. Number of galls were reduced by 42%, egg masses by 60%, and adult nematodes by 90% inside the tomato roots under greenhouse conditions. The strain was also reported to be a growth promoter, since there was an increase in the size of F1 plants. In another greenhouse study, Aseel et al. (2019) observed very interesting observations against tomato mosaic virus (ToMV). Significant reductions in the disease severity and incidence, as well as the viral accumulation level, were observed as a response to the mycorrhizal colonization of the infected plants after 21 days of inoculation. Enhanced values of growth parameters, photosynthetic pigments, and flavonoid content were reported by AMF colonization. Salt tolerant strains *Trichoderma asperellum* (F-01763), *Hypocrealixii* (F-01760), and *Hypocreavirens* (F-01775) were observed to be antagonistic against tomato root rot caused by *Rhizoctoniasolani* (AG-4) (Kashyap et al., 2020). Zhao et al. (2020) transferred flagellin gene from *Pseuodomonas syringae* pv. Tomato str. DC3000 to the antagonist yeast *Sacharomyces cerevisiae* (EBY100). The strain was very effective against postharvest disease caused by *Botrytis cinerea*.

15.2 Conclusion and Prospects

The present world is facing problems of clean and safe environment. In agriculture, due to the use of pesticides, crop protection strategies have been redefined by using new tools based on biological control agents (BCAs) and natural antimicrobial chemicals for disease control. A lot of research laboratories in India are now focusing on identification of the microorganisms which can either inhibit the growth or kill the plant pathogens. The mode of action of each BCA is different, so we need to focus on developing efficient BCAs. The isolation of compounds that affect plant metabolism may help to overcome the problems related to use of living microorganisms, through the introduction of new secondary metabolites-based biopesticides and biofertilizers for the desired beneficial effects on crops.

REFERENCES

Ahuja, D. B.; Ahuja, U. R.; Srinivas, P.; Singh, R.V.; Malik, M.; Sharma, P.; Bamawale, O. M. Development of Farmer-led Integrated Management of Major Pests of Cauliflower Cultivated in Rainy Season in India. J. Agric. Sci. 2012, 4(2), 79–90.

Aseel, D. G.; Rashad, Y. M.; Hammad, S. M. Arbuscular mycorrhizal fungi trigger transcriptional expression of flavonoid and chlorogenic acid biosynthetic pathways genes in tomato against tomato mosaic virus. Sci. Rep. 2019, *9*(1), 1–10.

Balaji, L. P.; Ahir, R. R. Evaluation of plant extracts and biocontrol agents against leaf spot disease of brinjal. *Indian Phytopathol.* 2011, *64* (4), 378–380.

Barnwal, M. K.; Maiti, D.; Pandey, A.C. Management of tomato wilt caused by *Fusarium oxysporum f. sp. lycopersici* by using bioagents and organic amendments. *Indian Phytopathol.* 2011, *64* (2), 194–196.

Basu, A. Employing Eco friendly potato disease management allows organic tropical Indian production systems to prosper. *Asian J. Food Agro-Ind.* 2009, 80–87.

Bunbury-Blanchette, A. L.; Walker, A. K. Trichoderma species show biocontrol potential in dual culture and greenhouse bioassays against Fusarium basal rot of onion. *Biol. Control.* 2019, *130*, 127–135.

Chanchaichaovivat, A.; Ruenwongsa, P.; Panijpan, B. Screening and identification of yeast strains from fruits and vegetables: Potential for biological control of postharvest chilli anthracnose (*Colletotrichum capsici*). *Biol. Control.* 2007, *42*(3), 326–335.

Chemeltorit, P. P.; Mutaqin, K. H.; Widodo, W. Combining *Trichoderma hamatum* THSW13 and *Pseudomonas aeruginosa* BJ10–86: A synergistic chili pepper seed treatment for *Phytophthora capsici* infested soil. *Eur. J. Plant Pathol.* 2017, *147*(1), 157–166.

Contina, J. B.; Dandurand, L. M.; Knudsen, G. R. Use of GFP-tagged *Trichoderma harzianum* as a tool to study the biological control of the potato cyst nematode *Globodera pallida. Appl. Soil. Ecol.* 2017, *115*, 31–37.

De Medeiros, H. A.; de Araújo Filho, J. V.; De Freitas, L. G.; Castillo, P.; Rubio, M. B.; Hermosa, R.; Monte, E. Tomato progeny inherit resistance to the nematode *Meloidogyne javanica* linked to plant growth induced by the biocontrol fungus *Trichoderma atroviride.* Sci. Rep. 2017, *7*(1), 1–13.

Dutta, P.; Das, B. C. Management of collar rot of tomato by *Trichoderma* spp. and chemicals. *Indian Phytopathol.* 2002, *55*(2), 235–237.

Dwivedi, R. Effect of growth products of staled fungal spp. on germination and yield of exotic potato cultures. *Indian Phytopathol.* 1988, *41*, 500–502.

El Khaldi, R.; Daami-Remadi, M.; Cherif, M. Biological control of stem canker and black scurf on potato by date palm compost and its associated fungi. *J. Phytopathol.* 2016, *164*(1), 40–51.

El-Mougy, N. S.; Abdel-Kader, M. M. Biocontrol measures against onion basal rot incidence under natural field conditions. *J. Plant. Pathol.* 2019, *101*(3), 579–586.

Elshahawy, I. E.; Saied, N. M.; Abd-El-Kareem, F.; Morsy, A. A. Field application of sclerotial mycoparasites as biocontrol agents to *Stromatiniacepivora*, the cause of onion white rot. *J. Plant Pathol.* 2017, *99*(2), 391–401.

Jadon, K. S. Eco-friendly management of brinjal collar rot caused by *Sclerotium rolfsii* Sacc.*Indian Phytopathol.* 2009, *62*(3), 345–347.

Jakatimath, S. P.; Mesta, R. K.; Mushrif, S. K.; Biradar, I. B.; Ajjappalavar, P. S. In vitro evaluation of fungicides, botanicals and bio-agents against *Phomopsis vexans*, the causal agent of fruit rot of brinjal. *J. Pure. Appl. Microbiol.* 2017, *11*(1), 229–235.

Jambhulkar, P. P.; Sharma, P.; Raja, M.; Lakshman D. K.; Rokadia, P.; Jambhulkar, N.Assessing synergism of combined applications of *Trichoderma harzianum* and *Pseudomonas fluorescens* to control blast and bacterial leaf blight of rice. *Eur. J. PlantPathol.* 2018, *152*(3), 747–757.

Jayaraj, J.; Radhakrishnan, N. V.; Velazhahan, R. Development of formulations of *Trichoderma harzianum* strain M1 for control of damping-off of tomato caused by *Pythium aphanidermatum. Arch. Phytopathol. Plant Prot.* 2006, *39*(1), 1–8.

Kapoor, A. S. Biocontrol potential of *Trichoderma* spp. against important soilborne diseases of vegetable crops. *Indian Phytopathol.* 2008, *61*(4): 492–498.

Kashyap, P. L.; Solanki, M. K.; Kushwaha, P.; Kumar, S.; Srivastava, A. K. Biocontrol potential of salt-tolerant Trichoderma and Hypocrea isolates for the management of tomato root rot under saline environment. *J. Soil Sci. Plant Nutr.* 2020, *20*(1), 160–176.

Kaur, M.; Sharma, O. P.; Sharma, P. N. *In vitro* effect of *Trichoderma* species on *Colletotrichum capiscic*a using fruit rot of chilli (*Capsicum annuum*L.). *Indian Phytopathol.* 2006, 59(2), 243–245.

Larkin, R. P.; Tavantzis, S. Use of biocontrol organisms and compost amendments for improved control of soilborne diseases and increased potato production. *Am. J. Potato.Res.* 2013, *90*(3), 261–270.

Leon,V. C.; Raja, M.; Pandian R. T. P.; Kumar, A.; Sharma, P. Opportunistic endophytism of *Trichoderma* species in rice Pusa Basmati-1 (PB1). *Indian. J. Exp. Biol.* 2018, *56*, 121–128.

Manoranitham, S. K.; Prakasam, V.; Rajappan, K.; Amutha, G. Effect of two antagonists on damping off disease of tomato. *Indian Phytopathol.* 2000, *53*(4), 441–443.

Mathur, K., Zacharia, S. Bohra, B. Rhizosphere competence of *Trichoderma harzianum* for biological control of root- rot pathogen, *Rhizoctonia solani* in different chilli cultivars.*Indian Phytopathol.* 2006, *59*(2), 246–248.

Mishra, R. K.; Gupta, R. P. *In vitro* evaluation of plant extracts, bio-agents and fungicides against Purple blotch and Stemphylium blight of onion. *J. Med. Plant. Res.* 2012, *6*(48), 5840–5843.

Muthukumar, A.; Eswaran, A.; Sanjeevkumar, K. Exploitation of *Trichoderma* species on the growth of *Pythium aphanidermatum* in chilli. *Braz. J. Microbiol.* 2011, *42*(4), 1598–1607.

Nahar, N., Islam, M. R., Uddin, M. M., de Jong, P., Struik, P. C., Stomph, T. J. Disease management in eggplant (*Solanum melongena* L.) nurseries also reduces wilt and fruit rot in subsequent plantings: A participatory testing in Bangladesh. *Crop Prot.* 2019, *120*, 113–124.

Osman, H. A.; Ameen, H. H.; Mohamed, M.; El-Mohamedy, R.; Elkelany, U. S. Field control of *Meloidogyne incognita* and root rot disease infecting eggplant using nematicide, fertilizers, and microbial agents. *Egypt. J. Biol. Pest Control.* 2018, *28*(1), 40.

Pandey, S.; Pundhir, V. S. Mycoparasitism of potato black scurf pathogen (*Rhizoctonia solani*Kuhn) by biological control agents to sustain production. *Indian J. Hortic.* 2013, 70(1),71–75.

Prakasam, V.; Sharma, P. *Trichoderma harzianum* (Th-3) a potential strain to manage the purple blotch of onion (*Allium cepa L.*) caused by *Alternariaporri* under North Indian Plains. *J. Agric. Sci.* 2012, 10, 266–272.

Raja, M.; ThavaPrakasa Pandian R.; Sharma, M.; Jambhulkar, P. P.; Sharma, P. Study of induced systemic resistance in tomato against *Fusarium oxysporum* f. sp. *lycopersici* causing wilt of tomato. *Indian Phytopathol.* 2016, *69*(4):539–542.

Ram, R. M.; Tripathi, R.; Birla, H.; Dilnashin, H.; Singh, S. P.; Keswani, C. Mixed PGPR consortium: an effective modulator of antioxidant network for management of collar rot in cauliflower, Arch. Phytopathol. Plant Protect. 2019, 52(7-8), 844–862.

Saxena, A.; Raghuwanshi, R.; Singh, H. B. Elevation of defense network in chilli against *Colletotrichum capsici* by phyllospheric *Trichoderma* strain. *J. Plant Growth Regul.* 2016, *35*(2), 377–389.

Shahnaz, E.; Razdan, V. K.; Rizvi, S. E. H.; Rather, T. R.; Gupta, S.; Andrabi, M. Integrated disease management of foliar blight disease of onion: A case study of application of confounded factorials. *J. Agric. Sci.* 2013, *5*(1), 17–22.

Sharma, P.; Kulshrestha, G.; Gopal, M.; Kadu, L. N. Integrated management of chilli die back and anthracnose in Delhi region. *Indian Phytopathol.* 2004, *57*(4), 427–434.

Sharma, P.; Sain, S. K. Induction of systemic resistance in tomato and cauliflower by *Trichoderma* species against stalk rot pathogen. (*Sclerotinia sclerotiorum*). *J. Biol. Control.* 2004, *18*(1), 21–28.

Sharma, P.; Sain, S. K. Use of biotic agents and abiotic compounds against damping off of cauliflower caused by *Pythium aphanidermatum. Indian Phytopathol.* 2005, *58*(4), 395–401.

Sharma, P.; Sain, S. K. James, S. Compatibility study of *Trichoderma* isolates with fungicides against damping-off of cauliflower and tomato caused by *Pythium aphanidermatum. Pestic. Res. J.* 2003, *15*(2), 133–138.

Sharma, P.; Sharma, M.; Raja, M.; Shanmugam, V. Status of *Trichoderma* research in India: A review. *Indian Phytopathol.* 2014, *67*(1), 1–19.

Sharma, P.; Sharma, M.; Raja, M.; Singh, D. V.; Srivastava, M. Use of *Trichoderma* spp. in biodegradation of Carbendazim. *Indian J. Agric. Sci.* 2016, *86*(7), 59–62.

Sharma, P.; Singh, L.; Adlakha, D. Antagonistic potential of *Trichoderma* and *Aspergillus*species on *Sclerotiniasclerotiorum*(Lib.) de Barry causing rots in cabbage and cauliflower. *Pestic. Info.* 2001, *2*, 41–44.

Sharma, P.; Zewain, Q.; Bahadur, P.; Sain, S. K. Effect of soil solarization on sclerotial viability of *Sclerotiniasclertiorum* (Lib.) de Bary of cauliflower (*Brassica oleracea* L. var. *botrytis* subvar, *cauliflower* D. C.). *Indian J. Agric. Sci.* 2005, *75*, 90–94.

Sriram, S.; Savitha, M. J.; Ramanujam, B. *Trichoderma*-enriched coco-peat for the management of *Phytophthora* and *Fusarium* diseases of chilli and tomato in nurseries. *J. Biol.Control.* 2010, *24*(4), 311–316.

Vasanthakumari, M. M.; Shivanna, M. B. Biological control of anthracnose of chilli with rhizosphere and rhizoplane fungal isolates from grasses. *Arch. Phytopathol. Plant Prot.* 2013, *46*(14), 1641–1666.

Yadav, M.; Rakholiya, K. B.; Pawar, D. M. Evaluation of bioagents for management of the onion purple blotch and bulb yield loss assessment under field conditions. *Bioscan.* 2014, *8*(4): 1295–1298.

Yao, Y.; Li, Y., Chen; Z.; Zheng, B.; Zhang, L.; Niu, B.; Wang, Q. Biological control of potato late blight using isolates of Trichoderma. *Am. J. Potato. Res.* 2016, *93*(1), 33–42.

Zapata-Sarmiento, D. H.; Palacios-Pala, E. F.; Rodríguez-Hernández, A. A.; Melchor, D. L. M.; Rodríguez-Monroy, M.; Sepúlveda-Jiménez, G. *Trichoderma asperellum*, a potential biological control agent of *Stemphylium vesicarium*, on onion (*Allium cepa*L.). *Biol. Control.* 2020, *140*, 104105.

Zewain, Q. K.; Bahadur, P.; Sharma, P. Effect of fungicides and neem extract on mycelial growth and myceliogenic germination of *Sclerotinia sclerotiorum. Indian Phytopathol.*2004, *57*(1), 101–103.

Zewain, Q. K.; Bahadur, P.; Sharma, P. Integrated disease management of stalk rot of cauliflower. *Indian Phytopathol.* 2005, *58*(2), 167–173.

Zhao, S.; Guo, Y.; Wang, Q.; Luo, H.; He, C.; An, B. Expression of flagellin at yeast surface increases biocontrol efficiency of yeast cells against postharvest disease of tomato caused by *Botrytis cinerea. Postharvest Biol. Technol.* 2020, *162*, 111112.

Section X

Management of Crop Diseases through Viral Biopesticides

16

Usage of Entomopathogenic Viruses for Insect Pest Control

Muhammad Sarwar, Sidra Sarwar, and Roohi Aslam

CONTENTS

16.1 Introduction

All known animal, insect, and bacterial viruses have an extracellular transport form that includes nucleic acid and a protein shell in which this nucleic acid is enclosed. A great variety of viruses attack and kill many insects. These viruses are called entomopathogenic viruses and have been found in many insect orders (Khalid et al., 2015; Sarwar, 2016a).

The nucleocapsid surrounded by a lipid bilayer envelope is called a virion (infective unit of a virus, while viroid is a single RNA molecule able to cause cell infection and replicate) (Ibarra and Rincon-Castro, 2008).

Some viruses are additionally occluded into a protein matrix that is named an occlusion body (OB). Once a viral particle gains entry into a permissive (susceptible) cell, its nucleic acid takes charge of the cell's metabolic system and profusely replicates into new virus particles, until the cell is normally depleted of all its content and dies. Viruses cannot replicate "in vitro" successfully on artificial media, where only organic material sources are provided (Sarwar, 2020a).

There are presently more than 20 known groups of insect viruses, which are classified into 19 families (primarily Ascoviridae, Baculoviridae, Nudivirus, Iridoviridae, Polydnaviridae, and Poxviridae), isolated from insects to date, and show as much variation in structural and organizational paradigms similar to viruses from most other animal classes (Fauquet et al., 2005). Some of these viruses have been used for the purposes of biocontrol of insect pests in agriculture (Sarwar, 2016b).

Most viruses that attack insects belong to a group called nuclear polyhedrosis viruses or NPVs. Their victims are usually young larvae of butterflies and moths, which become infected by eating of NPV particles and typically die within several weeks. Prevailing environmental factors heavily influence the insect-killing efficiencies of viruses. For example, they are adversely affected by sunlight, while the relatively slow speed at which they kill has also hindered their widespread acceptance for biocontrol (Sarwar, 2020b).

Many of the insect pests are susceptible to viral infections (Figure 16.1) and hence, these viruses can be used as biological control agents. Although there is a great diversity of insect viruses, only a few are frequently observed in insect populations, such as the ascovirus, iridovirus, polydna virus, baculovirus, cypovirus, entomopox virus. Among these, only a few show potential to be used as control agents, mainly the baculoviruses (Rincon-Castro and Ibarra, 2011).

It is determined that baculoviruses (nuclear polyhedrosis and granulosis viruses) have considerable potential as components of integrated forest insect pest management systems. The prime limiting factors in their large-scale use at present are the high cost of virus production and underdevelopment of application technology (Morris, 1980).

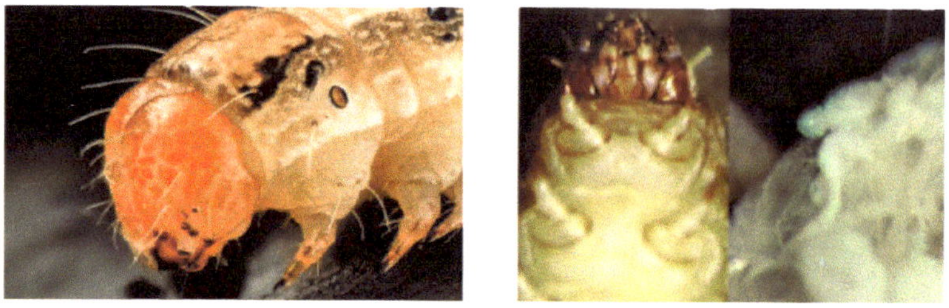

FIGURE 16.1 Viruses infected Lepidopteran caterpillars.

TABLE 16.1

The Most Diverse Group of Entomopathogenic Viruses Used on Insect Populations

Pathogen	Host Range	Use and Host Plant
Lymantria dispar nuclear plyhedrosis (NPV)	Gypsy moth caterpillars	Forest
Orgyialeuco stigma NPV	Tussock moth caterpillars	Forest
Neodiprion lecontei NPV	Pine sawfly larvae	Pine
Cydia pomonella Granulovirus (GV)	Codling moth caterpillars	Orchard
Helicoverpa zea single-enveloped NPV	Corn earworm	Maize, beet, tomato, tobacco, cotton, cabbage
Spodoptera exigua multi-enveloped NPV	Beet armyworm	Cotton, maize
Helicoverpa armigera NPV	Cotton bollworm	Cotton, pepper, tobacco
Periplaneta fuliginosa densonucleosis virus (DNV)	Smokybrown cockroach	House
Pieris brassicae GV	Cabbage butterfly	Cabbage
Plutella xylostella GV	Diamondback moth	Radish, cabbage, mustard
Dendrolimus punctatus cytoplasmic polyhedrosis virus (CPV)	Pine moths *D. punctatus*, *D. kikuchii*, *D. spectabilis*, *D. supreans*	Pine

16.2 Entomopathogenic Viruses for Usages

Most research is focused on baculoviruses that are important pathogens of some globally important pests for which control has become difficult due to either pesticide resistance or pressure to reduce pesticide residues. However, the absence of a practical in vitro mass production system, generally higher production costs, limited post-application persistence, slow rate of kill, and high host specificity currently contribute to restricted use in pest control (Sarwar et al., 2020a).

It is found that baculoviruses have considerable potential as components of integrated forest insect pest management systems. The prime limiting factors in their large-scale use at present are the high cost of virus production and underdevelopment of application technology (Sarwar et al., 2020b).

Several lepidopteron pests are important hosts of baculoviruses including nucleopolyhedro viruses (NPV) and granulo viruses (GV) (Table 16.1). Virus particles invade the nucleus of the midgut, fat body, or other tissue cells, compromising the integrity of the tissues and liquefying the cadavers. Before death, infected larvae climb higher in the plant canopy, which aids in the dissemination of virus particles from the cadavers to the lower parts of the canopy. This behavior aids in the spread of the virus to cause infection in healthy larvae (Boucias and Pendland, 1998).

16.3 Classification of Entomopathogenic Viruses

Entomopathogenic virus classification, similar to any other type of viruses, follows the same criteria to classify the diversity of viral groups that attack insects. However, the ultimate criterion is the sequencing of the genetic material which determines not only the discrimination between viral species, but also establishes the evolutionary relationship among viruses within the same group (Van Regenmortel et al., 2000).

Insect viruses are named in acronyms, according to their host and the viral group to which these belong. For example, all nucleopolyhedro viruses are named NPV, just as the granulo viruses are named GV, the entomopox viruses are EPV, the iridoviruses are IV, and the cytoplasmic polyhedrosis viruses (cypo viruses) are CPV.

16.4 Major Groups of Entomopathogenic Viruses

Insect viruses are highly diverse, and only a few groups are frequently found in insect populations and show potential to be used as biological control agents. Each of these groups possesses a distinctive strategy of replicating and expressing its genetic material in receptive host cells.

16.4.1 Baculovirus

Baculoviruses make up a family Baculoviridae (Figure 16.2) of insect viruses and are grouped into two main groups or genera; Nucleopolyhedro virus or NPVs (Figure 16.3), and Granulovirus or GVs (Figure 16.4). Both groups of the family Baculoviridae contain circular double-stranded DNA genome of approximately 80–180 kb (Kilobases—one thousand nucleic bases), which is condensed within nucleocapsids and are predicted to encode

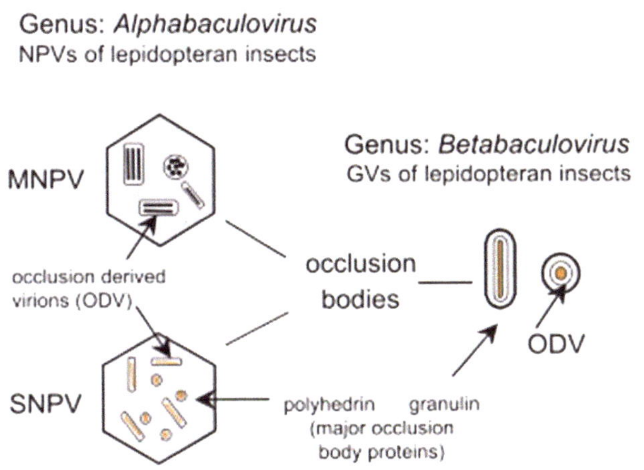

FIGURE 16.2 Baculoviruses.

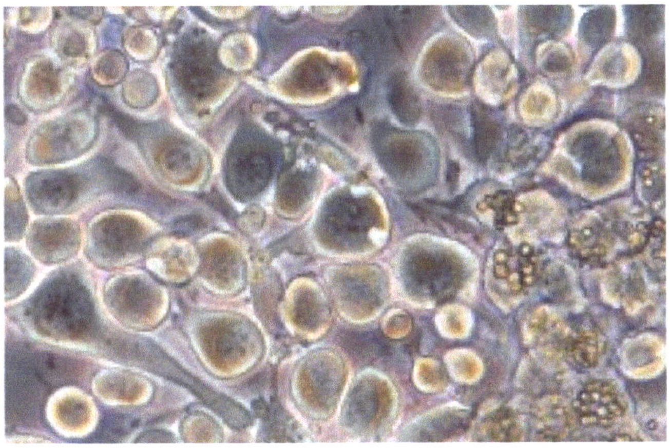

FIGURE 16.3 Nucleopolyhedrovirus.

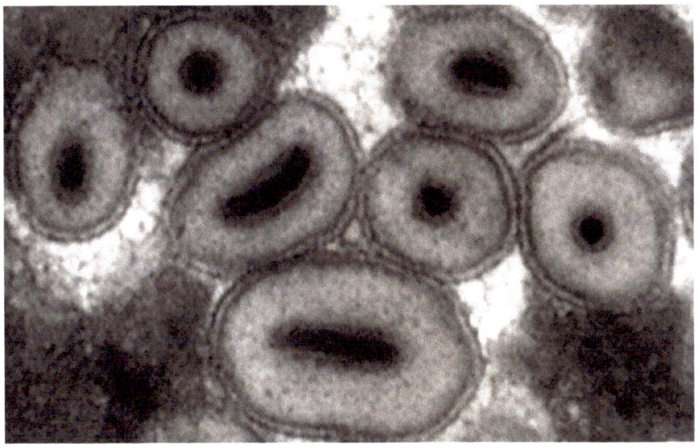

FIGURE 16.4 Granulovirus.

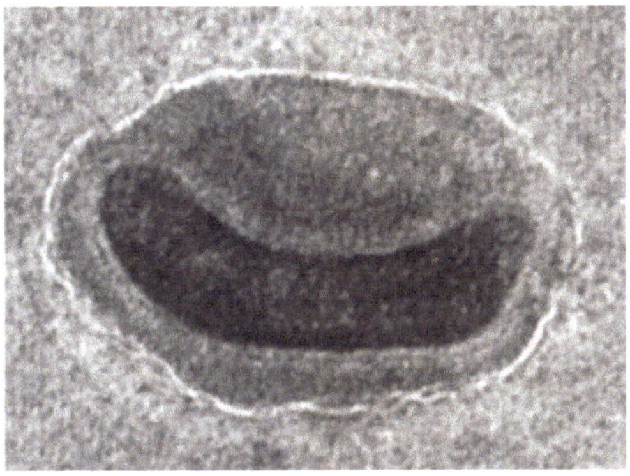

FIGURE 16.5 Entomopoxvirinae.

from about 90 to 180 genes (Okano et al., 2006; Van Oers and Flak, 2007). These are occluded either within polyhedral or granular OBs (2 μm–15 μm in size) mainly consisting of polyhedrin protein. Although there have been more than 600 isolates reported from a variety of insect species, 90% of these have lepidopteran hosts (Adams and McClintock, 1991; Vlak, 1992).

The SNPV type species (BmSNPV) infects silkworms (*Bombyx mori*), and the MNPV type species (AcMNPV) infects larvae of *Autographa californica* (Lepidoptera: Noctuidae). The NPV virions replicate only in the nuclei of susceptible cells and their OBs fluctuate between one and 15 pm in size, clearly visible under light microscopy. On the other hand, GV type species infects codling moth larvae (*Cydia pomonella*; Lepidoptera: Tortricidae). GV virions are always single within the envelope. OBs are very small (0.2–0.5 μm) as compared with NPV polyhedra, but this is because there is only one virion per OB. GV virions replicate in the cytoplasm of susceptible cells. Additionally, there is a group not recognized as belonging to the family Baculoviridae, but due to many similarities with baculoviruses and its use as a biological control agent, it is included in this chapter. This is the nudivirus or nonoccluded baculovirus (NOB) specific of the coconut rhinoceros beetle, *Oryctes rhinoceros* (Coleoptera: Scarabaeidae). The rod-shaped virions are very similar to those of baculoviruses, except that they are not occluded within an OB. NPV virions replicate only in the nuclei of susceptible cells and their OBs fluctuate between one and 15 μm in size (Tinsley and Kelly, 1985; Sarwar and Roohi, 2020).

16.4.2 Entomopoxvirus

Members of the family Poxviridae also show a wide host range, with vertebrate and invertebrate hosts. The Poxviridae family is separated into two subfamilies: the Entomopoxvirinae (Figure 16.5), which comprise poxviruses of insects; and the Chordopoxvirinae, which comprise poxviruses of vertebrates (Goodwin et al., 1991). The first subfamily or Entomopoxvirus (EPV) comprise three genera based on host insect and virion morphology. Virus genera are designated as Entomopoxvirus A, Entomopoxvirus B, and Entomopoxvirus C. The first one infects only coleopteran species and the type species infects *Melolontha melolontha*, the second one infects lepidopteran and coleopteran species and the type species infects *Amsacta moorei*, and the third one infects only dipteran species and the type species infects *Chironomus luridus*. A fourth group (D) has been proposed, which attacks hymenopterans. (Arif and Kurstak, 1991). They show allantoid- to brick-shaped virions, occluded within ovoid OBs called spheroids (Adams, 1991). Virions are some of the largest in the virus world, measuring up to 400 μm in length and 250 μm in width, and contain dsDNA ranging in size from 270–320 Kb. Virions replicate in the cytoplasm of susceptible cells. Entomopox viruses have been isolated from 27 species of orthopterans, lepidopterans, dipterans, and coleopterans (Arif, 1984).

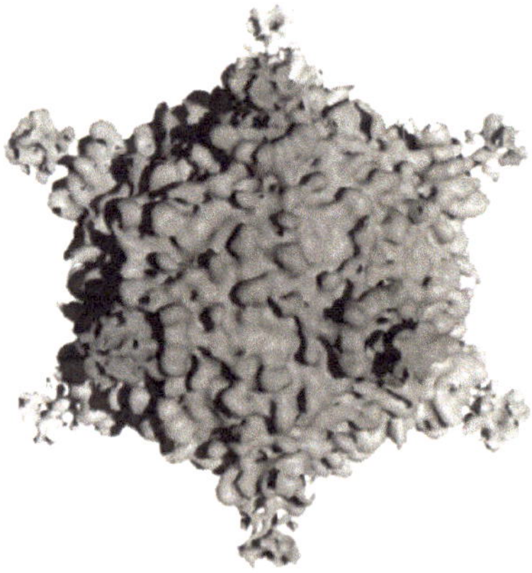

FIGURE 16.6 Cypovirus.

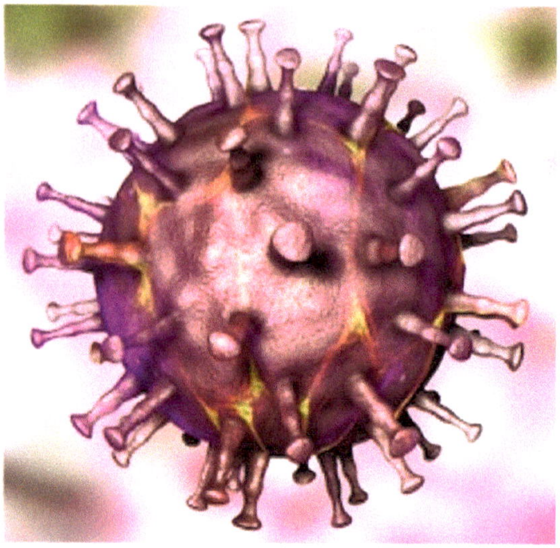

FIGURE 16.7 Iridovirus.

16.4.3 Cypovirus

Reoviridae is a family of segmented dsRNA viruses with 12 genera; some of them infect mammals (Evans and Entwistle, 1987). The viruses of this family isolated of insects are called cytoplasmic polyhedrosis viruses or cypo viruses or just CPVs (Figure 16.6). CPVs are six commonly isolated from insects and have only one genus, *Cypovirus*. These viruses have linear dsRNA genomes divided into 10–12 segments of about 12–32 kb in total. The size and number of fragments depend on the species. Icosahedral virions, not enveloped 60–80 µm in diameter, show 12 lateral projections and are occluded within large isometric crystal named polyhedra, of up to 10 µm in size (Hukuhara and Bonami, 1991). Virions replicate in the cytoplasm of the midgut epithelial cells. CPVs are mainly isolated from Lepidoptera insects, and occasionally from Diptera or Hymenoptera, and rarely from Coleoptera or Neuroptera. The *B. mori* cytoplasmic polyhedrosis virus or BmCPV is the first report of an occluded cytoplasmic insect virus (Hukuhara, 1985).

16.4.4 Iridovirus

Iridovirus (Figure 16.7) are large icosahedral viral particles of 120 to 300 µm in diameter, and consist of a central core of nucleic acid and proteins, virions that bud from the plasma membrane, and are not occluded in a protein matrix. Genomic DNA of iridovirus is a linear molecule fluctuating from 140 to 303 Kb. The replication of iridovirus includes nuclear and cytoplasmic stages, but virion assembly occurs exclusively in the cytoplasm (Goorha, 1982). Iridoviruses infecting insects belong to two genera. Genus *Iridovirus*, viral particles fluctuate between 120 and 130 µm in size and the type species is rice borer *Chilo suppressalis*; and *Chloriridovirus*, with a larger viral particle (180 µm) and the type species is isolated from mosquito larvae of *Aedes taeniorhinchus*. Iridovirus has been isolated from Diptera, Hemiptera, Lepidotera, Coleoptera, and Hymenoptera (Balange and Devauchelle, 1982).

FIGURE 16.8 Ascovirus.

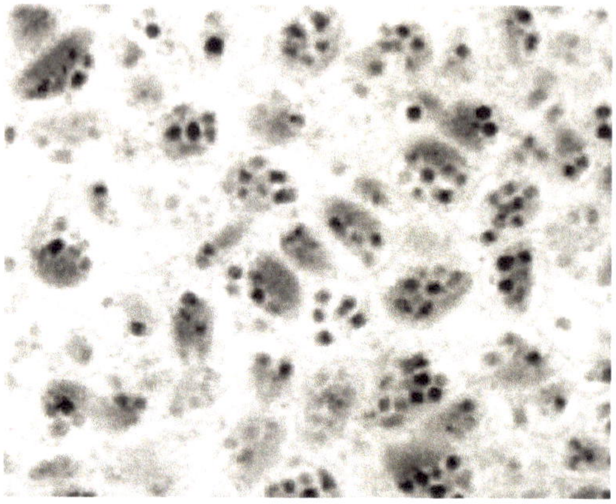

FIGURE 16.9 Polydnavirus.

16.4.5 Ascovirus

Ascoviridae members with double-strand DNA viruses primarily infect invertebrates, mainly noctuids and spodopteran species. It contains two genera: *Ascovirus* (Figure 16.8), which contains three species, and *Toursvirus*, with a single species. The type species of *Ascovirus* is *Spodoptera frugiperda ascovirus 1a*, which infects the army worm (*Spodoptera frugiperda*). Their genomes contain circular dsDNA of 100–180 Kb. So far, eight species have been identified from larvae of the same number of noctuid (Lepidoptera: Noctuidae) species (Xue and Cheng, 2011; Federici et al., 2009).

16.4.6 Polydnavirus

Polydnavirus (PDV) is a member of the family Polydnaviridae of insect viruses (Figure 16.9). Members of this family show

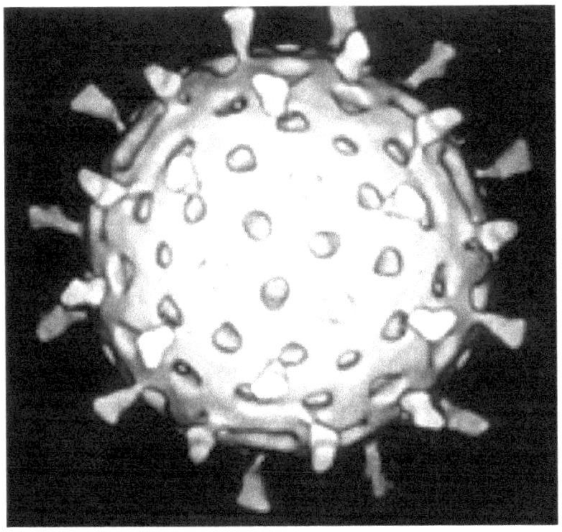

FIGURE 16.10 Mesoniviridae virus.

nonoccluded, ovoid virions, containing multipartite dsDNA, totaling 75-200 Kb when all the fragments are added. There are currently 53 species in this family, divided among two genera recognized within the family. *Ichneovirus* type species infect the ichneumonid wasp *Campoletis sonorensis*, and *Bracovirus* type species infect the braconid wasp *Cotesia melanocella*. Polydnaviruses form a symbiotic relationship with parasitoid wasps (ichnoviruses occur in ichneumonid wasps species and bracoviruses in braconid wasps), but these wasps are themselves parasitic on Lepidopteron moths and butterflies (Burke and Strand, 2012; Strand and Burke, 2015)

16.4.7 Mesoniviridae

Mesoniviridae is a family of positive-sense, single-stranded RNA insect viruses within the broad order Nidovirales. Prior to the proposal of the Mesoniviridae (Figure 16.10) family, the order Nidovirales is subdivided into three families: Arteriviridae, Coronaviridae, and Roniviridae, with the former family having 12.7–15.7kb genomes and the latter two having genomes ranging from 26.3 to 31.7kb, and being referred to as the "large nidovirales". Within the order Nidovirales, Cavally virus (CavV) has been isolated from pools of *Aedes*, *Culex*, *Anopheles*, and *Uranotaenia* mosquitoes, and Nam Dinh virus (NDiV), discovered only in *Culex* species. Within the family Mesoniviridae, a single viral genus hosts all known isolates: *Alphamesonivirus*, which is further subdivided into seven recognized viral species called *Alphamesoniviruses-1–7* (Vasilakis et al., 2014).

16.5 Virus Replication in Host Insect

Normally, the most usual mode of infection due to insect viruses is oral, by ingesting food previously contaminated with the inoculum (a certain amount of a pathogen able to cause infection). Other alternative routes of infection occur when adults transmit the infection to their progeny (vertical transmission), by inoculating the eggs either superficially or internally (transovarial transmission). Also, if an adult parasitoid oviposits eggs into an infected individual, inocula can be transmitted to healthy individuals by the parasitoid, during subsequent ovipositions (horizontal transmission). In infection by baculoviruses, once the inoculum (polyhedra or granules) are ingested, they reach the midgut (mesenteron) lumen, where pH is normally high (9.5–11.5). Due to these conditions and perhaps the joint action of proteolytic enzymes, OBs are degraded and the enveloped virions are released to the lumen. Virions envelopes have high affinity to the membrane of the gut epithelial cells (especially those of the microvilli) and attach to them, releasing the now naked virions (nucleocapsids) inside the cell nucleocapsids then move toward the cell nucleus where they replicate (Kurstak, 1991).

When insects inadvertently consume OB-contaminated food, the infection cycle of a virus starts. Viruses gain entry into a permissive host cell, using a variety of infective mechanisms; endocytosis, phagocytosis, pinocytosis, membrane fusion, etc. Subsequently, the viral genome is dissociated from the capsid and released into the host cell in a great variety of mechanisms. Once in the cell (either in the nucleus or in the cytoplasm), the viral genome replicates itself, frequently using the host cell's enzymatic machinery, creating a great number of identical copies. Also, structural, functional, and auxiliary viral genes are transcribed and translated into proteins, which will be used to assemble, part by part, new viral particles. The new copies of the viral genome are packed within the new capsids and the new viral progeny is released by a variety of procedures, which are distinctive of each viral group. Some viruses are released by inverted pinocytosis, which engulfs the viral particle within a membrane vesicle, creating what is known as "enveloped viruses" (Kamita et al., 2010).

16.6 External Symptoms Development in Insect by Virus Infection

The first detectable signs appear when the insect becomes slow in its movements, stops feeding, and growth practically ceases. In baculoviruses, infected insects become whitish in color because of the massive infection of the fat body, visible through a more translucent integument (exoskeleton), which turns thinner as the infection advances until it ruptures. A grayish to creamy liquid is released, where billions of OBs are suspended, after the larva crawled up and hang head down by its crochets in an inverted "V" shape position, which facilitates the spread of inocula in the field (Granados and Williams, 1986).

The insects infected with Entomopox virus show similar symptoms (Figure 16.11) to baculovirus infections, except that the whitish coloration of the fat body is located at the posterior part of the abdomen, which becomes swollen due to the accumulation of spheroids. Some coleopterans show a swollen rectal sac, which becomes irregular in form and whitish in color. The infection develops slowly in any case and, although common in nature, rarely develops into epizootic levels in insect populations. On the other hand, the Cipo virius only infects the midgut epithelial tissue, larvae develop diarrhea and vomit, stop feeding, and growth practically ceases. Usually, symptoms are apparent at the end of the larval stages and death overcomes just before pupation (Tanada and Kaya, 1993).

FIGURE 16.11 Signs and symptoms of viruses infected larvae.

The IV infections can be detected earlier as iridescent spots in legs, prolegs, and thorax. No behavioral change occurs until larvae are close to death and stop feeding. Polydnavirus infections, due to their mutualistic relationship with the wasp hosts, show neither lethality nor any detrimental effect on the wasps. Consequently, no signs or symptoms are apparent. The ascovirus infections show no symptoms during the first stages of the disease, until larvae stop growing and hemolymph becomes opaque and whitish, due to presence of free vesicles. Still, larvae can survive for two to six weeks, but it eventually dies (Miller and Ball, 1998).

16.7 Field Application of Insect Viruses

Entomopathogenic viruses have been applied as targeted biocontrol agents against forestry and agriculture pests.

16.7.1 Virus Production and Formulation

The production of baculoviruses as bioinsecticides has been carried out exclusively (Figure 16.12) for larva, by massively infecting susceptible individuals. Their strategy of production requires a highly reliable technique to maintain a big insect colony, under optimal rearing conditions, in such a way that it can provide massive numbers of susceptible individuals in a continuous and synchronized manner. Rearing mechanization is highly recommended, both to standardize the production and to reduce costs. This can be achieved by using artificial or semi-artificial diets, which not only expedites the production process, but also allows keeping a more thorough control of undesirable infections agents. Although the infection of larvae may be achieved by injection of EVs, the most practical method is based on the contamination of the insect diet with the OB's (Tanada and Kaya, 1993).

The optimum inoculum dose and the most adequate larval stage to be infected are determined experimentally, in order to ensure the maximum production of OBs per larva, using the minimum inoculum dose. One example is the use of one to five million OBs per container with 10 3rd instar larvae. This larval stage allows the development of the highest larval biomass the insect can accumulate at the moment of maximum infection of tissues. Infected insects are collected and immediately processed, or stored under freezing conditions. The most frequent processing of larvae consists of the disintegration of the infected tissues, usually by blending the dead larvae, followed by sieving through a wide mesh, which will eliminate large pieces of unblended tissues (normally, uninfected tissues) and large pieces of integument. The obtained OB concentrate should be then quantified and formulated, or else stored under freezing conditions. Because baculoviruses are highly virulent and infective, there is a high possibility that the whole insect colony become infected with it. This is why these two phases of production must be physically isolated and that the personnel involved with one phase, must be different from the other, these are among other safety measures. In addition, usually the insect colony used to produce the virus is the same species as the one to be controlled in the field; however, in some situations, the viral strain shows a somewhat extended host range. Therefore, sometimes the alterative host may be used in the production of the virus, if its rearing conditions are more practical and cheaper (Boming et al., 2006).

The adding of inert material as talc, diatomaceous, or earth to the OB concentrate is a very important step in the formulation process of virus products. Some additives, such as dispersants, adherents, and even UV-protectants and feeding stimulants, can be included in the formulation. All these products improve the application and persistence of the product in the field. Formulation by micro-encapsulation is also feasible. At the end of formulation, a product can show different presentations, such as wettable powders, dispersed granules, suspensions, etc. Some minerals are now added to powder formulations, such as bentonite, attapulgite, silica sand, among others. Likewise, some inert materials of organic origin have been tested, such as wheat bran, milled nutshells, and crushed corn cobs (Williams and Cisneros, 2001).

There is a wide variety of surfactants in the market, such as Adsee, Agral NN, Chevron, Rhoplex B60A, Tween 80, etc. A more persistent application in the field may render better control of the pest; therefore, some adhesive chemicals can be added to the formulations, such as Agral NN, polyvinyl alcohol, Hyvis 150, slim milk, etc. Because the efficiency of the product is based on the ingestion of the inoculum, baits, and phagostimulants such as sugar, cottonseed oil, glycerin, Gustol, etc., can be added, too. Also, because a major threat to the applied viruses in the field is sunlight, and more specifically, UV rays, some UV-protectants may be added, such as Tinopal LPW, Phorwite AR, Leucophor BS, activated carbon, folic acid, etc. (Williams and Cisneros, 2001).

Baculovirus-based bioinsecticides are formulated as wettable powders, which can be easily sprayed in the field, using the same equipment used for the application of chemical insecticides. However, some formulations have been developed recently, using micro-encapsulation. These are based on a mixture of

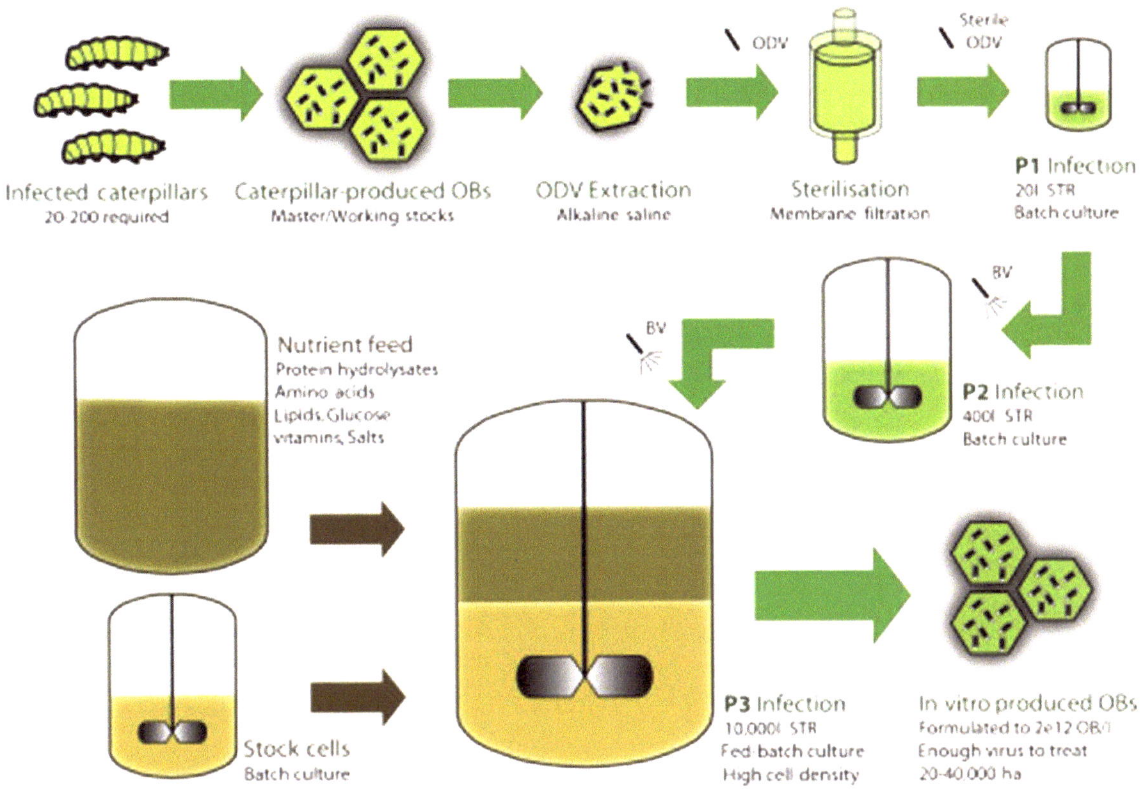

FIGURE 16.12 Production of entomopathogenic viruses.

the baculoviruses with some polymers such as gelatin, pectin, chitin, and calcium alginate, although starch is the most common and practical to be used. The viral product should show a shelf-life of at least 18 months. On the other hand, quality control measures should be also including an estimate of contaminant microorganisms in the product, especially if the microbial content includes some potential vertebrate pathogens. Quality tests may also include resistance to abrasive materials, susceptibility to UV rays from sunlight, possible synergistic or antagonistic factors, rain effect, etc. More detailed tests may comprise DNA tests to prove the genome integrity, serological tests of the capsid proteins, and even safety tests to vertebrates; although most of these tests are rather required during the registration process of the product (Burges and Jones, 1998).

The CL50 values indicate the virulence level of each lot of production and adjustments must be done during formulation, in order to standardize every lot to a certain level of virulence. Potency of the product is usually measured by the concentration of OBs per weight unit of the product (3 × 109 OB gr^{-1}). The in vitro production (in insect cells) of viral insecticides constitutes a very attractive alternative as it would substitute the laborious and expensive maintenance of the massive insect colony for a highly homogeneous, accurately controlled cell culture kept in large fermenters or bioreactors. This technique is based on the inoculation of the sterile culture medium kept in the fermenter, with a cell line highly susceptible to the virus to be produced, and, most importantly, adapted to the in vitro massive production. The fermenter must be kept under adequate conditions (temperature, agitation, aeration, pH, etc.), to ensure the optimum proliferation of the cell line, up to a density that is appropriate to infect. At this point, inoculation should be carried out with EVs extracted from previously inoculated cell cultures, at a density previously established to achieve an effective and rapid infection of the cells, as OBs must be produced as fast as possible. Once cells burst and release the OBs to the medium, these are collected and concentrated, usually by centrifugation, and the concentrate is formulated as described above (Hunter-Fujita, 1998; Reid et al., 2014).

One of the advantages of the in vitro production is the elimination of insect parts, tissue residues, other microorganisms, and any other contaminant, usually found in the larva production. Unfortunately, the in vitro production still faces important limitations that hamper the use of this technique at an industrial level, to date. The cost of rearing live hosts adds greatly to the final cost of the product. It is to be hoped that insect cell culture systems currently being developed for other uses may ultimately make viral pesticides more cost-effective.

16.7.2 Strategies of Virus Application in Field

Baculoviruses play an important ecological role in regulating the size of insect populations. Baculovirus-based biocontrol applications have been restricted to lepidopteran and hymenopteran pests. One of the advantages of using viral bioinsecticides in the field for the control of insect pest is that viruses are able to be formulated in such a way that the equipment normally used to spray chemical insecticides, can also be used in the application of these bioinsecticides. The classical biological control has been successfully used with baculoviruses. Introduction of the virus into an insect population is never subjected to the pressure of this factor before, and once inoculated and established, the virus would persist into the insect population without any further application (Hostetter et al., 1985; Sarwar, 2015a).

Still, viruses can be used in an augmentation strategy, where viruses in the insect populations are found in low levels, either because they constitute a minor natural factor of mortality in the insect population or because they are at the beginning of the epizootic cycle. Other cases of the inoculative strategy make use of a variety of methods to spread the pathogen. For example, the release of predators and parasites inoculated with the virus or the release of inoculated birds whose feces may spread out the virus. In any case, the application of the virus on the insect population increases the natural level of inoculum to trigger an epizootic that ideally, would cause the reduction of the pest population. The augmentation technique will be always more successful when the insect population is at low densities, and therefore used as a preventative measure (Podgwaite, 1985; Sarwar, 2015b).

16.8 Conclusion and Future Prospects

Baculoviruses are perhaps unique among viruses in the breadth of their biotechnological applications. The baculovirus is frequently studied and the granulosis virus, or granulo virus, appears to be the most widespread. These insect-specific viruses are used not only for insect pest management purposes, but also as laboratory research tools for production of recombinant proteins and for protein display, and as potential vectors for human gene therapy. The basic virus particle comprises a nucleic acid (DNA or RNA), which directs the virus replication and is encapsulated within a protein case known as "capsid," which plays an important role in the host cell infection process. Once a viral particle gains entry into a permissive (susceptible) cell, its nucleic acid takes charge of the cell's metabolic system and profusely replicates into new virus particles, until the cell is normally depleted of all its content and dies. Probably, the greatest potential of viruses will be realized when used as an indirect or preventative strategy with an opinion to prevent the occurrence of pest emergency situations.

REFERENCES

Adams, J. R. Introduction and classification of viruses of invertebrates. In: Adams, J. R. and Bonami, J. R. (Eds.), *Atlas of Invertebrate Viruses*. CRC Press, Boca Ratón, FL. 1991; pp. 1–8.

Adams, J. R.; McClintock, J. T. Baculoviridae. Nuclear polyhedrosis viruses. In: Adams, J. R. and Bonami, J. R. (Eds.), *Atlas of Invertebrate Viruses*. CRC Press, Boca Ratón, FL. 1991; pp. 87–204.

Arif, B. M. The entomopox viruses. *Adv. Virus Res.* 1984, 29: 195–201.

Arif, B. M.; Kurstak, E. Entomopox viruses. In: Kurstak, E. (Ed.), *Viruses of Invertebrates*. Marcel Dekker, Inc, New York. 1991; pp. 179–195.

Balange-Orange, N.; Devauchelle, G. Lipid composition of an Iridescent virus type 6 (CIV). *Archives of Virology*. 1982, 73: 363–367.

Boming, B. C.; Maramorosch, K.; Shatkin, A. J. Insect viruses: Biotechnological applications. *Advances in Virus Research*. 2006, 68: 552 pp.

Boucias, D. G.; Pendland, J. C. Major groups of insect viruses. In: *Principles of Insect Pathology*. Springer, Boston, MA. 1998; pp. 181–183.

Burges, H. D.; Jones, K. A. Formulation of bacteria, viruses and protozoa to control insects. In: Burges H. D. (Ed.), *Formulation of Microbial Biopesticides*. Springer, Dordrecht. 1998; pp. 33–127.

Burke, G. R.; Strand, M. R. Polydnaviruses of parasitic wasps: Domestication of viruses to act as gene delivery vectors. *Insects*. 2012, 3(1): 91–119.

Evans, H. F., Entwistle, P. Viral diseases. In: Fuxa, J. R. and Tanada, Y. (Eds.), Epizootiology of insect diseases. Wiley, New York. 1987; pp. 257–322.

Fauquet, C. M.; Mayo, M. A.; Maniloff, J.; Desselberger, U.; Ball, L. A. Virus taxonomy. In: 8th Report of the International Committee on Taxonomy of Viruses. Sydney, 1999 and Paris, 2002 Elsevier Academic Press, San Diego, CA, USA. 2005; 1259 pp.

Federici, B. A.; Bideshi, D. K.; Tan, Y.; Spears, T.; Bigot, Y. Ascoviruses: Superb manipulators of apoptosis for viral replication and transmission. In: *Lesser Known Large dsDNA Viruses. Current Topics in Microbiology and Immunology*. Springer, Berlin. 2009; Vol. 328, pp. 171–196.

Goodwin, R. H.; Milner, R. J.; Beaton, C.D. Entomopoxvirinae. In: Adams, J. R. and Bonami, J. R. (Eds.), *Atlas of invertebrate viruses*. CRC Press, Boca Ratón, FL. 1991; pp. 259–285.

Goorha, R. Frog virus 3 DNA replication occurs in two stages. *Journal of Virology*. 1982, 43: 519–528.

Granados, R. R.; Williams, K. A. *In vivo* infection and replication of baculoviruses. In: Granados, R. R. and Federici, B. A. (Eds.), *The Biology of Baculoviruses*. CRC Press, Boca Ratón, FL. 1986; Vol. 2, pp. 89–108.

Hostetter, D. L; Bell, M. R. Natural dispersal of baculovirus in the environment. In: Maramorosch, K. and Sherman, K. E. (Eds.), *Viral Insecticides for Biological Control*. Academic Press, Orlando, FL. 1985; pp. 249–284.

Hukuhara, T. Pathology associated with cytoplasmic polyhedrosis viruses. In: Maramorosh, K. and Sherman, K. E. (Eds.), *Viral Insecticides for Biological Control*. Academic Press, Orlando, FL. 1985; pp. 121–162.

Hukuhara, T.; Bonami, J. R. Reoviridae. In: Adams, J. R. and Bonami, J. R. (Eds.), *Atlas of Invertebrate Viruses*. CRC Press, Inc., Boca Ratón, FL. 1991; pp. 393–434.

Hunter-Fujita, F. R.; Entwistle, P. F.; Evans, H. F.; Crook, N. E. *Insect Viruses and Pest Management*. Wiley, West Sussex, UK.1998; 632 pp.

Ibarra, J. E.; Rincon-Castro, M. C. D. Insect viruses diversity, biology, and use as bioinsecticides. In: Claro, K. D. et al. (Eds.), *Tropical Biology and Conservation Management*. Encyclopedia of Life Support Systems, Oxford. 2008; Vol. VII, p. 22.

Kamita, S. G.; Davis, C. A.; Kang, K. D.; Inceoglu, A. B.; Hammock, B. D. Genetically modified baculoviruses for pest insect control. In: Insect Control: Biological and Synthetic Agents, Gilbert, L. I.; Gill, S. S. (Eds.). Academic Press, London. 2010, 383-386.

Khalid, P. A.; Hussain, M.; Hassan, M.; Sarwar, M. Sarwar, N. Evaluation of Bt-cotton genotypes for resistance to cotton leaf curl disease under high inoculum pressure in the field and using graft inoculation in glasshouse. *Plant Pathology Journal*. 2015, 31(2): 132–139.

Kurstak, E. *Viruses of Invertebrates*. Marcel Dekker, Inc., New York. 1991; 360 pp.

Miller L. K.; Ball, L. A. *The Insect Viruses (The Viruses)*. Springer, New York. 1998; 434 pp.

Morris, O. N. Entomopathogenic viruses: Strategies for use in forest insect pest management. *Canadian Entomologist*. 1980, 112(6): 573–584.

Morales-Ramos, J. A.; Guadalupe-Rojas, M.; Shapiro-Ilan, D. I. (Eds.). Elseiver, London.

Okano, K.; Vanarsdall, A. L.; Mikhailov, V. S.; Rohrmann, G. F. Conserved molecular systems of the Baculoviridae. *Virology*. 2006, 344: 77–87.

Podgwaite, J. D. Strategies for field use of baculoviruses. In: Maramorosch, K. and Sherman, K. E. (Eds.), *Viral Insecticides for Biological Control*. Academic Press, Orlando, FL. 1985; pp. 775–797.

Reid, S.; Chan, L.; Van Oers, M. M. Production of entomopathogenic viruses. In: *Invertebrates and Entomopathogens*. 2014; pp. 437–482.

Rincon-Castro, M. C. D.; Ibarra, J. E. Entomopathogenic viruses. In: Rosas, N. (Ed.), *Biological Control of Insect Pests*, 1st ed. Studium Press LLC, Houston, TX, USA. 2011; pp. 29–64.

Sarwar, M. Biopesticides: An effective and environmental friendly insect-pests inhibitor line of action. *International Journal of Engineering and Advanced Research Technology*. 2015b, 1(2): 10–15.

Sarwar, M. Insects as transport devices of plant viruses. In: Awasthi, L. P. (Ed.), *Applied Plant Virology: Advances, Detection, and Antiviral Strategies*. Elsevier Inc., London. 2020a; pp. 381–402.

Sarwar, M. Microbial insecticides: An ecofriendly effective line of attack for insect pests management. *International Journal of Engineering and Advanced Research Technology*. 2015a, 1(2): 4–9.

Sarwar, M. Mite (Acari, Acarina) vectors involved in transmission of plant viruses. In: Awasthi, L. P. (Ed.), *Applied Plant Virology: Advances, Detection, and Antiviral Strategies*. Elsevier Inc., London. 2020b; pp. 257–273.

Sarwar, M. Mosquito-borne viral infections and diseases among persons and interfering with the vector activities. *International Journal of Vaccines and Vaccination*. 2016b, 3(2): 00063.

Sarwar, M. Prevalence of multiple viral diseases associated with honey bees colony collapse and control of disorders. *International Journal of Zoology Studies*. 2016a, 1(2): 29–34.

Sarwar, M.; Roohi, A. New advances in insect vector biology and virus epidemiology. In: Awasthi, L. P. (Ed.), *Applied Plant Virology: Advances, Detection, and Antiviral Strategies*. Elsevier Inc., London. 2020; pp. 301–311.

Sarwar, M.; Mariya, A.; Sidra, S.; Romana, I. Different nematodes and plasmodiophorids as vectors of plant viruses. In: Awasthi, L. P. (Ed.), *Applied Plant Virology: Advances, Detection, and Antiviral Strategies*. Elsevier Inc., London. 2020a; pp. 275–290.

Sarwar, M.; Shad, N. A.; Batool. R. Integrated management of vectored viral diseases of plants. In: Awasthi, L. P. (Ed.), *Applied Plant Virology: Advances, Detection, and Antiviral Strategies*. Elsevier Inc., London. 2020b; pp. 707–724.

Strand, M. R.; Burke, G. R. Polydnaviruses: From discovery to current insights. *Virology*. 2015, 479–480: 393–402.

Tanada, Y.; Kaya, H. K. *Insect Pathology*. Academic Press, San Diego, CA. 1993; 666 p.

Tinsley, T. W.; Kelly, D. C. Taxonomy and nomenclature of insect pathogenic viruses. In: Maramorosh, K. and Sherman, K. E. (Eds.), *Viral Insecticides for Biological Control*. Academic Press, Orlando, FL. 1985; pp. 3–25.

Van Oers, M. M.; Vlak, J. M. Baculovirus genomics. *Current Drug Targets*. 2007, 8: 1051–1068.

Van Regenmortel, M. H. V.; Fauquet, C. M.; Bishop, D. H. L.; Cartens; E. B.; Estes; M. K.; Lemon, S. M.; Maniloff, J.; Mayo, M. A.; McGeoch, D. J.; Pringle, X. R.; Wickner, R. B. Virus taxonomy. In: 7th Report of the International Committee of Taxonomy of Viruses 1996, Jerusalem. Academic Press, San Diego. 2000; 1162 p.

Vasilakis, N.; Guzman, H.; Firth, C.; Forrester, N. L.; Widen, S. G.; Wood, T. G.; Rossi, S. L.; Ghedin, E.; Popov, V.; Blasdell, KR.; Walker, P. J.; Tesh, R. B. Mesoniviruses are mosquito-specific viruses with extensive geographic distribution and host range. *Virology Journal*. 2014, 11: 97.

Vlak, J. M. The biology of baculovirus in vivo and in cultured insect cells. In: Vlak, J. M., Schlaeger, E. J. and Bernard, A. R. (Eds.), *Baculovirus and Recombinant Protein Production Processes*. Editiones Roche, Interlaken, Switzerland. 1992; pp. 2–10.

Williams, T.; Cisneros, J. Formulacion y aplicacion de los baculovirusbioinsecticidas. In: Caballero, P., Lopez-Ferber, M. and Williams, T. (Eds.), *Los baculovirus y susaplicacionescomobioinsecticidas en el control biológico de plagas*. PHYTOMA-Universidad Publica de Navarra, Espana. 2001; pp. 313–372.

Xue, J. L.; Cheng, X. W. Comparative analysis of a highly variable region within the genomes of *Spodoptera frugiperda* ascovirus 1d (SfAV-1d) and SfAV-1a. *Journal of General Virology*. 2011, 92(12): 2797–2802.

17

Viral Biopesticides: An Ecofriendly Non-Chemical Approach for the Management of Insect Pests in Crops

Meenakshi Devi, Neha Upadhyay, Suman Devi, and Bab Li

CONTENTS

17.1 Introduction

A virus is a small infectious agent that can replicate only in the living cells of organisms. Viruses infect all types of organisms, from animals and plants to bacteria and insects. The study of viruses is known as virology, and the study of viruses causing diseases in insects is known as insect pathology. Inclusion viruses are submicroscopic, obligate, intracellular, and pathogenic organisms. Seven families of viruses, namely, Baculoviridae, Reoviridae, Iridoviridae, Poxviridae, Parvoviridae, Picornaviridae, and Rhabdoviridae, cause diseases in insects. But the viruses of families Baculoviridae and Reoviridae are the most important for their role as biopesticides because of their high virulence. The Baculoviridae family contained NPVs and GVs whereas, the NPVs are mass-produced in larval hosts grown on an artificial diet or host plant. Usually third to fourth instar larvae of *Helicoverpa armigera* are infected with the viral food. The perfect phase of viral disease occurs over a period of 5–10 days. Once the complete infection of the virus in the larvae is completed, the larvae start 'putrefying' releasing billions of polyhedra. In commercial production, larvae are being harvested before purification to keep bacteria at a low level in the final product. After the larval production phase is complete, the larvae are collected and formulated. NPVs are being used against *H. armigera* and *Spodoptera litura* on cotton, corn, sorghum, tomatoes, and chrysanthemum. It is also being used against *Anticarsia gemmatalis* of soybean. One of the most important successes in commercial production and use of a GV is *Cydia pomonella* GV (CpGV) on apples and pears.

17.2 Classification

Baculoviruses are different from vertebrate viruses and therefore safe to humans and other vertebrates. The classification of baculoviruses is represented in a flow chart.

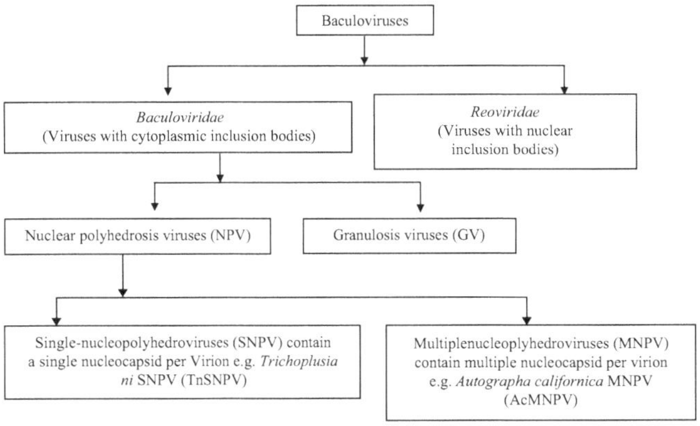

17.2.1 Mechanism of Action of Baculoviruses

Baculoviruses have double-stranded genome with rod-shaped nucleocapsids. The virus acts as a stomach poison. The infectious virus particles or virions are occluded in protein bodies called polyhedra (NPV) or granules (GV). The NPV particles are called as polyhedral inclusion bodies (PIBs). NPV polyhedra are larger than the virions (usually 1–15 μm) and may contain many virions.

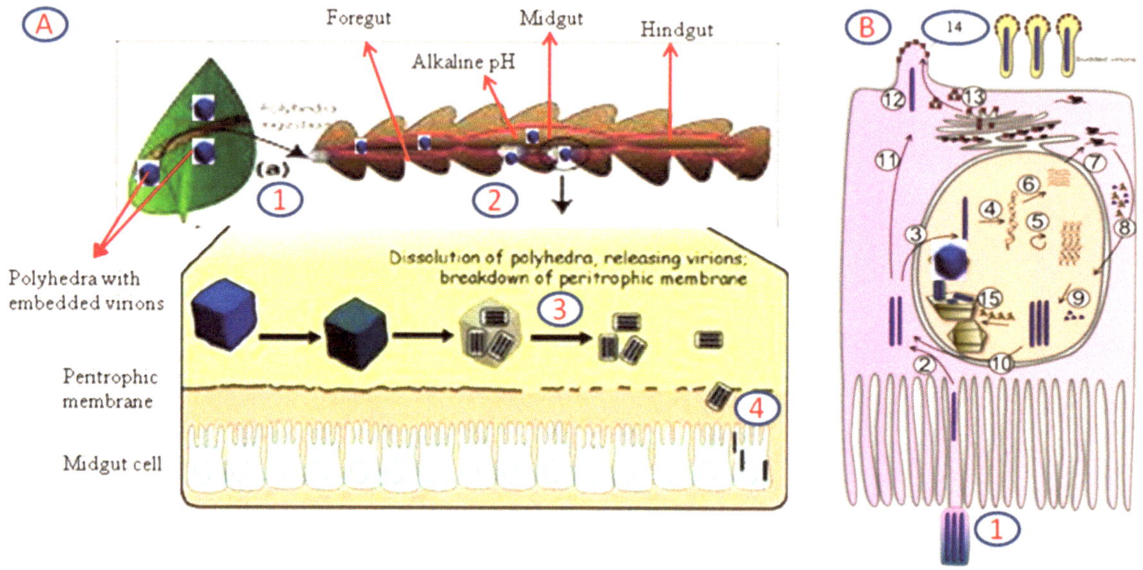

FIGURE 17.1 Cross-sectional diagram of an insect larva. (**A**) A baculovirus occlusion body (OB) ingested with infected food starts a new cycle of infections (1). As OBs pass the foregut and enter the alkaline midgut of insect the protein matrix is dissolved (2), releasing ODV (3). Virus and host encoded enzymes present in the OB degrade the peritrophic membrane, allowing the ODV to enter the cell (4). (**B**) Representation of a process of viral replication. ODV enters the cell by fusion with microvilli epithelial cell (1), nucleocapsids (NC) released into cytoplasm (2). NC can reach the nucleus (3), disassemble the genome and release it (4). Then early genes are transcribed (6) and then translated (7). Some of the proteins translocate to the nucleus (8), participate in transcription/replication of genomes, NC and virion assembly (9). NC is transferred to cytoplasm in the early stages of viral infection (10), reaches the basolateral cell membrane (CM) (11) emerges as budded virus (BV) (12) in the spots where the viral envelope fusion protein (EFP) (14) accumulates using the secretory pathway (13). NC is enveloped in the nucleus at the very late stages of infection and occluded in the polyhedral shaped protein matrix (OB) (15) (Haase et al. 2015).

The infection occurs after a susceptible host eats the plant foliage infected with PIBs then the virus enters the midgut of the insect larvae. The proteinaceous polyhedral comes in contact with the midgut alkaline pH and rapidly dissolves inside, thereby releasing the infectious virions. After the liberation of virus particles, the nucleocapsid envelope fuses with microvillar membrane of the gut wall cells. The nucleocapsids are released to enter the nucleus where viral DNA replicates and produce secondary infections, which invade the fat body and the haemolymph. The massive destruction of body tissue eventually kills the insect. The infection caused by baculoviruses is called "wilting disease" because the larva becomes wilted and tissues of the host liquefy; infection of the epidermis causes the host to appear to melt, releasing virus particles into the environment. Baculoviruses are considered to be the most beneficial of the insect viruses to man, because of their utility in insect control, their specificity to the arthropods, and their most recent use in fundamental biological studies using molecular techniques (Miller 1997) (Figure 17.1).

17.3 History of Viral Biopesticides

The earliest records of baculoviruses can be found in literature from the 16th century in reports of wilting disease infecting silkworm larvae. Paralysis and subsequent liquification occurring in the larvae affected with baculoviruses were found in many ancient literatures. In the 1930s and 1940s, rod-shaped virions were identified within the crystalline polyhedrin. During the same period, baculoviruses were observed to be an effective biological control agent of an insect pest. It was discovered that the spruce sawfly (accidentally introduced into North America) could be effectively controlled by the subsequent introduction of a baculovirus.

The first baculovirus to be registered as a pesticide (in 1975) was a commercial failure. However, the use of a baculovirus as a pest control agent that was a nucleopolyhedrovirus used to control the tussock moth in 1984 was a notable success. Baculoviruses are now being used for making them recombinant by utilizing genetic engineering for insect control.

The first viral insecticide Elcar™ was introduced by Sandoz Inc. in 1975 (Ignoffo and Couch 1981). Elcar™ was a preparation of *Heliothis zea* NPV, which is relatively a broad range baculovirus and infects many species belonging to genera *Helicoverpa* and *Heliothis*. HzSNPV provided control of not only cotton bollworm but also of pests belonging to genera attacking soybean, sorghum, maize, tomato, and beans. Broad spectrum biopesticide based on HaNPV is also used in India (Srinivasa et al. 2008). About 20,000 ha of maize annually were controlled with *Spodoptera frugiperda* NPV in Brazil (Moscardi 1999). The NPV for *Spodoptera litura* has been tested on cabbage crops in India (Kumari and Singh 2009).

Granulovirus CpGV is the active component of a number of biopesticides used for the protection of apple and pear orchards against the codling moth (*Cydia pomonella*). Some of the trademarks of CpGV based products are Granusal™ in Germany, Carpovirusine™ in France, Madex™ and Granupom™ in Switzerland, and Virin CyAP in Russia (Vincent et al. 2007). The GVs of *H. magnanima* and *A. honmai* were registered in 2003; however, the use of GVs has recently declined. One reason for the reduction in use of GVs in Japanese tea fields is the changing pattern of occurrence of other pests. Other important viruses that are currently employed to control insects include the tea tortricids, *Adoxophyes honmai* and *Homona magnanima* granuloviruses (GV) in Japan. The area sprayed with GVs comprised 5,850ha in Kagoshima in 1995, equivalent to 80% of all the tea fields in the prefecture (Nishi and Nonaka, 1996). The GVs of *H. magnanima* and *A. honmai* were registered in 2003; however, the use of GVs has recently declined.

In China, 12 baculoviruses have been authorized as commercial insecticides (Sun and Peng 2007), including *H. armigera* NPV (the most widely used virus in China for cotton, pepper, and tobacco protection), *S. litura* NPV (vegetables), *S. exigua* NPV (vegetables), *Buzura suppressaria* NPV (tea), *Pieris rapae* GV, and *Plutella xylostella* GV (vegetables). Sun and Peng (2007) also founded a cypovirus (CPV) produced in China for the control of *Dendrolimus punctatus*, an insect pest of pine forests.

17.4 Viral Biopesticides: Indian Scenario

Biopesticides fall under the Insecticide Act (1968) under which any microbial organism manufactured or sold for pest and disease control should be registered with the Central Insecticides Board (CIB) of the Ministry of Agriculture. The national agricultural research system, comprising of the many ICAR institutes as well as state agricultural universities, plays a leading role in promoting biopesticides. The Project Directorate of Biological Control is involved in testing the quality of biopesticides and training the officers of the state department of agriculture in quality control protocols. The National Centre for IPM routinely incorporates the use of pesticides in its IPM validation programs and demonstrations, as do the IPM centers of the Directorate of Plant Protection, Quarantine and Storage. Commodity research boards have also played a role in researching and developing biopesticides for pest control in key crops such as cotton, coffee, tea, and cardamom. In India, extensive research has been conducted on the use of NPVs for tackling two major pests, namely *Spodoptera litura* and *Helicoverpa armigera*. Nuclear polyhedrosis viruses like HaNPV and SlNPV are increasingly being used as alternatives to chemicals.

The Government of India allocates funds for IPM programs for all major crops. Major national research programs such as the National Agricultural Technology Project (2000–2006) and the current National Agricultural Innovation Project also contain important biopesticide research and development components. At the state level, 50% of the plant protection budget is allocated to ecofriendly agriculture (Singhal 2004) to cover both the training of farmers and the procurement of biopesticides for distribution. The DBT had a substantial funding program for the research and development of microbial pesticides since 1989, with over 200 projects funded (Wahab 2004). States such as Tamil Nadu, Gujarat, Andhra Pradesh, and Maharashtra have been particularly active in promoting microbial pesticide use.

17.5 Mass Production of HaNPV and SlNPV

The mass production of HaNPV and SlNPV involves three steps:

(1) Rearing of the adult gram pod borer and a tobacco caterpillar for mass production of eggs.

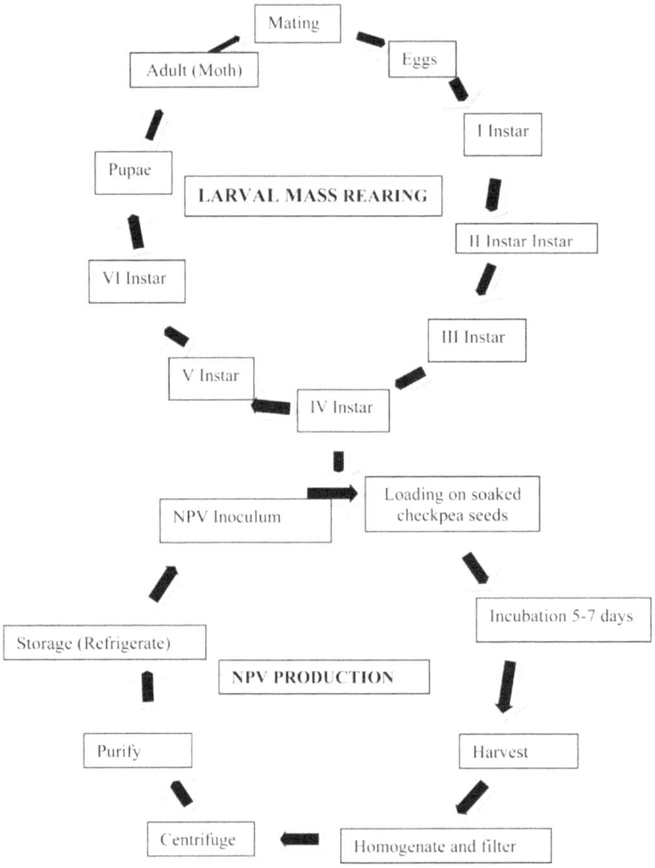

FIGURE 17.2 Mass production of NPV of *Helicoverpa armigera*.

(2) Rearing of larvae of the above species either on the host plants like chickpea and castor under semi-natural conditions or on a synthetic diet in laboratory conditions.

(3) The inoculation of HaNPV and SlNPV into the larvae of the gram pod borer and tobacco caterpillar, respectively, for mass multiplication of viruses and extraction of polyhedral occlusion bodies (POBs) from the diseased larvae, which are used as biopesticide on the crop plants.

17.5.1 Details of Mass Production and Major Equipment Required

Major equipment such as a centrifuge, laminar flow, magnetic shaker, microscopes, autoclave, coolers, refrigerators, incubator, distillation units, etc., are required in addition to glassware, plastic trays, basins, and iron racks for mass production of Ha NPV and Sl NPV.

17.5.1.1 Mass Production of NPV of Helicoverpa armigera

See Figure 17.2.

17.5.1.2 Mass Production of NPV of Spodoptera litura

Mass rearing of *Spodoptera litura* can be done on castor leaves in the laboratory. The steps involved in the mass production of SlNPV are shown in these figures.

Pre starve the fourth instar larvae

↓

Prepare virus suspension containing 10^8 POB/ml in water

↓

Dip clean castor leaves in virus suspension and shade dry

↓

Allow the caterpillar to feed for 2 days

↓

Collect the diseased larvae in distilled water

↓

Allow to putrefy

↓

Macerate the larvae in a blender

↓

Filter through muslin cloth

↓

Decant supernatant and discard

↓

Polyhedra settle at bottom as white layer

↓

Sediment contains POB

↓

Suspend in distilled water

↓

Centrifuge for 1 min at 500 RPM

↓

Discard the pellet

↓

Supernatant contains POBs

↓

Centrifuge at 2500 RPM for 15 minutes

↓

Discard supernatant

↓

Collect pellet

↓

Re-suspend in distilled water

↓

Repeat differential configuration

↓

Pure POB's

17.6 Environmental Factors Affecting the Action NPV

The environmental factors that affect the action of NPV are an ultraviolet component of sunlight, rainfall, temperature, humidity, and leaf surface compounds. The application of HaNPV in the evening hours provides better efficacy than the morning hours in the field. NPV degrades in the sunlight quickly, and hence adjuvants have to be added along with it while spraying. Solar radiation affects the field persistence of baculovirus. Ultraviolet radiation in the range of 280–310 nm inactivates baculovirus. Fluorescent brighteners can be used to increase its persistence in the field. Leucaena leaf extract, Eucalyptus leaf extract and Ranipal can be used with HaNPV to enhance its efficacy (Kalawate et al. 2005). Washing of the virus from the leaves by rain is a major factor affecting the persistence, and hence stickers have to be added with the viruses.

17.7 Advantages and Disadvantages of Using Viral Pesticides

Advantages of using viral pesticides include:

(1) A high degree of specificity, which makes them especially valuable for use in integrated pest management program.
(2) They are safe for humans and other warm-blooded animals.
(3) The residues present after application of viral pesticides pose no hazards to humans or other animals.

Disadvantages include the following:

(1) They are toxic to only a specific species or group of insects, each application may control only a portion of the pests present in a field, garden, or lawn. If other types of pests are present in the treated area, they will survive and may continue to cause damage.
(2) They are inactivated by heat, UV and rain can wash out the virus present on the foliage.
(3) Special formulation and storage procedures are necessary.

17.8 Conclusion and Future Aspects

Viral biopesticides provide a promising alternative approach to pest control. Available data suggest that the viruses are effective against insects and do not pose any deleterious effects on other components of the ecosystem (other invertebrates, plants, and vertebrates including humans). In India, the potential of baculovirus has not been utilized fully to control the economic insects. The new developments in this field depend upon the development of recombinant viral biopesticides and its commercial production. The most important aspect is to educate the farmers about the benefits of the NPVs. The inclusion of viral biopesticides in organic farming and integrated pest management has to be made understood by the farmers.

REFERENCES

Haase, S.; Ferrelli, M. L.; Pidre, M. L.; Romanowski, V. Genetic Engineering of Baculoviruses Molecular Virology Viral Genetics and Biotechnological Applications. Available online: http://www.intechopen.com/books/current issues in molecular virology viral genetics and biotechnological applications/ genetic engineering of baculoviruses. 2015.

Ignoffo, C. M.; Couch, T. L. The Nucleopolyhedrosis Virus of *Heliothis* Species as a Microbial Pesticide. In: Burges H. D. (ed.), *Microbial Control of Pests and Plant Diseases*. Academic, London, 1981, 329–362.

Kalawate, A.; Nachane, M. N.; Dhondge, A. S. Efficacy and retentivity of HaNPV under field conditions on cotton. *Plant Prot. Bull.* 2005, 57(1/2), 16–19.

Kumari, V.; Singh, N. P. *Spodoptera litura* nuclear polyhedrosis virus (NPV-S) as a component in integrated pest management (IPM) of *Spodoptera litura* (Fab.) on cabbage. *J. Biopest.* 2009, 2, 84–86.

Miller, L. K. Introduction to the Baculoviruses. In: Miller L. K. (ed.), *The Baculoviruses*. Plenum Press, New York, 1997.

Moscardi, F. Assessment of the application of baculoviruses for the control of lepidoptera. *Ann. Rev. Entomol.* 1999, 44, 257–289.

Nishi, Y.; Nonaka, T. Biological control of the tea tortrin using granulosis virus in the tea field. *Agrochem. Jpn.* 1996, 69, 7–10.

Singhal, V. Biopesticides for sustainable agriculture: Prospects and constraints. In: Kaushik E. (ed.), *Biopesticides in India*. All India Biotech Association, VIPPS Centre, New Delhi, 2004, 31–40.

Srinivasa, M.; Jagadeesh, Babu, C. S.; Anitha, C. N.; Girish, G. Laboratory evaluation of available commercial formulations of HaNPV against *Helicoverpa armigera* (Hub.). *J. Biopest.* 2008, 1, 138–139.

Sun, X. L., Peng, H. Recent advances in biological pest insects by using viruses. *Virol. Sin.* 2007, 22, 158–162.

Vincent, C.; Andermatt, M.; Valéro, J.; Virosoft, C. P. Viral biopesticides for coddling moth control. In: Vincent C., Goethel M. S., Lazarovits G. (eds.), *Biological Control: A Global Perspective*. CAB International, Oxfordshire/Cambridge, 2007, 336–343.

Wahab, S. The department of biotechnology initiates towards the development and use of biopesticides in India. In: Kaushik E. (ed.), *Biopesticides for Sustainable Agriculture*: Prospects and Constraints. The Energy and Resources Institute, New Delhi, 2004, 73–90.

18

A Modern Weapon in Organic Agriculture: Management of Crop Diseases through Viral Biopesticides

Upasana Mohapatra, Siddhartha Das, Sudeepta Pattanayak, and L. P. Awasthi

CONTENTS

18.1 Introduction

Due to destructive activities of various pests like insects, fungi, and weeds, global crop yield is reduced by 20 to 40% annually and has serious effects on feed production (Kachhawa, 2017). The green revolution has a remarkable role in managing many pests and diseases, but chemicalization in agriculture has created several problems like environmental pollution, pesticide residues in foodstuff, imbalance in ecological equilibrium, and resurgence of minor pests and pathogens. By applying chemical pesticides such as insecticides, herbicides, and fungicides, this food crisis was resolved to large extent. Uninhibited use, over-application, and unsustainability of chemical pesticides have necessitated development of an alternative as they create resurgence of pests, phytotoxicity on fruits, health hazards to human beings, destruction of beneficial microorganisms, and environmental pollution. In Sri Lanka, more people die of pesticide poisoning than of malaria. So the recent importance should be on the development of non-chemical/ecofriendly methods for the management of this disease, which has provided motivation for more substantial exploration of natural resources and identification of effective plant extracts/botanical biopesticides or viral biopesticides. Similarly, organic farming has appeared to be an important priority area globally in view of the growing demand for safe and healthy food, long-term sustainability, and concerns on environmental pollution associated with indiscriminate use of agrochemicals (Cao, 2010).

Although biopesticides make up around 1% of the total pesticides globally, their number and growth rate have been showing an increasing trend in the past two decades (Kachhawa, 2017).

18.2 Pesticides

Any substances used to repel, control, or sometimes kill certain forms of plants or animal life which are considered pests are known as pesticides. These pesticides include insecticides for controlling insects, herbicides for controlling weeds and other unwanted vegetation, fungicides to control molds and mildew, disinfectants for controlling the spread of bacteria, acaricides to control mites, and rodenticides to control mice and rats.

18.2.1 Harmful Effects of Pesticides

i) Use of huge amounts of agricultural chemicals in food production, a low level of pesticide residues enter the human body through diets.
ii) Pesticide residues are also found in the air and water, including drinking water.
iii) Pesticides cause nausea, headache, fatigue, cancer, reproductive organ harm, endocrine disruption, nerve, skin, and eye irritation, dizziness, nausea, and systemic poisoning.
iv) Pesticides cause environmental problems like water pollution and air pollution.
v) They have harmful effects on natural predators and beneficial microorganisms.

18.3 Biopesticides

The pesticides developed or derived from natural substances or living organisms such as plants, animals, and microorganisms (virus, fungus, bacteria) to control serious pests like insects, weeds, and fungi through a non-toxic ecofriendly mode of action are known as biopesticides. The biopesticides mainly include insect pheromones, plant extracts and oil, plant growth regulators, and insect growth regulators (Kumar, 2015).

18.3.1 Advantages of Biopesticides

i) They are highly specific and affect only the targeted pest or closely related pests.
ii) They do not harm humans or beneficial organisms, whereas chemical pesticides affect non-target organisms including predators and parasites, as well as humans, as they are broad-spectrum.
iii) These are environment friendly and biodegradable.
iv) They have lower pesticide residues and avoid pollution problems.
v) Biopesticides can be used as a component of integrated pest management (IPM) and can greatly reduce the use of conventional (chemical) pesticides, also achieving almost the same level of crop yield (Vaughn, 1993).

18.3.2 Types of Biopesticides

The Environment Protection Agency (EPA) separates biopesticides into three major classes based on the type of active ingredient used, namely microbial, biochemical, or plant incorporated protectants (GMOs).

18.3.2.1 Biochemical Pesticides

- Insect pheromones
- Plant extracts and oils
- Plant growth regulators
- Insect growth regulators

18.3.2.2 Microbial Pesticides

- Bacterial biopesticides
- Fungal biopesticides
- Viral biopesticides
- Other microbial biopesticides

18.3.2.3 Biopesticide Formulations

18.4 Viral Biopesticides

A virus enclosed in a protective coat protein consists of one or more nucleic acids, and replicates within suitable host cells (Kachhawa, 2017).

Around 1,000 species of insect virus from at least 13 different insects order have been isolated and classified based on the production of inclusion bodies (Figure 18.1, Table 18.1).

Baculoviruses are highly pathogenic, insect-specific, double-stranded DNA enveloped viruses with circular, supercoiled genomes. They are used in natural form as biocontrol agents against numerous serious insect pests.

18.4.1 Mode of Action

i) Baculoviruses are ingested by larvae and initiate infection.
ii) They enter the insect's body through midgut.
iii) They spread throughout the body from the midgut but they are limited to the insect midgut or the fat body in some insects.
iv) In the case of an occluded virus, the occlusion bodies are dissolved in insect midgut and the virus infects the insect gut cell and spreads in the body.
v) Because of infection the insect dies and disintegrates and the virus infects a new host.

18.4.2 Strategies for Pest Control through Virus

18.4.2.1 Introduction of Virus

The virus is introduced either deliberately or by accident and provides effective control of the pest.

Examples of the introduction of viruses include:

- Sawflies are controlled by introducing an NPV from Europe.
- The non-occlude BV of the coconut palm rhinoceros beetle from Malaysia was introduced to the Pacific Islands.

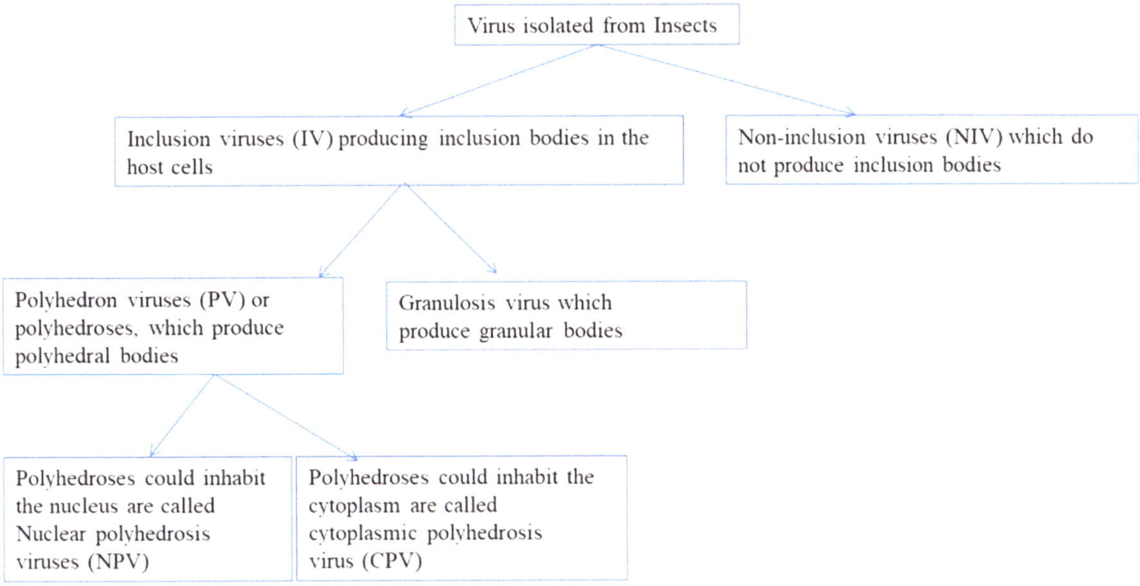

FIGURE 18.1 Classification of entomogenic virus.

TABLE 18.1

List of Entomogenic Virus Types and Their Hosts

S. No.	Nature of Virus	Infected Host
1	Nuclear polyhedrovirus	*Helicoverpa armigera* *Spodoptera litura* *Amsacta albistriga* *Spilosoma obliqua* *Pericallia ricini* *Pseudaletia separata* *Spodoptera mauritia* *Corcyra cephalonica* *Plusia chalcites* *Antheraea mylitta* *Dasychira mendosa* *Plusia peponis*
2	Cytoplasmic polyhedrosis virus	*Helicoverpa armigera*
3	Granulosis virus	*Cnaphalocrocis medinalis* *Pericallia ricini* *Achaea janata* *Phthorimaea operculella* *Chilo infuscatellus*
4	Pox virus	*Amsacta moorei*

18.4.2.2 Supplementation of Existing Virus

Increase in the amount of already present virus may cause an incremental rise in the infection of an insect population.

For example, a single NPV application to young trees may create protection from sawflies.

18.4.2.3 Manipulation of Existing Virus

The increase in level of virus disease by adoption of certain management practices can infect more insect pests.

For example, an NPV controls the pest *Wiseana cervinata* in New Zealand. Cultivation of pasture land beyond the cultivated land can control the pest because of increase in virus reservoir.

18.4.3 Mode of Application of Viral Pesticide

18.4.3.1 Release of Virus-Infected Insects

This method is highly suitable for the virus, which survives poorly outside its host.

For example, release of *O. rhinoceros* non-occluded BV to control the beetle in South Pacific Island was successful.

18.4.3.2 Spraying of Virus

Virus can be sprayed as aqueous sprays through insecticide sprayer with small droplets or through the overhead irrigation system.

Examples of virus spraying include:

- *P. apercu/ella* GV is sprayed through a sprayer to control the larvae.
- NPV of *H. zea* and *S. jrugiperda* is sprayed onto corn through an overhead irrigation system.
- Virus can be directly applied to the soil, e.g., the NPV of soybean looper *Pseudoplusia*.
- During planting time, *includens* is applied to the soil.

18.4.3.3 Through Baits

Application of virus through baits is advantageous because less insecticidal material is required. But the increased cost of application makes it a less usable method.

Examples include:

1. The application of viruses in baits includes the mixture of *H. zea* SNPV and cottonseed and soybean.
2. Wheat bran is also used as bait to apply NPV to corn.

18.4.3.4 Dipping of Seedlings

The inclusion body can be used as suspension for dipping the seedling before planting.

TABLE 18.2

List of Insects Registered for Field Application (Carter, 1984)

S. No.	Virus	Applied On	Trade Name
1	*Heliothis zea* SNPV	Cotton soybean	Elcar
2	*Orgyia pseudotsugata* MNPV	Fig	TM Biocontrol Virtuss
3	*Lymantria dispar* MNPV	Decidous trees	Gypchek SAN 404
4	*Neidiprion sertifer* SNPV	Pine trees	Neochek S
5	*Neidiprion lecontei*	Pine trees	Lecontvirus
6	*Dendrilimus spectobilis* CGV	Pine trees	Matsukemin

For example, before cabbage transplanting, the seedlings are dipped in the inclusion body suspension.

18.4.4 Insects under Trial or in Use

The viruses registered for field application are listed in Table 18.2.

18.4.5 Timing of Application

Till the larvae come to later instar the significant pest damage is not noticed. So a huge amount of viral insecticide is required to control the pest. And most insect viruses kill their host more slowly than chemical insecticides. So the virus should be applied before the pest damages the crop. The damaging level of the pest can be noticed by using a pest forecast system and the virus can be applied. In some cases the later instar becomes resistant to the virus. So application of virus after appearance of first instar larvae is done to control the pest (Carter,1984; Senthil-Nathan, 2015).

18.5 Viruses as Biocontrol Agents for Plant Pathogens

Generally, the viruses infecting bacteria, i.e., bacteriophages, are used as biocontrol agents for bacterial diseases.

Advantages of bacteriophages include:

- Highly specific
- Easy to produce
- Long shelf life
- Non-toxic to eukaryotic cells
- Readily isolated from the host plant

Disadvantages of bacteriophages include:

1. Development of resistance to virus in bacterial population.

The avirulent strain pathogens are used as a vehicle to produce the phages, which creates a self-perpetuating biocontrol system in the field.

Fungal viruses or mycoviruses are also used as biocontrol agent for controlling fungal diseases. The viruses are transmitted through hyphal connections. Viral infection does not allow to cause any symptoms by fungal pathogen or makes it a hypovirulent strain.

Application of virus for disease control

Examples include:

- Application of virus-infected hypovirulent strain to chestnut blight infected tress reduces the virulence of fungal pathogens and protects the chestnut tree.
- The foliage treatment with phages in peach reduces the infection level by *Xanthomonas campestris* pv. pruni (Xcp).
- Phage treatment to tomato plant reduces the gall size formed by *Agrobacterium tumefaciens.*
- Application of bacteriophages to paddy results in reduction of bacterial leaf blight in paddy.
- Biweekly spray application of phage suspension is effective against Xcp and reduces the leaf spot disease incidence.
- A mixture of three phages of *Erwinia amylovora* application reduces the fire blight incidence in apple, pear, and raspberry.
- Foliar application of mixture of phages reduces the bacterial blight incidence in potato.
- In onion, the *Xanthomonas* blight is reduced by the application of phages.
- Bacterial blotch of mushroom is reduced by application of phages.
- *Ralstonia* wilt of tomato is suppressed by using the phages.

18.6 Viruses as Bioherbicide to Control the Weeds

Viruses affecting the weeds are considered bioherbicides but some scientists consider them inappropriate candidates to be used as bioherbicides because they can be used only in a broader ecosystem to control the invasive species, and not in the specifically managed areas. Also, the greater genetic variability and lesser host specificity make the virus unsuitable for being used as herbicide. But in some cases the viruses are used to control the unwanted species in the field.

Examples include:

- Application of Tobacco Mild Green Mosaic Tobamo virus controls the tropical soda apple (*Solanum viarum*) in Florida.
- Application of Araujia Mosaic Virus reduces the incidence of moth plant (*Araujia hortorum*) in New Zealand.
- A virus resembling Tobacco Rattle Virus can be used as weedicide to control *Impatiens glandulifera* in Central and Western Europe.
- Pepino Mosaic Virus (PepMV) and Obuda Pepper Virus (ObPV) are used as biocontrol agents for weed *Solanum nigrum.*

18.7 Conclusion and Future Prospectives

Although many viruses are used as biopesticide, there are many more still to be developed as biopesticide. As timing of application in field is crucial, the knowledge of damage level, lifecycle of the pest, and disease forecasting should be improved.

The promotion of the virus must include the education of all involved with the sale and use to understand how the virus works, and what effects to expect, both in the short term and in the long term. If this is done viruses will become useful tools in the control of pests. Biopesticides are advantageous over chemical pesticides like low research expenditure, faster rate of product development, as well as flexible registration process, production, and commercialization. Biopesticides can and will play a significant role in creating a new generation of sustainable agriculture products. They are the source of alternatives to some of the most problematic chemical pesticides currently in use that are under ever-increasing scrutiny. Biopesticides may also offer solutions to concerns such as pest resistance to traditional chemical pesticides and public concern about the side effects of pesticides on the surrounding environment and, ultimately, on human health. Thus, the application of different biopesticides was found effective to control the various pests of agriculture crops. Moreover, the use of biopesticides will be promoted to sustain the yield of agricultural crops (Copping, 2000; Jarvis, 2001; Singh et al., 2009).

REFERENCES

Cao, C.; Park, S.; Gardener, B. B. M. Biopesticide controls of plant diseases: Resources and products for organic farmers in *Ohio. Fact Sheet*. 2010.

Carter, J. B. Viruses as pest-control agents. *Biotechnol. Genet. Eng. Rev.* 1984, 1(1), 375–419.

Copping, L. G.; Menn, J. J. Biopesticides: A review of their action, applications and efficacy. *Pest Manage. Sci.* 2000, 56(8), 651–676.

Jarvis, P. Biopesticides: Trends and opportunities. *Agrow Reports*. PJB Publications 2001, 18:20.

Kachhawa, D. Microorganisms as a biopesticides. *J. Entomol. Zool. Stud.* 2017, 5(3), 468–473.

Kumar, V. A review on efficacy of biopesticides to control the agricultural insect's pest. *Int. J. Agr. Sci. Res.* 2015, 4(9), 168–179.

Senthil-Nathan, S. A review of biopesticides and their mode of action against insect pests. In *Environmental Sustainability*. Springer 2015, pp. 49–63.

Singh, H. B.; Singh, B. N.; Singh, S. P.; Sarma, B. K.; Singh, S. R. Biological control of plant diseases: Current status and future prospects. In: Johri, J. K. (ed.) *Recent Advances in Biopesticides*: Biotechnological Applications. New Indian Publishing Agency, New Delhi, 2009, pp. 193–304.

Vaughn, J. L. Viral insecticides. 1993.

Section XI

Management of Crop Diseases through Algal Biopesticides

19

Crop Disease Control through Application of Algal Biopesticides: Alternate Route Map in Organic Farming

Sudeepta Pattanayak, Siddhartha Das, and Gayatri Biswal

CONTENTS

19.1 Introduction

The world's population is increasing at a fast rate and in order to feed the entire population, the production of food also increased. It is a mainstay of the Indian economy and contributes 18% toward GDP (Sharma et al., 2018). The green revolution has taken an important role in modernizing the techniques to meet the entire population's food demand. After the green revolution, excessive use of chemicals and high yielding varieties have increased dramatically, leading us to a path of environmental destruction and diverted from sustainable agriculture. The immense use of chemicals is misbalancing the ecosystem by polluting the environment, water bodies, and groundwater through leaching loss. The regular and high use of pesticides helps create new races of resistant pests (Sharma et al., 2018). The residue of chemicals can stay in the soil, air, water, and food year after year and make ill to the human, animal, and some beneficial insects by genetic disorders, population changes and gene erosion, etc. Additionally, plants are also affected by the high use of chemicals. The ill-effects of chemicals also remain in the food chain for a longer time.

Several biotic and abiotic factors are responsible for the decrease in crop yield. It was estimated that the crop loss due to plant disease is nearly 10–20%. To maintain the quality of feed, food and fiber, plant diseases need to be controlled. In the past 100 years, plant diseases have been controlled by several agrochemicals. Therefore, biological control measures are sustainable alternatives to chemical and synthetic pesticides. Several beneficial micro-organisms are used as a biological control agent to protect the plants in the biological control method.

19.2 Algal Biopesticide

Algal biopesticide stands out as an important biocontrol agent and playing an important part in controlling several diseases and pests. Algae can synthesize several metabolites having potential bio pesticidal action. In addition to plant disease pest control, it has the following advantages (Costa et al., 2019):

- They have simple nutritional requirements for their growth.
- They can produce higher biomass per unit area.
- It doesn't need arable land for growth and production.
- It can directly or indirectly improve the soil health like the availability of micro and macronutrients (Abdel, 2012), reducing soil erosion, formation of soil crust, etc.
- They can also be used as biofertilizer.

Among the several microalgae-based biopesticide, cyanobacteria are one of the important biocontrol agents. It can control numerous pathogenic fungal genera and soil-borne diseases by producing secondary metabolites having low molecular weight and biocidal activity (Berry et al., 2008; Rastogi et al., 2009). They are antibiotic, antifungal, andinst nematodes (Kulik et al., 1995; Kiviranta et al., 1993).

19.3 Production and Cultivation of Microalgal Biopesticide

The production and cultivation of microalgae are attributed to better sustainability, such as less dependent on soil fertility and type of water used like brackish water, seawater or wastewater (Zhou et al., 2017). Additionally, it can be produced by reusing industrial waste to reduce production cost, increase sustainability, and remediate wastewater and soil (Morais et al., 2014). The commercial scale of cultivation and production of microalgae is highly expensive and not yet explored properly. Therefore, to enhance the economic feasibility, we can use domestic, industrial and agricultural waste as an economical source of nutrients (Klein et al., 2018). The greatest difficulty of microalgae-based products is to reach the competitive price in the market, and it can overcome by using the waste effluents as a nutrient source (Klein et al., 2014).

Microalgae are referred as 'living biorefinery' due to their capability of generating a wide range of chemicals to be used as source of renewable energy, converting water and carbon dioxide into oxygen, Primary (carbohydrate, lipid and protein) and secondary metabolites (terpenes, phenolic substances, etc.) through photosynthesis process which play a crucial role in pest management.

Utilization of phosphorus, nitrogen and heavy metals rich waste watering cultivation of microalgae helps not only in reduction of production cost but also to treat waste effluents (Zhou et al., 2017). *Chlorella* sp. is observed to reduce the amount of phosphorus, nitrogen and heavy metals by 80% when added to crop runoff water of Metropolitan Wastewater Treatment Plant in Saint Paul. Microalgae use nitrogen and phosphorus, which have a crucial physiological role in plant growth and defense mechanism against several pests. In 2017, Chen et al. grew *Chlorella sorokiniana, Galdieria sulphuraria* and *C. vulgaris* by recycling nutrient from a hydrothermal liquefaction process; they then observed the high production of organic nitrogen, phosphate and polysaccharides (Chen et al., 2017). In 2015, Rosa et al. observed an increase of biomass up to 31.4% and 96% in carbohydrate when *Spirulina* sp. LEB 18 was cultivated with the recycling of culture media to fight against several pathogens. In 2017, Liu et al. used a solution of concentrated residuals, i.e (Rosa et al., 2015). sugarcane molasses after crystallization of sugar in industries, and concluded the increased production of lipids, biomass and astaxanthin in *C. zofingiensis* (Liu et al., 2017). Terpenes can be extracted from lipids and used as biopesticides. Stimulation of lipids synthesis and decrease of carbon source expenditure was being observed in *C. vulgaris* by using residual glycerol produced from biodiesel production (Ma et al., 2016).

Use of phytohormones has achieved high attention in algal production to influence growth and production of biomolecules by adjusting several metabolic pathways. The phytohormones are produced by microalgae both exogenously and endogenously. One herbicidal compound found in *Scytonema hofmanni*, which is not developed commercially till now.

19.4 Mechanism of Control

19.4.1 Against Pathogenic Microbes

For the first time, it was observed that *Chlorella*, the green algae, produces chlorellin that act as a biopesticide and effectively control the growth of both gram-positive as well as gram-negative bacteria (Gupta et al., 2013).

Microalgae can control by secreting proteins, carbohydrates, oils, phenolic compounds, allelochemicals, tocopherols, polyphenols and sesquiterpenes (Costa et al., 2019). These chemicals and enzymes control the pest by inactivating their enzymes, inhibiting their protein synthesis mechanism, modifying their structure and function (Swain et al., 2017). Additionally, the microalgae can also stimulate plant immune systems the pathogen through several metabolic processes such as activation of enzymes with defense characters (Renuka et al., 2018; Gupta et al., 2013). The microalgae protect the plant from several insect pest and pathogen attack by stimulating the plant defense mechanism by activating the action and production of several antioxidant compounds like catalase, polyphenol oxidase, chitinase, endoglucanase, peroxidase and polyphenol alanine (Renuka et al., 2018). Additionally, algae can produce biocidal compounds and hydrolytic enzymes like majusculonic acid, benzoic acid, etc. (Prassanna et al., 2008b; Gupta et al., 2013). The algae can persist and colonize in rhizosphere region through which it indirectly supplements antagonism on several pests by releasing antimicrobial compounds, hydrolytic enzymes, etc.

19.4.2 Against Insects

The cyanobacteria class plays a major role in plant protection by producing biofilm, polymeric and extracellular substances. This group is known to produce several toxins that act as an antifeeding substances for larvae. It also reduces mosquito population rice crop when applied it as a biofertilizer (Renuka et al., 2018).

Microalgae can also control many insects by secreting alkaloids, sesquiterpenes and amines. It is observed that algae primarily inhibit the early larval stages compared to adult larva in which the algae inhibit mainly population and fecundity. For example, microalgae *Calothrix* spp. is known to secret a sesquiterpene i.e. Eremophilone having insecticidal activity, which is used in rice crop pest control (Hernandez-Carlos et al., 2011).

19.4.3 Mechanism of Control Against Nematodes

The group microalgae decrease the usage of toxic chemicals imparted to boost pest resistance in plants such as nematodes. Algae produce nematicidal substances and peptide toxin to fight against nematodes.

19.4.4 Mechanism of Control Against Weeds

Microalgae secret allelochemical compounds help in controlling wide range of weeds, larvae of aquatic invertebrates (Pratt et al., 1945). One of the most important herbicides, Triketones were derived from triketone, a major allelochemical which restrict the growth of weeds through an enzymatic action of complex II of photosynthesis (Duke et al., 2010).

Table 19.1 shows the microalgae, its potential metabolite and the method of extraction.

19.5 Conclusion and Future Trends

The craving of human beings to avoid the use of conventional chemicals in crop plants to go in a trend of sustainable agriculture

TABLE 19.1
List of Microalgae Used as Potential Biopesticide

Activity	Microalgae	Metabolite produced	Method of extraction	Method of Quantification or Identification	Reference
Antimicrobial	*Anabaena laxa*	Laxaphycin C, Anatoxin-a, Laxaphycin B	Successive gel filtration	High performance liquid chromatography (HPLC)	Hernandez-Carlos et al. 2011, Frankmolle et al. 1992
	Fischerella sp.	Ambiguine isonitrile	Using water or methanol	HPLC	Raveh and Carmeli et al. 2007
Antimicrobial and molluscicidal	*Fischerella ambigua*	Ambigol A and B	Using ethyl acetate, ether, methanol and mixture of water, dichloromethane	HPLC	Falch et al. 1993
Antifungal	*Spirulina platensis*	Phenolic compounds	With Folinciocalteu reagent	Spectrophotometry	Hussein et al. 2009
	Chlorella vulgaris	Phenolic compounds	With Folinciocalteu reagent	Spectrophotometry	Hussein et al. 2009
Insecticidal	*Calothrix* sp.	Eremophilone	Essential oil obtained by vacuum distillation	Mass spectrometry and gas chromatography	Hockelmann et al. 2009
Larvicidal	*Lyngbya* spp.	Pahayokolides	Extraction using water and methanol	HPLC	Raveh and Carmeli et al. 2007
	Anabaena sp.	Anatoxin-a	Ultrasonic disintegration of algal cell wall	Mass spectrometry and gas chromatography	Berry et al. 2008
Herbicidal	*Nostoc* sp. *ATCC 53789*	Cryptophycin	Using methanol	Unrealized	Berry et al. 2008
	Microcystis aeruginosa	Microcystin	Free-thawed with water	Unrealized	Berry et al. 2008

and ecofriendly agriculture has created demand for new chemicals having a less toxic effect. Recent studies and research on controlling several pests by utilizing microalgae have proved potentially promising. Microalgae are antifungal, antibiotic, entomopathogenic, etc. Additionally, it can also be used as a biofertilizer, growth regulator, increase soil organic matter, helps in crust formation, etc. In contrast to this, microalgae tend to have poor storage, slow action, sometimes costly also. The success of algal biopesticides are mainly dependent on the production of biomass.

The development and discovery of biopesticides by using biocontrol micro-organisms is one of the foremost challenges for scientists because the metabolism of microalgae for secreting these metabolites and their mode of action is still unknown or poorly explored. There is high potential to discover new noble bioactive agents from algae and cyanobacteria group to use as biopesticide. Many scientists have done their research in the past, but more of those researches are subjected to only in lab conditions. Henceforth, it is required to do more investigation to discover the novel characters and substances unknown to us. This research will surely help us find the path of sustainable agriculture and lead us to ensure food safety. Thus, Algal biopesticide, an ecofriendly approach to agriculture as well as the foundation stone and base of food webs, a strong awareness among the farmers and growers by the government, more investment and development in researches, collaboration among research institutions need to be created so that we can undoubtedly achieve sustainable agriculture development in the near future.

REFERENCES

Abdel-Raouf, N. Agricultural importance of algae. *Afr. J. Biotechnol.* 2012, 11, 11648–11658.

Berry, J. P.; Gantar, M.; Perez, M. H.; Berry, G.; Noriega, F. G. Cyanobacterial toxins as allelochemicals with potential applications as algaecides, herbicides and Insecticides. Mar. Drugs. 2008, 6, 117–146.

Chen, L.; Zhu, T.; Fernandez, J. S. M.; Chen, S.; Li, D. Recycling nutrients from a sequential hydrothermal liquefaction process for microalgae culture. *Algal Res.* 2017, 27, 311–317.

Costa, J. A. V.; Freitas, B. C. B.; Cruz, C. G.; Silveira, J.; Morais, M. G. Potential of microalgae as biopesticides to contribute to sustainable agriculture and environmental development. *J. Environ. Sci. Health, Part B.* 2019, 54, 366–375.

Duke, S. O.; Cantrell, C. L.; Meepagala, K. M.; Wedge, D. E.; Tabanca, N.; Schrader, K. K. Natural toxins for use in pest management. *Toxins.* 2010, 2, 1943–1962.

Falch, B. S.; Koenig, G. M.; Wright, A. D.; Sticher, O.; Ruegger, H.; Bernardinelli, G. Ambigol A and B: New biologically active polychlorinated aromatic compounds from the terrestrial blue-green alga *Fischerella ambigua. J. Org. Chem.* 1993, 58, 6570–6575.

Frankmolle, W. P.; Knubel, G.; Moore, R. E.; Patterson, G. M. L. Antifungal cyclic peptides from the terrestrial blue green alga *Anabaena Laxa*. II. Structures of laxaphycins A, B, D and E. J. *Antibiot.* 1992, 45, 1458–1466.

Gupta, V.; Ratha, S. K.; Sood, A.; Chaudhary, V.; Prasanna, R. New insights into the biodiversity and applications of cyanobacteria (blue-green algae)—Prospects and challenges. *Algal Res.* 2013, 2, 79–97.

Hernandez-Carlos, B.; Gamboa-Angulo, M. M. Metabolites from freshwater aquatic microalgae and fungi as potential natural pesticides. *Phytochem. Rev.* 2011, 10, 261–286.

Hockelmann, C.; Becher, P. G.; Reu, S. H. V.; Juttner, F. Sesquiterpenes of the geosmin-producing cyanobacterium Calothrix PCC 7507 and their toxicity to invertebrates. *Z. Naturforsch. C.* 2009, 64, 49–55.

Hussein, M. Y.; Abd El-All, A. A. M.; Mostafa, S. S. M. Bioactivity of algal extracellular by products on cercospora leaf spot disease, growth performance and quality of sugar beet. Presented at the 4th Conference on Recent Technologies in Agriculture, Cairo, Egypt, 2009.

Kiviranta, J.; Abdel-Hameed, A.; Sivonen, K.; Niemela, S. I.; Carlberg, G. Toxicity of cyanobacteria to mosquito larvae screening of active compounds. *Environ. Toxicol. Water Qual.* 1993, 8, 63–71.

Klein, B. C.; Bonomi, A.; Maciel Filho, R. Integration of microalgae production with industrial biofuel facilities: A critical review. *Renewable Sustainable Energy Rev.* 2018, 82, 1376–1392.

Kulik, M. M. The potential for using cyanobacteria (blue-green algae) and algae in the biological control of plant pathogenic bacteria and fungi. *Eur. J. Plant Pathol.* 1995, 101, 585–599.

Liu, T.; Liu, F.; Wang, C.; Wang, Z.; Li, Y. The boosted biomass and lipid accumulation in *Chlorella Vulgaris* by supplementation of synthetic phytohormone analogs. Bioresour. *Chlorella Minutissima*. Bioresour. Technol. 2017, 238, 248–253.

Ma, X.; Zheng, H.; Addy, M.; Anderson, E.; Liu, Y.; Chen, P.; Ruan, R. Cultivation of chlorella vulgaris in wastewater with waste glycerol: Strategies for improving nutrients removal and enhancing lipid production. *Bioresour. Technol.* 2016, 207, 252–261.

Morais, M. G.; Morais, E. G.; Vaz, B. S.; Costa, J. A. V. Industrial effluents as a nutritional source in microalgae cultivation. *Mini-Rev. Org. Chem.* 2014, 17, 1–7.

Pratt, R.; Oneto, J. F.; Pratt, J.; American, S.; Jul, N.; Pratt, R.; Oneto, J. F.; Pratt, J. Studies on *Chlorella Vulgaris* influence of the age of the culture on the accumulation of chlorellin. *Am. J. Bot.* 1945, 32, 405–408.

Rastogi, R. P.; Sinha, R. P. Biotechnological and industrial significance of cyanobacterial secondary metabolites. *Biotechnol. Adv.* 2009, 27, 521–539.

Raveh, A.; Carmeli, S. Antimicrobial ambiguines from the cyanobacterium *Fischerella sp.* collected in Israel. *J. Nat. Prod.* 2007, 70, 196–201.

Renuka, N.; Guldhe, A.; Prasanna, R.; Singh, P.; Bux, F. Microalgae as multi-functional options in modern agriculture: current trends, prospects and challenges. *Biotechnol. Adv.* 2018, 36, 1255–1273.

Rosa, G. M.; Moraes, L.; Cardias, B. B.; Souza, M. R. A. Z.; Costa, J. A. V. Chemical absorption and CO_2 biofixation via the cultivation of *Spirulina* in semi continuous mode with nutrient recycle. *Bioresour. Technol.* 2015, 192, 321–327.

Sharma, K. R.; Raju S. V. S.; Jaiswal, D. K.; Thakur, S. Biopesticide: An effective tool for insect pest management and current scenario in India. *Indian J. Agric.* Allied Sci. 2018, 4(2), 59–62.

Swain, S. S.; Paidesetty, S. K.; Padhy, R. N. Antibacterial, antifungal and antimycobacterial compounds from cyanobacteria. *Biomed Pharmacother.* 2017, 90, 760–776.

Zhou, W.; Li, Y.; Gao, Y.; Zhao, H. Nutrients removal and recovery from saline wastewater by *Spirulina Platensis. Bioresour. Technol.* 2017, 245, 10–17.

Section XII

Management of the Diseases and Pests of Crops through Protozoan Biopesticides

20

Entomopathogenic Protozoa Roles in the Management of Insects Pest Populations

Muhammad Sarwar, Naveed Akhter Shad, and Riffat Batool

CONTENTS

20.1 Introduction

Most species are free-living, but all higher-level animals are infected with one or more species of protozoa. Infections range from asymptomatic to life-threatening, depending on the species and strain of the parasite and the resistance of the host (Sarwar, 2016).

Entomopathogenic protozoans are an extremely diverse group of organisms comprising thousands of species that attack invertebrates, including insect species, and are commonly referred to as microsporidians. Primary hosts are European corn borer and other caterpillars, locusts, grasshoppers, crickets, forest and tree crop pests, certain crop pests, and mosquito populations (Brooks, 1988).

Generally, protozoa (Figure 20.1) are host-specific and slow-acting, producing chronic infections with general debilitation of the host. The infection results in reduced feeding, vigor, fecundity, and longevity of the insect host as inundatively applied microbial control agents. However, only a few species have been moderately successful (Solter and Becnel, 2000). The *Nosema pyrausta* is another beneficial microsporidian that reduces fecundity and longevity of the adults and also causes mortality of the larvae of European corn borer (Lacey et al., 2015).

In addition to the differences in tissue specificity and size of spores, light microscopy studies revealed some stages of the pathogen (uninucleate bodies and plasmodia) apparently not previously observed in *Pyrgomorphae* (Lange and Brito, 1992).

Members of the subkingdom Protozoa represent a heterogenous group of protists that belong to the kingdom Protoctista. Comparative analysis of the ribosomal RNA sequences of protozoa has placed them in an evolutionary branch between the plant and animal kingdoms. These unicellular organisms are an ancient group that arose well before animals, helminthes, and yeasts. Protozoa are not monophyletic and diverged very early in evolutionary time and on a scale greater than that estimated for either the plant or animal kingdoms. Protozoa possess at least one motile stage powered by cilia, flagella, or pseudopodia (Williams et al., 2002). The vegetative cells of most protozoa are membrane-bound and lack the rigid cell wall structure common to both fungi and plants. The structure of the cell membrane of protozoa is similar to that found in other eukaryotes and generally fits the fluid mosaic model. In many cases, the outer membrane surface may be coated with a thin pellicle or glycocalyx (Lee et al., 1985; Englund and Sher, 1988).

Protozoan pathogens naturally infect a wide range of insect hosts (Table 20.1). Although these pathogens that can kill their insect hosts, many are more important for their chronic

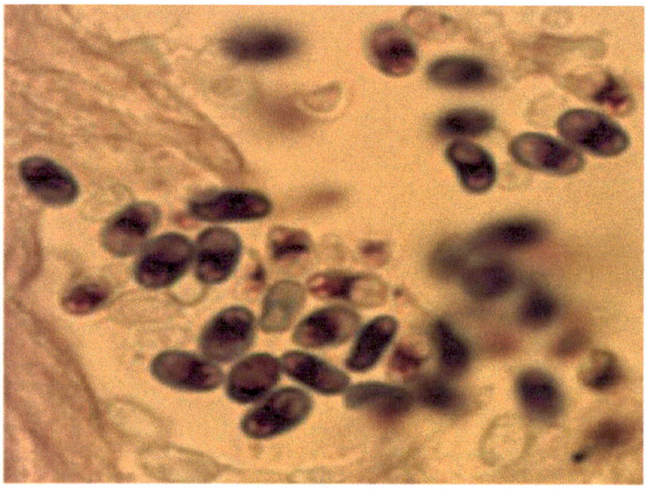

FIGURE 20.1 Entomopathogenic protozoa.

debilitating effects. For pathogenicity, the infectious stage spores made by protozoan upon ingestion by insect in the midgut germinate to release sporoplasm for invading the target cells for causing infection of the host.

20.2 Bioassays of Protozoa

To conduct a quantitative bioassay, spores must be counted and a dilution series made, so that a known concentration of spores can be fed to the host. To calculate a dosage or concentration for any bioassay, the spores, or other infectious stages must be accurately counted. Viability can be assessed without infecting host animals. Viable microsporidian spores must be capable of germination. Spores can be fed to the host on an artificial diet or their natural food. Experimentation is necessary in order to find the optimum dosage for best spore production. The *Nosema* and *Endoreticulatus* species in Lepidoptera develop more slowly than *Vairimorpha necatrix*, dictating that infections be initiated earlier in larval life than is recommended for *V. necatrix*. It is found that freeing the spores of host tissues and microbial contaminants is often necessary for storage.

20.3 Classification of Protozoa

On the basis of light and electron microscopic morphology, the protozoa are currently classified into six phyla. But most entomophilic protozoa occur in the microspora with fewer in Apicomplexa, Sarcomastigophora, and Ciliophora. Generally protozoa, particularly the microsporidia, cause debilitative diseases that result in subtle regulation of host populations (Yaeger, 1996). Overall, there are many protozoa that could be manipulated in some manner to increase their activity against noxious insects (Baker et al., 1990).

Microsporaorganisms have been previously considered "primitive" protozoa, however, molecular phylogenetic analysis has resulted in the insights that these organisms are not primitive, but instead degenerate, and that microsporidia are related to the fungi, either as a basal branch of the fungi or as a sister group (Capella-Gutierrez, 2012; Corsaro et al., 2016). It should be noted that discrepancies exist between phylogenies based on traditional characters and those based on molecular analysis and that these types of discrepancies have not been fully resolved (Vossbrinck and Debrunner-Vossbrinck, 2005). Here, microsporida or microsporidia will be treated as part of the protozoa.

20.3.1 Apicomplexan

The Apicomplexans are unicellular and spore-forming obligate endoparasites of animals, except *Nephromyces*, which is a symbiont in marine animals, originally classified as a chytrid fungus (Saffo et al., 2010). Most of them possess a unique form of organelle, which is an adaptation that the apicomplexan applies in penetration of a host cell. Apicomplexan parasites have complex life cycles involving both asexual and sexual replication in a variety of invertebrate and vertebrate hosts. Typically, a host is infected via an active invasion by the parasites (similar to entosis), which

TABLE 20.1

Protozoan Insecticides: A Summary of Products and Their Uses

Pathogen	Host Range	Uses and Comments
Protozoa		
Nosema locustae	European corn borer caterpillars, grasshoppers, and crickets	Useful for rangeland grasshopper control and is active only if ingested. Not recommended for use on a small scale, such as backyard gardens, because the disease is slow-acting and grasshoppers are very mobile. Also effective against caterpillars.
Gregarina garnhami	Desert locust *Schistocerca gregaria*	Reduces the weights of male fledglings (new adults) without causing observable effects on molting or feeding.
Lambornella clarki	Western treehole mosquito *Aedes sierrensis*	Biological control and manipulated biological control agent of treehole mosquitoes.
Chilodonella uncinata	Mosquitoes	Wide host range including Culex, Anopheles, and Aedes.
Microspora		
Anncaliia algerae	Mosquitoes	Manifestations of myositis, keratoconjunctivitis, and cellulitis.
Trachipleistophora anthropopthera	Insects	Encephalitis, keratitis, and disseminated infection.
Trachipleistophora hominis	Mosquitoes	Myositis, keratoconjunctivitis, sinusitis, and encephalitis.
Tubulinosema acridophagus	Fruit fly	Myositis and disseminated infection.

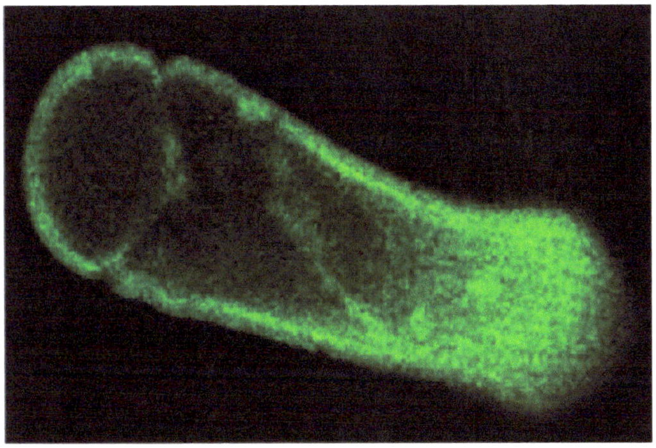

FIGURE 20.2 *Gregarina garnhami.*

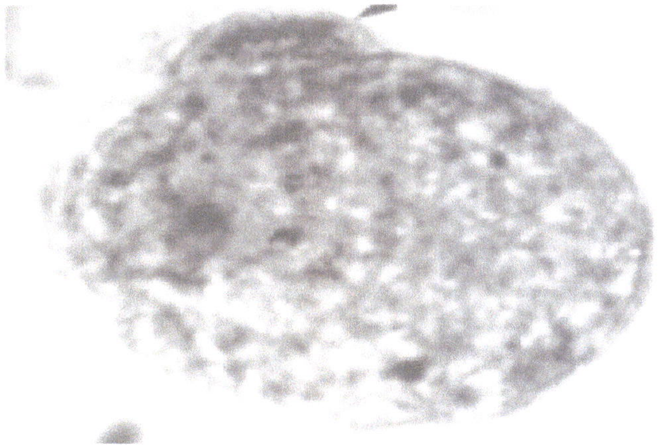

FIGURE 20.3 *Adelina tribolii.*

divide to produce sporozoites that enter its cells. Eventually, the cells burst by releasing merozoites, which infect new cells. The Apicomplexa is a diverse group that includes organisms such as the coccidia, gregarines, piroplasms, haemogregarines, and plasmodia (Stefan et al., 2004; Sibley, 2010).

Within this phylum, three groups (coccidians, gregarines, and haemosporidians) are important.

20.3.1.1 Gregarines

The gregarines (Apicomplexa: Eugregarinorida) are generally parasites of annelids, arthropods, and mollusks. They are often found in the guts of their hosts, but may invade the other tissues. Gregarines are relatively large protozoan parasites of the guts and body cavities of invertebrates including annelids, tunicates, sipunculids, and especially arthropods. During the preliminary study, 20 gregarine species have been recorded belonging to the taxa: Actinocephalus, Euspora, Gamocystis, Gregarina, Hirmocystis, Hyalospora, and Leidyana (Levine, 1988a).

20.3.1.1.1 Eugregarines

The eugregarines are probably the most frequently encountered protozoa in insects and the most innocuous. Moderate infestations of the gregarine *Gregarina garnhami* (Figure 20.2) reduce the weights of male ledglings (new adults) of desert locust *Schistocerca gregaria* without causing observable effects on molting or feeding. The larvae of deerfly *Chrysops fuliginosus* are infected with the eugregarine *Cometoides pechumani*. The six species in genus Trogoderma are infected with a eugregarine *Pyxinia* sp. The gregarines in the mosquito *Aedes aegypti* interfere with the development of dog heartworm *Dirofilaria immitis*, for which the mosquito serves as a vector. The gregarine trophozoites destroy epithelial cells of the digestive tract and Malpighian tubules, which result in fewer cells available for filarial growth. Also, the Malpighian tubules of infected mosquitoes are shorter and deformed, and the weights of infected pupae are significantly lower than those of uninfected pupae (Perkins et al., 2000).

20.3.1.1.2 Neogregarines

Some species exhibit high pathogenicity and control potential, which undoubtedly can be attributed to one or more reproductive cycles in the life of these organisms. The two neogregarines that have received the most attention are *Mattesia grandis*, described from the boll weevil *Anthonomus grandis*, and *Mattesia trogodermae*, described from the khapra beetle *Trogoderma granarium* (Levine, 1988b).

20.3.1.2 Coccidians

They are commonly parasites of the epithelial cells of the gut, but may infect other tissues. Coccidia *Adelina tribolii* (Figure 20.3), infects several stored product pests and causes striking epizootics in populations of the confused flour beetle *Tribolium confusum*. Also, an epizootic by a species of Adelina in a laboratory colony of the wax moth *Galleria mellonella*, suggests that the coccidian might be used to control wax moths in beehives (Duszynski et al., 2004).

20.3.1.3 Haemosporidians

The Haemosporidia have more complex lifecycles that alternate between an arthropod and a vertebrate host. The trophozoite parasitizes erythrocytes or other tissues in the vertebrate host. Microgametes and macrogametes are always found in the blood. The gametes are taken up by the insect vector during a blood meal. The microgametes migrate within the gut of the insect vector and fuse with the macrogametes. The fertilized macrogamete now becomes an ookinete, which penetrates the body of the vector. The ookinete then transforms into an oocyst and divides initially by meiosis and then by mitosis (haplontic lifecycle) to give rise the sporozoites. The sporozoites escape from the oocyst and migrate within the body of the vector to the salivary glands where they are injected into the new vertebrate host when the insect vector feeds again (Morrison, 2009).

20.3.2 Sarcomastigophora

The phylum Sarcomastigophora belongs to the Protista or protoctista kingdom and includes many unicellular or colonial, autotrophic, or heterotrophic organisms. It gets its name from the combination of "Sarcodina" (which is an older term used for amoeboids) and "Mastigophora" (an older term for flagellates). It is divided into three subphyla: the Mastigophora, the

Sarcodina, and the Opalinata. It can be described either as kingdom Protista—phylum Sarcomastigophora, or in older classifications as phylum Protozoa—subphylum Sarcomastigophora. It is characterized by flagellae, pseudopodia, or both, reproduction asexual, but when sexually, it is essentially by syngamy and spore formation is absent (Katherine and Dunster, 1996; Pawlowski and Burki, 2009).

20.3.2.1 Flagellates

The association of flagellates with insects varies, ranging from mutualism and commensalism to apparent parasitism. Many are digenetic and spend part of their life cycle in an insect, most often Hemiptera or Diptera, and part in other animals or plants. Flagellate causes infections by *Herpetomonas swainei* in a population of the sawfly *Neodiprion swainei* occurring in all larval instars, eonymphs, and adults. The flagellate *Blastocrithidia caliroae* is isolated from quarantined eonymphs of the pear-slug *Caliroa cerasi*. Also the flagellate *Herpetomonas chatoni* is observed in several noctuid adults (*Discestra trifolii*, *Agrotis exclamationis*, and *A. segetum*) caught in light traps (Boucias and Pendland, 1998).

20.3.2.2 Amoebidae

The *Malameba locustae* (Figure 20.4), which infects grasshoppers and some crickets, is probably the most important entomophilic amoeba from the standpoint of natural or applied control of a host. This has been demonstrated in reports of a high natural incidence of *M. locustae* in the brown locust *Locustana pardalina*. Infections by *M. locustae* also have been observed in silverfish (Thysanura) from a colony that is started with insects collected in a culture room containing desert locusts *Schistocerca gregaria* that are heavily infected with the amoeba.

20.3.3 Ciliophora

Members of the phylum Ciliophora are protistans possessing cilia in at least one stage of their life cycle, and having two different types of nuclei: one macronucleus and one or more micronuclei. Only one ciliophoran, *Balantidium* coli, infects humans that also parasitizes pigs and monkeys.

20.3.3.1 Ciliates

Most entomophilic ciliates are associated with dipterans, primarily mosquitoes and chironomids. Treeholes and similar container-type mosquito habitats represent breeding sites for some of the most effective mosquito vectors of important human pathogens (e.g., *Ae. aegypti* and yellow fever and dengue fever, *Ae. polynesiensis* and filariasis, *Ae. albopictus* and dengue, and *Ae. sierrensis* and its role as a vector of the dog health importance of *Ae. triseriatus* and La Crosse Virus). Biological control of mosquitoes with an endoparasite *Lambornella clarki* (Tetrahymenidae: Ciliophora), has potential as a manipulated biological control agent of the western treehole mosquito *Aedes sierrensis* (Jirivavra, 1997; Sarwar, 2014; 2015). A number of advancements can be made to improve production efficiency and biocontrol potential of *L. clarki* (Egerter et al., 1986).

One other microbial control agent *Chilodonella uncinata* (Figure 20.5) with many biological control properties is known to have control over mosquito vector of Japanese encephalitis. Bioassay indicated that a dose of this protozoan formulation can be used as a potential biopesticide to reduce vectors of malaria, dengue, Chikungu, and filariasis in the context of integrated vector management (Das, 2019).

20.3.4 Microspora

Microsporidia are ubiquitous in the environment and infect almost all invertebrates and vertebrates, as well as some protists. Microsporidia have been initially described with the identification of *Nosema bombycis* as the organism responsible for the disease pebrine in silkworms. Spores from microsporidia are commonly found in surface water. For insects, fish, laboratory rodents, and rabbits microsporidia are important pathogens; and they have been investigated as biological control agents for destructive species of insects (Izquierdo et al., 2011; Weiss and Becnel, 2014). Several species of microsporidia have caused significant agricultural economic losses including *Nosema apis* and *Nosema ceranae* in honeybees (Fries, 1993), *Loma salmonae* in salmonid fish (Kent et al., 1989), and *Thelohania* sp., in shrimp (Overstreet, 1973).

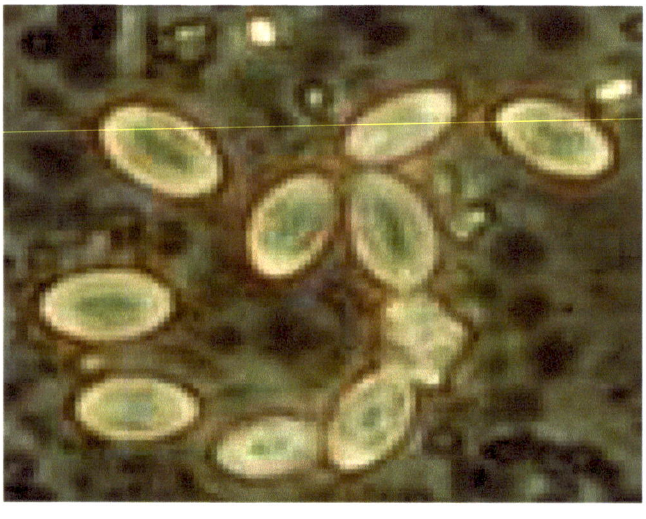

FIGURE 20.4 *Malameba locustae.*

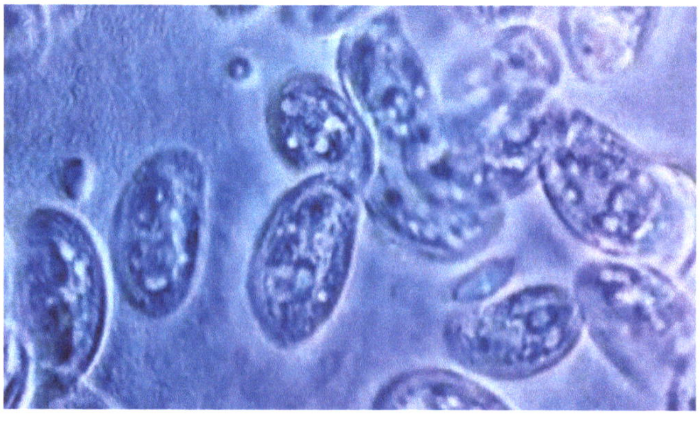

FIGURE 20.5 *Chilodonella uncinata.*

Some microsporidia are being investigated as microbial insecticides, and at least one is available commercially, but the technology is new and work is needed to perfect the use of these organisms (Usta, 2013).

20.3.4.1 Life Cycle

The general features of the life cycle are: (1) spores are ingested or inhaled and then germinate, resulting in extension of the polar tube, and injection of the sporoplasm into the host cell cytoplasm; (2) merogony ensues as injected sporoplasm develops into meronts (the proliferative stage), which multiply, depending on the species, by binary fission or multiple fission, forming multinucleate plasmodial forms; (3) sporogony follows merogony, and the meront cell membranes thicken forming sporonts; (4) after dividing sporonts give rise to sporoblasts that develop into mature spores without additional multiplication; and (5) the host cell becomes distended with mature spores, ruptures, and releases spores into the environment (Canning, 1991).

20.3.4.2 Mode of Action and Symptoms

Most microsporidia must be eaten to infect an insect, but there may also be some natural transmission within a pest population, for example by predators and parasitoids. The pathogen enters the insect body via the gut wall, spreads to various tissues and organs, and multiplies; sometimes causing tissue breakdown and septicemia. Infected insects may be sluggish and smaller than normal, sometimes with reduced feeding and reproduction, and difficulty in molting. Death may follow if the level of infection is high. One advantage by this type of infection is that the weakened insects are more likely to be susceptible to adverse weather and other mortality factors (Han and Weiss, 2017).

They have a very unique invasion organelle, the polar tube, which upon appropriate environmental stimulation rapidly discharges out of the spore, pierces a cell membrane, and serves as a conduit for sporoplasm passage into a host cell (Weiss and Becnel, 2014).

20.3.4.3 Some Beneficial Microsporidia

A microsporidium *Nosema pyrausta* (*Perezia pyraustae*) (Figure 20.6) infects several insect species, including European corn borer, for which it can be an important natural control agent. Infection can spread from diseased to healthy larvae via contaminated frass and by migration of infected larvae between plants.

Microsporidium *Nosema locustae* (Figure 20.7) is the commercially available species, and marketed under several labels for the control of grasshoppers and crickets.

Another microsporidium *Vairimorpha necatrix* (Figure 20.8) has great commercial potential. It has a wide host range among caterpillar pests, including corn earworm and European corn borer, various armyworms, fall webworm, and cabbage looper.

20.3.4.3.1 Forest and Tree Crop Pests

The host-pathogen relationship exists between the spruce budworm *Choristoneura fumiferana* and the protozoan *Nosema fumiferanae* in pine forests. Larvae that are infected early in their development from parental infections fail to mature and they die. Infection with common microsporidium *Nosema tortricis* is the most involved in the epizootic, along with *Octosporea viridanae*, *Thelohania weiseri*, and *Plistophora* sp., resulting reduction in tortrix densities. Occurrence of microsporidia in the codling moth, *Laspeyresia pomonella*, reveales that *Nosema carpocapsae* infections predispose the overwintering larvae to the fungus *Beauveria bassiana* (Tanada and Kaya, 1993).

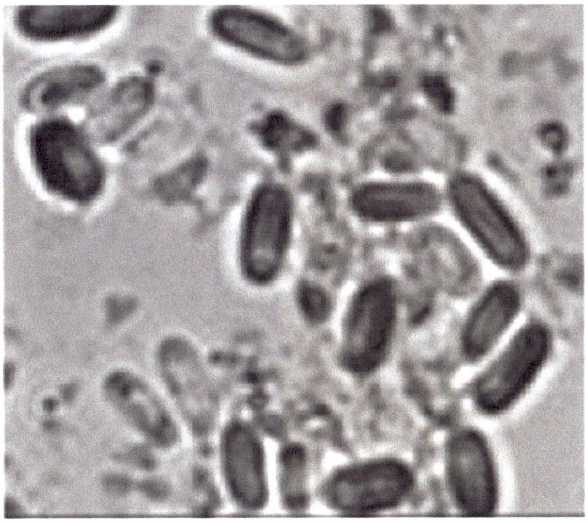

FIGURE 20.6 *Nosema pyrausta*.

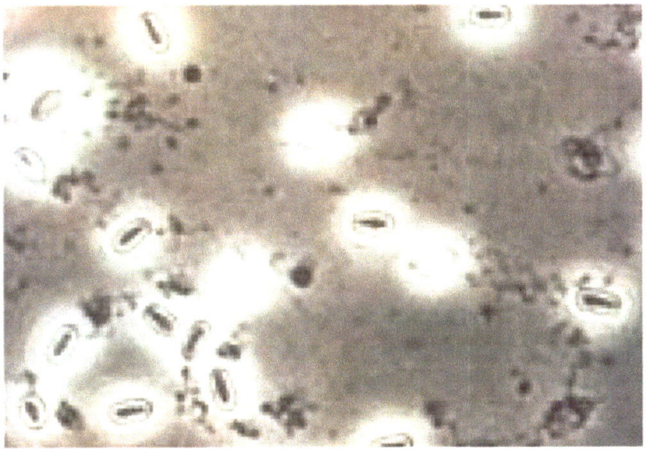

FIGURE 20.7 Spores of *Nosema locustae*.

Single applications of doubled spore dosages, 5.0×10^{10} spores per tree, increased infections. *P. schubergi* infects a wide variety of forest Lepidoptera and Hymenoptera, and it has been tested against larvae of *Anisota senatoria* and *Symmerista canicosta* on red oak trees resulting in more increased infectivity at four days postapplication. The *P. schubergi* has been combined with *Nosema disstriae* in equal concentrations and applied at the rate of 1.8×10^{11} spores per trembling aspen *Populus tremuloide* against the forest tent caterpillar *Malacosoma disstria* increased infection over the natural incidence (Weinzierl and Tess, 1989).

20.3.4.3.2 Crop and Storage Pests

There are natural host-pathogen relationships between *Nosema pyrausta* and the European corn borer *Ostrinia nubilalis*. The females from a disease-free population produce about twice as

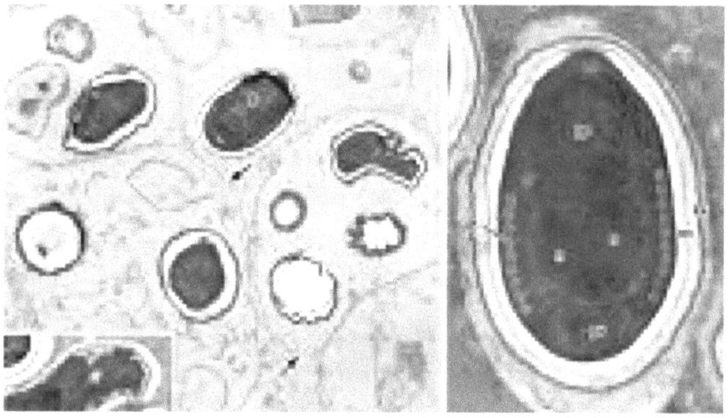

FIGURE 20.8 *Vairimorpha necatrix.*

many eggs as females from populations with infected larvae. The corn earworm *Heliothis zea* is commonly infected in nature by the microsporidium *Nosema heliothidis* that has been considered an important factor in the regulation of pest population. The presence of three grasshopper species, *Oedaleonotus enigma, Melanoplus bivittatus*, and *Melanopolus sanguinipes*, is important for the persistence and increase of infections by *N. locustae.* Generally, infections first appear in *O. enigma* and *M. bivittatus* early in the season; these species produce inoculum for infection of *M. sanguinipes*, which then produces inoculum for infection of the next generation (Hazard, 1981).

As with most of protozoans, *N. locustae* is not particularly virulent and therefore is most useful as a long-term control agent of rangeland grasshoppers, is compatible at low concentrations (10% actual ingredients) and its wheat bran formulations can be tank-mixed with chemical insecticides (Henry and Oma, 1981).

Studies establishe that *N. cuneatum* and *N. acridophagus* are infective for the corn earworm also reveale that more *N. acridophagus* can be produced per insect in the corn earworm than in grasshoppers, but both insects produced about equal concentrations of *N. cuneatum* (Ewen and Mukerji, 1980).

A dimorphic spore-forming microsporidium, *Vairimorpha necatrix*, previously named *Nosema necatrix* and *Thelohania diazoma*, has been isolated from at least 14 species of field-collected lepidopteran larvae. *V. necatrix* also infects other species of Lepidoptera, including such important agricultural pests as *Pseudaletia unipuncta, Heliothis zea, Heliothis virescens* (Sarwar, 2017a; Sarwar et al., 2009; 2011), *Autographa californica, Trichoplusia ni, Spodoptera frugiperda, Spodoptera exigua, Spodoptera ornithogalli* (Sarwar, 2017b), *Plathypena scabra, Pseudoplusia includens*, and *Hyphantria cunea*, which appear useful against the black cutworm *Agrotis ipsilon* (Sarwar, 2019). Also, the application of *V. necatrix* on cabbage infested with *Trichoplusia ni* and *Pieris rapae*, and soybeans or sorghum attacked by the green cloverworm *Plathypena scabra*, significantly provides crop protection. The storage of *V. necatrix* spores on dry corn meal is excellent, and survival is best when spores are stored at −15°C in water, water plus glycerin, and cadavers in glycerin (Hoffmann and Frodsham, 1993).

The larvae of the subsequent generation could become infected by consuming food contaminated by contact with infected adults, consuming cadavers of infected adults, and transovarial or transovum transmission (Maddox, 1981).

20.3.4.3.3 Mosquitoes

Microsporidia have a role in regulating natural mosquito populations, but when colonies of Aedes, Anopheles, and Culex are maintained, infections occur only in the anophelines. The incidence of infection among *Anopheles albimanus* is 75%, with few becoming adults (Sarwar; 2020). In one test, *Plistophora (Vavraia) culicis* of about 6,000 to 7,000 spores per ml has been applied to small pools of water inhabited with *Culex pipiens fatigans* (*C. quinquefasciatus*) that resulted in population suppression and microsporidium persisted for 24 months postinoculation. Microsporidium *N. algerae* exhibits a high pathogenicity for anopheline mosquitoes that can interfere with the ability of mosquitoes to vector malarial organisms. In a field test against *Anopheles albimanus, N. algerae* applied at rates of 2.15×10^7 to 2.15×10^9 spores per m^2 gave some mortality among larvae with massive spore concentrations (Henry, 1981).

20.4 Conclusion and Future Prospects

Entomopathogenic protozoans commonly referred to as microsporidians have holozoic nutrition. That is, they require organic materials, which may be preformed organic substances particulate or in solution. For insects, protozoa are important pathogens and they have been investigated as biological control agents for destructive species of pests. Frequently they reduce host reproduction or feeding rather than killing the pest outright. Microsporidia are generally host-specific, slow-acting, and produce chronic infections with general debilitation of the host insects. As the body of information expands, efficacy of protozoa when used in baits or on particulate carriers is becoming increasingly clear, so the aspects of their applied use require more attention. Laboratory mass production and field application of the mosquito pathogens could result in their successful use as biological control agents with excellent potential for self-persistence in treated treeholes.

Some microsporidia are being investigated as microbial insecticides, and a few are available commercially, but the technology is new and work is needed to perfect the use of these organisms. Development of new and specific formulations, application techniques, and equipment will be further required. As the course of protozoa use continues and more are brought into pests control operation, it will then be better to further evaluate the true potential of these organisms. There is a scope to utilize the biodiversity of the entomophathogen group of protozoa in the context of integrated pest management wherein simpler, easier, and cheaper mass production techniques have to be developed.

REFERENCES

Baker, C.; Henis, J.; Baker, R.; Dunn, P. Commercial production and formulation of microbial biocontrol agents. New directions in biological control. Alternatives for suppressing agricultural pests and diseases. Proceedings of a UCLA Colloquium, Frisco, CO, January 20–27, 1989. Alan R. Liss, Inc. 1990, pp. 333–344.

Boucias, D. G.; Pendland, J. C. Insect pathogenic protozoa. In: *Principles of Insect Pathology*. Springer, Boston, MA. 1998, p. 550.

Brooks, W. M. Entomogenous protozoa. In: *Handbook of Natural Pesticides, Microbial Insecticides, Part A*, Ignoffo, C. M., Mandava, N. B. (Eds.), Entomogenous Protozoa and Fungi. CRC Press, Baco Raton, FL. 1988, pp. 1–149.

Canning, E. U. *Phylum Microspora*. Jones and Bartlett, Boston, MA. 1991, pp. 53–72.

Capella-Gutierrez, S.; Marcet-Houben, M.; Gabaldon, T. Phylogenomics supports microsporidia as the earliest diverging clade of sequenced fungi. *BMC Biol*. 2012, 10: 47.

Corsaro, D.; Michel, R.; Walochnik, J.; Venditti, D.; Muller, K. D.; Hauroder, B.; Wylezichl, C. Molecular identification of *Nucleophaga terricolae* sp. nov. (Rozellomycota), and new insights on the origin of the Microsporidia. *Parasitol. Res*. 2016, 115: 3003–3011.

Das, B. P. *Chilodonella uncinata* – As potential protozoan biopesticide for mosquito vectors of human diseases. *Acta Sci. Microbiol*. 2019, 2 (12): 90–96.

Duszynski, D.W.; Upton, S.J.; Couch, L. *The Coccidia of the World*. Department of Biology, University of New Mexico, and Division of Biology, Kansas State University. 2004.

Egerter, D. E.; Anderson, J.R.; Washburn, J.O. Dispersal of the parasitic ciliate *Lambornella clarki*: Implications for ciliates in the biological control of mosquitoes. *Proc. Nati*. Acad. Sci. USA. 1986, 83: 7335–7339.

Englund, P. T.; Sher. A. The biology of parasitism. In: *A Molecular and Immunological Approach*. Alan R. Liss, New York. 1988, p. 499.

Ewen, A.B.; Mukerji, M.K. Evaluation of *Nosema locustae* as a control agent of grasshopper populations in Saskatchewan. *J. Invertebr. Pathol*. 1980, 35: 295–303.

Fries, I. *Nosemaapis*—A parasite in the honey bee colony. *Bee World*. 1993, 74: 5–19.

Han, B.; Weiss, L.M. Microsporidia: Obligate intracellular pathogens within the fungal kingdom. *Academic Press, Lond. Microbiol. Spectr*. 2017, 5 (2): 17.

Hazard, E. I. Identification of microsporidia. In: *Microbial Control of Pests and Plant Diseases, 1970–1980*, H. D. Burges (Ed.). 1981, 163–182.

Henry, J. E. Natural and applied control of insects by protozoa. *Ann. Rev. Entomol*. 1981, 26: 49–73.

Henry, J. E.; Oma, E. A. Protozoa: Pest control by *Nosema locustae*, a pathogen of grasshoppers and crickets. In: *Microbial Control of Pests and Plant Diseases, 1970–1980*, Burges, H. D. (Ed.). 1981, pp. 573–586.

Hoffmann, M. P.; Frodsham, A. C. *Natural Enemies of Vegetable Insect Pests*. Cooperative Extension, Cornell University, Ithaca, NY. 1993, 63 pp.

Izquierdo, F.; Castro-Hermida, J. A.; Fenoy, S.; Mezo, M.; Gonzalez-Warleta, M.; DelAguila, C. Detection of microsporidia in drinking water, wastewater and recreational rivers. *Water Res*. 2011, 45: 4837–4843.

Jirivavra, A. H. U. Research methods for entomopathogenic protozoa. In: *Manual of Techniques in Insect Pathology*. Academic Press, San Diego, California. 1997, 117–151.

Katherine, D.; Dunster, J. A. *Dictionary of Natural Resource Management*. UBC Press, Vancouver. 1996, p. 337.

Kent, M. L.; Elliott, D. G. Groff, J. M.; Hedrick, R. P. *Loma salmonae* (Protozoa: Microspora) infections in seawater reared coho salmon *Oncorhynchus kisutch*. *Aquaculture*. 1989, 80: 211–222.

Lacey, L. A.; Grzywacz, D.; Shapiro-Ilan, D. I.; Frutos, R.; Brownbridge, M.; Goettel, M. S. Insect pathogens as biological control agents: Back to the future. *J. Invertebr. Pathol*. 2015, 132: 1–41.

Lange, C. E.; Brito, J. M.; Henry, J. E. Characteristics of a microsporidium (protozoa: microspora) infecting grasshoppers (Orthoptera: Pyrgomorphidae) in Cape Verde, Africa. *J. Protozool*. 1992, 39 (4): 494–498.

Lee, J. J.; Hutner, S.H.; Bovee, E.C. *An Illustrated Guide to the Protozoa*. Society of Protozoologists, Lawrence, KS. 1985, pp. 16–17.

Levine, N. D. Progress in taxonomy of the Apicomplexan protozoa. *J. Protozool*. 1988a, 35 (4): 518–520.

Levine, N. D. *The Protozoan Phylum Apicomplexa*. CRC Press, Boca Raton, FL.1988b, pp. 1–2.

Maddox, J. Y.; Brooks, W. M.; Fuxa, J. R. *Vairimorpha necatrix*, a pathogen of agricultural pests. In: *Microbial Control of Pests and Plant Diseases, 1970–1980*, Burges, H. D. (Ed.). 1981, pp. 587–594.

Morrison, D. A. Evolution of the Apicomplexa: Where are we now? *Trends Parasitol*. 2009, 25 (8): 375–382.

Overstreet, R. M. Parasites of some penaeid shrimps with emphasis on reared hosts. *Aquaculture*. 1973, 2: 105–140.

Pawlowski, J.; Burki, F. Untangling the phylogeny of amoeboid protists. *J. Eukaryotic Microbiol*. 2009, 56 (1): 16–25.

Perkins, F. O.; Barta, J. R.; Clopton, R. E.; Peirce, M. A.; Upton, S. J. Phylum apicomplexa. In: *An Illustrated Guide to the Protozoa: Organisms Traditionally Referred to as Protozoa, or Newly Discovered Groups*, Lee, J. J., Leedale, G. F., Bradbury, P. (Eds.). 2nd ed. Society of Protozoologists, Lawrence,Kansas, USA. 2000, 190–369.

Saffo, M. B.; McCoy, A. M.; Rieken, C.; Slamovits, C. H. Nephromyces, a beneficial apicomplexan symbiont in marine animals. *Proc. Natl. Acad. Sci*. 2010, 107 (37): 16190–16195.

Sarwar, M. Characteristics of armyworm *Spodoptera litura* (Fabricius) (Lepidoptera: Noctuidae) occurrence and control on cotton. *Scholars J. Res. Agric. Biolo*. 2017a, 2 (1): 33–38.

Sarwar, M. Cotton bollworm *Helicoverpa armigera* (Hubner, 1809) (Lepidoptera: Noctuidae) and development of integrated pest management platform. *Scholars J. Res. Agric. Biol*. 2017b, 2 (1): 63–70.

Sarwar, M. A Guide to the honey bee's protozoan nosema disease and treatment recommendations for pest control. *Int. J. Res. Appl. Phys*. 2016, 1 (7): 1–11.

Sarwar, M. Multiple integrated pest management (IPM) and area-wide management (AWM) approaches for insect pests of chickpea *Cicer arietinum* (L.) Walp crop. In: *Handbook of Chickpeas: Nutritional Value, Health Benefits and Management*, Lund, A. T., Schultz, N. D. (Eds.). Nova Science Publishers, Inc., New York. 2019, pp. 237–263.

Sarwar, M. Proposing solutions for the control of dengue fever virus carrying mosquitoes (Diptera: Culicidae) *Aedes aegypti* (Linnaeus) and *Aedes albopictus* (Skuse). *J. Pharmacol. Toxicol. Stud*. 2014, 2 (1): 1–6.

Sarwar, M. Source reduction practices for mosquitoes (Diptera) management to prevent dengue, malaria and other arboborne diseases. *Am. J. Clin. Neurol. Neurosurg*. 2015, 1 (2): 110–116.

Sarwar M. Typical flies: Natural history, lifestyle and diversity of Diptera. In: *Typical Flies: Natural History, Lifestyle and Diversity of Diptera*, Sarwar, M. (Ed.). InTech, London, UK. 2020; p. 50.

Sarwar, M.; Ahmad, N.; Tofique, M. Host plant resistance relationships in chickpea (*Cicer arietinum* L.) against gram pod borer (*Helicoverpa armigera*) (Hubner). *Pak. J. Bot*. 2009, 41 (6): 3047–3052.

Sarwar, M.; Ahmad, N.; Tofique, M. Identification of susceptible and tolerant gram (*Cicer arietinum* L.) genotypes against gram pod borer (*Helicoverpa armigera*) (Hubner). *Pak. J. Bot*. 2011, 43 (2): 1265–1270.

Sibley, L. D. I. How apicomplexan parasites move in and out of cells. *Curr. Opin. Biotechnol.* 2010, 21 (5): 592–598.

Solter, L. F.; Becnel, J. J. Entomopathogenic microsporidia. Field manual of technique in invertebrate pathology. In: *Application and Evaluation of Pathogens for Control of Insects and other Invertebrate Pests*, Lacey, L. A., Kaya, H. K. (Eds.). Kluwer Academic, Dordrecht. 2000, 231–254.

Stefan, H. I.; Kappe, C. A.; Bergman, B. L. W.; Coppens, I.; Nussenzweig, V. Apicomplexan gliding motility and host cell invasion: overhauling the motor model. *Trends Parasitol.* 2004, 20 (1): 13–16.

Tanada, Y.; Kaya, H. K. *Insect Pathology*. Academic Press, Inc., San Diego. 1993; 666 pp.

Usta, C. Microorganisms in biological pest control—A review (bacterial toxin application and effect of environmental factors). In: *Current Progress in Biological Research*. InTech, London. 2013, pp. 287–317.

Vossbrinck, C. R.; Debrunner-Vossbrinck, B. A. Molecular phylogeny of the microsporidia: Ecological, ultrastructural and taxonomic considerations. *Folia Parasitol.* 2005, 52: 131–42.

Weinzierl, R.; Tess, H. *Alternatives in Insect Management: Microbial Insecticides*. Cooperative Extension, University of Illinois, Circular 1295. 1989, 12 pp.

Weiss, L. M.; Becnel, J. J. *Microsporidia: Pathogens of Opportunity*. Wiley-Blackwell, Oxford. 2014, 736 p.

Williams, B. A.; Hirt, R. P.; Lucocq, J. M.; Embley, T. M. A mitochondrial remnant in the microsporidian *Trachipleistophora hominis*. *Nature*. 2002, 418: 865–869.

Yaeger, R. G. Protozoa: Structure, classification, growth, and development. In: Medical Microbiology, Baron S. (Ed). 4th ed. University of Texas, Galveston. 1996.

21

Protozoan Biopesticides: A New Horizon in Organic Farming

Pradipta Banerjee, Siddhartha Das, Sudeepta Pattanayak, and L.P. Awasthi

CONTENTS

21.1 Introduction

Organic farming is an environment-friendly approach that involves techniques to achieve good and sustainable crop yields without affecting the natural flora and fauna. It is a modern approach followed by farmers all over the world, where crops and animals can co-exist together. Organic farming has a long-term benefit, which includes increasing soil fertility, controlling pests and diseases in an environment-friendly way, producing nutritious crops and feed to get the maximum benefit out of it. In organic farming, every insect should not be treated as a pest, every plant out of place should not be considered as a weed. Furthermore, all problems related to pests and weeds should not be solved by spraying artificial chemicals. The objective of organic farming is to check the growth of weeds and pests to a threshold level so as to utilize maximum benefits that they may provide, rather than to eradicate them completely. To achieve this, biofertilizers and biopesticides are widely used by farmers all over the world. In our present topic, we will focus on the importance of biopesticides with special reference to protozoan biopesticide, which is a hot topic of research in recent times.

The inorganic, commercial pesticides that are generally available in the market have several adverse effects. They can kill the beneficial insects, which can eat the pests and with a single spray, the ecological balance between the pests and the predators will be disturbed. The chemicals used as pesticides often possess long-term effects, as they are not biodegradable and stays in the environment for an infinite period of time. This may lead to biomagnification of these chemicals and sometimes they can enter into the animal system through food and create health hazards. Moreover, the pests are capable to resist a particular chemical pesticide over a few breeding cycles via genetic modification. If a particular pest becomes resistant to a pesticide, a farmer has to use stronger chemical agents to control the pest, which in turn will damage the natural environment. Thus, the use of natural control measures to check the pests has become popular in modern-day agriculture.

21.2 Biopesticides: A Modern Approach to Control Pests

A variety of insects, bacteria, fungi, nematodes, viruses, etc. which can damage agriculture crops are regarded as pests. Often it is found that the chemical pesticides that are used to control pest populations are detrimental to natural microorganisms and indiscriminate use of these chemicals may cause various types of environmental problems (Arora et al., 2012). Today, biopesticides are widely used because they are more effective, eco-friendly, biodegradable and generally no harmful residues are left on the environment (De and De, 2019). Cheng et al. (2010) reported that production of chemical pesticides had decreased by 2% per year whereas production of biopesticides had increased by 20% per annum.

The concept of biocontrol agent is not new. In China, nests of ant *Oecophylla smaragdina* were sold near Canton in the third century for use in control of citrus insect pests such as *Tesseratoma papillosa* (Lepidoptera). In Europe (1776), the control of bedbugs (*Cimex lectularius*) was successfully achieved by releases of the predatory pentatomidae *Picromerus bidens*. It was reported that *Trichogramma* sp. were shipped from the United States to Canada for control of lepidopterous pests in

1882. Plant extracts were probably the first biopesticide used in agriculture in the 17th century. In the 19th century, research was carried out with mineral oils as plant protectant. In the first half of the 20th century, biopesticide came to the limelight and spores of *Bacillus thuringiensis* were the most widely used biopesticide. In the latter part of the 20th century, there were many chemical pesticides gained popularity, but over the course of time most of them showed adverse effects on nature.

Plants and microorganisms serve as the primary sources of biopesticides due to the presence of bioactive compounds and antimicrobial agents (Nefzi et al., 2016). The bioactive compounds present in plants include quinones, phenols, steroids, terpenes, alkaloids, and saponins (Mizubuti, 2007). Generally, biopesticides are made up of living organisms which are extremely pathogenic for only one target species. Bacterial species such as *Pseudomonas fluorescens*, *Bacillus popilliae*, *Bacillus thuringiensis*, etc., fungal species such as *Beauveria bassiana*, *Verticillium lecanii*, *Paecilomyces fumosoroseus*, etc., and viruses such as *Neodiprion sertifer*, *Lymantria dispar*, etc., are some popular biopesticides (De and De, 2019). The pests cannot easily develop resistance against these biopesticides, unlike chemical pesticides.

As per the published reports so far, more than 1,500 species of fungi, viruses, bacteria, and protozoa can infect arthropod species (Fountain and Wratten, 2013). The first attempt to use them as bioinsecticides was done by a Russian scientist called Elie Metchnikoff, in 1884. He used *Metarhizium anisopliae* to infect two invertebrate pests, namely, cereal cockchafer, *Anisoplia austriaca*, and the sugar-beet weevil, *Bothynoderes punctiventris*. The first commercially available bioinsecticide was Bt, produced in the 1930s.

21.3 Types of Biopesticides

Biopesticides are categorized into three important classes, as per the US Environmental Protection Agency:

1. Microbial pesticides
2. Plant-incorporated protectants
3. Biochemical pesticides

21.3.1 Microbial Pesticides

These constitute the largest group of broad-spectrum pest-specific biopesticides. Microbial biopesticides are products derived from microorganisms themselves or the metabolites they produce. They are capable of controlling different kinds of pests, although each individual active ingredient is relatively specific for its target pest(s). For instance, there are certain fungi that can control some weeds. Again, there are some other fungi that are capable of killing specific insects. Microbial biopesticides can be sub-categorized as:

- Fungal biopesticides
- Bacterial biopesticides
- Viral biopesticides
- Protozoan biopesticides

21.3.2 Plant-Incorporated Protectants (PIPS)

These pesticides are natural substances that plants can produce from genetic material that has been added to the plant. By means of genetic engineering, the gene for the BT pesticidal protein can be taken out and introduced into the plant's own genetic material. The genetically modified plant is then capable of producing the elements that destroy the pest.

21.3.3 Biochemical Pesticides

The modes of action of these naturally occurring substances are non-toxic to the environment. Sex hormone, scented plant extracts, secondary metabolites of plants, alkaloids, suberins, phenols terpenes, flavonoids, carotenoids, mineral oils, etc. can act as biopesticides. Insect sex pheromones interfere with mating of insects. Scented plant extracts can attract insect pests to trap them. The Environmental Protection Agency (EPA) has put forward a committee to describe whether a substance meets the criteria for a biochemical pesticide. A biochemical biopesticide is mainly involved in developing plant immune system and defending them against both microbial pathogens and invertebrate pests (Golob et al., 1999).

In areas where niche markets, pesticide resistance, and environmental concerns act as limiting factors for the use of conventional chemical pesticide products, biopesticides play a significant role in providing pest management. Broadly speaking, biopesticides have a diverse spectrum of applications in controlling not only pests, but also weeds, plant diseases, and nematodes. Some examples of classical biopesticides include:

Insect Control

Bacteria	*Bacillus thuringiensis*, *B. sphaericus*, *Paenibacillus popilliae*, *Serratia entomophila*
Fungi	*Beauveria* spp., *Metarhizium*, *Entomophaga*, *Zoopthora*, *Paecilomyces fumosoroseus*, *Normuraea*, *Lecanicillium lecanii*
Virus	Nuclear polyhedrosis viruses, granulosis viruses, non-occluded baculoviruses
Protozoa	*Nosema*, *Thelohania*, *Vairimorpha*
Entomopathogenic nematodes	*Steinernema* spp., *Heterorhabditis* spp.
Others	Pheromones, parasitoids, predators, microbial by-products

Weed Control

Fungi	*Colletotrichum gloeosporioides*, *Chondrostereum purpureum*, *Cylindrobasidium laeve*
Bacteria	*Xanthomonas campestris* pv. *Poannua*

Plant Disease Control

Fungi	*Ampelomyces quisqualis*, *Candida* spp., *Clonostachys rosea f. catenulate*, *Coniothyrium minitans*, *Pseudozyma flocculosa*, *Trichoderma* spp.
Competitive and Soil Inoculants	*Bacillus pumilus*, *B. subtilis*, *Pseudomonas* spp., *Streptomyces griseoviridis*, *Burkholderia cepacia*

Nematicides

Bacteria	*Bacillus firmus, Pasteuria penetrans*
Mollusc parasitic nematode	*Phasmarhabditis hermaphrodita*
Nematode trapping fungi	*Myrothecium verrucaria, Paecilomyces lilacinus*

21.4 Recent Trends in Biopesticides Using Protozoa

As is evident from the above discussions, biopesticides play an important role in controlling pests, diseases, weeds, and nematodes. Microbial, biochemical, and plant-incorporated protectants are three common biopesticide types used widely. Microbial biopesticide, especially, protozoan biopesticide is a popular research topic in recent times. There are only a handful of reports available regarding uses of protozoa as biopesticide. Thus, this is an interesting topic of research for the microbiologists working on biopesticide formulation.

Protozoans are primitive organisms that are available widely in soil, water and can be cultured in easy, inexpensive methods.

21.4.1 Biology of Protozoa in General

Protozoa are single-celled eukaryotes existing in both water and soil. Some species of protozoa are responsible for serious human diseases, such as malaria, vectored by mosquitoes. Most species of entomopathic protozoa cause chronic infections that weaken but do not kill their host. Protozoans are small usually microscopic and can be seen with naked eyes. They are the simplest and most primitive organism with protoplasmic grade of organization. Their body is unicellular with one or more nuclei. Protozoan body is naked or covered by pellicle. In some forms, the body is covered with shells and often provided with an internal skeleton. They can be found alone or form loose colonies in which individuals remain alike and independent. Protozoa can be symmetrical, non-symmetrical, bilateral, radial, and spherical with body form usually constant, varied in some, while changing with environment or age in many. Protoplasm is differentiated into an outer ectoplasm and inner endoplasm. Locomotary organelles are pseudopodia, flagella, and cilia, or sometimes locomotary organs are absent. Nutrition may be holozoic (animal-like), holophytic (plant-like), saprozoic, or parasitic. Digestion is intracellular, and takes place inside the food vacuoles. Respiration occurs by diffusion through general body surface. Excretion occurs through contractile vacuoles or generally through body surface. Reproduction may take place asexually (binary or multiple fission) or sexually (conjugation of adults, hologamy, or by fusion of gametes, syngamy). Alternation of generation observed, i.e., life cycle includes asexual and sexual phases. Cyst formation is common to resist unfavorable conditions. Protozoans are ubiquitous or cosmopolitan, free-living species are usually aquatic (marine or freshwater). Several protozoans are commensal, symbiotic, and parasitic species. About 50,000 known species of protozoa have been reported, of which some are fossil and approximately 10,000 are parasitic.

21.4.2 Protozoan Biopesticides

There are not many studies regarding protozoan biopesticides, although research is ongoing for development of more efficient biopesticides using protozoa. Microsporidians or entomopathogenic protozoans form an extremely diverse group comprising of nearly 1,000 species that can attack invertebrates including insect species (Kachhawa, 2017; Solter et al., 2000). Microsporan protozoans are used as possible components of integrated pest management (IPM) (Anonymous, 2020). They are ubiquitous, obligate intracellular parasites. One of the more well-known protozoan microsporan protozoa is *Nosema locustae,* which a natural biocontrol agent of many grasshopper species. *Nosema* infects at least 90 species of grasshoppers, although it is non-toxic to humans and other mammals. It infects and weakens young grasshoppers and adversely affects female grasshoppers' reproductive ability.

Another species of these genera is *Nosema pyrausta*, which is a biocontrol agent for larvae of European corn borer. (Lacey and Goetell, 1995; Siegel et al., 1986). They can be characterized as slow-acting and host-specific, capable of producing chronic infections with general debilitation of the host. These protozoans form infectious spores and if these spores are ingested by the host insect, they can cause severe pathogenicity. The spore germinates in the host's midgut and sporoplasm is eventually released, invading the target cells and causing infection. Nosema can infect beetles, lepidoptera, locusts, and other orthoptera orally. Spores of size 3 × 5 µm, germinate in the gut of the host and reside in cells of fat body, gut, and other tissues. The spores then replicate and sporulate, resulting in organ destruction of the host insect. The infection is often chronic, devastating, and transmitted by contaminated food, cannibalism, and transovarially (Burges and Jones, 1998). Mites (*Hypoaspis miles*) are very effective against Sciarid larva and other invertebrates while *Malameba locustae* and *Mattesia* spp. are known to be antagonistic to insects such as lepidoptera, grasshoppers, and coleopteran (Kumari et al., 2014). Vairimorpha and Malamoeba are genuses of microsporidian parasites which have proven their potential against Lepidoptera and locusts, respectively. When ingested, these protozoans interfere with the pests' feeding behavior and starve the pests to death.

21.5 Conclusion

Most protozoan biopesticides are highly specific for their host and show persistent stages. The only drawback is that they cause chronic infection and often slow-acting compared to bacterial and fungal biopesticides. The choice of biological control agents of plant should be appropriate to get the maximum yield of a particular crop. The process of pest control should be fast, inexpensive, and must completely control natural enemies. To get consistent control of pests, the inherent complex relationship between the agent, the pest, the timing, and the environment is critical (Fountain and Wratten, 2013). Warm, humid conditions in the tropics are best suited for biopesticide application. Change in environmental conditions affects both the pest and the biological control agent. Moreover, the health condition of the pest impacts on resistance to fungal and protozoan attacks, with respect to pest age, nutrition, genetics, and physical damage through various mechanical, chemical, or biological means. Researchers can influence pest health and increase the susceptibility of the pest to biopesticides, by simply using resistant crop cultivars or manipulating the habitat to alter pest nutrition. The application of biopesticides should be optimized considering the environmental factors.

REFERENCES

Anonymous. 2020. Society for invertebrate pathology—Microbial control division biopesticides. http://www.sipweb.org/microspor/div_micro.html (Accessed Jan 5, 2020).

Arora, N. K.; Tewari, S.; Singh, S.; Lal, N.; Maheshwari, D. K. PGPR for protection of plant health under saline conditions. In *Bacteria in Agrobiology: Stress Management*; Maheshwari, D. K., ed.; Springer, Berlin, 2012, pp. 239–258.

Burges, H. D.; Jones, K. A. Formulation of bacteria, viruses and protozoa to control insects. In *Formulation of Microbial Biopesticides*; Burges, H. D., ed.; Springer, Dordrecht, 1998, pp. 31–127.

Cheng, X. L.; Liu, C. J.; Yao, J. W. The current status, development trend and strategy of the bio-pesticide industry in China. *Hubei Agricultural Sciences*. 2010, 49, 2287–2290.

De, T.; De, L. C. Bio-pesticides: A viable tool for organic farming. *International Journal of Microbiology Research*. 2019, 7, 1660–1664.

Fountain, E. D.; Wratten, S. D. Conservation biological control and biopesticides in agricultural. In *Encyclopedia of Ecology (2nd Edition)*; Elsevier, USA, 2013, Vol. 1, pp. 377–381.

Kachhawa, D. Microorganisms as a biopesticide. *Journal of Entomology and Zoology Studies*. 2017, 5 (3), 468–473.

Kumari, S.; Chourasia, S. K.; Jha, M. N.; Kant, R.; Singh, U.; Kumar, P. Microbial pesticide: A boom for sustainable agriculture. *International Journal of Scientific & Engineering Research*. 2014, 5 (6), 1394.

Lacey, L. A.; Goettel, M. S. Current developments in microbial of insect pests and prospects for early 21st Century. *Entomophaga*. 1995, 40, 3–27.

Mizubuti G. S. E.; Junior V. L.; Forbes G. A. Management of late blight with alternative products. *Pest Technology*. 2007, 2, 106–116.

Nefzi, A.; Abdallah, B. A. R.; Jabnoun-Khiareddine, H.; Saidiana-Medimagh, S.; Haouala, R.; Danmi-Remadi, M. Antifungal activity of aqueous and organic extracts from *Withania somnifera* L. against *Fusarium oxysporum* f. sp. *radicis-lycopersici*. *Journal of Microbial and Biochemical Technology*. 2016, 8, 144–150.

Siegel, J. P.; Maddox, J. V.; Ruesink, W. G. Lethal and sublethal effects of *Nosema pyrausta* on the European corn borer in Central Illinois. *Journal of Invertebrate Pathology*.1986, 48, 167–173.

Solter, L. F.; Becnel, J. J. Entomopathogenic microsporidia. Field manual of technique in invertebrate pathology. In *Application and Evaluation of Pathogens for Control of Insects and other Invertebrate Pests*; Lacey, L. A., Kaya, H. K., eds.; Kluwer Academic, Dordrecht, 2000, pp. 231–254.

Section XIII

Management of Crop Diseases through Botanical Biopesticides

22

Management of Fungal Diseases of Crops through Plant-Based Natural Products and Biopesticides

Deeksha Joshi, Monika Upadhyay, and Raghvendra Tiwari

CONTENTS

22.1 Introduction

Majority of more than 100,000 known fungi survive as saprophytes; around 10,000 fungal species are known to attack plants and are responsible for causing the great majority, almost two-thirds, of infectious plant diseases (Agrios, 2005). On average, world agriculture sustains losses of around 16% annually as a result of microbial plant diseases, with 70–80% of these losses occurring due to fungal infections alone (Oerke, 2006).

22.2 Disease Control in Organic Farming

Organic farming, as a cultivation system relies primarily on the use of various agronomic, biological, and mechanical methods, as opposed to using synthetic materials, to fulfill any specific function within the system (FAO, 1999).

As more and more land and crops are being brought under organic cultivation, effective management of diseases and insect pests poses a major challenge. Disease development in any crop occurs as a result of the interaction of three components *viz.* a virulent pathogen, a susceptible host, and favorable environmental conditions; also referred to as a "disease triangle" (Agrios, 2005). Any effective disease management strategy should aim to break this triangle by targeting one or more of these components. In this scenario, use of natural plant products and biological control agents present the most feasible alternatives before organic growers.

22.3 Plant-Based Extracts for Fungal Disease Control

Plants have the ability to synthesize a range of secondary metabolites, like phenols, phenolic acids, quinones, flavones, flavonoids, flavonols, tannins, coumarins, etc., that can be toxic to various pathogens. To date, thousands of phytochemicals with inhibitory effects on various types of plant pathogenic microbes have been reported (Cowan, 1999; Das et al., 2010). To date, hundreds of botanicals with antifungal properties have been reported based on *in vitro* and *in vivo* studies.

22.3.1 Neem and Its Derivatives

Neem (*Azadirachta indica*) and its various derivatives (oil, seed extract, kernel extract, leaf extract) are the most common botanical pesticides currently being used in organic farming. Neem is a rich source of various types of antimicrobial compounds with most important constituents being azadirachtin, nimbolinin, nimbin, nimbidin, nimbidol, sodium nimbinate, gedunin, salannin, and quercetin. The seeds, in particular, are a very rich source of neem-derived botanical pesticides and contain numerous azadirachtin analogs, but the major form is azadirachtin A. Besides direct antimicrobial activity, induction of resistance in plants by neem extract has been suggested as a mechanism for suppression of plant diseases.

Although neem extracts have mainly gained recognition for their insecticidal properties; their efficacy against plant

TABLE 22.1
Plant Extracts Effective against Different Fungal Plant Diseases

S. No.	Plant Extract	Part Used	Disease	References
1	Neem	Leaf, Seed	Anthracnose of pepper, early blight of tomato, powdery mildew of pea, rice blast, leaf spot disease of sesame, wilt disease of tomato, fruit rot in chilies	Nduagu et al., 2008; Dharam et al., 1985; Surwase et al., 2009; Kandari et al., 2007; Arroyo et al., 2007; Amadioha 2000; Prithiviraj et al., 1998; Guleria et al., 2006; Sundaramoorthy et al., 2014
2	Garlic	Bulb	Foot rot of *Phaseolus vulgaris*, citrus green and blue molds, fruit rot in chilies	Russel et al., 1997; Obagwu et al., 2003; Sundaramoorthy et al., 2014
3	Ginger	Leaf	Root rot disease of cow pea, yam tuber rot	Suleiman et al., 2009; Okigbo et al., 2005
4	Lemongrass	Leaf	Black mold disease on onion bulbs	Abdullah et al., 2006
5	Onion	Bulb	Fruit rot in chilies	Sundaramoorthy et al., 2014
6	Extracts of clove, cinnamon, and coriander	Leaves/ bark		Carmello et al., 2018
7	*Artemessiaannua*, *Eucalyptus globulus*, *Ocimum sanctum*, and *Rheum emodi*	Leaves	Wilt disease of brinjal	Joseph et al., 2008

pathogenic fungi, bacteria, and nematodes have also been documented (Locke, 1995). Soil application and sprays of neem oil have been shown to control phytopathogenic fungi like *Fusarium* sp., *Rhizoctonia solani*, *Sclerotium rolfsii*, *Sclerotinia sclerotiorum*, *Botrytis cinerea*, *Puccinia* spp., *Magnaporthe grisea*, etc., which cause diseases like rusts, scabs, powdery mildew, blast, and rots (Pasini et al., 1997). Aqueous extracts of neem leaves and seed kernel extracts showed effective control of leaf stripe of barley caused by *Drechslera graminea* (Paul and Sharma 2002). Sprays of neem extracts were are reported effective in suppressing diseases of tomato, cowpea, pea, and rice (Table 22.1).

22.3.2 Garlic

Garlic (*Allium sativum* L.) is used worldwide as a food additive, spice, and medicine. Garlic extracts are extensively recommended for disease control in India (Curtis et al., 2004) (Table 22.1). On slicing or crushing, garlic develops its characteristic odor because cellular damage leads to mixing of the vacuolar enzyme alliinlyase and its cystosolic substrate alliin which results in production of thiosulphenic acid, which undergoes spontaneous dimerization to diallylthiosulphinate or allicin. This allicin has been shown to be the major antimicrobial substance in garlic. It works by crossing the cell membrane and undergoes thiol-disulphide exchange reactions with free thiol groups in cell proteins and these proteins then form the basis of its antimicrobial action (Slusarenko et al., 2008).

Besides neem and garlic, extracts from eucalyptus (*Eucalyptus globulus*, Labill.), lemongrass, onion, and ginger (*Zingiber officinale* Rosc) have also been exploited for disease control under organic farming conditions (Table 22.1).

22.4 Microbial Pesticides for Biological Control of Diseases

Biopesticides based on the fungal bioagent *Trichoderma* spp. and bacterial bioagent *Pseudomonas fluorescens* have been used most extensively to date. Bioagents mainly work by altering the biotic and abiotic environment from one that favors disease/pathogen to one that discourages the activity of the pathogen by employing mechanisms like mycoparasitism, antibiosis, competition, and induction of systemic resistance in plants against diseases.

22.4.1 *Trichoderma* spp.

Trichoderma is a genus of filamentous deuteromycetes with only one known teleomorph, *Hypocrea* Fr. *Trichoderma* colonies display rapid growth, abundant powdery green conidia, and ill-defined conidiophores. *Trichoderma*-based biopesticides have been classified as Class IV, which indicates low hazard (green label) indicating that they are extremely safe to mammals, man, and other non-target organisms like pollinators, bees, fishes, etc., have no phytotoxicity and are exempt from residual analysis. *Trichoderma* spp. have been effectively used for management of various diseases in several crops in India including cereals, vegetables, pulses, oilseeds, cash crops, and fruit crops (Sharma et al., 2014) (Table 22.2).

22.4.2 *Pseudomonas fluorescens*

Fluorescent pseudomonas are gram-negative, chemoheterotrophic motile rods with polar flagella, and are grouped in r-RNA homology group I (Palleroni et al., 1973) known to produce a large number of secondary metabolites, which may exert antagonistic activity against plant pathogens (Table 22.2). The formulations of this bacteria are also classified under Class IV indicating low hazard, no phytotoxicity, and exempt from residual analysis.

22.5 Application of Biocontrol Agents

Seed treatment, seedling treatment, and soil application are the most common methods for application of bioagents for disease management. Seed coating with biocontrol agents is the easiest method of delivering the bioagents.

TABLE 22.2

Management of Fungal Diseases of Different Crops by Biocontrol Agents

S. No.	Crop	Diseases	Bioagent	References
1	Rice	Sheath blight, brown spot, rice blast	*Trichoderma* spp. and *P. fluorescens*	Mathivanana et al., 2005, Khalili et al., 2012
3	Wheat	Talk all on wheat, root rot	*T. koningii*, *P. fluorescens*, *T. viride*, *T. harzianum*	Duffy et al., 1996, Salehpour et al., 2005
4	Maize	Damping off, ear rot of maize	*T. harzianum*, *P. fluorescens*	Chen et al., 2005, Nayaka et al., 2009
5	Black gram	Root rot, dry root rot	*T. harzianum*, *P. fluorescens*	Indra et al., 2003, Karthikeyan et al., 2005
6	Sugarcane	Red rot	*Trichoderma* spp. *Pseudomonas* spp.	Joshi et al., 2019, Vishwanathan et al., 2000
7	Mango	Anthracnose	*T. asperellum*, *P. fluorescens*	Villalobus et al., 2013, Vivekananthan et al., 2004
8	Pigeon pea	Wilt	*Trichoderma* spp., *P. fluorescens*	Kumar et al., 2009, Siddiqui et al., 2009
9	Chickpea	Wilt	*T. harzianum*, *P. aruginosa*	Dubey et al., 2007, Anjaiah et al., 2003
11	Tomato	Wilt, foot rot	*Trichoderma* spp.	Verma et al., 2017
12	Potato	Dry rot, soft rot	*T. viride*, *T. harzianum* *Pseudomonas* spp.	Ayed et al., 2006, Cronin et al., 1997
13	Onion	Basal rot disease, wilt	*T. harzianum*, *P. fluorescens*	Coşkuntuna et al., 2008, Tehrani et al., 2003

Application of bioagents directly to soil is the most effective method to suppress soil-borne pathogens like *Pythium* sp., *Phytophthora* sp., *Rhizoctonia solani*, *Sclerotium rolfsii*, *Sclerotinia sclerotiorum*, *Verticillium*, *Fusarium*, etc., which cause a wide array of diseases like root, stem, and collar rots, damping off, and wilts. The bioagents may be applied directly to soil or delivered through any organic substrate like farmyard manure or composts.

22.6 Panchgavya and Dashagavya for Disease Control

In addition to botanicals and bioesticides, panchgavya and dashagavya are two preparations that haven approved for use in organic farming for disease management. Panchgavya refers to a blend of five products obtained from cows, namely fresh cow dung (5 kg), cow urine (3 l), cow milk (2 l), curd (2 l), and cow ghee (1 kg). Dashagavya is prepared by adding ingredients like sugarcane juice (3 l), tender coconut water (3 l), banana paste (12 fruits), and grape juice (2 l) to panchgavya. To prepare the formulations, cow dung and ghee are initially mixed in a container and fermented for three days with intermittent stirring. The rest of the ingredients are added on the fourth day and allowed to ferment for 15 days with stirring twice daily. The final formulation is ready in 18 days. 500 g jaggery can be used in place of sugarcane juice while 100 g yeast powder mixed with 100 g jaggery and 2 l of warm water can be used in place of grape juice. These mixtures can be applied as seed treatment, soil drench (50 l/ha), or foliar sprays (3–4 l formulation diluted with 100 l water). Vedic literature has emphasized the role of panchgavya both in soil and plant health promotion as well as in according protection to plants against diseases. With the growing popularity of organic farming and organic foods, there has been a renewed interest in the use of such ancient tools for disease control and several studies are being carried out to scientifically validate the potential of panchgavya against diseases and understand the underlying mechanisms. Sireesha (2013) observed that foliar sprays of panchgavya gave effective control of rice blast caused by *Magnaporthe grisea*. Similarly, seed treatment with panchgavya showed significant reduction in *Curvularia lunata* causing grain discoloration in rice, Phytophthora blight, and anthracnose of bell pepper (Ashlesha & Paul, 2014) and early blight of tomato (Ramesh et al., 2015). Sarkar et al. (2014) had observed enhanced production of defense-related enzyme polyphenol oxidase in plants treated with panchgavya.

22.7 Conclusion

Despite certain doubts regarding the potential of organic farming to meet the world food requirement, organic farming is now a growing and booming sector worldwide as well as in India and as such, there is an urgent need to address the concerns of the organic farmers especially with respect to the management of various pests and diseases. Use of plant extracts and biocontrol agents offer a viable and effective option for organic growers to successfully manage the disease problems in field conditions. These techniques offer farmers protection against a diverse array of pathogens in various crops. However, there is a need to further standardize and refine these technologies to improve the level of control accorded in field.

REFERENCES

Abdulla, M. A.; El Mohamedi, R. S. R.; Badeaa, R. I. Effect on some volatile compounds on black mould disease on onions bulbs during storage. *Res. J. Agric. Biol. Sci.* 2006, *2*, 384–390.

Agrios, G. N. *Plant Pathology* (5th edition). Elsevier-Academic Press, San Diego, CA, 2005.

Amadioha, A. C. Controlling rice blast in vitro and in vivo with extracts of *Azadirachta indica*. *Crop Prot.* 2000, *19*(5), 287–290.

Anjaiah, V.; Cornelis, P.; Koedam, N. Effect of genotype and root colonization in biological control of fusarium wilts in pigeon pea and chickpea by *Pseudomonas aeruginosa* PNA1. *Can. J. Microbiol.* 2003, *49*(2), 85–91.

Arroyo, F. T.; Moreno, J.; Daza, P.; Boianova, L.; Romero, F. Antifungal activity of strawberry fruit volatile compounds against *Colletotrichum acutatum*. *J. Agric. Food Chem.* 2007, *14*, 5701–5707.

Ashlesha, P. Y. S. Antifungal bioefficacy of organic inputs against fungal pathogens of bell pepper. *Paripex-Indian J. Res.* 2014, *3*(6), 4–9.

Ayed, F.; Remadi, M. D.; Khiareddine, H. J.; Mahjoub, M. Potato vascular fusarium wilt in Tunisia: Incidence and biocontrol by *Trichoderma* spp. *Plant Pathol. J.* 2006, *5*(1), 92–98.

Carmelloa, C. R.; Cardosob, J. C. Effects of plant extracts and sodium hypochlorite on lettuce germination and inhibition of *Cercospora longissima in vitro. Sci. Hortic.* 2018, *234*, 245–249.

Chen, J.; Harman, G. E.; Comis, A.; Cheng, G. W. Proteins related to the biocontrol of *Pythium* damping-off in maize with *Trichoderma harzianum* Rifai. *J. Integr. Plant Biol.* 2005, *47*, 988–997.

Coskuntuna, A.; Özer, N. Biological control of onion basal rot disease using *Trichoderma harzianum* and induction of antifungal compounds in onion set following seed treatment. *Crop Prot.* 2008, *27*(3–5), 330–336.

Cowan, M. M. Plant products as antimicrobial agents. *Clin. Microbiol. Rev.* 1999, *12*, 564–582.

Cronin, D.; Loccoz, Y. M.; Fenton, A.; Dunne, C.; Dowling, D. N.; O'Gara, F. Ecological interaction of a biocontrol *Pseudomonas fluorescens* strain producing 2,4-diacetylphloroglucinol with the soft rot potato pathogen *Erwinia carotovora* subsp. *atroseptica. FEMS Microbiol. Ecol.* 1997, *23*(2), 95–106.

Curtis, H.; Noll, U.; Stormann, J.; Slusarenko, A. J. Broad-spectrum activity of the volatile phytoanticipinallicin in extracts of garlic (*Allium sativum* L.) against plant pathogenic bacteria, fungi and Oomycetes. *Physiol. Mol. Plant Pathol.* 2004, *65*(2), 79–89.

Das, K.; Tiwari, R. K. S.; Shrivastava, D. K. Techniques for evaluation of medicinal plant products as antimicrobial agent: Current methods and future trends. *J. Med. Plant Res.* 2010, *4*, 104–111.

Dharam, V.; Sharma, R. K. Efficacy of fungicides XXIX studies on the fungicidal properties of neem oil. Indian J. Plant Pathol. 1985, *3*, 241–242.

Dubey, S. C.; Suresh, M.; Singh, B. Evaluation of *Trichoderma* species against *Fusarium oxysporum* f. sp. *ciceris* for integrated management of chickpea wilt. *Biol. Control.* 2007, *40*(1), 1187–127.

Duffy, B. K.; Simon, B. K.; Weller, D. M. Combination of *Trichoderma koningii* with fluorescent pseudomonads for control of talk-all on wheat. *Phytopathology.* 1996, *86*, 188–194.

FAO (Food and Agriculture Organization of the United Nations). *Organic Agriculture.* Food and Agriculture Organization of the United Nations, Rome, 1999.

Guleria, S.; Kumar, A. *Azadirachta indica* leaf extract induces resistance in sesame against *Alternaria* leaf spot disease. *J. Cell Mol. Biol.* 2006, *5*, 81–86.

Indra, N.; Gayathri, S. Management of blackgram root rot caused by Macrophomina phaseolina by antagonistic microorganisms. *Madras Agric. J.* 2003, *90*, 490–494.

Joseph, B.; Dar, M. A.; Kumar, V. Bioefficacy of plant extracts to control *Fusarium solani* f. sp. *melongenae* incitant of brinjal wilt. *Global J. Biotechnol. Biochem.* 2008, *3*(2), 56–59.

Joshi, D.; Singh, P.; Holkar, S.; Kumar, S. *Trichoderma* mediated suppression of red rot of sugarcane under field conditions in sub-tropical India. *Sugar Tech.* 2019, *21* (3), 496–504.

Kandhari, J. Management of sheath blight of rice through fungicides and botanicals. *Indian Phytopathol.* 2007, *60*, 214–217.

Karthikeyan, M.; Bhaskaran, R.; Radhika, K.; Mathiyazhagan, S.; Jayakumar, V.; Sandosskumar, R. Endophytic *Pseudomonas fluorescens* Endo2 and Endo35 induce resistance in black gram (*Vigna mungo* L. Hepper) to the pathogen *Macrophomina phaseolina. J. Plant Interact.* 2005, *1*(3), 135–143.

Khalili, E.; Sadravi, M.; Naeimi, S.; Khosravi, V. Biological control of rice brown spot with native isolates of three *Trichoderma* species. *Braz. J. Microbiol.* 2012, *43*(1), 297–305.

Kumar, S.; Upadhyay, J. P.; Rani, A. Evaluation of *Trichoderma* species against *Fusarium udum* Butler causing wilt of pigeon pea. *J. Biol. Control.* 2009, *23*(3), 329–332.

Locke, J. C. Fungi. In Schmutterer, H. (Ed.), *The Neem Tree.* VHC, Weiheim, Germany, 1995, pp. 118–126.

Mathivanan, N.; Prabavathy, V. R.; Vijayanandraj, V. R. Application of talc formulations of *Pseudomonas fluorescens* Migula and *Trichoderma viride* Pers. ex S.F. Gray decrease the sheath blight disease and enhance the plant growth and yield in rice. *J. Phytopathol.* 2005, *153*(11–12), 697–701.

Nayaka, S. C.; Shankar, A. C.; Reddy, M. S.; Niranjana, S. R.; Prakash, H. S.; Shetty, H. S.; Mortensen, C. N. Control of *Fusarium verticillioides*, cause of ear rot of maize, by *Pseudomonas fluorescens. Pest Manage. Sci.* 2009, *65*(7), 769–775.

Nduagu, C.; Ekefan, E. J.; Nwankiti, A. O. Effect of some crude plant extracts on growth of *Colletotrichum capsici* (Synd) Butler & Bisby, causal agent of pepper anthracnose. *J. Appl. Biosci.* 2008, *6*, 184–190.

Obagwu, J.; Korsten, L. Control of citrus green and blue molds with garlic extracts. *Eur. J. Plant Pathol.* 2003, *109*(3), 221–225.

Oerke, E. C. Crop losses to pests. *J Agric. Sci.* 2006, *144*, 31–43.

Okigbo, R. N.; Nmeka, I. A. Control of yam tuber rot with leaf extracts of *Xylopia aethiopica* and *Zingiber officinale. Afr. J. Biotechnol.* 2005, *4*(8), 804–807.

Palleroni, N. J.; Kunisawa, R.; Contopoulou, R.; Doudoroff, M. Nucleic acid homologies in the genus *Pseudomonas. Int. J. Syst. Bacteriol.* 1973, *23*, 333–339.

Pasini, C.; Aquila, F.; Curir, P.; Gullino, M. L. Effectiveness of antifungal compounds against rose powdery mildew (*Sphaerothecapannosa* var. *rosae*) in glasshouses. *Crop Prot.*1997, *16*(3), 251–256.

Paul, P. K.; Sharma, P. D. *Azadirachta indica* leaf extract induces resistance in barley against leaf stripe disease. *Physiol. Mol. Plant Pathol.* 2002, *61*, 3–13.

Prithiviraj, B.; Singh, U. P.; Singh, K. P.; Plank-schumacher, K. Field evaluation of ajoene, a constituent of garlic (*Allium sativum*) and neemazal, a product of neem (*Azadirachta indica*) for the control of powdery mildew (*Erysiphepisi*) of pea (*Pisumsativum*). *J. Plant Dis. Prot.* 1998, *105*(3), 274–278.

Ramesh, G.; Ajithkumar, K.; Savitha, A. S.; Patil, S. G. Integrated influence of organic manures in addition to inorganic fertilizers on growth, yield parameters and early blight disease of tomato (*Lycopersiconesculentum*L.). *Int. J. Biol. Pharm.* 2015, *6*(6), 478–483.

Russell, P.; Mussa, A. E. A. The use of garlic (*Allium sativum*) extracts to control foot rot of *Phaseolus vulgaris* caused by *Fusarium solani* f. sp. *phaseoli. Ann. Appl. Biol.* 1997, *86*(3), 369–372.

Salehpour, M.; Etebarian, H. R.; Roustaei, A.; Aminian, H. Biological control of common root rot of wheat (*Bipolarissorokinian*) by *Trichoderma* isolates. Plant Pathol. J. 2005, *4*, 85–90.

Sarkar, S.; Kundu, S. S.; Ghorai, D. Validation of ancient liquid organics- Panchagavya and Kunapajala as plant growth promoters. *Indian J. Tradit. Knowl.* 2014, *13*(2), 398–403.

Sharma, P.; Sharma, M.; Raja, M.; Shanmugam, V. Status of *Trichoderma* research in India: A review. *Indian Phytopathol.* 2014, *67*(1), 1–19.

Siddiqui, Z. A.; Shakeel, U. Biocontrol of wilt disease complex of pigeon pea (*Cajanus cajan* (L.) Millsp.) by isolates of *Pseudomonas* spp. *Afr. J. Plant Sci.* 2009, *3*(1), 001–002.

Sireesha, O. Effect of plant products, Panchagavya and bio-control agents on rice blast disease of paddy and yield parameters. *Int. J. Res. Biol. Sci.* 2013, *3*(1), 48–50.

Slusarenko, A. J.; Patel, A.; Portz, D. Control of plant diseases by natural products: Allicin from garlic as a case study. *Eur. J. Plant Pathol.* 2008, *121*(3), 313–322.

Suleiman, M. W.; Emua, S. A. Efficacy of four plant extracts in the control of root rot disease of cowpea (*Vigna unguiculata* (L.) Walp). *Afr. J. Biotechnol.* 2009, *8*(16), 3806–3808.

Sundaramoorthy, S.; Usharani, S.; George, A. P. Antifungal activity of plant products for the management of fruit rot infection in chillies. *Plant Pathol.* 2014, *13*(2), 87–99.

Surwase, A. G.; Badgire, D. R.; Suryawanshi, A. P. Management of pea powdery mildew by fungicides, botanicals and bio-agents. *Ann. Plant Prot. Sci.* 2009, *17*, 384–388.

Tehrani, S.; Ramezani, M. Biological control of *Fusarium oxysporum*, the causal agent of onion wilt by antagonistic bacteria. *Commun. Agric. Appl. Biol. Sci.* 2003, *68*(4), 543–547.

Verma, N. P.; Kaur, I.; Masih, H.; Singh, A. K.; Singla, A. Efficacy of *Trichoderma* in controlling *Fusarium* wilt in tomato (*Solanum lycopersicum* L.) Res. *Environ.* Life Sci. 2017, *10*(7), 636–639.

Villalobos, S. S.; Doralinda, A.; Miguel, A.; Delano Frier, J. P.; deFolter, S.; Garcia, P.; Pena-Cabriales, J. J. Potential use of *Trichoderma asperellum* T8a as a biological control agent against anthracnose in mango (*Mangifera indica* L.). *Biol. Control.* 2013, *64*(1), 37–44.

Viswanathan, R.; Samiyappan, R Efficacy of *Pseudomonas* spp. strains against soil borne and sett borne inoculum of *Colletotrichum falcatum* causing red rot disease in sugarcane. *Sugar Tech.* 2000, *2*(3), 26–29.

Vivekananthan, R.; Ravi, M.; Ramanathan, A.; Samiyappan, R. Lytic enzymes induced by *Pseudomonas fluorescens* and other biocontrol organisms mediate defence against the anthracnose pathogen in mango. *World J. Microbiol. Biotechnol.* 2004, *20*(3), 235–244.

23

Management of Fungal Diseases of Crop Plants through Biopesticides

B. Meena

CONTENTS

23.1 Introduction

Biological control of plant diseases opened an era of new technology to manage crop diseases and received the attention of researchers throughout the world, which will enhance the sustainability of agricultural production systems and to reduce the use of chemical pesticides. Fluorescent pseudomonads have emerged as the largest and potentially most promising group of plant growth-promoting rhizobacteria (PGPR) for biocontrol of plant diseases. Fluorescent pseudomonads associated with plants include *Pseudomonas fluorescens*, *P. putida*, *P. aeruignosa*, and *P. aureofaciens*. Many of the fluorescent pseudomonads, predominantly *P. fluorescens* have been isolated from suppressive soil for the management of soil-borne and foliar diseases. *P. fluorescens* and *P. putida* are easy to isolate and grow in the laboratory and they are not fastidious in their requirement of nutrition.

Several studies have indicated that PGPR may stimulate the production of biochemical compounds associated with host defense, massive accumulation of phytoalexins, phenolic compounds, increase in the activities of PR-proteins, defense enzymes, and transcripts, and enhanced lignification. The induction of SAR using various ISR inducers has been of recent interest with quite reasonable success. Induced resistance against Fusarial wilt of carnation caused by *Fusarium oxysporum* f. sp. *dianthi* was found by prior application of *Pseudomonas* spp. strain WCS417r (Van Peer et al., 1991). Verhagen et al. (2009) studied the ability of *P. fluorescens* CHAO to induce resistance in grapevine against *Botrytis cinerea* and highlighted the importance of salicylic acid, pyochelin, and pyoverdin in priming phytoalexin responses and induced resistance. Vleesschauwer et al. (2008) demonstrated the ability of *P. fluorescens* WCS374r to trigger ISR in rice (*Oryza sativa*) against the leaf blast pathogen, *Magnoporthe oryzae* and found that WCS374r-induced resistance is regulated by an SA-independent but jasmonic acid/ethylene-modulated signal transduction pathway.

23.2 Characteristics of an Ideal Biocontrol Agent

- High rhizosphere competence
- High competitive saprophytic ability
- Enhanced plant growth
- Ease for mass multiplication
- Broad spectrum of action
- Reliable control
- Safe to environment
- Compatible with other rhizobacteria
- Should tolerate desiccation, heat, oxidizing agents, and UV radiation

23.3 Plant Growth-Promoting Activity

The bacteria that provide some benefit to plants are of two general types; those that form a symbiotic relationship with them and those that are free-living in the soil, but are often found near, on, or even within the roots of plants. Free-living soil bacteria are usually referred to as plant growth-promoting rhizobacteria or PGPR. The beneficial effects of these bacteria have been variously attributed to their ability to produce various compounds including phytohormones, organic acids, and siderophores, to fix atmospheric nitrogen, to solubilize soil phosphate, to produce antibiotics that suppress deleterious rhizobacteria, or to show some other unidentified mechanisms (Glick, 1995).

23.4 Formulation Development

Major research on biocontrol is centered on the use of cell suspensions of PGPR directly to seed. Technologies become viable only when the research findings are transferred from lab to field. Bacterial cell suspension cannot be used for large-scale field use due to difficulty in storage, transport, and handling. Commercial application of PGPR either to increase crop health or to manage plant diseases depend on the development of commercial formulations with suitable carriers that support the survival of bacteria for a considerable length of time. Carriers should be economical and easily available. The carriers with smaller particle size increased the surface area and thereby increased the resistance to desiccation of the bacteria by the increased coverage of the bacterial cells (Dandurand et al., 1994).

A powder formulation with a longer shelf-life would be beneficial. Talc is chemically referred to as magnesium silicate $\{Mg_3Si_4O_{10}(OH_2)\}$ and available as powder form from industries suited for a wide range of applications.

23.5 Delivery Systems

PGPR are delivered through several means based on nature of survival and mode of infection of pathogens. It is delivered through seed, soil, foliage, rhizomes, setts, or through a combination of several methods of delivery.

23.5.1 Seed Treatment

Seed bacterization means treating seeds with bacterial cultures that will improve plant growth; such bacterial cultures that improve plant growth are called as bacterial fertilizers. Transfer of technology for commercial use could be possible if PGPR strains are available as a product. Seed bacterization with peat-based formulation of *P. fluorescens* strain 1 at the rate of 10 g kg^{-1} seed reduced rice blast and sheath blight disease (Rabindran & Vidhyasekaran, 1996). Treatment of tomato seeds with powder formulation of PGPR (*Bacillus subtilis*, *B. pumilus*) reduced symptom severity of tomato mosaic virus and increased the fruit yield (Murphy et al., 2000).

23.5.2 Soil Application

Soil being as the repertoire of both beneficial and pathogenic microbes, delivering of PGPR strains to soil will increase the population dynamics of augmented bacterial antagonists and thereby would suppress the establishment of pathogenic microbes onto the infection court. Hagedorn et al. (1993) highlighted that furrow application of *Pseudomonas* spp. at the rate of 14 ml m^{-1} increased the seedling stand in cotton. The improved seedling stand was due to the suppression of seedling disease in cotton by the antagonistic bacteria.

23.5.3 Foliar application

The efficacy of biocontrol agents for foliar diseases is greatly influenced by microclimate. The concentration of nutrients like amino acids, organic acids, and sugars exuded through stomata, lenticels, hydathodes, and wounds varies highly. It affects the efficacy and survival of antagonists in phylloplane. The dosage and frequency of application have to be standardized based on the crop value, which could be a reliable and practical approach. Selected strains from many genera of bacteria isolated from these suppressive soils have the potential to reduce plant diseases when applied to the plant root environment (Weller et al., 2002). The biological control of plant pests and diseases using a single organism has been reported to give inconsistent and poor performance. Bioformulations combining *P. fluorescens* Migula strains Pf1 and AH1 and *Beauveria bassiana* (Balsamo) Vuill. isolate B2 effectively reduced the incidence of leaf folder insect and sheath blight disease on rice plants and showed the possibility of controlling both pest and disease using a single bioformulation (Karthiba, 2010).

23.6 Mechanisms of Biocontrol

Besides the capacity to colonize roots intensively for an extended period of time, other mechanisms are involved that makes the fluorescent pseudomonads as an effective biocontrol agent. PGPR that indirectly enhance plant growth via suppression of phytopathogens do so by a variety of mechanisms. These include the ability to produce siderophores that chelate iron, making it unavailable to pathogens; the ability to synthesize antifungal metabolites such as antibiotics, fungal cell wall-lysing enzymes, or hydrogen cyanide, which suppress the growth of fungal pathogens; the ability to successfully compete with pathogens for nutrients or specific niches on the root; and the ability to induce systemic resistance. Biochemical and molecular approaches are providing new insight into the genetic basis of these traits, the biosynthetic pathways involved, their regulation, and importance for biological control in laboratory and field studies (Nelson, 2004; Harish et al., 2008).

23.6.1 Antibiosis

Antibiosis is now often implicated as an important mechanism of biological control, resulting from the fact that it is an attractive mechanism to study and can provide a highly effective mode of action. Inhibition of pathogens of several crops by the release of diffusible or volatile metabolites such as pyrrolnitrin, pyoluteorin, phenazine, or cyanide was confirmed by application of DNA technology and biochemical reaction techniques (Gutterson, 1990). The compound 2,4-Diacetylphloroglucinol (DAPG) is a phenolic molecule produced by certain plant-associated fluorescent pseudomonads of worldwide origin (Thomashow et al., 1997). It has antifungal, antibacterial, antihelminthic, and phytotoxic properties.

Phenazine comprises a large family of heterocyclic nitrogen containing brightly colored pigments with broad-spectrum antibiotic activity. Several strains of fluorescent pseudomonad produce antifungal metabolites namely phenazines (Thomashow et al., 1997). Pyrrolnitrin (3-chloro-4-(2'-nitro-3'chloro-phenyl) pyrrole) is a broad-spectrum antifungal metabolite produced by many fluorescent strains of the genus *Pseudomonas*. Pyoluteorin is an aromatic polyketide antibiotic consisting of a resorcinol ring, which is derived through polyketide biosynthesis. It is produced by several *Pseudomonas* sp., including strains that suppress plant diseases caused by phytopathogenic fungi (Maurhofer et al., 1994).

23.6.2 HCN Production

Hydrogen cyanide production by certain fluorescent pseudomonads was found to influence the plant root pathogens. Wei et al. (1991) reported that four PGPR strains of *P. fluorescens viz.*, G8-4, *P. aureofaciens* 28-9 and 36-5 and *P. putida* 34-13 produced HCN *in vitro*, whereas two strains *P. aureofaciens* 25-33 and *Serratia plymuthica* 2-67 that induced resistance in the host showed no HCN production.

23.6.3 Competition for Nutrients and Space

Competition between pathogenic and saprophytic microorganisms for organic materials released from the roots can reduce growth and/or pathogenic activity of the pathogens. Many successful antagonists that do not produce antibiotics are able to grow rapidly at the wound sites and are better to extreme nutrients and environmental conditions compared with postharvest pathogens.

23.6.4 Siderophore Production

Siderophores are low-molecular-weight molecules that are secreted by microorganisms to take up iron from the environment and their modes of action in suppression of disease were thought to be solely based on competition for iron with the pathogen. These siderophores have been classified as either pyoverdins or pseuobactins. Production by certain fluorescent pseudomonads of extracellular, water-soluble, yellow and green pigments in KB medium that fluoresce in UV light is known for a hundred years. All such pigment (or siderophore) producing pseudomonads *P. aeruginosa*, *P. fluorescens*, and *P. putida* belong to one intrageneric homology group. Pseudobactin, the first siderophore was isolated, purified, and characterized by X-ray crystallography form *P. fluorescens* strain B10.

23.6.5 Induced Systemic Resistance

Induced resistance is a state of enhanced defensive capacity developed by a plant when appropriately stimulated (Audenaert et al. 2002). Induced resistance results from perception of rhizobacteria by plant roots which give rise to an increased level of resistance that is expressed upon subsequent infection by a pathogen. Application of PGPR strains showed enhanced resistance to bacterial speck and spot of tomato (Kavitha & Umesha, 2007).

Several studies have indicated that PGPR may stimulate the production of biochemical compounds associated with host defense, massive accumulation of phytoalexins, phenolic compounds, increase in the activities of PR-proteins, defense enzymes, and transcripts, and enhanced lignification. Peroxidase (PO) catalyzes the last step in the biosynthesis of lignin and other oxidative phenols and it is associated with disease resistance in plants. In groundnut, increased activity of PO was observed due to application of *P. fluorescens* and PO isoforms were expressed at higher levels (Meena et al., 2000). Phenylalanine ammonia lyase (PAL) is the first enzyme involved in phenylpropanoid pathway and plays a key role in the biosynthesis of phenolics and phytoalexins.

Induction of higher PPO activity was noticed in tomato and hot pepper pretreated with fluorescent pseudomonads strain against *Pythium* diseases (Ramamoorthy et al., 2002). Recently reports on mechanisms of biological control revealed that several microbial strains protect plants from various pests, diseases, and phyto-nematodes in several crops by activating defense genes encoding chitinase, glucanase, peroxidase, and synthesis of phytoalexins and inducing physiological changes (Rajendran et al., 2007).

23.7 Conclusion and Future Prospects

PGPR have gained worldwide importance and acceptance for agricultural benefits. These microorganisms are the potential tools for sustainable agriculture and the trend for the future. Scientific researches involve multidisciplinary approaches to understand adaptation of PGPR to the rhizosphere, mechanisms of root colonization, effects on plant physiology and growth, biofertilization, induced systemic resistance, biocontrol of plant pathogens, production of determinants, etc. The inconsistency in performance of these PGPR strains is a major constraint to their widespread use as biocontrol agent in commercial agriculture. However, genetic manipulation of PGPR has the potential to construct significantly better strains with improved biocontrol efficacy. The applications of mixtures of biocontrol agents may be a more ecologically sound approach because it may result in better colonization and better adaptation to the environmental changes occurring throughout the growing season. Future strategies are required to clone genes involved in the production of antibiotics, siderophores, and other metabolites, and to transfer these cloned genes into the strains having the good colonization potential along with other beneficial characters.

REFERENCES

Audenaert, K.; Pattery, T.; Cornelis, P.; Hofte, M. Induction of systemic resistance to *Botrytis cinerea* in tomato by *Pseudomonas aeruginosa* 7NSK2: Role of salicylic acid, pyochelin and pyocyanin. *Molecular Plant-Microbe Interaction* 2002, 15, 1147–1156.

Dandurand, L. M.; Morra, M. J.; Chaverra, M. H.; Orser, C. S. Survival of *Pseudomonas* spp. in air dried mineral powders. *Soil Biology and Biochemistry* 1994, 26, 1423–1430.

Glick, R. B. The enhancement of plant growth promotion by free living bacteria. *Canadian Journal of Microbiology* 1995, 41, 109–117.

Gutterson, N. Microbial fungicides: Recent approaches to elucidating mechanisms. *Critical Review of Biotechnology* 1990, 10, 69–91.

Hagedorn, C.; Gould, W. D.; Bardinelli, T. R. Field evaluation of bacterial inoculants to control seedling disease pathogens on cotton. *Plant Disease* 1993, 77, 278–282.

Harish, S.; Kavino, M.; Kumar, N.; Saravanakumar, D.; Soorianathasundaram, K.; Samiyappan, R. Biohardening with plant growth promoting rhizosphere and endophytic bacteria induces systemic resistance against banana bunchy top virus. *Applied Soil Ecology* 2008, 39, 187–200.

Karthiba, L.; Saveetha, K.; Suresh, S.; Raguchander, T.; Saravanakumar, D.; Samiyappan, R. PGPR and entomopathogenic fungus bioformulation for the synchronous management of leaf folder pest and sheath blight disease of rice. *Pest Management Science* 2010, 66, 555–564.

Kavitha, R.; Umesha, S. Prevalence of bacterial spot in tomato fields of Karnataka and effect of biological seed treatment on disease incidence. *Crop Protection* 2007, 26, 991–997.

Maurhofer, M.; Keel, C.; Haas, D.; Defago, G. Pyoluteorin production by *Pseudomonas fluorescens* strain CHAO is involved in the suppression of *Pythium* damping-off of cress but rot of cucumber. *European Journal of Plant Pathology* 1994, 100, 221.

Meena, B.; Radhajeyalakshmi, R.; Marimuthu, T.; Vidhyasekaran, P.; Doraiswamy, S.; Velazhahan, R. Induction of pathogenesis-related proteins, phenolics and phenylalanine ammonia-lyase in groundnut by *Pseudomonas fluorescens. Journal of Plant Diseases and Protection* 2000, 107, 514–527.

Murphy, J. F.; Zehnder, G. W.; Schuster, D. J.; Sikora, E. J.; Polston, J. E.; Kloepper, J. W. Plant growth promoting rhizobacterial mediated protection in tomato against Tomato mottle virus. *Plant Disease* 2000, 84, 779–784.

Nelson, L. M. Plant growth promoting rhizobacteria (PGPR): Prospects for new inoculants. *Plant Management Network* 2004, 10, 301–305.

Rabindran, R.; Vidhyasekaran, P. Development of a formulation of *Pseudomonas fluorescens* PfALR2 for management of rice sheath blight. *Crop Protection* 1996, 15, 715–721.

Rajendran, L.; Samiyappan, R.; Raguchander, T.; Saravanakumar, D. Endophytic bacteria mediate plant resistance against cotton bollworm. *Journal of Plant Interactions* 2007, 2, 1–10.

Ramamoorthy, V.; Raguchander, T.; Samiyappan, R. Enhancing resistance of tomato and hot pepper to *Pythium* diseases by seed treatment with fluorescent pseudomonads. *European Journal of Plant Pathology* 2002, 108, 429–441.

Thomashow, L. S.; Bonsall, R. F.; Weller, D. M. Antibiotic production by soil and rhizosphere microbes in situ. In *Manual of Environmental Microbiology*; Hurst, C. J., Knudsen, G. R., McInerney, M. J., Stetzenbach, L. D., Walter, M. V., Eds., ASM Press, Washington, DC, 1997, pp. 493–499.

Van Peer, R.; Niemann, G. J.; Schippers, B. Induced resistance and phytoalexin accumulation in biological control of Fusarium wilt of carnation by *Pseudomonas* sp. strain WCS417r. *Phytopathology* 1991, 81, 728–734.

Verhagen, B. W. M.; Aziz, P. T.; Couderchet, M.; Hofte, M.; Aziz, A. *Pseudomonas* spp.-induced systemic resistance to *Botrytis cinerea* is associated with induction and priming of defence responses in grapevine. *Journal of Experimental Botany* 2009, 61, 249–260.

Vleesschauwer, D.; Djavaheri, M.; Bakker, P. A. H. M.; Hofte, M. *Pseudomonas fluorescens* WCS374r-Induced systemic resistance against *Magnaporthe oryzae* is based on pseudobactin-mediated priming for a salicylic acid-repressible multifaceted defense response. *Plant Physiology* 2008, 148, 1996–2012.

Wei, G.; Kloepper, J. W.; Tuzun, S. Induction of systemic resistance of cucumber to *Colletotrichum orbiculare* by select strains of plant growth promoting rhizobacteria. *Phytopathology* 1991, 81, 1508–1512.

Weller, D. M.; Raaijmakers, J. M.; McSpadden Gardener, B. B.; Thomashow, L. S. Microbial populations responsible for specific soil suppressiveness to plant pathogens. *Annual Review of Phytopathology* 2002, 40, 309–348.

24

Management of Bacterial Diseases of Crops through Plant-Based Natural Biopesticides

Prem Chandra and Enespa

CONTENTS

24.1 Introduction

Food demand is increasing due to the overgrowth of the human population worldwide and is expected to reach 9.6 billion by 2050 (Deetman et al., 2020). Hazards connected with water, soil, and air pollution, and health risks to humans, plants, animals, and other wildlife make pesticides environmentally offensive (Atreya et al., 2011). Some common bacterial diseases and pathogens affect crops and their symptoms are given in Table 24.1. The bioformulation-based biopesticides of *Trichoderma* spp., *Pseudomonas* spp., *Bacillus* spp., *Agrobacterium radiobacter*, and *Aspergillus niger*, *Bacillus thuringiensis* (Bt), *Metarhizium* spp., *Beauveria bassiana*, and nuclear polyhedrosis virus (NPVs) are used in plant protection. Globally, more than 225 microbial biopesticides are industrialized in the Organization of Economic Development and Cooperation. The United States, Canada, and Mexico are NAFTA countries used about 45% of all biopesticides sold, 20% used by Oceanic countries, 20 % used by European Union, 10% used by South and Latin Americas, and only 5% used by Asian countries (Rischer et al., 2020).

24.2 Management of Bacterial Plant Diseases

The development of phytopathogenic resistant bacteria due to the limited amount of chemical production for their control the problem is further compounded (de Oliveira Pires et al., 2020). In terms of biological control, more emphasis is laid nowadays on the management of bacterial plant pathogens using bacteriophages and bacteriocins (Thind, 2015). Through the use of biological control such as bacteriophages, bacteriocins and siderophores, chemicals, and heat therapy, a brief account of the management of bacterial plant pathogens achieved is given below (Narayanasamy, 2013).

24.2.1 Biological Control

Streptomycin is a well-known resistant against the *E. amylovora* strain caused fireblight of pome fruits and *X. vesicatoria* caused bacterial spot of tomato and pepper (Gusberti et al., 2015). For strains of *X. citri* subsp. *citri* from Argentina and strains of *X. alfalfae* subsp. *citrumelonis* from Florida, multiple and independent introductions of copper resistance genes took place (Behlau et al., 2011). Therefore, an attractive alternative to chemical use offers a biological control in applied agriculture (Vidogbéna et al., 2015). Plant diseases are gaining importance as biological control (Köhl et al., 2019). Biocontrol agent must colonize the host surface for a successful biological control. A biocontrol agent establishment is necessary before arrival of the pathogen when the competitor is not a violent settler. For biological control, a general requirement is that an antagonist colonizes (Von Itzstein, 2007). *P. fluorescens* is a soilborne bacteria that can be hostile in pear flowers and give control of fireblight and it is known as a good commercial bactericide (Stockwell et al., 2010).

24.2.2 Plant Growth-Promoting Rhizobacteria

The PGPR employed the main mechanism of biocontrol such as antagonism for an ecological niche or a substrate, inhibitory

TABLE 24.1

Common Bacterial Diseases That Affect Crops and Symptoms

Bacterial Disease	Pathogen	Factors Conductive to Spread	Affected Crops	Symptoms	References
Black rot	*Xanthomonas campestris* pv. *campestris*	Warm, wet conditions	Brassicas	Light brown to yellow V-shaped lesions on the leaf, vein blackening	Vicente and Holub, 2013
Bacterial canker	*Clavibacter michiganensis* pv. *michiganensis*	Moderate temperatures and high humidity	Tomato; capsicum; chili	Seedlings may die and older plants may wilt and die eventually	Lamichhane et al., 2015
Bacterial soft rot	*Pseudomonas* spp., *Erwinia* spp.	Warm, wet conditions	Wide range of vegetables such as lettuce; brassicas; cucurbits	Wet, slimy, soft rot that affects any part of vegetable crops including heads curds	Banerjee et al., 2018
Bacterial leaf spot	*Xanthomonas campestris* – various strains	Overhead irrigation and windy conditions	Range of vegetables including lettuce; cucurbits	Lettuce: large brown to black circular areas that begin as small translucent spots	Tsitsigiannis et al., 2008
Bacterial wilt	*Ralstonia solanacearum*	High temperature, more moisture	Potato; tomato; capsicum; eggplant	Wilting, yellowing, and stunting of plants but they may wilt rapidly	Badillo-Vargas et al., 2019
Bacterial blight	*Pseudomonas syringae* pv. *pisi*	Cool, wet, windy conditions	Peas	Water-soaked spots on leaves and stipules which become dark-brown	Sandipan et al., 2019
Bacterial speck	*Pseudomonas syringae*	Humidity and overhead irrigation	Tomato	Small dark spots surrounded by a yellow halo on leaves	Rekanović et al., 2019
Bacterial brown spot	*Pseudomonas syringae*	Cool, wet, windy conditions	Beans	Tan to reddish-brown spots on leaves	Volpiano et al., 2019

metabolites production, and induction of systemic resistance in host plants to a broad spectrum of pathogens (Chandra & Enespa, 2019). Synthesis of allelochemicals by PGPB such as iron-chelating siderophores, antibiotics, biocidal volatiles, lytic enzymes, and detoxification of chemical compounds controls the phytopathogens. Pseudomonads produced antibiotics such as amphisin, oomycin A, phenazine, HCN, tropolone, pyoluteorin, pyrrolnitrin, tensin, 2, 4-diacetylphloroglucinol (DAPG), and cyclic lipopeptides (Chandra & Enespa, 2017). *Bacillus*, *Streptomyces*, and *Stenotrophomonas* spp. produce oligomycin A, zwittermicin A, xanthobaccin, and kanosamine compounds. *Pantoea dispersa* produces an esterase which causes an irretrievable detoxification of albicidin, thereby causing the leaf scald of sugarcane produced by *X. albilineans* (Champoiseau et al., 2006).

24.2.3 Induced Systemic Resistance

The induced systemic resistance (ISR) is a PGPB trigger a phenomenon to protect the plants from several types of phytopathogens, i.e., fungal, bacterial, and viral. A systemic acquired resistance (SAR) phenomenon is similar to ISR, and initiated in plants by phytopathogens. The ISR expression is dependent on the host plant and bacterial strain combination (Gunatilaka et al., 2006). ISR and SAR are dependent on the gathering of salicylic acid (SA). PGPB do trigger a SA-dependent signaling pathway triggered by producing very small amounts (nonograms) of SA in the rhizosphere (Veselova et al., 2015). In Arabidopsis thaliana the way of SA-independent ISR pathway has been observed, which depend on ethylene signaling and Jasmonic Acid (JA). ISR and SAR can upturn protection beside pathogens that are resisted through both pathways besides comprehensive shield to a broader spectrum of pathogens than ISR/SAR alone (Choudhary et al., 2007).

24.2.4 Bacteriocins

Various strains of bacteria-produced bacteriocins are active against some other strains of the same or closely related species. The synthesis of DNA inhibited specifically by Agrocin 84 because of its being a structural analog of adenine. The crown gall of stone fruits controls with *A. radiobacter* K 84/agrocin 84 covered already (Mukerji, 2000). In beans, a purified syringacin 4-A has been tried to control by *P. syringae* pv. *phaseolicola*. *P. syringae pv. ciccaronei* at 6000 arbitrary units mL^{-1} produced crude bacteriocin reduced the severity of olive knots caused by *P. savastanoi* pv. *savastanoi* using the reduced knot weight 81 and 51% on leaves and leaf scar sites, respectively (Lavermicocca et al., 2002).

24.2.5 Siderophores

Siderophores are high-affinity iron (III) chelators released by bacteria, fungi, and many plants. At neutral pH iron Fe^{3+} ions have very low solubility, so, they cannot be utilized by organisms. By the chelation as soluble Fe^{3+} complexes ions dissolves by the siderophores that can be taken up by vigorous transference mechanisms. Ferric iron chelates hydroxamic acid as a derivative of siderophore group very strongly (Chandra & Singh, 2016). The limited supply of iron (III) in the rhizosphere as siderophores sequester in habited its availability to the pathogens and also reduce their growth. Certain soilborne phytopathogens reduced their growth due to biocontrol action of *Fluorescent pseudomonads*. The production of pyoverdines of a yellow-green pigment worked as siderophores fluoresce under UV light (Chandra & Enespa, 2019).

24.2.6 Bacteriophages

Viruses that kill bacteria, known as bacteriophages, and their bacterial pathogen killing capability inspired scientists to use them for the control of phytopathogenic bacteria in plants. The combined inoculation of *E. carotovora* sub sp. *carotovora* and its phage-controlled rotting of carrot discs (Buttimer et al., 2017). The phage-resistant bacterial strains control with host range (h) mutant phage. Additionally to achieving highly operative disease control, some potential advantages in using phages in disease control are given:

- One application will take care of this in many cases; there is no need for constant applications because the phages are self-replicating.
- The phages are usual constituents of the biosphere; so they can be effortlessly isolated from wherever their congregations (bacteria) are present.
- On bacteria the phage receptors known as pathogenicity determining factors; phage-resistant mutant development usually is central to tempering in bacterial virulence.
- To eukaryotic cells the phages are non-toxic where use of chemicals is prohibited they can be used in cases such as before harvesting the treatment of peach fruits.
- In contrast to broad-spectrum antibiotics, they eliminate only target bacteria in specific phages. So, they do not injure to indigenous beneficial flora.
- Most agrochemicals are not affected by phages so they can be used with a combination of these chemicals and also with biocontrol agent.

24.2.7 Heat Therapy

In order to control the bacterial diseases of plants, heat therapy has been successfully used. To manage the ratoon-stunting disease of sugarcane, hot water, hot air, and moist hot air treatment are used, but among them, the moist hot air treatment method is more effective and hence widely used. The recommended temperature is 54°C for 4 h in India are followed, while the treatment is done at 51°C for 4 h in the United States for the moist hot air treatment methods (Yulianti & Wijayanti, 2020). Treatment with hot water at 50°C for 30 min for cowpea and at 52°C for 30 min green gram has been suggested for the control of bacterial blight of cowpea and bacterial leaf spot of green gram, respectively.

24.2.8 Through Plant-Based Natural Products

In plant protection, the use of natural products might also be dissimilar. A living organism synthesized a natural or biogenic substance directly or by its derivatives from the biogenic origin substances through chemical reactions such as the decomposition of biological materials (Yáñez-Mó et al., 2015). Plants, animals, and microbes synthesized a physiologically active chemical known as a natural product. More than 12,000 secondary metabolites have been isolated from the plant, which is less than 10% of the total. These synthesized substances served as plant defense mechanisms in contradiction of microorganisms, insects, and herbivores (Cowan, 1999; Gyawali & Ibrahim, 2014).

24.3 Conclusions

Biopesticides can form conventional packages more easily, and there is increased quality and productivity when compared with chemical pesticides. Biopesticides meet consumer demand for health and wellness. Chemical pesticide restrictions are estimated to continue and develop resistance in the distribution and sustainability of GM crops. Therefore, biopesticides are an advanced technology that can enhance the quality, quantity, sustainability, and durability of IPM programs.

REFERENCES

Atreya, K.; Sitaula, B. K.; Johnsen, F. H.; Bajracharya, R. M. Continuing issues in the limitations of pesticide use in developing countries. *J. Agric. Environ. Ethics*, 2011, *24*(1), 49–62.

Badillo-Vargas, I. E.; Chen, Y.; Martin, K. M.; Rotenberg, D.; Whitfield, A. E. Discovery of novel thrips vector proteins that bind to the viral attachment protein of the plant bunyavirus tomato spotted wilt virus. *J. Virol.*, 2019, *93*(21), e00699-19.

Banerjee, J.; Gantait, S.; Sarkar, S.; Bhattacharyya, P. K. Transgenic research on tomato: Problems, strategies, and achievements. In *Biotechnologies of Crop Improvement*, 2018 (pp. 287–334). Springer, Cham.

Behlau, F.; Canteros, B. I.; Minsavage, G. V.; Jones, J. B.; Graham, J. H. Molecular characterization of copper resistance genes from *Xanthomonas citri* subsp. *citri* and *Xanthomonas alfalfae* subsp. *citrumelonis*. *Appl. Environ. Microbiol.*, 2011, *77*(12), 4089–4096.

Buttimer, C.; McAuliffe, O.; Ross, R. P.; Hill, C.; O'Mahony, J.; Coffey, A. Bacteriophages and bacterial plant diseases. *Front. Microbiol.*, 2017, *8*, 34.

Champoiseau, P.; Daugrois, J. H.; Girard, J. C.; Royer, M.; Rott, P. C. Variation in albicidin biosynthesis genes and in pathogenicity of *Xanthomonas albilineans*, the sugarcane leaf scald pathogen. *Phytopathol.*, 2006, *96*(1), 33–45.

Chandra, P.; Singh, E. Applications and mechanisms of plant growth-stimulating rhizobacteria. In *Plant-Microbe Interaction: An Approach to Sustainable Agriculture*, 2016 (pp. 37–62). Springer, Singapore.

Chandra, P.; Singh, E. Microbial volatiles as chemical weapons against pathogenic fungi. In *Volatiles Food Security*, 2017 (pp. 227–254). Springer, Singapore.

Chandra, P.; Singh, E. Soil–microbes–plants: Interactions and ecological diversity. In *Plant Microbe Interface*, 2019 (pp. 145–176. Springer, Cham.

Choudhary, D.K.; Prakash, A.; Johri, B.N. Induced systemic resistance (ISR) in plants: mechanism of action. Ind J Microbiol., 2007, 47 (4), 289-297.

Cowan, M. M. Plant products as antimicrobial agents. *Clin. Microbiol. Rev.*, 1999, *12*(4), 564–582.

de Oliveira Pires, M. S.; de Carvalho Alves, M.; Pozza, E. A. Multispectral radiometric characterization of coffee rust epidemic in different irrigation management systems. *Int. J. Appl. Earth Obs. Geoinf.*, 2020, *86*, 102016.

Deetman, S.; Marinova, S.; van der Voet, E.; van Vuuren, D. P.; Edelenbosch, O.; Heijungs, R. Modelling global material stocks and flows for residential and service sector buildings towards 2050. *J. Cleaner Prod.*, 2020, *245*, 118658.

Gunatilaka, A. L. Natural products from plant-associated microorganisms: distribution, structural diversity, bioactivity, and implications of their occurrence. *J. Nat. Prod.*, 2006, *69*(3), 509–526.

Gusberti, M.; Klemm, U.; Meier, M. S.; Maurhofer, M.; Hunger-Glaser, I. Fire blight control: The struggle goes on. A comparison of different fire blight control methods in Switzerland with respect to biosafety, efficacy and durability. *Int. J. Environ. Res. Public Health*, 2015, *12*(9), 11422–11447.

Gyawali, R.; Ibrahim, S. A. Natural products as antimicrobial agents. *Food Control*, 2014, *46*, 412–429.

Köhl, J.; Kolnaar, R.; Ravensberg, W. J. Mode of action of microbial biological control agents against plant diseases: Relevance beyond efficacy. *Front. Plant Sci.* 2019, *10*, 845.

Lamichhane, J. R.; Barzman, M.; Booij, K.; Boonekamp, P.; Desneux, N.; Huber, L.; Sarah, J. L. Robust cropping systems to tackle pests under climate change. A review. *Agron. Sustainable Dev.*, 2015, *35*(2), 443–459.

Lavermicocca, P.; Lonigro, S. L.; Valerio, F.; Evidente, A.; Visconti, A. Reduction of olive knot disease by a bacteriocin from *Pseudomonas syringae* pv. *ciccaronei*. *Appl. Environ. Microbiol.*, 2002, *68*(3), 1403–1407.

Mukerji, K. G. Exploitation of protoplast fusion technology in improving biocontrol potential. In *Biocontrol Potential and Its Exploitation in Sustainable Agriculture*, 2000 (pp. 39–48). Springer, Boston, MA.

Narayanasamy, P. Mechanisms of action of bacterial biological control agents. In *Biological Management of Diseases of Crops*, 2013 (pp. 295–429). Springer, Dordrecht.

Rekanović, E.; Potočnik, I.; Stepanović, M.; Todorović, B.; Mihajlović, M.; Milijašević-Marčić, S. Control of pepper bacterial spot and tomato bacterial speck using bactericides in combination with a plant activator. *Pesticidi i fitomedicina*, 2019, *34*(1), 47–52.

Rischer, H.; Szilvay, G. R.; Oksman-Caldentey, K. M. Cellular agriculture—Industrial biotechnology for food and materials. *Curr. Opin. Biotechnol.*, 2020, *61*, 128–134.

Sandipan, P. B.; Patel, R. K.; Faldu, G. O.; Patel, D. M. Status of different diseases of cotton under South Gujarat Region of India. *Int. J. Curr. Microbiol. App. Sci.*, 2019, *8*(10), 2651–2657.

Stockwell, V. O.; Johnson, K. B.; Sugar, D.; Loper, J. E. Control of fire blight by *Pseudomonas fluorescens* A506 and *Pantoea vagans* C9-1 applied as single strains and mixed inocula. *Phytopathol.*, 2010, *100*(12), 1330–1339.

Thind, B. S. Diagnosis and Management of Bacterial Plant Diseases. In *Recent Advances in the Diagnosis and Management of Plant Diseases*, 2015 (pp. 101–117). Springer, New Delhi.

Tsitsigiannis, D. I.; Antoniou, P. P.; Tjamos, S. E.; Paplomatas, E. J. Major diseases of tomato, pepper and egg plant in green houses. *Eur. J. Plant Sci. Biotechnol.*, 2008, *2*(1), 106–124.

Veselova, S. V.; Nuzhnaya, T.V.; Maksimov. I.V. Role of jasmonic acid in interaction of plants with plant growth promoting rhizobacteria during fungal pathogenesis. In *Jasmonic Acid: Biosynthesis, Functions and Role in Plant Development*, 2015 (pp. 33-66). Nova Science, NY, USA.

Vicente, J. G.; Holub, E. B. *Xanthomonas campestris* pv. *campestris* (cause of black rot of crucifers) in the genomic era is still a worldwide threat to brassica crops. *Mol. Plant Pathol.*, 2013, *14*(1), 2–18.

Vidogbéna, F.; Adégbidi, A.; Tossou, R.; Assogba-Komlan, F.; Ngouajio, M.; Martin, T.; Zander, K. K. Control of vegetable pests in Benin–Farmers' preferences for eco-friendly nets as an alternative to insecticides. *J. Environ. Manage.*, 2015, *147*, 95–107.

Volpiano, C.G.; Lisboa, B.B.; Granada, C.E.; São José, J.F.B.; de Oliveira, A.M.R.; Beneduzi, A.; Vargas, L.K. *Rhizobia* for biological control of plant diseases. In *Microbiome In Plant Health and Disease*, 2019 (pp. 315–336). Springer, Singapore.

Von Itzstein, M. The war against influenza: discovery and development of sialidase inhibitors. *Nat. Rev. Drug Discovery*, 2007, *6*(12), 967–974.

Yáñez-Mó, M.; Siljander, P. R. M.; Andreu, Z.; Bedina Zavec, A.; Borràs, F. E.; Buzas, E. I.; Colás, E. Biological properties of extracellular vesicles and their physiological functions. *J. Extracell. Vesicle*, 2015, 4(1), 27066.

Yulianti, T.; Wijayanti, K. S. Management of ratoon stunting disease by hot water treatment to provide healthy cane seed. In *IOP Conf. Ser.: Earth Environ. Sci.*, 2020, *418*(1), 012063. IOP Publishing.

25

Management of Insect Pests of Cola nitida *and* C. acumunata *Schott and Endl Nuts (*Balanogastris kolae *and* Sophrorhinus *spp. – Coleoptera: Curculionidae) with Plant-Based Natural Products*

Asogwa, Evarestus Uche, and Theophilus Chinyere Nkasiobi Ndubuaku

CONTENTS

25.1 Introduction

The genus *Cola*, especially *Cola nitida* (Vent.) and *Cola acuminata* Scott and Endl. (1832) are important economic crops in West and Central Africa, the Caribbean Islands, Mauritius, Sri Lanka, and Malaysia (Eijnatten, 1969; Oladokun, 1982). Both are the only edible species of kola grown on a commercial scale. Kola belongs to the family Sterculiaceae but unlike cocoa, kola is native to West Africa.

The kola trees usually bear nuts that are uniform in color. The most common nut color is red followed by pink, while white is rare. Some trees bear nuts of one color, while others bear nuts of different colors (Figure 25.1). The white nuts are the most highly prized, probably because of the rarity (Eijnateen, 1969).

The *Cola* species of economic importance are *Cola acuminata* and *C. nitida* (Quarco, 1973; Daramola, 1978). *Cola nitida*, which is referred to as "the true kola of commerce," has featured in the internal trade of West Africa for a number of centuries (Nzekwu, 1961; Eijnatten, 1969). Kolanuts are a common sight in African markets, cities, and villages. Many Africans consume kolanuts regularly, even daily, for the medicinal, stimulating, and sustaining properties. Kolanuts contain caffeine (2 to 3%) and smaller amounts of theobromin and kolanin, which dispel sleep, thirst, and hunger and act as a stimulant and anti-depressant. They are also thought to reduce fatigue, aid digestion, and work as an aphrodisiac.

It contains caffeine (2–3%) and smaller amounts of theobromine and kolatin (a phenolic substance). The nuts, when chewed, dispel sleep, thirst, hunger, and fatigue. It is nutritious and contains 10.0% protein, 1.6% fats, and 53.1% carbohydrate in addition to 3.9% tannins and 3.5% ash (Eijnatten, 1969). *Cola* has numerous uses in social and ceremonial engagements by the natives in the forest regions of Africa (Nzekwu, 1961; Daramola, 1978). There is also an increasing demand for its industrial usage in pharmaceutical industries and for the production of *Cola*-flavored soft drinks, wines of varying alcoholic contents, candies, and dyes (Beattie, 1970; Ogutuga, 1975). Its uses have inevitably created a high demand in excess of its production (Oladokun, 1985). The kola pod husk has also been utilized for the production of liquid soap (Yahaya et al., 2001). The most recent and remarkable advancement in kola by-product utilization is the use of kola pod husk in the replacement of a very high proportion of known concentrates (corn-bran, wheat-bran, rice-bran, maize) used in animal feed formulations (Hamzat et al., 2007).

The kola weevils, *Balanogastris kolae* (Desbr) and *Sophrorhinus* spp. (Coleoptera: Curculionidae) (Figure 25.2) are the most destructive pests of kolanut in West Africa. They attack nuts from the field to storage and infestation ranges from 30% to 100% depending on the sanitary condition on the farm at time of harvest and in storage (Ndubuaku, 1989; Idowu & Ojelade, 1995). The larva feeds extensively on the nuts thereby reducing the kola to brown powdery mass, which drastically reduces their market value. The average period from oviposition to the emergence of the adults of *B. kolae* is 29 days and 31 days for *Sophrorhinus* spp. (NRI, 1996).

The larvae (Figure 25.3) feed and pupate inside the nuts with the adults emerging through circular holes. The weevil infestation also predisposes the nuts to secondary invasion by other micro-organisms, especially fungi which altogether lowers the market value and may eventually result in total destruction of the nuts (Figures 25.4 and 25.5). *Balanogastris kolae* and species of *Sophrorhinus* have been found infesting the same nut (Ivbijaro, 1976; Daramola, 1981; Odebode, 1990).

All the known chemical methods for the control of weevils in kola nuts are considered to be unsafe because the farmers and traders cannot easily apply most of them. In the absence of any recommendations on the use of pesticides for storage of kola nuts, the use of solutions of Lindane (Gammalin 20 Emulsifiable Concentrate) has become a common practice among farmers and

(a) (b)

FIGURE 25.1 Healthy whole white (A) and red (B) nuts.

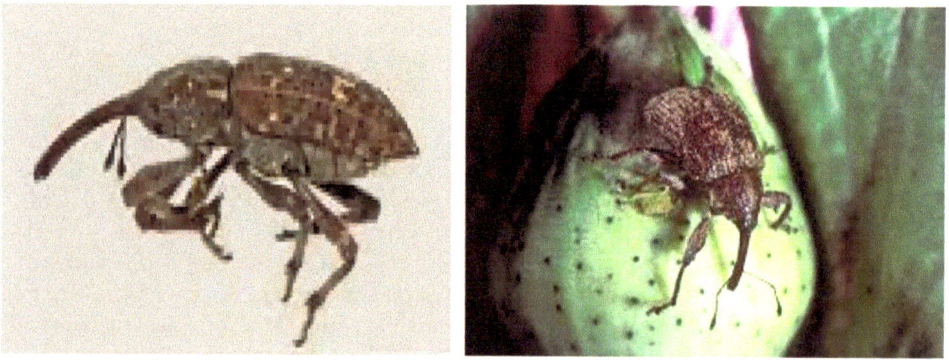

FIGURE 25.2 Adults of *Balanogastris kolae*.

FIGURE 25.3 Larvae of *Balanogastris kolae*.

FIGURE 25.4 Whole white nuts infested by *Balanogastris kolae*.

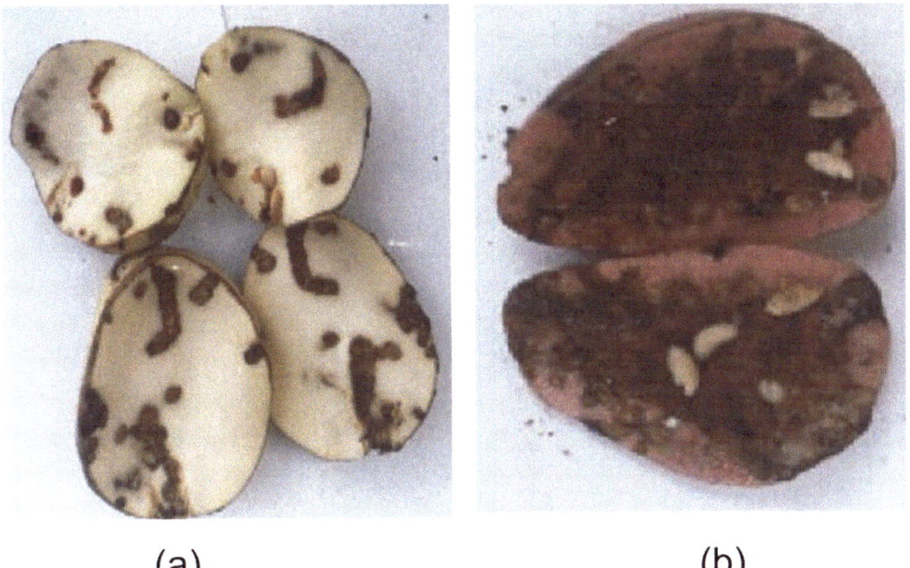

FIGURE 25.5 Heavily infested spilt white (A) and red (B) nuts.

traders in all kola-producing areas in Nigeria and Ghana which is one of the major kola nut exporters to Nigeria. This practice has persisted because no alternative safe and efficient methods of control of weevils in storage have been developed (Daramola, 1978; Ndubuaku, 1989). However, the huge losses resulting from kola weevil infestation during storage have caused kola farmers and traders to resort to the usage of hazardous pesticides for the control of the weevils. Unfortunately, since kolanuts are consumed in raw form without any further post-harvest processing, chemical control of weevils is highly unacceptable due to the probable high level of residue in the treated nuts and the possible long-term effect of these residues on consumers. This leaves researchers and investors with no other choice than to resort to the development and use of natural botanicals as the pest control alternative.

This book chapter gives an insight into the prospects of using botanicals (natural plant products) from known local plants for the protection of kolanuts against the major destructive weevils—*Balanogastris kolae* and *Sophrorhinus* spp.

25.2 Botanical Mortality against Weevils

The effect of botanicals on control of the weevil population in kolanuts during storage showed that *Capsicum longum* and *C. frutescens* powders caused significant high mortality of weevils in treated kolanuts (Ndubuaku, 2014). Adedire & Ajayi (1996) reported similar effectiveness of acetone extracts of *C. frutescens*, which caused 97.5% mortality of *S. zeamais* in stored maize.

The treatment of stored kolanuts with *Cymbopogon citratus* resulted in significantly high *B. kolae* adult mortality (Ndubuaku, 2014). Dike & Mba (1992) obtained similar results whereby cowpea grains were significantly protected from damage by *C. maculatus* by treatment with the stem powder of *C. citratus*. The effectiveness of lemon grass stem powder especially in protecting cowpea seeds from oviposition, progeny development, and damage by *C. maculatus* while at the same time ensuring the preservation of seed texture, color, and viability show its insecticidal potentials (Dike & Mba, 1992).

Results obtained from the treatment of stored kolanuts with *Vernonia amygdalina stem ash* showed that the ash was effective in causing adult *B. kolae* mortality (Ndubuaku, 2014). In a similar trial, Enobakhare et al. (2002), showed that the leaf powders of *V. amygdalina* applied to maize grain significantly reduced progeny development of *S. zeamai and wer*e effective in causing adult maize weevil mortality. The insecticidal properties might be attributed to one or more of the following: mechanical action, repellency stomach poison, starvation or desiccation as reported by Dales (1996).

The treatment of kolanuts with ash of *Azadirachta indica* stem resulted in significant mortality of weevils during storage (Ndubuaku, 2014). The efficacy of neem plant materials in the control of stored products weevils has been reported by many workers. Ivbijaro (1983a; b) reported that neem seed powder does not affect the survival of adult *Callosobruchus maculatus* but significantly reduces its fecundity, prolongs the development period, and considerably reduces the emergence of progeny. Popoola, (2004) reported that complete protection from *Prostephanus truncatus* was conferred on cassava chips treated with crude leaves extract of *A. indica* and that the inability of *P. truncates* to emerge on treated dried cassava chips may have been due to the oviposition, fertility, and reproductive deterrent effects.

Treatment of stored kolanuts with *Tetrapleura tetraptera* powder resulted in high mortality of weevils compared with control (Ndubuaku, 2014). This observation was in line with Popoola (2004) who reported pesticide properties of *T. tetraptera* on *Prostephanus truncatus* infesting dried cassava chips. These observations contradict the observation of Jackai (1993), that *T. tetraptera* has no insecticidal properties apart from its molluscidal property and the ability to emulsify neem oil.

Available literature on *T. tetraptera* showed that it is mostly used for medicinal purposes and as spice in black Africa. Adesina and Sofowora (1979) isolated Oleanolic acid, glycosides, scopoletin, and coumarin from the fruit of *T. tetraptera* and reported that scopoletin possesses antibroncho constrictor and antiarhythic effect.

The significant reduction of weevil population in stored kolanut treated with *Jatropha curcas* seed powder shows effective contact insecticidal activity (Ndubuaku, 2014). Similar result was reported by Anikwe (2000), who observed high mortality of *Sitophlus zeamais* in maize treated with seed oil extract of *Jatropha curcas*, but no mortality of adult weevils of *S. zeamais* treated with crude leaf, bark, and wood extracts, which is an indication that the allelochemicals responsible for toxicity to the weevils is contained in the seed (Fansworth, 1990).

In the experiment with *Carica papaya* (leaf extract), mortality of adult *B. kolae* was significantly higher ($p < 0.05$) than control (Ndubuaku, 2014). However, it was previously reported that papaw juice, along with a cold infusion of the bark of *Rauwolfa vomitoria* is used by Hausa traders in Sierra Leone as a preservative for kolanuts against weevils. All parts of the pawpaw tree are used for several medicinal purposes. The milky sap of the whole plant contains protoelytic ferment, papayotin, with an action like animal pepsin. This has the property of coagulating milk and of softening or digesting the fibrous tissue of flesh and is popularly used to make milk tender. The pesticide property of *C. papaya* is very likely due to the active principle it contains (Dalziel, 1937).

Also according to Ndubuaku, 2014, the significant *B. kolae* adult mortality ($p < 0.05$) recorded in the treatment of stored kolanuts with *Afromomum melegueta* powder corroborates a similar trial on *Sitophilus zeamais* in stored maize by Adedire and Ajayi (1996). It was observed that acetone extract of *A. melegueta* caused total mortality of weevils in 28 days after treatment which showed that the insecticidal strength of the extract increased with the length of storage (Ndubuaku, 2014). Lajide et al. (1998), also reported the effectiveness of *A. melegueta* fruit powder in controlling *S. zeamais* in maize by causing adult mortality of the insect, while Lale (1992), linked the effectiveness of *A. melegueta* with the presence of paradol in the seeds.

25.3 Discussion

There is a very high potential for the development of effective and environment-friendly plant-derived protectants in developing countries like Nigeria for the control of kola storage weevils. Generally, plants have evolved highly elaborate chemical defenses against insect attack, and they have therefore provided a rich source of biologically active chemical compound, which may be used as crop protecting agents. Fortunately, the tropics are well endowed with numerous flora which are currently not fully exploited. Essential or volatile oils extracted from plants have been shown to possess good potential for use as fumigants against stored products insects, including storage bruchids (Papachristos & Stamopolous, 2002).

The minimal protection of the kolanuts observed in most of the trials was due to lower weevil oviposition and reduced progeny emergence due to the effects of the plant extracts. The seed quality (color) was slightly affected by the extracts at all the treatment levels, suggesting that such treated produce will have general acceptability by consumers. Therefore, the color changes observed could have been a result of physiological factors associated with storing kolanuts in polythene bags and not from the plant extracts used. Most of the plants screened in this toxicity test against kola weevils had earlier been reported to be efficacious to one species of insect pest or the other. For instance, *Nicotiana tabacum* aqueous extracts (tobacco decoction) controls insects like caterpillars, beetles; stem borers, leaf miners, aphids, and thrips. In addition, aqueous extracts of *Nicotiana tabacum*, *Cymbopogon citratus*, and other plants have been found effective in reducing the number of *Podagrica* beetles damage on okra (Folorunsho et al., 2007).

There are vast amounts of secondary metabolites in higher plants: acids, alcohols, aldehydes, alkaloids, esters, fatty acids, flavones, glycosides, hydrocarbons, lactones, nitrogen-containing compounds, sterols, phenols, and terpenoids, which confer pesticidal activity on them (Dales, 1996). Therefore, it is logical to suggest that the reduction in population of kola weevil with these aqueous extracts may have been imparted by their water-soluble bioactive chemical components.

Crude preparations of neem pesticides can easily be prepared by farmers for pest management, and for the treatment of malaria and other disorders, for which reasons the tree is widely referred to as "Village Pharmacy" (NRC, 1992). Additionally, some farmers already use it to control grasshopper damage on cowpea farms during the dry season in a number of the northern states in Nigeria. The procedure for the crude oil extraction (UNIFEM, 1987) is simple and inexpensive, and requires no sophisticated equipment or skills.

The sudden increase in the number of emerging weevils after 56 days in storage coincided with the onset of another weevil generation and is indicative of the fact that plant materials degrade easily over time (Ndubuaku, 2014). It is therefore proposed that plant materials that are efficacious for protection of kolanuts should be applied intermittently to sustain the control of the weevils in storage. The loss in quality of kolanuts due to weevil infestation may be related both to the time of harvest and storage. According to Daramola (1976), the longer it takes for mature kola pods to be harvested, the higher the proportion of infested pods and the higher the intensity of the infestation in individual nuts. The farmers usually wait for the kola pods to mature and fall to the ground before they extract the nuts and sell in the market as unskinned nuts. A damage level of 38% to 56% by these weevils reported by Okunade and Lale (1998) conforms to the level of infestation recorded by earlier workers in Nigeria (Daramola, 1974; Idowu & Ojelade, 1995). The high number of weevil progeny development recorded in this present study further corroborated this.

25.4 Conclusion and Future Prospects

Botanical insecticides have long been touted as attractive alternatives to synthetic chemical insecticides for pest management because botanicals reputably pose little threat to the environment and human health. The body of scientific literature documenting bioactivity of plant derivatives to arthropod pests continues to expand, yet only a handful of botanicals are currently used in agriculture in the industrialized world, and there are prospects for commercial development of new botanical products. It is estimated that only about 10% of the over 250,000 different plant species in the world today have been examined chemically. Therefore, potential sources of plant chemicals still exist in the tropical forest which needs exploitation.

Most of the plant materials screened for their efficacy in the control of kola weevils are usually readily available to resource-poor farmers in areas where other forms of protection are difficult to obtain. Their extracts after application can easily be seen and washed off kolanuts before consumption. The additional benefits of using the plant materials are the low cost and minimal health risks. There is need for research to be intensified to incorporate these plant materials identified in this study to have strong protectant effect as components of integrated storage pest management programs to prevent post-harvest storage losses in kola.

Finally, it is obvious from the various studies that *B. kolae* is the most destructive field store pests of kolanut. There is therefore an urgent need to develop a comprehensive control package for the pest especially as kolanut does not require any other form of further post storage processing before consumption. This has become so imperative as to preserve the nuts and make them safe for consumers.

REFERENCES

Adedire, C.O.; Ajayi, T.S. Assesment of insecticidal properties of some plant extracts as grain protectants against maize weevil, *Sitophilus zeamais*. Niger. J. Entomol. 1996, 13:93–101.

Adesina, S.K.; Sofowora, E.A. Tetrapleura tetraptera. *Plantamedica*. 1979, 36: 270.

Anikwe, J.C. Efficacy of extracts of *Jatropha curcas* and *Jatropha gossypiifolia* as protectants of *maize grains against Sitophlus zeamais (Mostchulsky)* in Storage. *M.sc.* thesis, University of Ibadan, 2000.

Beattie, G.B. *Soft Drinks Flavours. Their History and Characteristics. 1. Cola or Kola Flavours.* The Flavour Industry, 1970, Pp. 390–394, Cammon House.

Dales, M.J. *A Review of Plant Materials Used for Controlling Insect Pest of Stored Products.* (NRI Bulletin No. 65). Natural Resource Institute, Chatham, Kent, 1996, 84pp.

Dalziel, J.M. *The Useful Plants of West Tropical Africa.* The Crown Agents, London, 1937, Pp. 100–107.

Daramola, A.M. The biology of kola weevils *Balanogastris kolae* (Desbr) on *Cola acuminata* and *Cola verticillata. Insect Sci. Appl.* 1981, 4:201–204.

Daramola, A.M. Effect of harvesting regime on weevil infestation of kolanuts in Southern Nigeria. Niger. *J.* Plant Prot. 1976, 2:78–88.

Daramola, A.M. *Insect Pest of Kola in Nigeria.* Research Bulletin. *No. 3.* Cocoa Research Institute of Nigeria, Ibadan, 1978, 33pp.

Daramola, A.M. (1974). Studies on the survival of kola weevils between seasons of kola production in Southern Nigeria. *Turialba*; 20(3): 309–310.

Dike, M.C.; Mbah, O.I. Evaluation of the lemon-grass, *Cymbopogon citratus* Staph. products in the control of *Callosobruchus maculatus* F, (Coleoptera. Bruchidae) on stored cowpea. *Niger.* J. Plant Prot. 1992, 14:88–91.

Eijnatten, C.L.; Van. M. *Kola, its botany and cultivation. Department of Agricultural Research Communication No.* 59. Royal Tropical Institute, Amsterdam, 1969, 100pp.

Enobakhare, D.A.; Law-Ogbomo, K.E. Efficacy of leaf powders of *Vernonia amygdalina* and *Ocimum gratissimum* as graln protectants against *Sitophilus zeamais* (Coleoptera: Curculionidae) in three maize varieties. *J. Agric. For. Fish.* 2002, 3: 22–27.

Farnsworth, N. R. (1990). In: Bioactive compounds from plants. *Ciba Foundations Symposium.* No. 154. John Willey & Sons. Pp. 2–7.

FDA (Federal Department of Agriculture). *National Agricultural Development Committee Report of the Study-Group on Cocoa, Kola, Coffee and Cashew.* FDA, Lagos, 1973, Pp. 63–120.

Folorunso, D.O.; Ofuya, T.I.; Adewumi, M.A. Effect of some aqueous medicinal plant extracts on *Podagrica* (Coleoptera: Chrysomelidae) infestation and yield of okra in a rain forest area of Nigeria. In: *Medicinal Plants in Agriculture: The Nigerian Experience.* Proceedings of the 3rd Annual Conference of School of Agricultural and Agricultural Technology (eds. Onibi, G.E.; Angele, S.O.; Adenkule, V.A.J.; Olufayo, M.O.). FUTA, Nigeria, April 16–19, 2007, Pp. 52–55.

Hamzat, R.A.; Omole, A.J.; Babatunde, B.B.; Adejumo, M.O.; Longe, O.G. Comparative value of kola testa, corn-bran and rice-bran with maize for mature African gaint land snail (*Archachatina marginata*). In: African Crop Science Conference Proceedings, Egypt, 2007, Vol. 8: Pp. 2127–2132.

Idowu, O.L.; Ojelade, K.T.M. Effects of timeliness of kola pod harvesting on weevil infestation in the field. *Ann. Rep. CRIN*, 1995, 1: 28–30.

Ivbijaro, M.F. Preservation of cowpea, *Vigna unguiculata* (L) Walp with the neem seed, *Azadirachta indica* A. Juss. *Prot. Ecol.* 1983a, 5:177–182.

Ivbijaro, M.F. The susceptibility of the immature and adult stages of the kolanut weevils *Balanogastris kolae* (Desbr) (Coleoptera: Curculionideae) to phosphine. *Niger. J. Entomol.* 1976, 3:53–56.

Ivbijaro, M.F. Toxicity of neem seed, *Azadirachta indica* A. Juss to *Sitophilus oryzae* (L) in stored maize. *Prot. Ecol.* 1983b, 5:353–357.

Jackai, L.E.N. The use of neem in controlling cowpea pests. *IITA Res.* 1993, 7:5–11.

Lajide, L., Adedire, C.O., Muse, W.A.; Agele, S.O. Insecticidal activity of powders of some Nigerian plants against the maize weevil *Sitophilus zeamais* Motsch. *E.S.N. Occas. Publ.* 1998, 31:227–235.

Lale, N.E.S. Oviposition—Deterrent and repellent effects of Products from dry Chilli pepper fruits, Capsicum species on *C. maculatus. Postharvest Biol. Technol.* 1992, 1:343–348.

Ndubuaku, T.C.N. Economic insect pests of kola. In: *Progress in Tree Crop Research.* 2nd ed. CRIN, Ibadan. 1989, Pp. 115–126.

Ndubuaku, T.C.N. Evaluation of aluminium phosphide, selected botanicals and cultural methods for the control of the kolanut weevil, *Balanogastris kolae* (Desbrochers, 1895) (Coleoptera: Curculionidae). PhD thesis. Department of Zoology, Faculty of Science, University of Ibadan, 2014, 136pp.

NRC (National Research Council). *Neem*: A Tree for Solving Global Problems. National Academy Press, Washington, DC, 1992, 141pp.

NRI (Natural Resources Institute). *A Guide to Insect Pests of Nigerian Crops, Identification, Biology and Control.* Federal Ministry of Agriculture & Natural Research, Nigeria and The Overseas Development & Administration, UK, 1996, 253pp.

Nzekwu, O. Kolanut. Niger. Mag. 1961, 71:298–305.

Odebode, A.C. Post-harvest rot of kola nuts caused by *Botryodiplodia theobroma* and *Fusarium pallidoroseum.* PhD thesis, University of Ibadan, Nigeria, 1990, 180pp.

Ogutuga, D.B.A. Chemical composition and potential commercial uses of kolanuts, *Cola nitida* (Vent). Scott and Endlicher. *Ghana J. Agric. Sci.* 1975, 8:121–125.

Okunade, S.O.; Lale, N.E.S. Assessment of damage to and loss of kolanut (Cola nitida Schott & Endl) caused by kola weevils *Balanogastris kolae* & *Sophrorhinus* spp in Maiduguri, Nigeria. *Entomol. Soc. Niger. Occas.* Publ. 1998, 31:215–219.

Oladokun, M.A.O. Morpho-physiological aspect of germination, rooting and seedling growth in kola (*Cola* spp). PhD thesis, University of Ibadan, Nigeria, 1982, 230pp.

Oladokun, M.A.O. Objectives and achievements in kola propagation research. Paper presented at Symposium to Mark the 25th Anniversary of Establishment of CRIN. 1985, Ibadan.

Papachristos, D.P.; Stamopoulos, D.C. Toxicity of three essential oils to immature stages of *Acanthoscelides obectus* (Say) (Coleoptera: Bruchidae). *J. Stored Prod. Res.* 2002, 38:365–373.

Popoola K.O.K. Pest status and control strategies of *Prostephanus truncatus* Horn, 1878 on dried cassava chips in Southwestern Nigeria. PhD thesis, University of Ibadan.

Quarco, T. *A Handbook on Kola CRIN,* Ibadan, Nigeria, 1973, 90pp.

UNIFEM (United Nations Development Fund for Women). Traditional methods of oil extraction and processing in oil extraction. *Food Cycle Technology Sourcebook No. 1.* UNIFEM, New York, USA, 1987, Pp. 9–12.

Yahaya, L.E.; Hamzat, R.A.; Aroyeun, S.O. Utilization of kola pod husk in liquid soap production. *Moor J.* Agric. Res. 2001, 3(2):252–256.

26

Natural Plant Products as Potential Synthetic Insecticide Alternatives in Developing African Countries

Asogwa, Evarestus Uche, and Theophilus Chinyere Nkasiobi Ndubuaku

CONTENTS

26.1 Introduction

Research into plant-derived natural products for agriculture went into decline for a number of years, but this trend is now being reversed as it becomes evident that plant natural products still have enormous potential to inspire and influence modern agrochemical research. A distinct propensity towards a "trek back to nature" has become evident in the last three decades especially in the fields of pesticides and pharmaceuticals. The need for safer, more natural pesticides is now accepted without any serious contention. The natural products, specifically plants in this context, are cheap, safe, effective, easily produced, and easily used.

This chapter gives an insight into the prospects of using natural plant products (botanicals) from known local plants as potential synthetic insecticide alternatives in developing African countries.

26.2 Pesticides of Plant Origins

Natural pesticides are active principles derived from plants for the management of human and animal pest organisms or it can be said to be biologically active ingredients, principally derived from plants, for the management of human and animal pest organisms (Ivbijaro, 1990). With the growing global demand for environmentally sound pest management strategies; there is a need to develop alternative pesticides with minimal or non-ecological hazards. Botanical pesticides are biodegradable (Devlin & Zettel, 1999) and their use in crop protection is a practical sustainable alternative. It maintains biological diversity of predators (Grange & Ahmed, 1988) and reduces environmental contamination and human health hazards.

Botanical insecticides are natural insecticides/toxicants derived from plants. Historically, the plant materials have been in use longer than any other group, with the possible exception of sulfur. Tobacco, pyrethrum, derris, hellebore, quassia, camphor, and turpentine were some of the more important plant products in use before the organized search for insecticides began in the early 1940s (Ware & Whitacre, 2004).

Botanicals as protectants have a great deal of advantages over synthetic protectants. This is because they are readily available and do not lead to the buildup of resistance by insect pests. They do not pose health hazards to the local farmers, as most of them are either medicinal plants or vegetables that are consumed as food. It also can help in rural development and employment for the rural populace through the establishment of cottage industries for the processing of selected plant materials for pest control (Ahmed et al., 1984).

The use of botanicals is known to have no likely effects on beneficial organisms such as parasites and predators. Workers with long experience with these plant extracts have not found any serious impact on beneficial fauna associated with most crops. For instance, the wolf spider, *Lycosa pseudoannulata*, which preys on leaf hoppers associated with rice in Asia, is not affected by neem oil or other extracts. There are reports of cases where parasitization was actually higher after application of neem, than when unsprayed. Also, *Telenomus remus* Nixon, an egg parasite of *Spodoptera litura*, was unaffected by a water suspension of neem seed kernel (Saxena, 1998; Schmutterer, 1990). The development of resistance to natural plant pesticides has been extremely slow and this can be further maintained by applying the botanicals in combination with biopesticides. Cases of resistance to synthetic insecticides are becoming common and the consequences include more frequent application of pesticides, higher doses, and more environmental contamination. Many synthetic insecticides rely on a single active ingredient, and once a pest population becomes resistant to it, the pesticide becomes useless. In contrast, botanical preparations contain a highly variable mixture of different plant chemicals, reducing the chances of provoking pest resistance.

26.3 Common Botanical Formulations Available for Use

The parts of the plants used for insecticide formulations from natural plants are the vegetative parts (bark, wood, leaves, fruits, and seeds) or the underground part (roots). The formulations can either be in the form of powders (dusts), oil extracts, ethanol extracts or aqueous extracts. Botanicals are processed into various forms; dusts and powders made from ground and dried plant parts; pure chemicals isolated from plants; and extracts or resins. These extracts and resins are formulated as liquid concentrates or impregnated onto dusts or wettable powders. The details of the commonly available formulations for use are as follows:

26.3.1 Powder Formulation

These are prepared by harvesting the appropriate plant part, which is then sun-dried or oven-dried under controlled temperature and then pulverized into fine powders using pestle and mortar or electric mill and sieved with a fine mesh (0.25 mm diameter sieve) (Adedire & Lajide, 2000).

26.3.2 Oil Formulation

The plant materials were pulverized and extracted by soaking in an organic solvent or extracted in a Soxhlet extractor for a minimum of 8 hours using organic solvent (ethanol, ethyl ether, or petroleum ether) (Egwunyenga et al., 1998). Another method utilized mainly for crude extraction of oil from seeds is by pounding it lightly in a mortar to obtain the kernel by removing the shell or outer coat. The kernel is ground into a paste, which is transferred to a pot and heated. The paste is heated alone briefly before the addition of water. The mixture is allowed to cool down after boiling. The oil settles on top of the cooled mixture and is easily scooped-off for use. This method is very effective for extracting neem seed oil (Jackai, 1993). The extraction is best done in an open area with good ventilation to avoid any hazard from the very strong and sometimes-offensive odor emitted during such boiling. It is better to gradually add a small amount of water to the paste while boiling rather than adding a large quantity of water at once.

26.3.3 Alcoholic or Ethanolic Extracts

The selected plant part was chopped into small pieces and ground in a mortar. The resulting paste is mixed with a known quantity of absolute ethyl alcohol. The solution is allowed to stand for 72 hours, after which the alcohol was evaporated off (recovered) using a rotary evaporator. The paste was then mixed with a measured quantity of distilled water and sieved through a fine nylon mesh to obtain the alcoholic extract solution (Padi et al., 2002).

26.3.4 Aqueous Formulation

In this case, the pulverized plant materials are extracted by using water as the solvent. The aqueous solution was obtained by pressing out fresh juice and diluting in water or through maceration (i.e. steeping in water for prolonged periods). It can also be got by infusion (the immersion of plants in already boiled water for prolonged periods) (Jackai, 1993; Lale, 1995).

26.3.5 Commercial Formulation

The plant materials were pulverized and extracted in a gaint Soxhlet extractor for a minimum of 8 hours using organic solvent (ethanol, ethyl ether or petroleum ether) (Egwunyenga et al., 1998). The extraction can be done in standard laboratories or in a small-scale industry. Some examples of commercial formulations of neem based insecticides (NBIs) products are Neemol, Super Neemol, Neem Azal, and Potenised Neem Systemic Oil (Padi et al., 2002). In some countries, especially India, NBIs are available in the open market (e.g. Repelin, Wellgro, Nimbosol, and Neemark (Schmutterer, 1990). *Azadirachta indica* (neem tree) has also been commercialized in the United States where it is registered as "Margosan O" and "Azatin E.C." (Larson, 1989). In Nigeria it is marketed as Neembecidine, Biocatch, Biomagic, and Green Magic. The Pyrethrum Board of Kenya factory in Nakuru, Northern Kenya has successfully extracted and formulated various insecticides from pyrethrum plants *Chrysanthemum cinerariaefolium* Trev. Schultz—Bip. Kenya has so far remained the major supplier of pyrethrum powder, aerosol, bait, and insecticides in the world today (Olaifa, 2009).

26.4 Limitations of Pesticides of Plant Origin (Botanicals)

There is the problem of local applicators for botanicals irrespective of their attractive qualities and great promise as alternatives to synthetic chemicals. So far there have been no locally developed applicators for the application of the aqueous and oil extracts of the plant materials. Motorized or Knapsack sprayers could be used but the initial cost of acquiring such sprayers by local farmers is a serious limitation. Also, fatty oils contained in the botanicals reduce the cleanliness and visual appeal of the treated plant materials/products.

The application of aqueous plant extract solutions is restricted mainly to growing crops. Their use as protectants of stored product increases the moisture content of the commodities, which results in mold infestation.

Pesticide industries in Africa have refused to take up the production and marketing of the various formulated botanical pesticides from plants. Unfortunately, multinational Agrochemical Companies (Novartis, Zeneca, Excel, Bayer, Af-Chem sofaco, Dow Chemicals, etc.) view phytochemicals as useful leads molecules for the synthesis of new classes of insecticides with novel mode of action but are not interested in developing botanical insecticides because of the required dependency on the natural resources often in a foreign country. Also, developed nations are not really interested in the formulation and production of purely botanical pesticides because they do not have abundant raw materials for them in their native homes. They do not want to create another OPEC situation, where they can be held to ransom for the supply of the raw materials. For example, for over 25 years (1980–2005) only a single new botanical insecticide from neem received registration for use in the United States and parts of Europe. A second group of botanical insecticides based on plant essential oils have recently been commercialized in America and India, but in terms of presence in the market they are currently in their infancy (Ahmed et al., 1984; Larson, 1989).

The formation of botanicals relies solely on forest resources, but uncontrolled harvesting of the plant parts impacts negatively on the diversity of forest species, reducing genetic and species richness. At present, plants are over-exploited for their medical and pesticidal use (Fasola, 2006). In developing countries there is no control of the rate and mode of collection of plants from the wild. These plant parts are continually being harvested without replanting, which is dangerous to their diversity. The intensive annual harvesting of plant materials (fruits, bark oil seeds, leaves, corns bulb, and rhizome) could gradually eliminate such species from the forest.

26.5 Policies for Sustainable Use of Botanicals

Great concerns by man for himself and his environment have necessitated the search for the use of pesticidal plants, which are believed to be more humanly safe and ecologically tolerable than synthetic pesticides. Recent efforts in sustainable use of natural pesticides of plant origin have centered on the production of plant powders, oils, and extracts using a single active ingredient of bioactive plant materials. For example, Margosan-O^R and Azatin^R formulations were based purely on "Azadirachtin" found in neem seeds.

Therefore, policies for sustainable production and use of botanicals for crop protection should evolve along the following principles:

1. Government in developing countries should encourage the development and use of botanicals by relaxing the laws governing the processing and use of botanical pesticides. At present many developing nations still follow the highly regulatory procedure employed by the United States of America for search and registration of pesticides for protection of food crops (Lale, 2002).
2. There should be high level funding of research into the identification, breeding, and propagation of prospect plants that could serve as ingredients for production of botanicals. This will involve the establishment of germplasm banks (seed banks) and seed gardens.
3. Farmers should be encouraged to use parts of their farmlands to propagate these useful plant materials.
4. The required parts of these plants (seeds, leaf, root, or stem bark) should attract economic and profitable rates from farmers to provide incentives for their production instead of planting food crops.
5. The policy should protect and encourage indigenous companies involved in processing and marketing of plant-derived pesticides such that local people will derive maximum possible benefits from these resources.
6. A policy should be put in place to prevent indiscriminate harvesting of plant materials with pesticidal or medicinal activities in order to prevent environmental degradation and loss of biodiversity.
7. There should be appropriate legislation by governments in developing nations to protect the intellectual property rights of discovered pesticides.

26.6 Discussion

Botanical insecticides have long been touted as attractive alternatives to synthetic chemical insecticides for pest management because botanicals reputably pose little threat to the environment and human health. A number of plant substances have been considered for use as insect antifeedants or repellents, but apart from some natural mosquito repellents, little commercial success has ensued for plant substances that modify arthropod behavior. Several factors appear to limit the success of botanicals, most notably regulatory barriers and the availability of competing products (newer synthetics, fermentation products, microbials) that are cost-effective and relatively safe compared with their predecessors.

The Forestry Research Institute of Nigeria (FRIN), Ibadan has the mandate for most of the plant biodiversity utilized in the formulation of botanicals. The only place and perhaps the best market for botanical pesticides are the developing countries in the Tropical African region. Nigeria must therefore seize this opportunity of apparent lack of interest from the developed world, to develop its own pesticide industry based on botanical pesticides. The country must therefore not waste further time to put in place regulations for the registration and patenting of botanical pesticides.

26.7 Conclusions and Suggestions for the Future

The following recommendations are proposed to increase agricultural yield and reduce the cost of agriculture and impact of chemicals on health.

(a) The private sector and nongovernmental organization (NGOs) should be encouraged by the abundant diverse flora in the forests and savannahs of Nigeria to fund research on botanicals.

(b) Periodic update of research and development on natural pesticides through workshops and conferences should be funded by government in collaboration with industries and NGOs to ensure increased food production in a world-friendly environment.

REFERENCES

Adedire, C.O.; Lajide, L. Toxicity and oviposition deterence activities of powders and extracts of pepper fruit plant *Dennettia tripetala* Baker to the pulse bettle *C. maculatus* F. E.S.N. Occasional Publication. 2000, 32: 199–206.

Adedire, C.O.; Adebowale, K.O.; Dansu, O.M. Chemical composition and insecticidal properties of Monodora tenuifolia seed oil (Annonaceae). *Journal of Tropical Fruit Products*. 2003, 9: 15–25.

Ahmed, S.; Wallace, C.; Mitchel, W.C.; Saxena, R. (eds.). Workshop report on the use of indigenous plant materials for pest control by limited—Resource farmers, an EWC/UH database. University of Hawaii, USA. 1984, August 6–10, 1984.

Devlin, J.F.; Zettel, T. *Ecoagriculture: Initiatives in Eastern and Southern Africa*. Weaver Press, Harare. 1999.

Egwunyenga, O.A.; Alo, E.B.; Nmorsi, O.P.G. Laboratory evaluation of the repellency of *Dennittia tripetale* to *Dermestes maculatus* (F). *Journal of Stored Products Research*. 1998, 34: 195–199.

Grange, N.; Ahmed, S. *Handbook of Plants with Pest Control Properties*. Wiley, New York. 1988.

Ivbijaro, M.F. Natural pesticides: Role and production potential in Nigeria. In National Workshop on the Pesticide Industry in Nigeria University of Ibadan. 1990, September 24–27, 1990, 24 pp.

Jackai, L.E.N. The use of neem in controlling cowpea pests. *IITA Research*. 1993, 7: 5–11.

Lale, N.E.S. An overview of the use of plant products in the management of stored product coleoptera in the tropics. *Postharvest News and Information*. 1995, 6(6): 69–75.

Lale, N.E.S. (2002). Stored products entomology and acarology in Tropical Africa. Mole Publications, Maiduguri, Nigeria 240pp

Larson, R.O. (1989). The commercialization of neem. In: M. Jacobson (ed). The neem tree, focus on photochemical pesticides. CRR Press: Boca Raton, Florida. Pp. 155-168.

Olaifa, J.I. Formulation of bioactive plants for plant and animal production in Nigeria: Realities and challenges. Key note address, Proceedings of the Humboldt Kellog/5th SAAT Annual Conference on Formulation of Medicinal for Crop and Animal Production in Nigeria. FUTA. 2009, April 20–23, 2009, pp X–XVII, Federal University of Technology Akure (FUTA), Nigeria.

Padi, B.; Adu-Acheampong, R.; Nkansah, A. Botanical pesticides for the control of cocoa capsids (Heteroptera: Miridae). In Proceedings of the 13th International Cocoa Research Conference. 2002, Kota Kinabalu, Sabah, Malaysia, pp. 403–409.

Saxena R.C. Green revolution without blues: Botanicals for pest management. In G.S. Dhaliwal, N.S., Randhawa, R. Arora and A.K. Dhawan (eds.) *Ecological Agriculture and Sustainable Development*, Vol. 2, Indian Ecological Society. Punjab Agricultural University, Ludhiana and Centre for Research in Rural and Industrial Development, Chandigarh. 1998.

Schmutterer, H. Properties and potentials of natural pesticides from the neem tree. Annual Review of Entomol. 1990, 35: 271–298.

Ware, G.W.; Whitacre, D.M. *The Pesticide Book*. 6th Ed. Meister Media Worldwide, Willoughby, OH (ISBN 1892829-11-8). 2004, 496pp.

27

Traditional Use of Store Grain Pest Control through Biopesticides

D.K. Kulkarni, R.B. Bhagat, P.V. Patil, and S.P. Taware

CONTENTS

27.1 Introduction

Historically, humans have tried to understand the availability of natural resources and their use to progress their lives. Agricultural scientists are diverting their attention towards traditional or indigenous technology for exploring possibilities of use in the future. Our age-old wisdoms were scientific and almost ecofriendly. It was the reason that all the plant protection practices followed in the medieval period were based on organic materials of both plant and animal origin, which included honey, clarified butter, milk and milk products, cow dung, and a number of plant species. The biochemical analysis of these organic materials has clearly indicated their anti-bacterial, anti-viral, anti-fungal, antiseptic, nematicidal, antifeedent, insect repellent, and insecticidal properties. It indicates that control of store grain pests using botanicals is economical, effective, and environmentally safe.

27.2 Indigenous Knowledge in Agricultural Pest Control

The Atharvaveda, the fourth Veda, describes the life of the ancient Indian agriculturist community. It is interesting to note that in Vedic times agriculture was considered the most honorable of human activities. The farmer is an authority in the knowledge and acquirement of food and is therefore highly respected. The common people choose their king amongst the agriculturists. The king is given the honor of being the first to plow the land. Everyone must cultivate the land and grow nutritious food. The farm is compared to the divine cow fulfilling all the desires of the hard-working farmer.

Farmers in ancient India developed nature-friendly farming systems and practices. Agricultural activities were designed to ensure ecological sustainability. India developed its own holistic scientific knowledge. It has a number of classical texts related to agricultural science. Kautilya's Arthashastra, Patanjali's Mahabhasya, Krishi-Parashara, Varahmihira's Bruhat Samhita, and Surapala's Vrikshayurveda are some of the manuscripts that contain valuable information about agricultural implements, selection of seeds, land preparation, pest control, storage, plant nutrients, grafting, soil selection, plant propagation, diseases and plant protection, mixed cropping, crop rotation, intercropping, shifting cultivation, terrace farming, etc.

Traditional knowledge exists worldwide in all communities covering varied areas including health, agriculture, and natural resource management. In most states, leaves of Neem (*Azadirachta indica*), Nirgudi (*Vitex negundo*), Karanj (*Pongamia pinnata*), Supla (*Mundulea sericea*), Tun (*Toona ciliate*), Teak (*Tectona grandis*), young leaves of Komal (*Koelzella padularia*), Bhang (*Cannabis sativa*), and Metho Dodi (*Leptodenia reticulate*) have been used. In addition, wood ash of some plants such as *Acacia nilotica*, Suru (*Casurina equisitifolia*), Mango (*Mangifera indica*), and Tamarand (*Tamarindas indica*) have been used. In some places, powdered rhizome of sweet flag (*Acorus calamus*) or turmeric (*Curcuma longa*) is used to protect stored paddy or rice.

In Bengal, leaves of Nishinda (*Vitex negundo*) and branches with green leaves are kept below and around gunny bags to repel insects/pests. (Naskar, 1998).

The Mahadeo koli tribe from Maharashtra utilizes leaves of Kulith (*Dolichos uniflorus*) and Sag (*Tectona grandis*) as preservatives for seed storage (Kulkarni and Kumbhojkar 2003). Effective treatments include salt powder, kitchen ash, *Vitex* leaves powder, cow dung ash, turmeric powder, and a mixture of neem leaves.

In the Konkan area, fruits of *Catunaregam spinosa* (Thumb) Tirveng mixed with grains to preserve it from infestation by insect pests. Plants infected by pulse beetle, are prohibited by applying Undi oil (*Calophyllum inophyllum*). The laboratory experiments showed strong toxicity of Undi oil on the pulse beetle. In 9 ml/kg of concentration, 100% mortality was observed after 24 hours. The oil, even at a low concentration, paralyzes the adults. Prolonged exposure leads to death (Patil et al. 2011, 2014). Many scientists have validated traditional knowledge of food grain pest control (Nene, 2005). Castor oil 10 ml/kg seed gives complete protection to mung bean against the bruchids without impairing germination. Oils such as sesamum, peanut, and sunflower had no deleterious effects on viability, palatability, cooking quality, or physical appearance of pulses.

27.3 Store Grain Pest Control: Experimental Work

Insect pests have been damaging stored grains in large quantities for years, resulting in major economic losses in India. Pulses are an important source of protein, fats, carbohydrates, sugars and vitamin B. *Callosbruchus chinensis* L. (Coleoptera: Bruchidae) is one of the most destructive and serious stored grain pest in tropical and subtropical region of the world (Khanam et al., 2008). Grain loss caused by storage pests such as cow pea beetle, (*Callosobruchus maculatus* F) also threatens food security.

There is serious need to find effective biological material to protect stored produce that is readily available, affordable, relatively less poisonous, and less damaging to the environment. It has stimulated interest in the development of alternative control strategies and the revitalization of traditional botanical insect pest control methods (Niber, 1994; Talukder and Howse, 1995, Ahuja et al. 2010). Singh et al. (2006) reported different oils as a grain protectant against *Callosobruchus chinensis* in green gram.

Farmers in developing countries use different plant materials to protect stored grains pest infestation by mixing grains with protectants made up of plant products. Local and tribal people from Western Maharashtra have traditional knowledge of plant resources for controlling different types of pests (Kulkarni and Kumbhojkar, 1996). Some farmers from Maharashtra use crushed leaves of *Combretum ovalifolium* Roxb to control beetles (*Carabus nemoralis*) in millet fields and the leaves of *Aristolochia bracteata* Retz for driving the beetles away. Farmers in Tamil Nadu use the liquid extracted from the latex of *Ficus* sp. mixed together to drive away the rhinoceros beetle (*Oryctes rhinoceros*) which is one of the most dangerous pests, causing mortality of coconut (*Cocos nucifera*) trees (Sinha and Bhatia, 1997). With this background, present study on documentation of natural resources for food grain pest control has been carried out in the Bhor and Mahad regions. An experimental part of bioassays to evaluate the efficacies of ten local species from villages of the Bhor and Mahad regions, respectively (Table 27.1).

Traditionally, local people used plant leaf powders for control of insects. The same method was used in the laboratory evaluation. Culture of pulse beetle, *Callosobruchus maculates* F., was maintained in the laboratory on cow pea grains. Leaf powders of ten plant species were mixed with 10 g of healthy noninfested cow pea grains at 1, 3, and 5% w/w. Three replicates were maintained in plastic bottles. Ten one-day-old unsexed pairs of adults were released in each plastic vial. A control without any treatment was maintained in three replicates and ten one-day-old unsexed adults were released. Adult mortality was recorded at 24, 48, and 72 hrs after the release of adults. The number of eggs laid was counted at 10 days after release of adults (sterile and fertile eggs were counted separately).

The average percentage mortality recorded 72 hrs after release of adults indicates 33.33% adult mortality in *Gnidia glauca* (Friesen.) Gilg. powder at 5% level, followed by *Eucalyptus globules* Labillis and *Catunaregam spinosa* (Thunb.) Tirveng. leaf powder at 3% and 5% levels. *Vitex negundo* L. powder at 3% level did not show any toxicity against bruchids. *Madhuca longifolia* (Koen.)Macbr. at 3% and 5% and *Pogostemon benghalensis* Burm. f. at 3% showed minimum adult mortality of 3.33%. *Azadirachta indica* A. Juss. at all three levels shows remarkable mortality.

There is significant lowering of egg numbers and adult emergence compared to the control. In the present laboratory experiment adult emergence, mortality, loss in weight were studied and results are encouraging (Figure 27.1).

On this background leaf powders of *Vitex negundo*, *Annona squamosa*, *Nicotiana tabacum*, and leaf, bark, and seed kernel of *Azadirachta indica* were tested against the pulse beetle. The eggs laid by both the species of bruchids were significantly decreased in *A. indica* seed kernel power. Very few adults emerged at seed kernel treatment (Khalequzzaman and Osmangoni, 2009).

Observations recorded on the average number of eggs laid on different treatments indicated the minimum number of eggs on

TABLE 27.1
List of Plants Used for Laboratory Testing

S.No.	Local Name	Botanical Name	Family	Plant Parts Used	Locality
1	Bakana limb	*Meliadubia* Cav.	Meliaceae	leaf	Bholawade
2	Gelphal	*Catunaregam spinosa* (Thunb.) Tirveng.	Rubiaceae	fruits	Taliye
3	Kadulimb	*Azadirachta indica* A. Juss.	Meliaceae	leaf	Velvand
4	Karanj	*Pongamia pinnata* (L.) Pierre	Fabaceae	leaf	Ambeshivthar
5	Moha	*Madhuca longifolia* (Koen.) Macbr.	Sapotaceae	leaf	Vadgaon
6	Nilgiri	*Eucalyptus globulus* Labillis	Myrtaceae	leaf	Nigudghar
7	Nirgudi	*Vitexnegundo* L.	Verbenaceae	leaf	Natambi
8	Phangali	*Pogostemon benghalensis* Burm. f.	Lamiaceae	leaf	Waki
9	Rametha	*Gnidia glauca* (Friesen.) Gilg.	Thymelaeceae	leaf	Kankwadi
10	Sag	*Tectona grandis* L.	Lamiaceae	leaf	Ambade

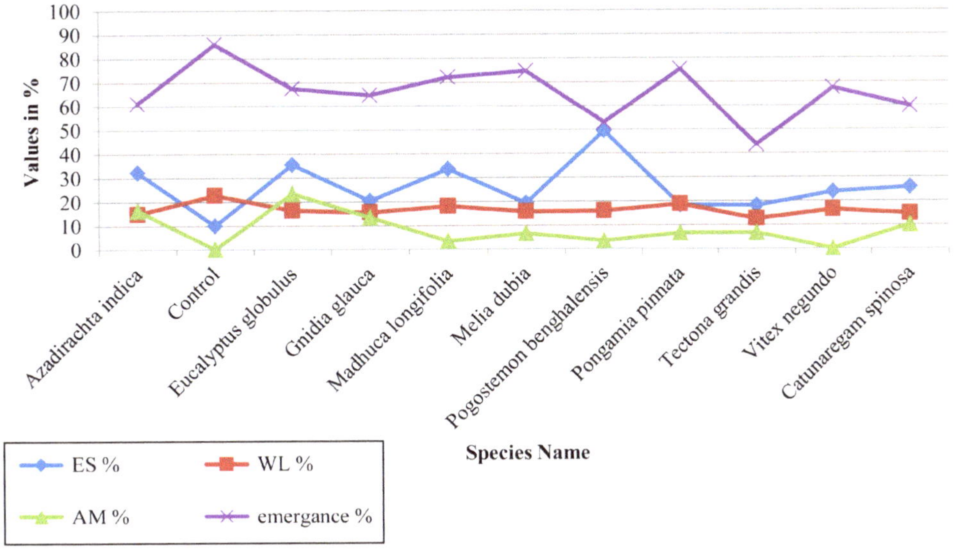

FIGURE 27.1 Anova analysis of 1% treatment.

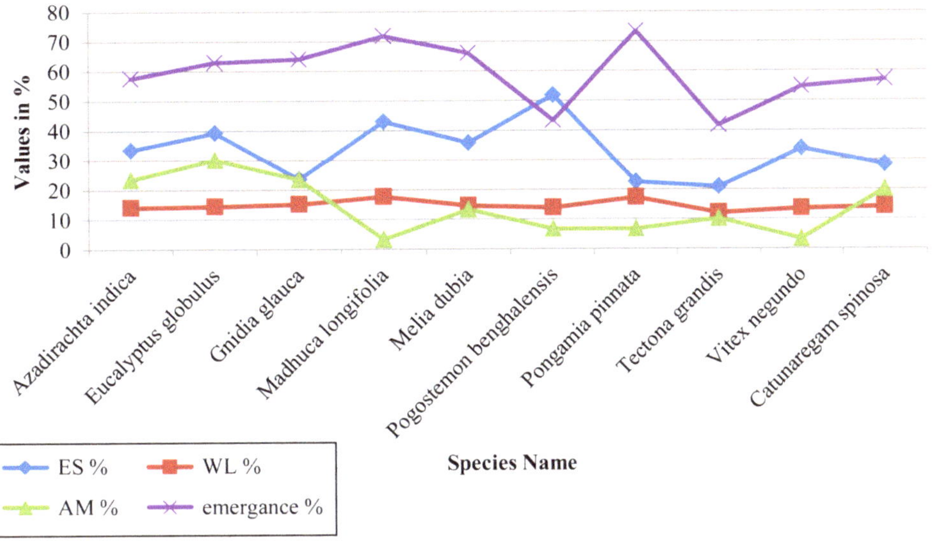

FIGURE 27.2 Anova analysis 3% treatment.

the seeds treated with 5% *Azadirachta indica* leaf powder (32.33 eggs/10 gm seed). The maximum number of eggs was recorded on the *Pongamia pinnata* leaf powder treatment at the 3% level (175 eggs/10 gm seed). Similarly, leaf powder of *Madhuca longifolia*, *Melia dubia*, *Pogostemon benghalensis*, *Tectona grandis*, *Vitex negundo*, and fruit powder of *Catunaregam spinosa* also showed a significant reduction in egg-laying. A less number of eggs laid in the treatment having with high adult mortality. These findings in respect of egg-laying are in correlation with that of the dosage applied (Figure 27.2).

An average number of adults emerged was found to be significantly reduced in the different treatments. The adult emergence was considerably inhibited in *Tectona grandis* L. and *Pogostemon benghalensis* leaf powder treatment. Low emergence of adults was observed in *Azadirachta indica* leaf powder treatments. *Melia dubia*, *Pongomia pinnata*, and *Madhuca longifolia* leaf powder fail to protect cowpea from bruchid attack, though some reduction was observed in *Catunaregam spinosa*, *Eucalyptus globulus*, and *Gnidia glauca* leaf powder treatment.

The average percentage of loss in grain weight due to attack of *Callosobruchus maculatus* was lowest in *Tectona grandis* leaf powder while the maximum percentage of loss was observed in *Pogostemon benghalensis* treatment (11.58%) at 5% level. The percentage of loss in grain weight in *Vitex negundo* and *Gnidia glauca* leaf powder treatment was found to be less than fifty percent of that observed in the control. Figure 27.3 gives detailed results of treatments.

Thus, laboratory experimental studies showed that leaf powders of *Gnidia glauca*, *Eucalyptus globulus*, *Catunaregam spinosa*, *Vitex negundo*, *Azadirachta indica*, *Madhuca longfolia*, and *Pogostemon benghalensis* showed the mortality of adults, and also reduction in oviposition and suppress adult emergence. Insecticidal properties of three *Jatropha* species against store grain pest (Bhagat et al., 2011).

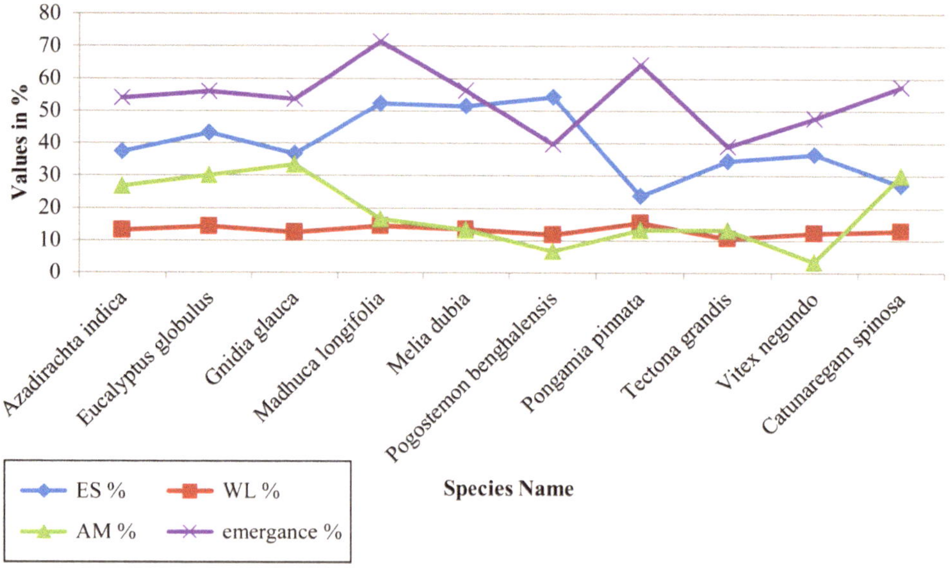

FIGURE 27.3 Anova analysis of 5% treatment.

27.4 *Jatropha* Species for Insect Control

Laboratory experiments were conducted to study the effect of *Jatropha* seed oil, root and leaf extracts of three species viz. *Jatropha nana* (JN), *J. gossypifolia* (JGO), and *J. glandulifera* (JGL). Four concentrations (20, 15, 10, and 5 mg/ml) of each dry extract were obtained by dissolving the dry extract in 1 ml of acetone. The stored grain pests, *Corcyra cephalonica* and *Tribolium castaneum*, were used for investigations. *Jatropha* species exhibited 50% (LC_{50}) and 95% lethal concentration (LC_{95}) values against *Corcyra* larvae and *Tribolium* adults. The activity of oils against *Corcyra* were found in the range of JGO < JGL < JN. Leaf pet ether extract activity observed was JN < JGL < JGO and leaf methanol extract was JGL < JN < JGO. While in the case of root pet extract these were JN < JGL and JGO does not show any activity up to 20 mg concentration. Root-methanol extract showed JGL < JN < JGO. The activity of oil against *Tribolium* adults showed that JGL <JN < JGO, while ether extracts were found in a range of JGO > JN > JGL and methanol extract JN < JGL < JGO. Phytochemical study showed almost all primary and secondary metabolites are found in leaf and root extracts. The pet ether extract of root showed that JGO < JN < JGL and methanol extracts were in a range of JN < JGO < JGL. TLC/HPTLC study showed presence of β-sitosterol in all extracts. *J. gossypifolia* oil was found to be most toxic to the *C. cephalonica* and *T. castaneum* with LC_{50} values of 2.17 and 1.08 mg, respectively (Bhagat Shinde and Kulkarni 2011). Bhagat and Kulkarni (2013) conducted a research experiment on stored grain pest *Callosobruchus chinensi s*L. for three *Jatropha* species for antifeedant and oviposition deterrent activity.

27.5 Conclusion

The extracts were used to search active components from various plants used by farmers and reported different results. It was found that though *Vitex negundo* leaves are used for paddy protection, leaf extract did not show this property. Similarly, Karanj (*Pongamia pinnata*) leaves and oil showed antifeedant property but cakes did not show significant grain protection. Products from several plants such as *Eclipta alba*, *Justicia adhadoda*, *Bassia latifolia*, *Mentha spicata*, *Chenopodium ambroidoises*, *Calotropis procera*, etc. have been tested. The active components have been identified and found effective but due to their high cost of synthesis were not viable commercially. Studies by the Indian Council of Agricultural Research (ICAR) have reported protection of rice by adding dried walnut and *Lantana* leaves. A number of species have been used as storage insecticides in various countries as food preservatives but these cannot be used for grain storage because of the high cost involved.

Acknowledgments

The authors are thankful to Officer-In-Charge-Agharkar Research Institute for facilities and In-charge, Botany group, ARI, Pune. The authors are thankful to Mr. G. G. Sohani, President of the BAIF Development Research Foundation, for his encouragement.

REFERENCES

Ahuja U., Ahuja, S.C. and Thakarar, R. 2010.Tradition Paddy storage. *Asian Agri-History*. **14**(3): 245–260.

Bhagat, R. and Kulkarni, D.K. 2013. Antifeedant and oviposition deterrent activity of three *Jatropha* species against stored grain pest *Callosobruchus chinensis* L. (Coleoptera: Bruchidae) *International Journal of Multidisciplinary Research*. **2**(7): 86–91.

Bhagat, R., Shinde, V. and Kulkarni, D.K. 2011. Insecticidal properties of three *Jatropha* species against stored grain pests. *Indian Journal of Entomology*. **73**(1): 30–33.

Khalequazzaman, M. and Osmangoni, H.M. 2009. Toxic potentials of some plant powders on survival and development of *Callosobruchus maculatus* (F.) and *Callosobruchus chinensis* L. *Journal of Life and Earth Science*. **3**(4): 1–6.

Khanam Muslima, L.A., Khan, A.R., Khalequzzamanb M. and Rahmanb, S.M. 2008. Effect of *Sapium indicum*, *Thevetia neriifolia* and *Jatropha gossypifolia* seed extract on the fecundity and fertility of *Tribolium castaneumand Tribolium confusum*. *Bangladesh Journal of Science and Industrial Research*. **43**: 55–66.

Kulkarni, D. K. and Kumbhojkar, M.S. 2003. Ethno-agricultrual study of Mahadeokolis in Maharashtra, India. *Asian Agri-History*. **7**(4): 295–312.

Kulkarni, D.K. and Kumbhojkar, M.S. 1996. Pest control in tribal area: An ethnobotanical approach. *Ethnobotany*. **8**: 56–59.

Naskar G.C. 1998. Rice seed preservation with leaves of nishinda. *Honey Bee* 9(1):13.

Niber, T.B. 1994.The ability of powders and slurries from ten plant species to protect stored grain from attack by *Prostephanus truncaus* Horn (Coleoptera: Bostrichidae) and *Sitophilus oryzae* L. (Coleoptera: Curculionidae). *Journal of Stored Products Research*. **30**(4): 297–301.

Sinha, R.K. and Bhatia, S. 1997. Indigenous knowledge and traditional wisdom in the farming communities of Asia with special reference to India. *Asian Agri-History*. **1**(4), 275–287.

Talukder, F.A. and Howse, P.E 1995. Evaluation of *Aphanamaxis polystachya* as a source of repellents, antifeedants, toxicants and protectants in storage against *Tribolium castaneum* (Herbst). *Journal of Stored Products Research*. **31**: 55–61.

Patil, P.V., Taware, S.P., Punalekar, S. and Kulkarni, D. K. 2011. Traditional plant resources used to control store food grains pests and validation of *Calophyllum inophyllum* L. oil. *Bioscience Discovery*. **2**(2):181–184.

Patil, P.V. Kulkarni, D.K. and Taware, S.P. 2014. Evaluation of traditional knowledge of plant resources to control store food grain pest- *Callosbruchus maculatus* F. *Indian Journal of Fundamental and Applied Life Sciences*. **4**(2): 1–5.

28

Botanical Pesticides for Organic Farming and Sustainable Agriculture

Desam Nagarjuna Reddy and Abdul Jabbar Al-Rajab

CONTENTS

28.1 Introduction

The use of botanical pesticides may present an alternative to chemical pesticides and be considered in agricultural practices to ensure agricultural production sustainability (Gurr et al. 2004; Médiène et al. 2011; Perfecto and Vandermeer 2010; Srinivasan 2006).

Historically, plant derivatives or botanical pesticides were used in ancient Egypt, India, China, and Greece to protect plants and food products (Thacker 2002). In the mid-1930s–1950s, however, the use of synthetic pesticides caused numerous problems that were unexpected at the time of their introduction: chronic poisoning of farmers and consumers; danger to the wildlife such as fish and birds, disturbance of natural biological control and pollination, groundwater pollution, potentially threatening human and environmental health (El-Wakeil 2013).

Isolated secondary metabolites from plants show feeding deterrent or toxic effects to insects in laboratory bioassays, and botanical pesticides have been the subject of several recent studies (Regnault-Roger et al. 2008). Also, it takes several steps to get commercial products as shown in Figure 28.1. There are many factors determining the use of botanical pesticides for pest control (Ali et al. 2017; Deng et al. 2009; Isman 2008). Instead of synthetic pesticides, botanical pesticides are used in organic gardens. In general, they respond and degrade rapidly with few exceptions like mammalian toxicity. Plant-derived products and its derivate are considered to be botanical pesticides.

Botanical pesticides may increase crop production and control different pests such as weeds, nematodes, insects, pathogens (Narwal 2010). Botanical pesticides are naturally occurring and extracted from the plants. Botanical pesticides are alternative to synthetic pesticides, but they are not necessarily less toxic to human beings (Boeke et al. 2004a, 2004b; Ei-Wakeil 2013; Philogene 2005; Stevenson et al. 2010).

28.2 Botanical Pesticides

28.2.1 Limitations of Botanical Pesticides Use

There are different factors limiting the use of artificial botanical pesticides such as the molecule's formulation, active ingredient, exposure time, contact time, concentration, climate, and application season (Dosemeci et al. 2002).

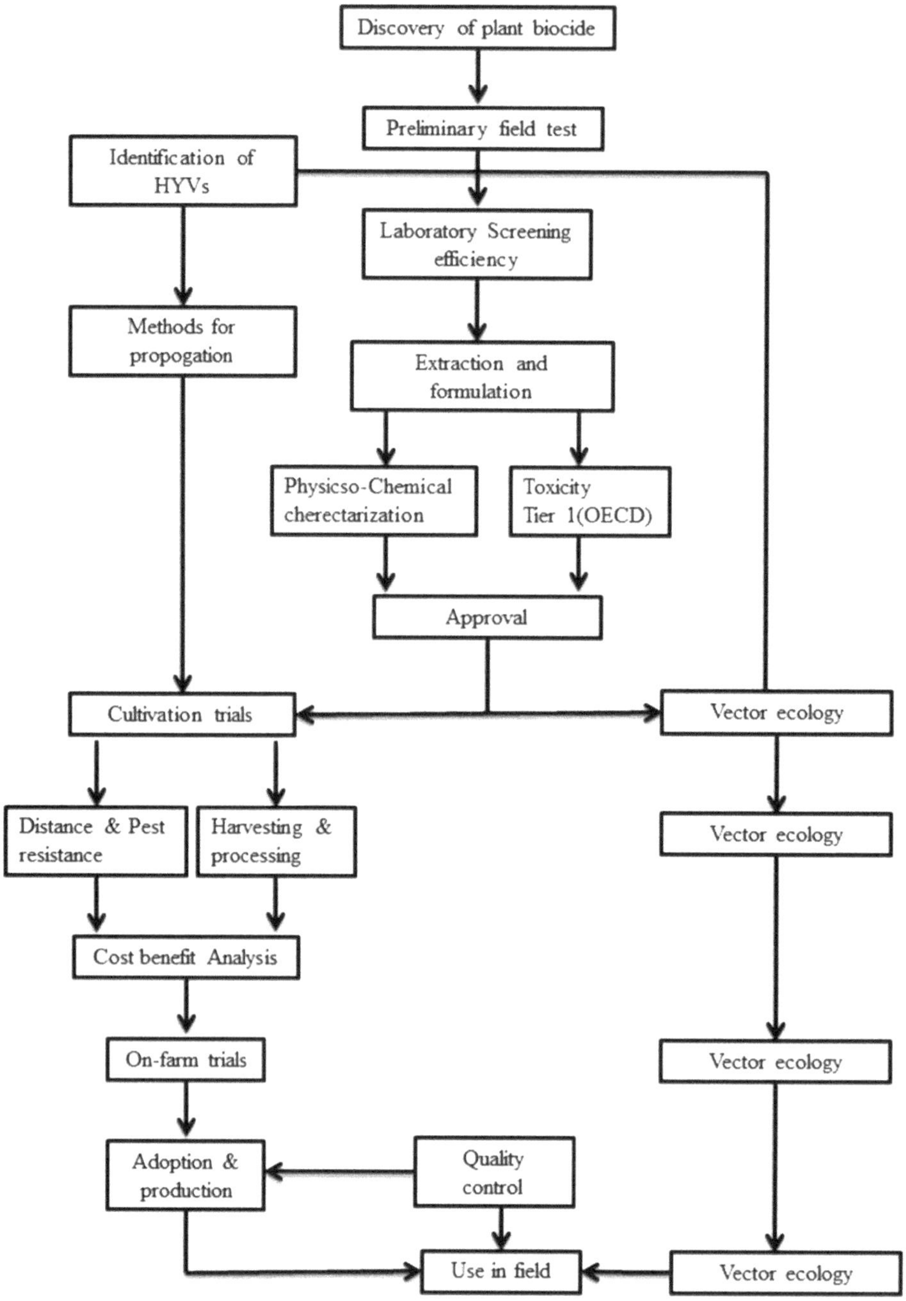

FIGURE 28.1 Steps and interactions between stages from discovery of a biopesticide to approved field application.

28.2.2 Resources of Botanical Pesticides

Currently, there are four major botanical insecticides that are used commercially, such as neem, rotenone, essential oils, and pyrethrum. Some other compounds are less often used, such as nicotine, sabadilla, and ryania.

28.2.3 Current Use of Botanical Pesticides

Botanical pesticides generally are classified in two generations. The first generation is of plant origin such as nicotine, rotenone, ryania, pyrethrum, sabadilla, and plant essential oils, while the second generation presents synthetic compounds such as Pyrethroids, Azadirachtin, and new botanical pesticides (Regnault-Roger et al. 2005).

28.2.3.1 First-Generation Botanical Pesticides

An alkaloid obtained from leaves of tobacco (*Nicotina tabacum*) and associated plant species, has a long history used as insecticide. As such organophosphate and carbonate insecticides are similar symptoms of poison (Regnault-Roger et al. 2008). Nicotine (Figure 28.2) should purely extreme toxicity of mammals (rat oral LD_{50} is 50 mg/kg) and its rapid cuticular absorption in humans, nicotine has seen reduced to use, primarily as a fumigant in greenhouses against soft-bodied pests (Casanova et al. 2002; El-Shazly et al. 2005).

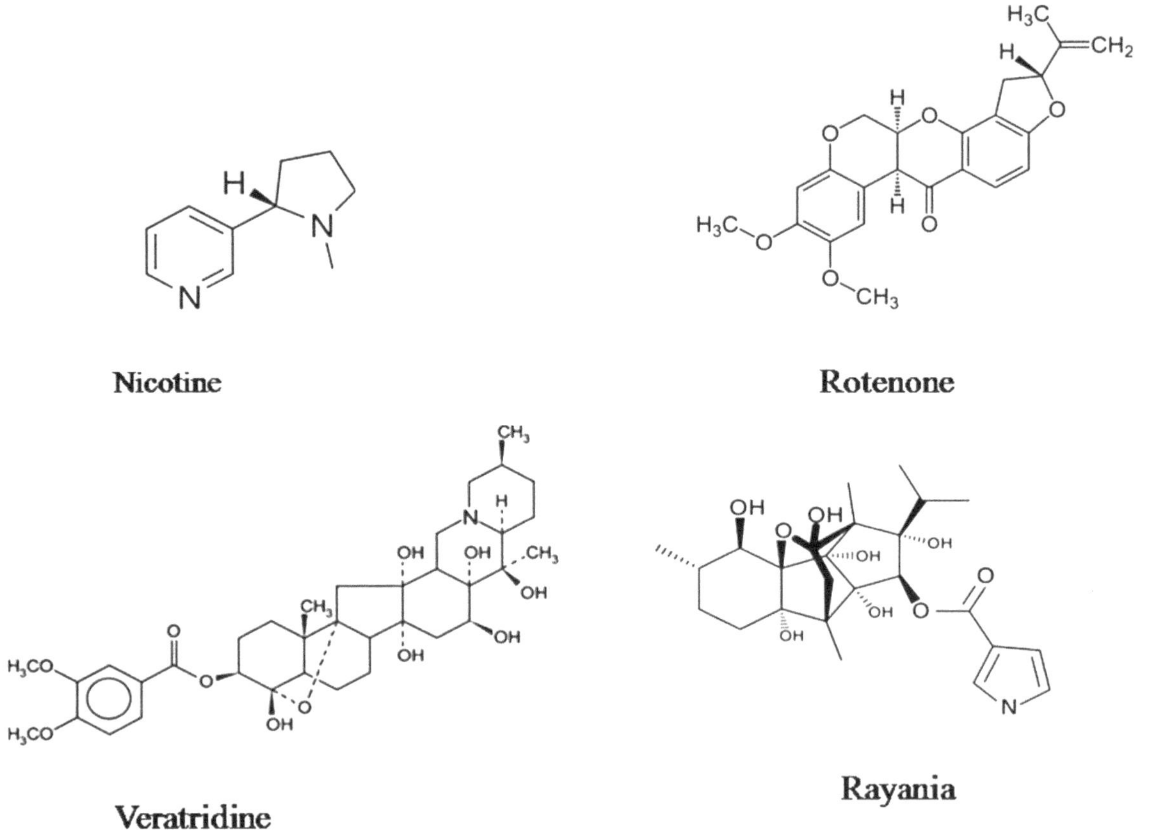

FIGURE 28.2 First-generation botanical pesticides.

Rotenone is one of the isoflavonoids obtained from the roots of the tropical legumes *Derris*, *Lonchocarpus*, and *Tephrosia* usually grown in Venezuela and Peru and is often called cube root. Rotenone (Figure 28.2) has been using insecticide for more than 150 years, but it is poisonous for fish. Extraction with roots from organic solvents resin contains more than 45% of total rotenoids; but literature indicates that the rotenone (44%) and deguelin (22%) major chemical constituents (Betarbet et al. 2000; Cabizza et al. 2004; Cabras et al. 2002), respectively. Liquid rotenone formulations contain as much as 8% rotenone and 15% total rotenoids.

Sabadilla is a botanical pesticide obtained naturally from *Schoenocaulon officinale* (South American lily). It contains cevedine-type alkaloids which are extremely toxic to mammals (rat oral LD_{50} is 13 mg/kg), while the commercial preparation contain < 1% of active ingredient, providing margin safety (Figure 28.2). These alkaloids are used primarily for citrus crops and avocado (Isman 2006) for organic farming in California about 100 kg is used annually. Sabadilla is used as botanical pesticides for insecticides. These alkaloids are extremely toxic, and affect nerve cell membrane action. Many insects are affected immediately after application, while other insects survive with paralysis for a few days. This compound is effectively synthesized by MGK 264 or PBO.

Rayania is a botanical pesticide obtained from the wood Caribbean shrub *Ryania speciosa*. The wood powder contains less than 1% ryanodine (Figure 28.2). Ryanodine effectively works on calcium and releases muscle tissue (National research council 2000). It would generally be used in limited extend by apple growers to control the *Cydia pomonella* (codling moth). It usually works on stomach poison and its slowly. Although it does not produce knockdown paralysis. This alkaloid causes to stop feeding the insects. This alkaloid is effectively synthesized by a PBO such as sabadilla alkaloids.

Pyrethrum alkaloids are obtained from *Chrysanthemum cinerariaefolin* dried flower. These flowers are dried and ground to a fine powder and it is extracted with n-hexane, after removing the solvents, the organic colored liquid contains the active botanical pesticides (Glynne-Jones, 2001). Among the six esters, three belong to chrysanthemic acid of esters and three are esters of pyrethric acid, namely pyrethrins I and II, which are the most naturally abundant and have most effective pesticidial activity (Figure 28.3). Pyrethrins work effectively on flying insects, producing hyperactivity and convulsions in most insects, due to the result of neurotoxic action of the pyrethins, which close the sodium channels in the nerve system. As such, mechanisms on insects, oreganochlorine and DDT are very similar to the pyrethrins. These are moderate effect to mammals (rat LD_{50} ranges from 350–500 mg/kg) pure pyrethrins and less toxic to mammals with technical grade (rat LD_{50} 1500 mg/kg) (Antonious, 2004; CDPR, 2005; El-Wakeil, 2013; Isman 2006; Rattan 2010). Essential oils are extracted from aromatic medicinal plants by stem distillation or normal distillation method using Clevenger apparatus. Essential oils are generally used as flavoring, fragrances in perfumes, and food industries, and recently in aromatherapy and herbal medicines (Buckle, 2003).

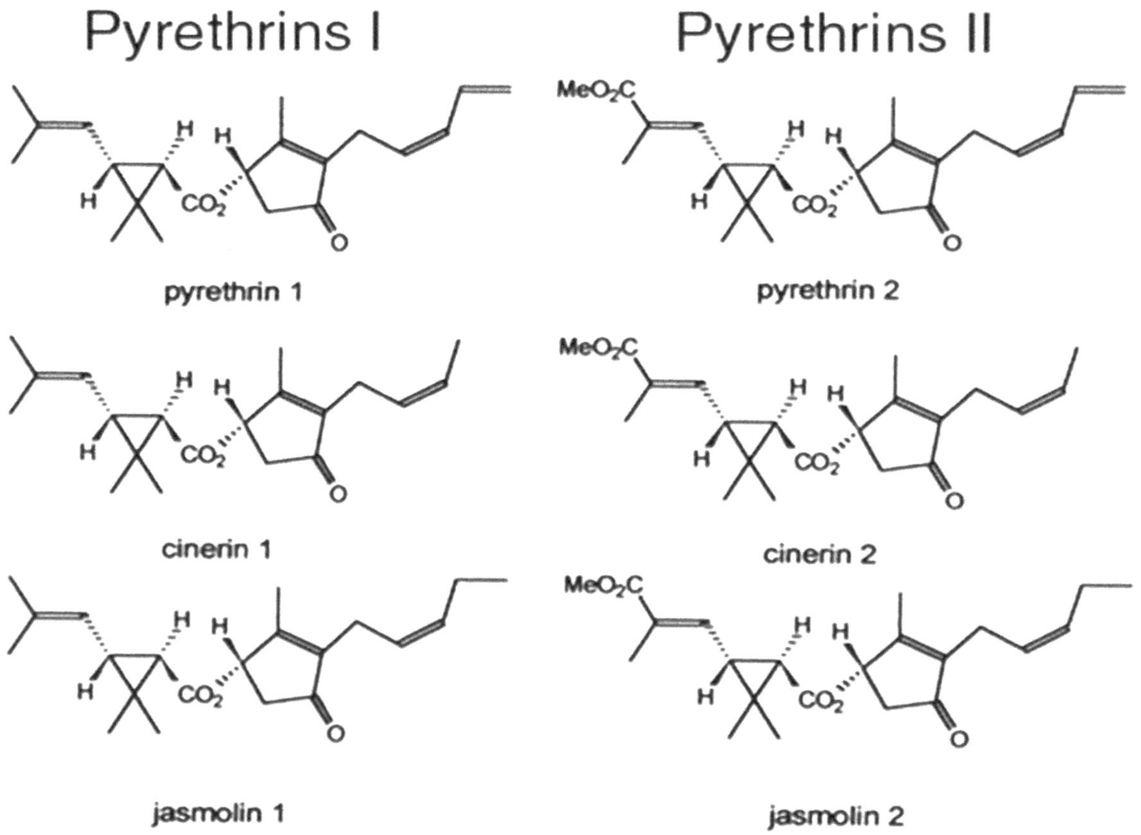

FIGURE 28.3 The six insecticidal constituents of pyrethrum extract.

Many essential oils are produced from botanical sources, many of which are members of the Lamiaceae family. Essential oils compose of mixtures of monoterpenes, sesquiterpenes, and phenols. 1, 8-cineole, eucalyptus, eugenol, thymol, and menthol are major components from *Rosmarinus officinale*, *Eucalyptus globus*, *Syzygium aromaticum*, *Thymus vulgaris*, and *Mentha* species respectively (Reddy, 2019) (Figure 28.4). Traditionally, essential oils have been used to protect the stored food products in the Mediterranean and North West Asia (Abdallah et al. 2004). Essential oils have a fast effect on some pests and interfere with the neuromodulator octopamine (Abdelgaleil et al. 2010; Adel et al. 2010; El-Hosary et al. 2011; El-Hawary et al. 2006; El-Sebai et al. 2005; Isman 2006; Rattan 2010; Sammour et al. 2011).

28.2.3.2 Second-Generation Botanical Pesticides

Neem and its derivatives are naturally occurring from the Indian neem tree *Azadirachta indica* (Dimetry et al. 2010, 2012). From Indian neem tree, two types of botanical pesticides sre present. These are obtained from neem oil, could effectively work against some phytopathogens and mites. Along with neem oil physical effects on pests and fungi. Neem oils are extracted with moderate polarity extracts of seed residue after removal of the oil as these extracts contain the complex mixture triterpenes Azadirachtin (Figure 28.5). These seeds contain more than ten Azadirachtin analog, among these Azadirachtin is a major component and the remainder are minor components, but they contribute little efficiency to the extract. Seed extracts contain contemplate quantities of other terpinoids like salanin, nimbin, and their derivatives, but these substances are controversial regarding role (Caboni et al. 2002; El-Wakeil 2013; Isman 2004; Weinzierl 2000). Azadirachtin blocks the release of molting hormone and works as antifeedant of several insects. In addition, it also is used in insect pest, medicinal applications, and also used in other areas (Figure 28.6).

Melia extracts are obtained from *Melia* seeds (*Melia azedarach*) contain a complex number of triterpenoids of the meliacarpins (Figure 28.5). But these terpenes are similar but not identical to the Azadirachtins. They act as insect growth regulating bioactivities (Ascher et al. 2002; Carpinella et al. 2003; Charleston 2004; Kraus 2002). China in 1990s botanical pesticide toosendanin (Figure 28.5) was extracted from bark of *Melia toosendan*. It is considered by most taxonomists to be synonymous with *Melia azedarach*. The extract contains complex mixture of triterpenoids. Toosendan was reported as a stomach poison for chewing insects (El-Wakeil 2013). It is relatively non-toxic to mammals and also not clear whether this substance in production and crop protection. East Africa, triterpenoids extracted from seeds of *Melia volkensii*. The active principles in *M. volkensii* include triterpinod salannin and also a major constituent of neem seed extracts, and some novel triterpenoids such as volkensin (Figure 28.5).

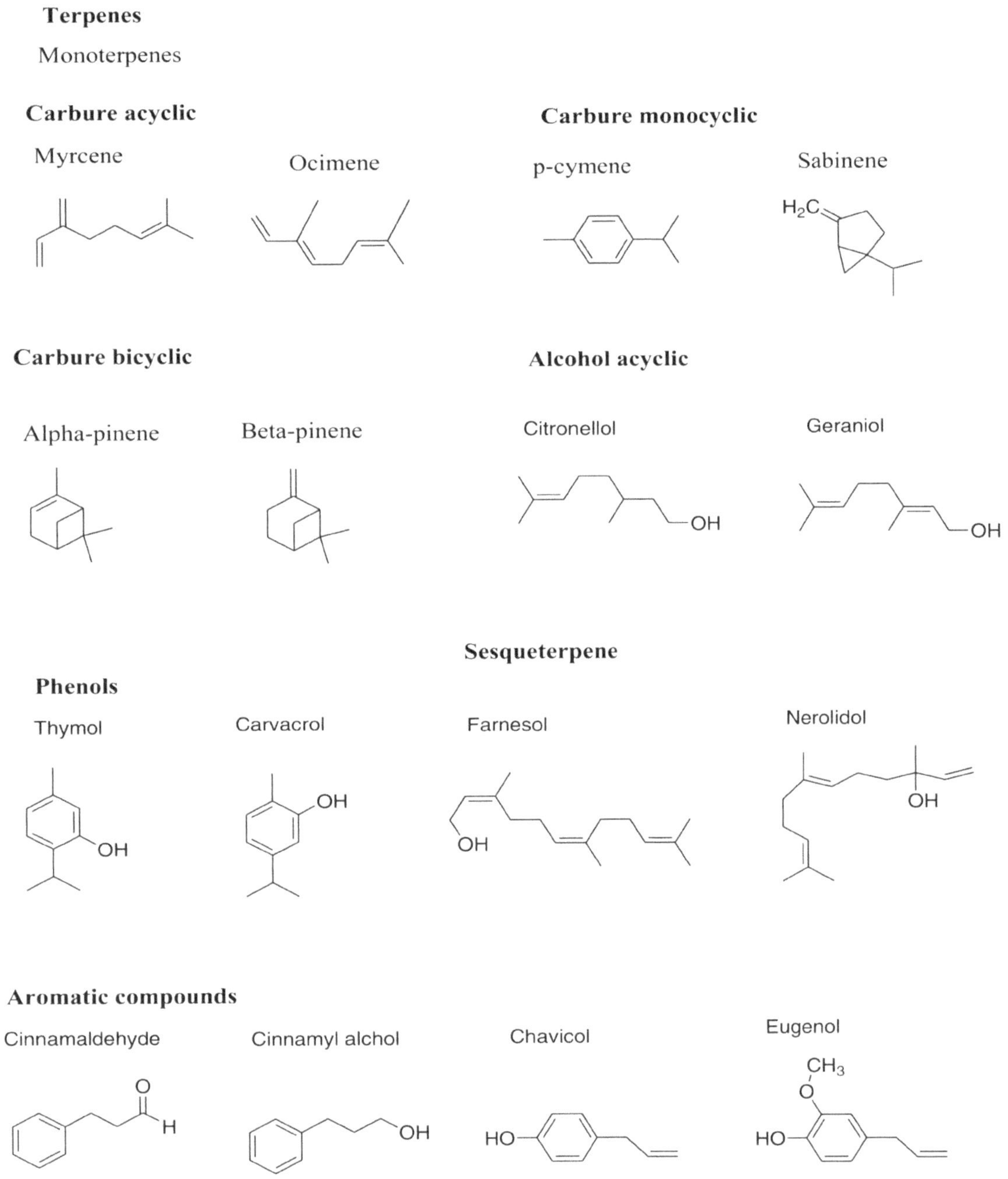

FIGURE 28.4 Chemical structures of selected components of essential oils.

28.2.4 Botanical Pesticides Advantages and Disadvantages

28.2.4.1 Advantages

- Botanical pesticides are produced from plants and known to the formers, because most of the time plants grown in the same region.
- The plant extracts also useful for household insect repellents and medical applications.
- Excellent degradation of active chemical constituents and risk residues decreases on food.
- Many active chemical compounds act rapidly.
- Most of the active components have stomach action and rapid degradation. They might be less aggressive with natural enemies.
- Most of the active chemical constituents are photo-toxic.
- Most of these active constituents are not developed for resistance as quickly as with synthetic compounds.
- Most of them are less toxic for wildlife.

28.2.4.2 Disadvantages

- Plant extracted chemical constituents are not truly pesticides.

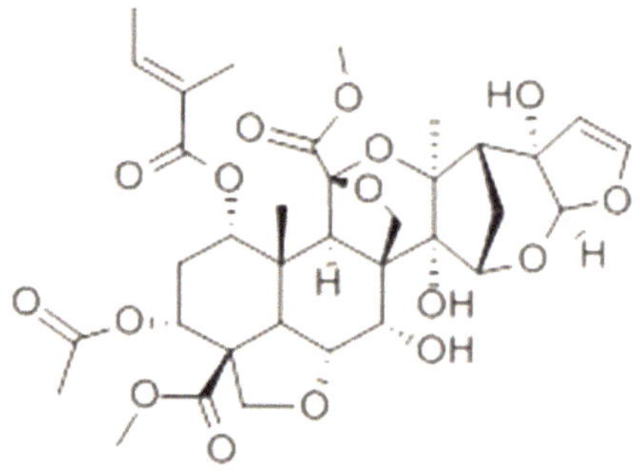

Azadirachtin

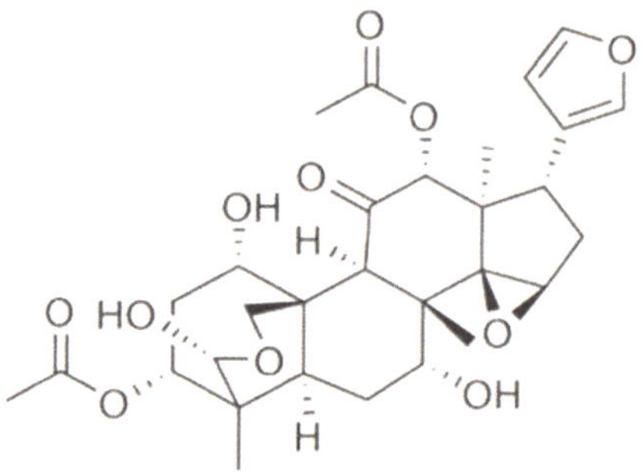

Toosendanin

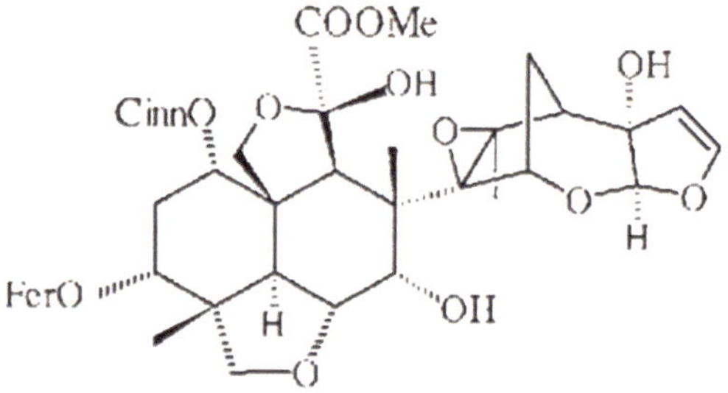

1-cinnamoyl-3-feruoyl-11-hydroxymeliacarpin

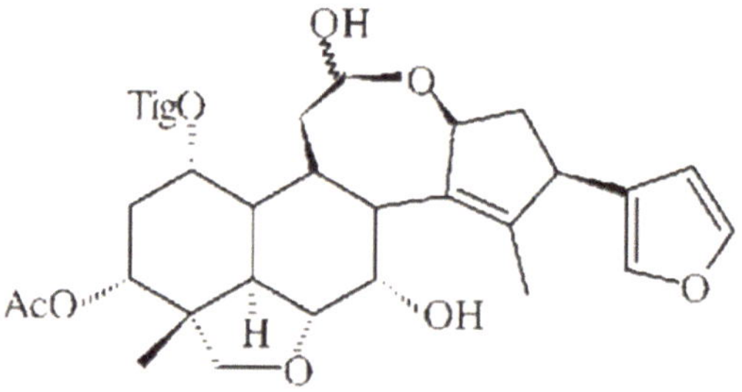

volkensin

FIGURE 28.5 Second-generation botanical pesticides.

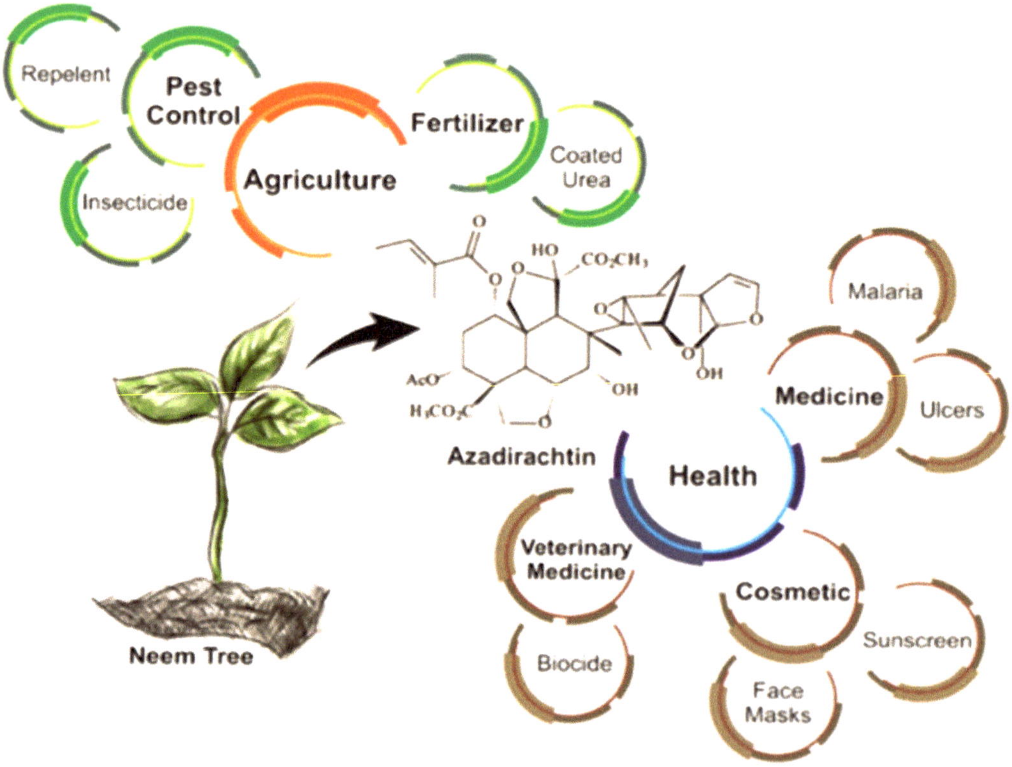

FIGURE 28.6 Potential applications of Azadirachtin in different areas.

- Many plant extracts are used only for insect deterrents and their action is short.
- Botanical pesticides are not necessarily available season long.
- There is no established residue tolerance for many botanical pesticides.
- In the presence of UV light, botanical pesticides are easily degradable and also their residue action also short.

28.2.5 The Role of Botanical Pesticides in the Future

Botanical pesticides prepared from Annona (Custered apple) species include *A. squamosal* (sweetsop) and *A. muricata* (soursop), source of pesticides from fruit juices from Asia. In 1980s clear investigations, identification and isolation of long-chain fatty acids and acetogenins are am important role of bioactivity. Acetogenins are insecticidal bioactivity and the annonin I and are the similar constituent asimicin (Figure 28.7) major source from seeds of *A. squamosa* and *Asimina triloba* (Amirican pawpaw), respectively (Johnson et al. 2000). Acetogenins are slow stomach poison, especially more effective against chewing insects like lepidopterans and Colorado potato beetle. Acetogenins have a similar mode of action to rotenone, which blocks the energy production in the mitochondria in insects (Johnson et al. 2000). There is no commercial development of this product as insecticide except for head lice shampoo. *Annona* seeds are more useful in tropical countries, commonly consumed as fruit juice. *A. Squamosa* seeds were reported to be effective against the diamondback moth (Isman 2006). Asimicin reduces the rate of oxygen consumption by fourth instar *Ostrinia nubilalis* as measured with constant manometer.

In the early 1990s, sucrose esters were discovered by the United States of Agriculture Department USAD from wild tobacco *Nicotina gossei*; they showed botanical pesticides' effect on soft-bodied insects and mites (United Nations 2011). Sucrose esters extraction from plant biomass commercially is not practical. Sucrose esters were registered in the United States as insecticide and miticide based on C_8 and C_{10} mono and di fatty acids and esters of sucrose octanoate and dioctanoate (Farone et al. 2002) (Figure 28.7). These types of compounds can be synthesized into a new class of botanical pesticides (Isman 2006).

Botanical pesticides are expected to play a major role in agriculture. In Europe and North America, organic food production has been estimated to be 8 to 15% of the annual total food production. On commercial scale, in markets botanical pesticides face the fewest competitors (National Council of Research 2000). Crop rotations botanical pesticidial products are placed in a better position among all; especially the use of spinosad resistance of diamondback moth to *B. thuringienis* (Zhao et al. 2002). In California, between 1998 and 2003 use of pesticide reduce risk increased than threefold from 138 to 483 t, whereas botanical pesticides use decreased from 652 to 472 t (CDPR 2005). Traditional plants have more attention in west Africa against insects (Belmain 2001, Bocke et al. 2004a, b). Some plants are more effective and active; some of these are volatile components and acts as fumigants, kill adult pests and their progeny (Jayasankara et al. 2005). As a result, developing known botanical pesticides rather than screening further novel bioactive substances is necessary to satisfy human curiosity for sustainable agriculture.

28.3 Weed Management

In agriculture, there are more than 250 major species of weeds that reduce the yield and crop growth (Narwal 2010). The continuous use of herbicides and pesticides for weed control generated many problems such as environment contamination, crop injury, herbicide resistance, etc. Consequently, weed control using non-chemical methods is preferred for sustainable organic agriculture.

28.3.1 Cover Crops

For sustainable agriculture weed management to use phytotoxic protect and cover crops is very effective. The herbicide Dicamba has been used effectively to control the broadleaf weeds in lands covered by rye, wheat, oats, barley, grain sorghum, and Sudan grass (Laxmishree et al. 2017).

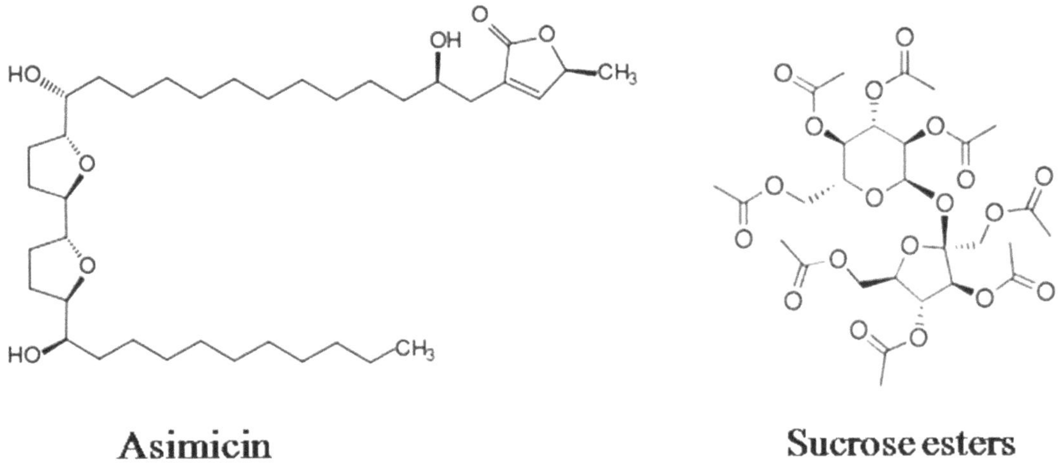

FIGURE 28.7 Future botanical pesticides.

The conceal effect of botanical pesticides of rye mulch on weds in the field is outstanding. From rye decaying (Thakore 2006) extracted nine acids, inhibited the growth of embassy plants. Before plant soybean, rye residues 30–60% inhibit the lambsquarters and red root pigweed, because from rye residue extracted phenylacetic acid and hydroxybutyric acid (Narwal 2010). From sunflower and tobacco (Batish et al. 2008; Laxmishree et al. 2017; Lucantoni et al. 2006; Scudeler et al. 2014) isolated 2,4-dihydroxy-1,4-benzoxazinone and 2-benzoxazolinone with phytotoxicity on some tested weeds. But botanical pesticides might be explained as herbicidal weed suppression under field conditions. Certain botanical pesticides discovering microbial transformation from plants residues such as wheat and rye, which may increase the phytotoxicity to weeds and also effectively reduce weed by intercropping (Awasthi and Avasthi 2017; Isman 2006; Isman 2008; Narwal 2010; Narwal 2017).

28.4 Nematodes Management

There are over 2200 plant species root-knot nematodes known worldwide (Sarmah et al. 1992). Some agricultural practices, plant resistance to nematodes, and use of botanical nematicides would reduce the damage of nematodes to crops and contribute effectively to the sustainable agriculture.

28.4.1 Nematicides from Plant Materials

Alfalfa, lespedeza, oats, and flax matured dried residues have been applied into the infested with root-knot remarkable reduced the incidence of root-knot in tomato (Narwal 2010; Narwal 2013). Incidence of *Tylenchulus semipenetrans* reduced by soil amended cotton waste and Lucerne pellets. B. longicaudatus and plant parasitic nematodes are reduced by rice straw (Karani et al. 2017; Khan et al. 2017; Narwal 2010). Flax and orchard grass reduce 75–90% incidence of root-knot in tomato field. Parasitic nematodes on wheat and mungbean were decreased by applying wheat straw, Neem cake, and fertilizers NPK (Narwal 2010) and also wheat straw and paddy husk reduced incidence of *M. incognita* (Narwal 2010). Okra might increase the plant growth. Infestation of root-knot is reduced by *A. mexicana* seed oil when applied soil drench or foliar spray. Infestation of okra is reduced by *A. pinnata* and biofertilizers.

28.5 Insects Pest Management

Naturally most pest species are in a regulated ecosystem process viz., for food competition, survival problems with natural enemies. In most cases the populations of pests are stable and the damage relatively insignificant. An artificial pesticide is more dangerous to the environment, human health and livestock, and does not easily degrade; it causes damage to natural resources, causes pest recovery by killing natural enemies of target pests, and could produce new pests, which were not previously present (e.g., cotton in white fly). Plant origin botanical pesticides (allelochemicals) are antibiotic, so it could be used as insect pest resistance in cropping systems include crop rotation, crop mixture, or intercropping (Batish 2006; Dutta 2015; Mahadevappa and Kulkami 1994; Narwal 2010) growing resistant verities (Bansal 1994; Prusty et al. 1994) and insecticidal botanical pesticides (Batish 2006; Lucantoni 2006; Singh et al. 2012).

28.6 Disease Management

The uninterrupted cultivation of the same or related crop leads increases pathogens' intensity in soil. Consequently, crop rotation is considered a natural method of disease prevention. Botanically unrelated crops are affected by different pathogens. Crop rotations are especially effective decreases the population of soil-borne pathogens (Narwal 2010). Intercropping species provide disease delaying the onset of disease, control spore distribution, and modify micro-environmental conditions. Sometimes plant organic matter and environment of plants indirectly increase the soil fungistatis and antibiosis. Intercropping systems and resistant plants prevent the spread of pathogens.

Many plants produce botanical pesticides either previous or after infections by certain pathogens, plants reduce the diseases (Zhao et al. 2002). Green manure with odor also controls this fungus (Narwal 2010). Before planting potatoes, soybean residue mixed in soil controlled by potato scab due to antibiotics and antagonistic production by *B. subtilis* and *S. scabies* respectively. Decomposition of plant residues releases botanical pesticides and it justifies the attention of plant pathogens (Council on Scientific Affairs 1997).

28.7 Soil Fertility

Worldwide tropical and subtropical climates are suitable for two to four crops grown in a year. Therefore, it is assumed that irrigated or heavy rainfall in a calendar year leads to multiple crop systems (MCSs). Accordingly, planned crop rotations, crop mixtures, and intercropping aim to make successfully ecological and sustainable agriculture, and also have a deleterious effect on environment conditions and maintain soil productivity over a long period of time. Soil sickness problems and pest checks are useful for maintaining soil fertility. To improve soil fertility farmers should follow both crop rotation and intercropping systems and also legumes maintains.

Soil fertility could be maintained or improved, when well-designed crop rotation and intercropping systems are present, which also prevent the development of pest and soil sickness when compared to monoculture (Anver et al. 1994).

28.8 Conclusion and Future Prospects

The plant kingdom provides a wide range of natural resources secondary metabolites (botanical pesticides) and structural biodiversity. Nowadays, developed and developing countries are using synthetic pesticides d to reduce fungal and insect pests, weeds, and diseases, which are extremely toxic to the environment and living organisms. So we should introduce and use new botanical pesticides for sustainable agriculture to reduce fungal and insect pests, and help disease management and environmental and living organisms. In this book, chapters present several different approaches. Because of these approaches, inhibition of larval growth, pupation, enzymes involved in the key process of life apolysis during molting, sclerotization, and emergence.

In soil, botanical pesticides released (organic and inorganic compounds), because of the crop, weed, and tree residues. These botanical pesticides are normally inhibited by the growth of crop or other plants depending on the amount of residue, type of residue, and time of decomposition period. There are some plant residues (mungbean, soybean, alfalfa, and leucaena) that increase crop growth and release botanical pesticides and biotransformation products decreases the incidence of weed, nematodes, insect pests, and disease. Intercropping, crop rotations, phytotoxic crop varities, natural herbicides are effective methods for ecological and sustainable agriculture.

REFERENCES

Abdallah, S. A.: Barakat, A. A.: Badawy, H. M. A.: Soliman, M. M. M. Insecticidal activity of different wild plant extracts against Aphis craccivora Koch. *Egypt J Biol Pest Control* 2004, 14, 165–173.

Abdelgaleil, S. A. M. Molluscicidal and insecticidal potential of monoterpenes on the white gardensnail, *Theba pisana* (Muller) and the cotton leaf worm, *Spodoptera littoralis* (Boisd.). *Appl Entomol Zool.* 2010, 45, 425–433.

Adel, M. M.: El-Hawary, F. M.: Abdel-Aziz, N. F.: Sammour, E. A. Some physiological, biochemical and histopathological effects of Artemisia monosperma against the cotton leafworm, Spodoptera littoralis. *Arch Phytopathol Plant Prot.* 2010, 43, 098–1110.

Ali, S. S.: Ahmed, S. S.: Rizwana, H.: Bhatti, F.: Khoso, A. G.: Mengal, M. I.: Jatoi, J. H.: Bugti, A.: Rind, M. A.: Shahwani, S. A. Efficacy of different biopesticides against major sucking pests on brinjal under field conditions. *J Basic Appl Sci.* 2017, 13, 133–138.

Antonious, G. F. Residues and halflives of pyrethrins on field grown pepper and tomato. *J Environ Sci Health.* 2004, B39, 491–503.

Ascher, K. R. S.: Schmutterer, H.: Mazor, M.: Zebitz, C. P. W.: Naqvi, S. N. H. *The Persian Lilac or Chinaberry Tree: Melia azedarach* L. Neem Foundation, Mumbai. 2002, 770–820.

Awasthi, A.: Avasthi, S. Role of some ecofriendly biopesticides to control the pest *Heliothis armigera* on tomato. *Green Chem Tehnol Lett.* 2017, 3(1), 14–17.

Bansal, G. L. Allelopathy of buttercups and their alleviation. *See Ref No.* 1994, 7, 16.

Batish, D. R.: Singh, H. P.: Setia, N.: Kaur, S.: Kohli, R. K. Chemical composition and phytotoxicity of volatile essential oils from intact and fallen leaves of *Eucalyptus citriodora. Z Naturforsch C.* 2006, c61, 465–471.

Belmain, S. R.: Neal, G. E.: Ray, D. E.: Golob, P. Insecticidal and vertebrate toxicity associated with ethnobotanicals used as post-harvest protectants in Ghana. *Food Chem Toxicol.* 2001, 39, 287–291.

Betarbet, R.: Sherer, T. B.: MacKenzie, G.: Garcia-Osuna, M.: Panov, A. V.: Greenamyre, J. T. Chronic systematic pesticide exposure reproduces features of Parkinson's disease. *Nat Neurosci.* 2000, 3, 1301–1306.

Boeke, S. J.: Baumgart, I. R.: Van Loon, J. J. A.: Dicke, M. Field trials with plant products to protect stored cowpea against insect damage. *Int J Pest Manage.* 2004b, 50, 1–9.

Boeke, S. J.: Baumgart, I. R.: Van Loon, J. J. A.: van Huis, A.: Dicke, M.: Kossou, D. K.: Toxicity and repellence of African plants traditionally used for the protection of stored cowpea against *Callosobruchus maculatus. J Stored Prod Res.* 2004a, 40, 423–438.

Buckle, J. *Clinical Aromatherapy: Essential Oils in Practice.* Churchill Livingstone, Edinburgh. 2003, p. 416.

Caboni, P.: Cabras, M.: Angioni, A.: Russo, M.: Cabras, P. Persistence of azadirachtin residues on olives after field treatment. *J Agric Food Chem.* 2002, 50, 3491–3494.

Carpinella, M. C.: Defago, M. T.: Valladares, G.: Palacios, S. M. Antifeedant and insecticide properties of a limonoid from Melia azedarach (Meliaceae) with potential use for pest management. *J Agric Food Chem.* 2003, 51, 369–374.

CDPR (California Department of Pesticide Regulation). Summary of pesticide use report data 2003, indexed by chemical. 2005. http://www.cdpr.ca.gov/ (Accessed: 25 March 2013).

Charleston, D. S. Integrating biological control and botanical pesticides for management of Plutella xylostella. Ph.D. thesis. Wageningen University, Wageningen, The Netherlands. 2004, p. 176.

Council on Scientific Affairs. *Education and Information Strategies to Reduce Pesticide Risks.* American Medical Association, Chicago, IL. 1997.

Deng, A. L.: Ogendo, J. O.: Owuor, G.: Bett, P. K.: Omolo, E. O.: Mugisha-Kamatenesi, M.: Mihale, J. M. Factors determining the use of botanical insect pest control methods by smallholder farmers in the Lake Victoria basin, Kenya. *African J Environ Sci Technol.* 2009, 3, 108–115.

Dimetry, N. Z. Prospects of botanical pesticides for the future in integrated pest management programme (IPM) with special reference to neem uses in Egypt. *Arch Phytopathol Plant Prot.* 2012, 45, 138–1161.

Dosemeci, M.: Alavanja, M. C.: Rowland, A. S.: Mage, D.: Zahm, S. H.: Rothman, N.: Lubin, J. H.: Hoppin, J. A.: Sandler, D. P.: Blair, A. A quantitative approach for estimating exposure to pesticides in the agricultural health study. *Ann Occup Hyp.* 2002, 46, 245–260.

Dutta, S. Biopesticide: An eco friendly approach for pest control. *World J Pharm Pharm Sci.* 2015, 4(6), 250–265.

El-Hawary, F. M. A.: Sammour, E. A. The bioactivity and mechanism of action of some wild plant extracts on *Aphis craccivora.* Bull NRC, Egypt. 2006, Vol. 31, pp. 545–556.

El-Hosary, R. A. Evaluation of some essential oils against *Sesamia cretica* Led. under field conditions. *J Am Sci.* 2011, 7, 563–568.

El-Sebai, T. N.: El-Wakeil, N. E.: Abdallah, S. A. Efficacy of certain mineral, natural oils and insecticides against the Whitefly, *Bemisia tabaci* on cucumber plants and their side effects on the associated predators. *Bull Entomol Soc Egypt, Econ Ser.* 2005, 31, 229–241.

El-Shazly, A. M.: Dora, G.: Wink, M. Alkaloids of Haloxylon salicornicum (Moq.) Bunge ex Boiss. (Chenopodiaceae). *Pharmazie.* 2005, 60, 949–952.

El-Wakeil, N. E. Botanical pesticides and their mode of action. *Gesunde Pflanzen.* 2013, 65(4), 125–149.

Farone, W. A.: Palmer, T.: Puterka, J. Polyol ester insecticides and method of synthesis. U.S. Patent. 2002, 6, 419, 941.

Gurr, G. M.: Wratten, S. D.: Altieri, M. A.: Ecological engineering for pest management. In *Advances in Habitat Manipulation for Arthropods.* CSIRO, Australia. ISBN-13:978–0-444-52241-2, 2004, Chapter 1, pp. 1–27.

Isman, M. B. Botanical insecticides, deterrents, and repellents in modern agriculture and an increasingly regulated world. *Annu Rev Entomol.* 2006, 51, 45–66.

Isman, M. B. Botanical insecticides: For richer, for poorer. *Pest Manage Sci.* 2008, 64, 8–11.

Isman, M. B. Factors limiting commercial success of neem insecticides in North America and Western Europe. In: Koul, O., Wahab, S. (eds.), *Neem: Today and in the New Millennium.* Kluwer Academic, Dordrecht. 2004, pp. 33–41.

Jayasekara, T. K.: Stevenson, P. C.: Hall, D. R.: Belmain, S. R. Effect of volatile constituents from *Securidaca longepedunculata* on insect pests of stored grain. *J Chem Ecol.* 2005, 31, 303–313.

Johnson, H. A.: Oberlies, N. H.: Alali, F. Q.: McLaughlin, J. E. Thwarting resistance: annonaceous acetogenins as new pesticidal and antitumor agents. In Cutler S. J., Cutler J. G. (eds.), *Biological Active Natural Products: Pharmaceuticals.* CRC Press, Boca Raton, FL. 2000, pp. 173–183.

Karani, A. O.: Ndakidemi, P. A.: Mbega, E. R. Botanical pesticides in management of common bean pests: Importance and possibilities for adoption by small scale farmers in Africa. *J Appl Life Sci Int.* 2017, 12(1), 1–10.

Khan, S.: Taning, C. N. T.: Bonneure, E.: Mangelinckx, S.: Smagghe, G.: Shah, M. M. Insecticidal activity of plant derived extracts against different economically important pest insects. *Phytoparasitica.* 2017, 45, 113–124. DOI:10.1007/s12600-017-0569-y.

Kraus, W. *Azadirachtin and Other Triterpenoids.* Neem Foundation, Mumbai. 2002, pp. 39–111.

Laxmishree, C.: Singh, N. Botanical pesticides-a major alternative to chemical pesticides: A review, *Int J Life Sci.* 2017, 5(4), 722–729.

Lucantoni, L.: Giusti, F.: Cristofaro, M.: Pasqualini, L.: Esposito, F.: Lupetti, P.; Habluetzel, A. Effects of a neem extract on blood feeding, oviposition and oocyte ultra structure in *Anopheles stephensi* Liston (Diptera: Culicidae). *Tissue Cell.* 2006, 38, 361–371.

Mahadevappa, M.: Kulkami, R. S. Management of pathenium through allelopathic interaction. In: Narwal, S. S., Tauro, P. (eds.), *Allelopathy in Pest Management for Sustainable Agriculture.* Scientific Publishers, Jodhpur, India. 1994, pp. 67–76.

Médiène, S.: Valantin-Morison, M.: Sarthou, J. P.: de Tourdonnet, S.: Gosme, M.: Bertrand, M.: Roger-Estrade, J.: Aubertot, J. N.: Rusch, A.: Motisi, N.: Pelosi, C.: Doré, T. Agroecosystem management and biotic interactions: A review. *Agron Sustainable Dev.* 2011, 31, 491–514. doi:10.1007/s13593-011-0009-1.

Narwal, S. S. Allelopathy in ecological sustainable organic agriculture. *Allelopathy J.* 2010, 25(1), 51–72.

National Council of Research. *The Future Role of Pesticides in US Agriculture.* National Academy Press, Washington, DC. 2000, p. 301.

Perfecto, I.: Vandermeer, J. The agroecological matrix as alternative to the land-sparing/agriculture intensification model. Proc Natl Acad Sci. U.S.A. 2010, 107, 5786–5791. doi:10.1073/pnas.0905455107.

Philogène, B. J. R.: Regnault-Roger, C.: Vincent, C. Botanicals: Yesterday's and today's promises. In: Regnault-Roger, C., Philogène, B. J. R., Vincent, C. (eds.), *Biopesticides of Plant Origin.* Lavoisier, Andover, Paris. 2005, pp. 1–15.

Prusty, L. C.: Mohanty, S. K.: Behera, B. Allelopathic impact of sunflower on the growth of succeeding crops and associated weeds. *See Ref No.* 1994, 7, 39.

Rattan, R. S. Mechanism of action of insecticidal secondary metabolites of plant origin. *Crop Prot.* 2010, 29, 913–920.

Reddy, D. N. Essential oils extracted from medicinal plants and their applications. In: Akhtar, M., Swamy, M., Sinniah, U. (eds.), *Natural Bio-active Compounds.* Springer, Singapore. 2019, Vol. 1, pp. 237–283.

Regnault-Roger, C.: Philogène, B. J. R. Past and current prospects for the use of botanicals and plant allelochemicals in integrated pest management. *Pharmac Biol.* 2008, 46, 41–52.

Regnault-Roger, C.: Philogène, B. J. R.: Vincent, C. Biopesticides of plant origin. Lavoisier, Paris. 2005, p. 313.

Sammour, E. A.: El-Hawary, F. M.: Abdel-Aziz, N. F. Comparative study on the efficacy of neemix and basil oil formulations on the cowpea aphid *Aphis craccivora* Koch. *Arch Phytopathol Plant Prot.* 2011, 44, 655–670.

Scudeler, E. L.: Padovani, C. R.: dos Santos, D. C. Effects of neem oil (*Azadirachta indica* A. Juss) on the replacement of the midgut epithelium in the lacewing *Ceraeochrysa claveri* during larval–pupal metamorphosis. *Acta Histochem.* 2014, 116, 771–780.

Singh, A.: Kataria, R.: Kumar, D. Repellence property of traditional plant leaf extracts against *Aphis gossypii* Glover and *Phenacoccus solenopsis* Tinsley. *Afr J Agric Res.* 2012, 7(11), 1623–1628.

Srinivasan, A. *Handbook of Precision Agriculture: Principles and Applications.* Haworth Press, New York, USA. 2006.

Stevenson, P. C.: Nyirenda, S. P.: Veitch, N. C. Highly glycosylated flavonoid glycosides from *Bobgunnia madagascariensis.* *Tetrahedron Lett.* 2010, 51, 4727–4730.

Thacker, J. M. R.: *An Introduction to Arthropod Pest Control.* Cambridge University Press, Cambridge. 2002, p. 343.

Thakore, Y. The biopesticide market for global agricultural use. *Ind Biotechnol.* 2006, 194–208.

United Nations. World population prospects: the 2008 revision. Population Division of the Department of Economic and Social Affairs of the United Nations Secretariat. 2009. https://population.un.org/wpp/.

Weinzier, R. A. Botanical insecticides, soaps, and oils. In: Rechcigl, J. E., Rechcigl, N. A. (eds.), Biological and biotechnological control of insect pests. Lewis Publishers, Boca Raton, FL. 2000, pp. 101–121.

Zhao, J. Z.: Li, Y. X.: Collins, H. L.: Gusukuma-Minuto, L.: Mau, R. F. L.: Thompson, G. D.; Shelton, A. M. Monitoring and characterization of diamondback moth resistance to spinosad. *J Econ Entomol.* 2002, 95, 430–436.

29

Management of Crop Diseases through Botanical Biopesticides: An Easy Arsenal of Crop Protection in Organic Agriculture

Siddhartha Das, Sudeepta Pattanayak, Pradipta Banerjee, and L.P. Awasthi

CONTENTS

29.1 Introduction

Random and haphazard uses of chemical pesticides with immeasurable, unauthenticated doses in food and soil system not only break the balance of ecosystems but also present a serious challenge for our future generation. Need grows in this stage to think about alternatives that are effective, non-hazardous, non-toxic, and ecofriendly forever, and use of botanicals are the only way to mitigate the problem in all directions. The chemical based agro-environmental system can also easily change into common environment friendly organic material (Gnanamanickam, 2002). Several plant extracts like essential oils, gums, resins have been found to employ biological activity against a wide range of fungal plant pathogens in laboratory as well as field conditions (Romanazzi et al., 2012). Such products are normally considered to be less hazardous and more acceptable for balancing our environment and could be used for substitute remedies for plant disease treatment (Chuang et al., 2007).

29.2 Types of Botanical Biopesticides

29.2.1 Essential Oil Components

Plant secondary metabolites are regarded as natural chemicals derived from plants and plant parts and used as an essential substitute for chemical or synthetic pesticides. Essential oils are volatile, natural, complex compounds with a strong aromatic odor and are formed from different plant families belonging to Lamiaceae, Asteraceae, Lauraceae, and Myrtaceae. These volatile chemicals can act as acaricide, insecticide, and antibiotics, and can also control the growth and colonization of other plant pathogens. In addition to this, they can attract herbivores or natural enemies (Bakali et al., 2008). This volatile oil contains tarpenes, terpenoids, and other alipathic and aromatic components. Terpenes can form diverse classes of compounds by combining various numbers of isoprene units (5-carbon base, C5) which differ structurally and functionally, while terpenoids are characterized by terpenes which carry oxygen. Structurally terpenes are diversified under different classes such as hemiterpenes (C5), monoterpenes (C10), sesquiterpenes (C15), diterpenes (C20), triterpenes (C30), and tetraterpenes (C40) (Aharoni et al., 2005). The composition, bioactivity, and toxicity of the extracts are different and it is believed that the possible source of the variation is due to factors such as humidity contest after harvesting, plant age, and extraction method (Lahlou, 2004).

29.2.2 Triterpenoids

Limonoids are triterpenes that are metabolically altered but structurally similar. These are known to have a precursor source with a 4,4,8-trimethyl 17-furanylsteroid skeleton and mainly been extracted from different plants, most belonging from the plant families Meliaceae and Rutaceae, Cneoraceae, Simaroubaceae (*Harrisonia* sp.), etc. (Manners, 2007).

29.2.3 Glucosinolates and Isothiocyanates (Brassicaceae)

Glucosinolates are very important sulfur- and nitrogen-containing secondary metabolites produced by mustards (*Brassica* sp. and *Sinapis* sp.) and some other members of the Capparales order. Glucosinolates are one type of complex secondary metabolites which contain b-D thioglucose and sulfonated oxime in their structure. This group is under thioglucosides, which are characterized by their variable side chains with aliphatic, heteroaromatic, and aromatic carbon moieties. Glucosinolates by inversion produce several degradation products like indole, thiocyanates, and isothiocyanates. When glucosinolates containing vegetables are cut or chewed, they come into contact with the enzyme myrosinase which hydrolyzes the aforesaid end product.

29.2.4 Cyanogenic Glycosides

Cyanogenic glucosides contain a few amino acids, which are derived from secondary metabolites (SM) of different plant origins (approximately 2500–3000 plant species). This kind of crucial secondary metabolites plays an important role in plant defense mechanisms for their characteristic bitter taste and secretion of poisonous hydrogen cyanide. In the case of insects chewing plant tissue when they eat and breakdown the plant tissue, cyanogenic glucosides are released from the vacuoles. The cyanogenic glucoside hydrolyzes to produce glucose, an aldehyde/ketone and toxic HCN. The process is termed cyanogenesis and used against different kinds of insects, fungi, nematodes, and weeds (Carlsen and Fomsgaard, 2008). The mode of action of cyanogenic glycosides is interesting as it produces a substantial amount of cyanide, which makes a barrier against oxygen transport through inhibition of cytochrome oxidase (Majak, 1992).

29.2.5 Alkaloids

Alkaloids are nitrogenous secondary metabolites extracted from diverse botanical families. Nicotine is considered to be the most primitive alkaloid in agriculture and used for its insecticidal property. Varatrine and cevatrine are two different alkaloids under the same class. Sabdilla (*Schoenocaulon officinale* Grey) seeds are used for the control of thrips though resistance phenomenan in thrips control is also mentioned in some literature (Humeres and Morse, 2006).

29.2.6 Phenolics-Flavonoids

Phenols are poisonous for fungi, bacteria, nematodes, weeds, and insects (Koul, 2008). Flavonoids are under this class and extracted from various kinds of vascular plants and bryophytes. Iwashina (2003) reported approximately 5000 different kinds of flavonoids are present and specialized for its feeding attractant and deterrent properties. Rotenone is extracted from Derris or Lonchocarpus (Leguminosae) and is considered to be the most effective secondary metabolite under this category.

29.2.7 Polyacetylenes and Polythienyls

Different sets of secondary metabolites are target-specific, which belongs to the family of Asteraceae, The two major components Polyacetylenes and Polythienyls, associated with the family Asteraceae are characterized for their fungicidal, insecticidal, and nematicidal properties (Chitwood, 2002).

29.2.8 Pyrethrum

Pyrethrum is the powdered form component extracted from members of the Asteraceae family such as *Pyrethrum* spp., *Chrysanthemum* spp., and *Tenacetum* spp. Structurally, it is made up of six esters: Jasmoline I and II, pyrethrins I and II, and cinerin I and II. The mode of action of pyrethrins is unique in insect and mite control through attachment with Na^+ channels. Hyperactivity followed consecutively by contraction is found in those insects, which finally die. Pyrethrins activity is known as knockdown effect.

29.2.9 Organic Acids

Vegetable oils are known to have large amounts of saturated or unsaturated fatty acids, which are heterogenous in nature. Additionally, they contain esterified long to medium carbon chains and high molecular weight esters of fatty acids. They act mainly on respiratory organs by creating an impermeable film on the insect cuticle so that the insect will suffocate after inhaling the toxic materials. Some toxic chemicals enter through the cuticle, break down oxidative phosphorylation, and destroy the cell membrane.

29.3 Types of Plant Parts Used

Different types of plant parts have been used for a long time in the traditional way as botanical biopesticides (Subramaniam, 1993; Tripathi, 1998 and Dhaliwal & Koul, 2007). Some of them are discussed in Tables 29.1, 29.2, 29.3, and 29.4.

29.4 Advantages and Disadvantages

Using botanicals has several advantages and disadvantages compared to chemicals, which are depicted in Table 29.5.

29.5 Conclusion and Future Prospects

At present in plant protection, using of chemotherapeutants is a common practice in different parts of the world. Pest management will be easier for its characteristic nature of non-phytotoxicity, systematicity, easy biodegradability, and stimulatory nature of host metabolism (Mishra and Dubey, 1994). Embryophyte plants contain a wide spectrum of secondary metabolites such as phenols, flavonoids, quinones, tannins, essential oils, alkaloids, saponins, and sterols. Such plant-derived chemicals may be exploited for their different biological properties. The widely used plant-based components are pyrethrum and neem-based

TABLE 29.1

Leaves Used as a Botanical Biopesticide

Sl. No.	Common Names	Scientific Names	Active Chemical Constituents	Property
1.	Nimbu	*Citrus limon*	Limonin, nomilin, obacunone	Antifeedant, toxicant
2.	Akanda	*Calotropis procera*	Latex containing poisonous constituents	Antifeedant
3.	Kalmegh	*Andrographis paniculata*	Andrographolide	Insecticidal
4.	Basak, Adusa	*Adhatoda vasica*	Vasicine, vasicinone, vasicinol, limonene	Insecticidal, antifeedant
5.	Pyaj	*Allium cepa*	Quercetin and phenolic compounds	Insecticidal
6.	Tulsi	*Ocimum sanctum*	Linalool, chavicol, eugenol, eugenol methyl ether, cineole, caryophyllene.	Insecticidal repellent
7.	Dill	*Anethum sowa*	Carvone, dillapiole	Insecticidal
8.	Bhang	*Cannabis sativa*	Resinoid, tetrahydrocannabinol, a phenolic type substance	Protectant
9.	Satyanashi, Sialkata	*Argemone maxicana*	Protopine nitrate, Berberine nitrate, Ceryl alcohol, β-sitostero	Protection
10.	Raimuniya	*Lantana camera*	Cayophyllene, cineol and β-pinene	Protectant
11.	Seeta-bani	*Artemisia capillaris*	Bornyl acetate, capillarin, capillen	Feeding deterrent
12.	Neem	*Azadirachta indica*	Limonoids, azadirechtins, salamin, nimbin	Insecticidal, Hormonal (JH), antifeedant, multifacial
13.	Bachh	*Acorus calamus*	Trans-asarone, cis-asarone, isoasarone	Repellent, antifeedant
14.	Camellia	*Camellia* spp.	Shikinic acid, caeffin and tannins	Insecticidal repellent
15.	Papaya	*Carica papaya*	Carpaine	Insecticidal
16.	Jangli Tamatar	*Lycopersicon hirsutum*	2-tridecanone, trans-caryophyllene	Repellency, toxicity
17.	Chakunda, Kasonda	*Cassia occidentalis*	Emodin	Insecticidal
18.	Charota	*Cassia tora*	Chrysophanic-9-anthrone	Antifeedant
19.	Balsam fir	*Abies balsamea*	Juvabione, dehyojuvabione	Hormonal (JH)
20.	Nimbu	*Citrus* spp.	Citropin, di-limonens, linalool, glucosides, terpenes, etc.	Insecticidal

TABLE 29.2

Shoot, Flower, Fruits, and Seeds Used as Botanical Biopesticides

Sl. No.	Common Names	Scientific Names	Plant Parts Used	Active Chemical Constituents	Property
1.	Ritha	*Sapindus mukorossi*	Seeds	Saponins	Insecticidal
2.	Kaju	*Anacardium occidentale*	Cashew nut shell oil	Phenolic constituents	Insecticidal
3.	Ramphal	*Annona squamosa*	Seeds	Anonaine, liriodenine, reticuline, norushinsunine	Insecticidal
4.	Kapur	*Cinnamomum camphora*	Shoot	Camphor oil	Insecticidal
5.	Bamboo	*Bambusa arundinacea*	Fresh and young shoots	Benzoic acid, cyanogenic glucoside	Insecticidal
6.	Latkan, Annatto	*Bixa orellana*	Seed coats	Bixin	Repellent
7.	Dill	*Anethum sowa*	Seeds	Cavone, dillapiole	Insecticidal
8.	Sharifa	*Annona reticulata*	Fruit and seed extracts	Annonacin, annonin, annonelliptine, asimicin, annonidines	Antifeedant, repellent
9.	Sarisha	*Brassica comprastis*	Seeds	2-Phenylethyl isothiocyanate	Fecundity reducing
10.	Genda	*Tagetes minuta*	Flowers	Terthienyl-2,2,5, 2 terthiophene), E-ocimenone	Larvicidal, repellent

products, which have powerful potential to act as fungicide, bactericide, insecticide, nematicide, etc. Many factors are barriers to the use of plant-derived products, particularly the easy availability and cost-effectiveness of newly synthesized chemicals. Plant essential oils, which maximally use plants for their self-defense or attractants, are very effectively used in disease and pest management for their characteristic antifungal and antimicrobial properties (Oxenham, 2003). Essential oils are gaining more attention for their safer use, comparatively higher production rate, worldwide acceptance, and their multi-purpose activity in agricultural scenarios (Table 29.6). The complication of gaining resistant strains by insect pests and fungi may be overcome by using fumigants of essential oils extracted from higher plants for controlling the stored grain pests, as essential oils contain combinations of several components (Dubey et al., 2007). Although various essential oils have been screened for their pesticidal activity against various pests, detailed studies viz. antifungal, insecticidal, repellency, oviposition, ovicidal, and antiaflatoxigenic activity, phytochemistry and safety profiles have not been done properly with most of the oils. The high activity of the phenolic components is observed in terms of the alkyl substitution into the phenol nucleus, which is known to

TABLE 29.3

Roots, Rhizomes, Bark, and Bulb Used as Botanical Biopesticides

Sl. No.	Common Names (Parts Used)	Scientific Names	Parts Used	Active Chemical Constituents	Property
1.	Khas	*Vetiveria zizanioides*	Roots	β-vetivene, azulene, zizanene, leavojujenol	Growth disrupter, repellent
2.	Pyaj	*Allium cepa*	Bulbs	Quercetin and phenolic compounds	Insecticidal
3.	Agar wood	*Aquilaria malacensis*	Agarwood dust	A-guaiene, caryophellene oxide, eudesmol	Protection, repellent
4.	Cinchona	*Cinchona officinalis*	Bark	Quinine, quinidine, cinchonine and cinchonidine	Insecticidal
5.	Rasun	*Allium sativum*	Bulbs	Diallyl di-sulfide, diallyl tri-sulfide	Insecticidal
6.	Adrak	*Zanthoxylum monophyllum*	Rhizomes	Gingerdione, paradol	Antifeedant
7.	Derris	*Derris elliptica*	Roots	Rotenone and dihydrorotenone	Insecticidal
8.	Safed til	*Sesamum indicum*	Roots	Sesamolin, Sesamin, sesangolin	Antifeedant
9.	Haldi	*Curcuma longa*	Rhizome	α-Phellandrene, curcumene, termerone, dehydro-termerone	Protectant, repellent
10.	Yellow prickle	*Zanthoxylum monophyllum*	Bark	Zanthophylline	Feeding deterrent

TABLE 29.4

Secondary Metabolites, Latex Used as Botanical Biopesticides

Sl. No.	Common Names	Scientific Names	Active Chemical Constituents	Property
1.	Tridhara	*Euphorbia antiquorum*	Latex contents 4.0–6.4% caoutchouc	Antifeedant
2.	Murry red gum	*Eucalyptus rostrata*	1,8-Cineole, α-phellandrene	Anti-Fecundity

TABLE 29.5

Whole Plant Used as Botanical Biopesticides

Sl. No.	Common Names	Scientific Names	Active Chemical Constituents	Property
1.	Nimbu	*Citrus* spp.	Linalool, glucosides, terpenes, citropin etc.	Insecticidal
2.	Indian Aconite, Bishnag	*Aconitum ferox*	Psuedaconitine, diacetyl pseudoaconitine, chasmaconitine, indaconitine, bikhaconitine	Aphicidal toxic to beetles

TABLE 29.6

Comparison of Botanicals and Chemicals

Particulars	Botanicals	Chemicals
Promote sustainable agriculture?	Yes. They decrease ill effects on the crop plants, soil health, and environment	They are a major barrier to sustainable agriculture
Crop losses?	No phytotoxicity	High phytotoxicity
Ecofriendly?	Yes	No
Biodegradable?	Yes	No
Organic farming?	Yes	No
Cost-effectivity?	Cheaper and more cost-effective	Less cost-effective
Integrated disease management (IDM)?	Incorporated under the framework of IDM	Though chemical management also comes under IDM but it is suggested to avoid this practice for its hazardous property

enhance the antimicrobial activity of phenols (Pelczar et al., 1988). As food preservatives, volatile oils may have their greatest potential use. Spices, which are used as integral ingredients in cuisine or added as flavoring agents to foods, are present in insufficient quantities for their antimicrobial properties to be significant. However, spices are often contaminated with bacterial and fungal spores due to their volatile oil content, often with antimicrobial activity, being enclosed within oil glands and not being released onto the surface of the spice matter. In many different herbs and spices, volatile oils are used as main attractant component and contain various aromatic compounds. If added to foodstuffs, they would cause no loss of organoleptic properties, but would hinder microbial contamination and therefore reduce the onset of spoilage.

REFERENCES

Aharoni, A.; Jongsma, M.A.; Bouwmeester, H.J. Volatile science? Metabolic engineering of terpenoids in plants. *Trends in Plant Science*. 2005, 10(12):594–602.

Bakkali, F.; Averbeck, S.; Averbeck, D.; Idaomar, M. Biological effects of essential oils- A review. *Food Chem. Toxicol.* 2008, 46:446–475, ISSN: 0278–6915.

Carlsen, S.C.K; Fomsgaard, I.S. Biologically active secondary metabolites in white clover (*Trifolium repens* L.): A review focusing on contents in the plant, plant–pest interactions and transformation, *Chemoecology*. 2008, 18:129–170.

Chitwood, D.J. Phytochemical based strategies for nematode control. *Annual Review of Phytopathology*. 2002, 40:221–249.

Chuang, P.H.; Lee, C.W.; Chou, J.Y.; Murugan, M.; Shieh, B.J.; Chen, H.M. Antifungal activity of crude extracts and essential oil of *Moringa Oleifera* Lam. *Bioresource Technology*. 2007, 98:232–236.

Dhaliwal, G.S.; Koul, O. *Biopesticides and Pest Management: Conventional and Biotechnological Approaches*. Kalyani Publishers, Ludhiana, India, 2007.

Dubey, R.K.; Kumar, R.; Dubey, N.K. Evaluation of *Eupatorium cannabinum* Linn. oil in enhancement of shelf life of mango fruits from fungal rotting. *World Journal of Microbiology and Biotechnology*. 2007, 23:467–473.

Gnanamanickam, S.S. *Biological Control of Crop Diseases*. Marcel Dekker Inc., New York, USA, 2002, 468 p.

Humeres, E; Morse, J.C. Resistance of avocado thrips (Thysanoptera: Thripidae) to sabadilla, a botanically derived bait. *Pest Management Science*. 2006, 62:886–889.

Iwashina, T. Flavonoid function and activity to plants and other organisms. *Biological Sciences in Space*. 2003, 17:24–44.

Koul, O. Phytochemistry and Insect control: An antifeedant approach. Critical reviews in plant sciences. *Critical Reviews in Plant Sciences*, 2008, 27:1–24

Lohlou, M. Essential oils and fragnance compounds: bioactivity and mechanism of action. *Flavour Fragr. J.* 2004, 19: 159–165

Majak, W. Mammalian metabolism of toxic glycosides from plants. *Journal of Toxicology: Toxin Reviews*. 1992, 11(1):1–40.

Manners, G.D. Citrus limonoids: Analysis, bioactivity and biomedical prospects. *Journal of Agricultural and Food Chemistry*. 2007, 55:8285–8294.

Mishra, A. K.; Dubey, N. K. Evaluation of some essential oils for their toxicity against fungi causing deterioration of stored food commodities. *Applied and Environmental Microbiology*. 1994, 60: 1101–1105.

Oxenham, S.K. Classification of an *Ocimum basilicum* germplasm collection and examination of the antifungal effects of the essential oil of basil. Ph.D. thesis. University of Glasgow, Glasgow, UK, 2003.

Pelczar, M.J.; Chan, E.C.S.; Krieg, N.R. Control of microorganisms, the control of microorganisms by physical agents. In *Microbiology*. McGraw Hill International, New York. 1988, pp. 469–509.

Romanazzi, G.; Lichter, A.; Gabler, F.M.; Smilanick, J.L. Recent advances on the use of natural and safe alternatives to conventional methods to control post harvest gray mold of table grapes. *Postharvest Biology and Technology*. 2012, 63:141–147.

Subramaniam, S. Potential pesticidal plants of India. In: *Botanicals and Biopesticides*, B.S. Pamar and C. Devakumar (eds.). Westvill Publishing House, New Delhi, India, 1993, pp. 74–100.

Tripathi, Y.C. Phytopesticides—An ecofriendly approach for pest control. *Indian Journal of Environmental Sciences*. 1998, 2(2):63–71.

Section XIV

Management of Insect Pests and Diseases of Crops through Biopesticides of Insect Sex Hormone

30

Management of Insect Pests through Hormones

Muhammad Sarwar and Naveed Akhter Shad

CONTENTS

30.1 Introduction

In order to enjoy biological success, all living organisms must feed, mate, and reproduce (Sarwar, 2019a; Sarwar, 2020a; Sarwar et al., 2020a). Insects impact human health through vector-borne diseases, and cause major economic losses by damaging crops and stored agricultural products. Moreover, several mainstay insecticides are being limited or withdrawn due to tightening of legislation. Thus, "greener" more selective insecticides are required so that the pests do not become further resistant (Sarwar, 2020b).

Hormones are blood-borne, signalling molecules that act as molecular messengers between the brain and peripheral tissues.

Insect hormones appear to be unique regulators of such vital processes as molting, reproduction, and diapause, and as metabolic regulators. The high biological activity and specificity of these hormones, together with their low toxicity for higher animals, hold promise for their utilization for insect control (Sarwar, 2020c; 2020d). Hormones regulate key processes in all multicellular organisms, and are critical for normal development, metabolism, reproduction, and behavior. All these processes can potentially be targeted to affect the survival of specific insects (Sarwar and Roohi, 2020). They may inhibit the formation of the chitin required to make a new cuticle at each molt, or they may replace or disrupt the production of the juvenile hormone (JH), which controls the molting process (Sarwar, 2015; Sarwar and Salman, 2015). By studying insect hormone systems, there is hope to learn how insects regulate different critical processes and how researchers may influence these systems by using synthetic, hormone-like substances to diminish specific insect populations (Sarwar, 2016).

Constituting by far the largest group of insects, beetles are a particularly attractive target for pest control. Knowledge on beetle hormone systems will eventually be used to develop artificial, chemically stable insect hormones (hormone mimetics) that can be used to destabilize the insects' ability to regulate their body fluid composition, and thereby introduce new, more specific, and environmentally safe pest control (Sarwar, 2013a).

30.2 Endocrine Regulation of Insect Growth and Development

In the adult stage, steroid 20-hydroxyecdysone and the sesquiterpenoid JHs are involved in the regulation of reproductive maturation (Riddiford, 1994). The epidermal cells reorganize for massive protein synthesis, as well as secretion of the new epicuticle and cuticle. Once 20E titer begins to decline, enzymes in the molting fluid are activated to start digestion of the procuticle. After this process is completed, the molting fluid is reabsorbed and pre-ecdysial tanning of the new cuticle takes place (Reynolds, 1987).

The larval-pupal molt is then induced with molting levels of ecdysteroids in the presence of juvenile hormone (JH). JH is required to prevent precocious development of the imaginal disks in holometabolous insects (Dhadialla et al., 1998).

The morphological and ultrastructural changes that occur in the epidermis during insect growth and development are dependent upon the regulation of gene expression with different titers of 20E in the absence or presence of JH (Riddiford, 1996).

FIGURE 30.1 *Spodoptera littoralis.*

FIGURE 30.2 *Spodoptera frugiperda.*

Toxic and anti-developmental effects of the JHA insecticides methoprene, fenoxycarb, and pyriproxyfen have been investigated on the larval and pupal stages of *Spodoptera littoralis* (Figure 30.1) and *Spodoptera frugiperda* (Figure 30.2). Bioassays showed that fenoxycarb has the highest toxicity and fastest speed of kill in second instar *S. littoralis.* All three JHAs affected the development of sixth instar (final instar) and pupal *S. frugiperda.* JH esterase (JHE) is a critical enzyme that helps to regulate JH levels during insect development. JHE activity in the last instar *S. littoralis* and *S. frugiperda* is 11 and 23 μmol min^{-1} ml^{-1} hemolymph, respectively. Methoprene and pyriproxyfen show poor inhibition of JHE activity from these insects, whereas fenoxycarb displays stronger inhibition. The inhibitory activity of fenoxycarb, however, is more than 1000-fold lower than that of OTFP, a highly potent inhibitor of JHEs. Surprisingly, topical application of methoprene, fenoxycarb, or pyriproxyfen on sixth instars of *S. littoralis* and *S. frugiperda* prevented the dramatic reduction in JHE activity that is found in control insects. The findings suggest that JHAs may function as JH agonists that play a disruptive role or a hormonal replacement role in *S. littoralis* and *S. frugiperda* (El-Sheikh et al., 2016).

The ecdysones are necessary for all molting to occur, but when the insect reaches the point of metamorphosis to the adult reproductive stage, it molts to a pupal form and finally to the adult. Both of these molts must take place in the absence of JH, but in the presence of molting hormone (MH). Insect growth regulators (IGRs) are compounds which interfere with insect metabolism in a manner that affects growth. IGRs act in several ways, they may inhibit the formation of the chitin required to make a new cuticle at each molt, or they may replace or disrupt the production of the juvenile hormone, which controls the molting process. Although compounds that mimic the action of JH (JH analogues/agonists) are efficient, they have a limited scope of application. Development of potent compounds counteracting JH action (JH antagonists) would find a wider range of control applications (Bengston et al., 1991; Mian et al., 1990).

30.2.1 Chitin Inhibitors

Chitin inhibitors disrupt the synthesis of chitin. They act against the larval stages, which when affected, usually fail to survive the next molt. Death is caused by incomplete ecdysis and cuticle malformation. There are several known groups of substances that act as chitin inhibitors. The best known are probably the benzoylphenylure, which include diflubenzuron and its analogues teflubenzuron, flufenoxuron, and triflumuron.

30.2.2 Juvenile Hormone and Juvenile Hormone Analogues

The best known of these juvenile hormone analogues are methoprene, hydroprene, and fenoxycarb. These substances do not kill adult insects, but they prevent juvenile stages from completing their development. Control of an insect population is therefore a gradual process.

30.2.3 Anti-Juvenile Hormones

The extracts from the plant *Ageratum houstonianum* cause premature metamorphosis in some Hemiptera. It is subsequently shown that the active molecules destroy the glands (corpora allata), which produce JH after first being converted to highly reactive metabolites by tissue-specific enzymes within the glands (Menn et al., 1989). Screening led to the discovery of anti-JHs such as fluoromevalonate (FMev). FMev showed activity against virtually all the Lepidoptera in which it is tested, but little in other insect orders (Staal, 1986).

30.3 Natural Substances with Molt-Inhibiting Activities

The principal hormones involved in the life processes of insects include neurohormones (neuropeptides), ecdysteroids (molting hormones), and the sesquiterpenoid juvenile hormones (Hoffmann and Lorenz, 1998).

The most efficient natural substance with molt-inhibiting activity is azadirachtin, a tetranortriterpenoid plant (neem tree, Meliaceae) limonoid with ecdysteroid-like structure (Mordue and Blackwell, 1993; Schmutterer, 1990). Azadirachtin-containing products have been formulated as liquids or dusts from ethanol extracts of neem tree seeds (Figure 30.3).

The family Meliaceae is the source of numerous other insecticidal compounds and members of the genus Aglaia (Figure 30.4) are shown to contain novel insecticidal constituents of the rocaglamide type (benzofurans), which evince strong insecticidal activity towards neonate larvae of Spodoptera (Gtissregen et al., 1997). The effect of

FIGURE 30.3 Neem seeds.

FIGURE 30.4 *Aglaia* genus.

FIGURE 30.5 *Lindera erythrocarpa.*

FIGURE 30.6 *Solidago serotina.*

brassinosteroids on insects is by their competition with molting hormones at the binding site of the hormone receptor, which results in delayed molting (Richter and Koolman, 1991). Very recently, two triterpenoids isolated from seeds of a cruciferous plant, cucurbitacins B and D, have been demonstrated to be insect steroid hormone antagonists acting at the ecdysteroid receptor (Dinah et al., 1997).

Five diterpene secondary metabolites with juvenile hormone antagonist activity from two plants; *Lindera erythrocarpa* (Figure 30.5) and *Solidago serotina* (Figure 30.6) have been isolated. Topical application of two diterpenes caused reduction in the expression of Met target genes and retardation of follicle development in mosquito ovaries (Lee et al., 2015).

30.4 Insect Hormones as Pesticides

Availability of a number of combination products involving hormone analogues and other IGRs and insecticides has enhanced the utility of IGRs in diverse pest management programs (Subramanian and Shankarganesh, 2016).

Insect hormones have been the target site for IGRs, viz., JH analogues, chitin synthesis inhibitors, and ecdysone agonists. Rigorous efforts by chemists have shaped the development of improved analogues of JHA, chitin synthesis inhibitors (CSIs), and ecdysone agonists to cater to diverse pest management programs. The efficacy and photostability of early isoprenoid juvenoids have been improved through structural modifications in the basic juvenoid structure to develop newer analogues such as fenoxycarb and pyriproxyfen. Besides the BPU derivatives of chitin synthesis inhibitors, newer classes of CSIs including thiadiazines (buprofezin) and oxazolines (etoxazole) have been developed in the recent past with improved efficacy against sucking pests and mites. Understanding the molecular basis of hormones and elucidation of the biochemical pathway of chitin synthesis will help to identify new target sites for novel IGR compounds with improved potency and spectrum of action for sustainable pest management. There is a scope for developing of chemicals with anti-JH and anti-ecdysteroid activities.

FIGURE 30.7 Mosquito immatures.

FIGURE 30.8 German cockroach.

FIGURE 30.9 Termite.

30.4.1 Insect Growth Regulators (IGRs)

Insect growth regulators (IGRs) used by the pest management industry include juvenile hormone analogues and chitin synthesis inhibitors. IGRs do not act on the nervous system and they are insecticides that disrupt critical physiological functions associated with normal insect growth, development, and reproduction (egg production). However, like all pesticides, IGRs should be handled safely and applied with a great deal of care and consideration for non-target organisms.

30.4.1.1 Chemical Class: Juvenile Hormone Analogues (JHAs)

The active ingredients of this chemical class are hydroprene, methoprene, pyriproxyfen, fenoxycarb). The first category of insect growth regulators important in managing urban and structural pests are the JHAs, which mimic a naturally occurring chemical in immature insects called juvenile hormone. As immature insects progress through their life cycle, however, the level of juvenile hormone in the blood is reduced through a decrease in its production and by juvenile hormone degrading enzymes. With less juvenile hormone present, the insect can then proceed naturally toward adulthood.

The developmental physiology of immature mosquitoes (Figure 30.7) and fleas exposed to methoprene is severely altered, resulting in death or severe developmental abnormalities that eventually lead to death. German cockroaches (Figure 30.8) exposed to JHAs during the last-instar molt into either adult males are physically incapable of mating or adult females with deformed ovaries. The result is that adult cockroaches in the population are sterile. As reproduction ceases, the population slowly declines as sterilized adults die of natural causes and are not replaced by nymphal cockroaches.

Methoprene is a synthetic analogue of naturally occurring insect JH. Methoprene has been registered for the control of specific insects including *Ephestia cautella*, *Plodia interpunctella*, *Lasioderma serricorne*, *Rhyzopertha dominica*, and *Oryzeephilus surinamensis*. Unfortunately, it does not control all pests and is unsuitable for control of *Callosobruchus maculatus*, *Liposcelis bostrychophilus*, and Sitophilus species (Arthur et al., 1990; Bengston et al., 1991).

30.4.1.2 Chemical Class: Chitin Synthesis Inhibitors (CSIs)

The active ingredients of this chemical class are diflubenzuron, hexaflumuron, noviflumuron, and lufenuron. The second category of insect growth regulators are the chitin synthesis inhibitors. Like the juvenile hormone analogues, chitin synthesis inhibitors do not act on the insect's nervous system. They disrupt an important biochemical pathway responsible for the synthesis of chitin. In the absence of chitin synthase, chitin cannot be synthesized. The prevention of chitin synthesis is fatal for the affected insect (Bennett et al., 2003a).

Chitin synthesis inhibitors used by the structural pest management industry include diflubenzuron, developed decades ago for the control of agricultural pests and now used in several baits for termite (Figure 30.9) control, and hexaflumuron and noviflumuron, developed specifically for the control of termites in a baiting system. Termite baits that contain chitin synthesis inhibiting insecticides block chitin formation in molting termites exposed to the active ingredient (Bennett et al., 2003b).

Lufenuron is a chitin synthesis inhibitor used for flea (Figure 30.10) control. It is delivered orally and absorbed directly into the animal's bloodstream. Because fleas are obligate blood-feeders (blood proteins are required for female fleas to produce eggs), consumption of lufenuron-tainted animal blood by adult female fleas results in the production of eggs that fail to hatch

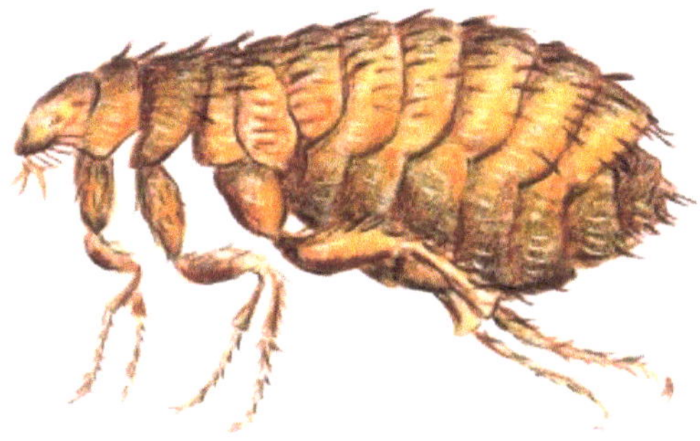

FIGURE 30.10 Flea.

(or first instar larvae that die soon after hatch) since insect eggs contain chitin. In addition, flea larvae are also killed by lufenuron. Adult fleas excrete large quantities of partially digested host blood. This high protein excrement is a primary food source of flea larvae during their development. Consumption of this lufenuron-tainted, dried blood is lethal to larval fleas when they molt (Braness, 2004).

Chlorfluazuron exhibits insecticidal and embryocidal activities against Lepidoptera and Coleoptera, and is used to control Noctuidae larvae on cotton (Fox, 1990). Babu et al. (1991) compared the effectiveness of triflumuron, diflubenzuron, and flucycloxuron at 20 ppm against *C. chinensis* on mung bean seed, which prevented oviposition and adult emergence for some times.

The potential for storage of insect pests to develop resistance or cross-resistance across the IGR range should not be ignored. Thus, the development of integrated pest management practices, which would optimize the potential benefits of IGRs, should seriously be considered (Sarwar, 2012; 2013b; 2017; 2019b; 2019c; Sarwar et al., 2020b).

30.5 Applying Insect Growth Regulators

The methods used to apply IGRs can be classified into the subsequent three groups: topical application, admixture to food media, and application to surfaces. For topical application, 1-100 µg of IGR are applied in solvent to the insect body. For the admixture of IGRs to insect food media, widely differing volumes of various solvents have been used. Generally, the treatment is applied directly to the food medium, although the compound is occasionally applied in solvent to the glass above the commodity in the treatment vessel. An IGR in solvent may also be directly applied to paper or cardboard, surface but this is a method which is rarely used (Dales, 1994).

Chitin synthesis inhibitors may be toxic to other arthropods such as Crustacea, and IGR metabolites may have adverse effects on vertebrates because of their ability to bind to certain members of the nuclear hormone receptor family (Couillaud and Peypelut, 1995). So, in open-field application of a potential new insecticide requires evaluation of its efficacy, selectivity, specificity, stability, compatibility, and innocuousness for other organisms.

30.6 Conclusions and Future Prospects

This chapter gives information about hormones in pest management covering aspects including physiology of insect hormones; evolution of hormone-based insecticides including juvenile hormone analogues, chitin synthesis inhibitors and ecdysone agonists; utility of insect growth regulators (IGRs) in management of crop pests, stored products, public health and urban pest management, termite management, their impact on natural enemies of crop pests, bees, and other pollinators; and development of resistance in insects to IGRS. Insect-specific growth regulators represent attractive control agents because of their safety to the environment and humans.

The significant advances in the use of combinatorial chemistry and highthroughput target site-directed assays will allow the agrochemical industry to make use of these technologies to discover new insecticides with insect hormone action. Certainly, strategies based on effects on the insect endocrine system will help to find new, promising substances that can be used in agricultural applications together with other approaches, for instance, on insect chemical communication, use of natural insect parasites, and plant genetic engineering.

REFERENCES

Arthur, F. H.; Throne, J. E.; Simonaitis, R. A.; Zehner, J. M. Evaluation of chlorpyrifos-methyl and chlorpyrifos-methyl plus methoprene as protectants of stored corn: small bin tests. *Journal of Economic Entomology*. 1990, 83: 1114–1121.

Babu, R. T.; Hussaini, S. H.; Sriramulu, M.; Siddiqui, M. K. H. Effect of inert clay and insect growth regulators on the development of *Callosobruchus chinensis* L. and the germination of mungbean seed [*Vigna radiate* (L.) Wilczek]. *Tropical Science*.1991, 31: 217–221.

Bengston, M.; Koch, K.; Strange, A. C. Development of insect growth regulators as grain protectants in Australia and South-East Asia. In: Proceedings of the 5th International Working Conference on Stored Product Protection, Bordeaux, France, September 1990. 1991, pp. 485–490.

Bennett, G. W.; Corrigan, R. M.; Owens, J. M. Pesticides. In *Truman's Scientific Guide to Pest Management Operations*, 6th ed. Advanstar Communications, Cleveland, OH. 2003a, pp. 39–62.

Bennett, G. W.; Corrigan, R. M.; Owens, J. M. Safety and the environment. In *Truman's Scientific Guide to Pest Management Operations*, 6th ed. Advanstar Communications, Cleveland, OH. 2003b, pp. 65–90.

Braness, G. A. Insecticides and pesticide safety. In: Hedges, S. A., Moreland, D. (eds.), *Handbook of Pest Control: The Behavior, Life History, and Control of Household Pests*, 9th ed. GIE Media, Richfield, OH. 2004, pp. 1098–1163.

Couillaud, E.; Peypelut, L. Endocrine-based opportunities for insect pest control. In: Muraleedharan, D. (ed.), *Recent Advances in Insect Endocrine Research*, Trivandrum, Ind. 1995, 117–153.

Dales, M. J. *Controlling Insect Pests of Stored Products Using Insect Growth Regulators and Insecticides of Microbial Origin*. Natural Resources Institute Bulletin 64. Natural Resources Institute, Chatham, UK. 1994, p. 58.

Dhadialla, T. S.; Carlson, G. R.; Le, D. P. New insecticides with ecdysteroidal and juvenile hormone activity. *Annual Review of Entomology*. 1998, 43: 545–569.

Dinah, L.; Whiting, E.; Girault, J. P.; Lafont, R.; Dhadialla, T. S.; Cress, D. E.; Mugat, B.; Antoniewski, C.; Lepesant, J. A. Cucurbitacins are insect steroid hormone antagonists acting at the ecdysteroid receptor. *Biochemistry Journal*. 1997, 327: 643–650.

El-Sheikh, E. S. A.; Kamita, S. G.; Hammock, B. D. Effects of juvenile hormone (JH) analog insecticides on larval development and JH esterase activity in two spodopterans. *Pesticide Biochemistry and Physiology*. 2016, 128: 30–36.

Fox, P. *Insect Growth Regulators*. P.J.B. Publications Ltd., Richmond, UK. 1990, 108 pp.

Gtissregen, B.; Fuhr, M.; Nugroho, B. W.; Wray, V.; Witte, L.; Proksch, R. New insecticidal rocaglamide derivatives from flowers of *Aglaia odorata. Zeitschrift fur Naturforschung C: Journal of Biosciences*. 1997, 52c: 339–344.

Hoffmann, K. H.; Lorenz, M. W. Recent advances in hormones in insect pest control. *Phytoparasitica*. 1998, 26(4): 323–330.

Lee, S. H.; Oh, H. W.; Fang, Y.; An, S. B.; Park, D. S.; Song, H. H.; Oh, S. R.; Kim, S. Y.; Kim, S.; Kim, N.; Raikhel, A. S.; Je, Y. H.; Shin, S. W. Identification of plant compounds that disrupt the insect juvenile hormone receptor complex. *Proceedings of the National Academy of Sciences of the United States of America*. 2015, 112 (6): 1733–1738.

Menn, J. J.; Raina, A. K.; Edwards J. P. Juvenoids and neuropeptides as insect control agents: Retrospect and prospects. In: Mcfarlane, N. R. (Ed.), *Progress and Prospects in Insect Control*. British Crop Protection Monograph. Proceedings of an International Conference, Reading, UK, September 1989. 1989, (43): 87–105.

Mian, L. S.; Mulla, M. S.; Hussain, N. Insect growth regulators as control agents against stored product insects. *Sarhad Journal of Agriculture*.1990, 6(3): 287–298.

Mordue, A. J.; Blackwell, A. Azadirachtin: An update. *Journal of Insect Physiology*. 1993, 39: 903–924.

Reynolds, S. E. The cuticle, growth and moulting in insects: the essential background to the action of acylurea insecticides. *Pesticide Science*. 1987, 20: 131–146.

Richter, K.; Koolman, J. Antiecdysteroid effects of brassinosteroids in insects. In: Cutler, H. G., Yokota, T., Adam, G. (eds.), *Brassinosteroids: Chemistry*, Bioactivity, and Application. American Chemical Society, Washington, DC. 1991, pp. 265–278.

Riddiford, L. M. Cellular and molecular actions of juvenile hormone I. General considerations and premetamorphic actions. *Advances in Insect Physiology*. 1994, 24: 213–274.

Riddiford, L. M. Molecular aspects of juvenile hormone action in insect metamorphosis. In: Gilbert, L. I. (Ed.), *Metamorphosis*. Academic, London. 1996, pp. 223–251.

Sarwar, M. Biology and ecology of some predaceous and herbivorous mites important from the agricultural perception. In: Haouas, D., Hufnagel, L. (Eds.), *Pests Control and Acarology*. IntechOpen Ltd., London, UK. 2019a, p. 29.

Sarwar, M. The dangers of pesticides associated with public health and preventing of the risks. *International Journal of Bioinformatics and Biomedical Engineering*. 2015, 1(2): 130–136.

Sarwar, M. Development and boosting of integrated insect pests management in stored grains. *Journal of Agriculture and Allied Sciences*. 2013a, 2(4): 16–20.

Sarwar, M. Frequency of insect and mite fauna in Chilies *Capsicum annum* L., onion *Allium cepa* L. and garlic *Allium sativum* L. cultivated areas, and their integrated management. *International Journal of Agronomy and Plant Production*. 2012, 3(5): 173–178.

Sarwar, M. House dust mites: Ecology, biology, prevalence, epidemiology and elimination. In: *Parasitology and Microbiology Research*, G. A. B. Pacheco; A. A. Kamboh(Eds.). InTech, London, UK. 2020a, p. 26.

Sarwar, M. Indoor risks of pesticide uses are significantly linked to hazards of the family members. *Cogent Medicine*. 2016, 3: 1155373.

Sarwar, M. Insects as transport devices of plant viruses. In: Awasthi, L. P. (Ed.), *Applied Plant Virology: Advances, Detection, and Antiviral Strategies*. Elsevier Inc., London. 2020b, pp. 381–402.

Sarwar, M. Integrated control of insect pests on canola and other brassica oilseed crops in Pakistan. In: Reddy, G. V. P. (Ed.), *Integrated Management of Insect Pests on Canola and Other Brassica Oilseed Crops*. CABI, UK. 2017, pp. 193–221.

Sarwar, M. Integrated pest management (IPM): A constructive utensil to manage plant fatalities. *Journal of Agriculture and Allied Sciences*. 2013b, 2(3): 1–4.

Sarwar, M. Mite (Acari, Acarina) vectors involved in transmission of plant viruses. In: Awasthi, L. P. (Ed.), *Applied Plant Virology: Advances, Detection, and Antiviral Strategies*. Elsevier Inc., London. 2020c, pp. 257–273.

Sarwar, M. Multiple insect pests management options for chickpea (*Cicer arietinum* L.) Walp grain with reference to storage. In: Lund, A. T., Schultz, D. (Eds), *Handbook of Chickpeas: Nutritional Value, Health Benefits and Management*. Nova Science Publishers, Inc., New York. 2019b, pp. 265–302.

Sarwar, M. Multiple integrated pest management (IPM) and area-wide management (AWM) approaches for insect pests of chickpea *Cicer arietinum* (L.) Walp crop. In: Lund, A. T., Schultz, N. D. (Eds.), *Handbook of Chickpeas: Nutritional Value, Health Benefits and Management*. Nova Science Publishers, Inc., New York. 2019c, pp. 237–263.

Sarwar M. Typical flies: Natural history, lifestyle and diversity of diptera. In: Sarwar, M. (Ed.), *Typical Flies: Natural History, Lifestyle and Diversity of Diptera*. InTech, London, UK. 2020d, p. 50.

Sarwar, M.; Roohi, A. New advances in insect vector biology and virus epidemiology. In: Awasthi, L. P. (Ed.), *Applied Plant Virology: Advances, Detection, and Antiviral Strategies*. Elsevier Inc., London. 2020, pp. 301–311.

Sarwar, M.; Salman, M. The paramount benefits of using insecticides and their worldwide importance in food production. *International Journal of Bioinformatics and Biomedical Engineering*. 2015, 1(3): 359–365.

Sarwar, M.; Mariya, A.; Sidra, S.;Romana, I. Different nematodes and plasmodiophorids as vectors of plant viruses. In: *Applied Plant Virology: Advances, Detection, and Antiviral Strategies*, L. P. Awasthi (Ed.). Elsevier Inc., London. 2020a; 275–290.

Sarwar, M.; Shad, N. A.; Batool. R. Integrated management of vectored viral diseases of plants. In: Awasthi, L. P. (Ed.), *Applied Plant Virology: Advances, Detection, and Antiviral Strategies*. Elsevier Inc., London. 2020b, pp. 707–724.

Schmutterer, H. Properties and potential of natural pesticides from the neem tree, *Azadirachta indica. Annual Reviews of Entomology*. 1990, 35: 271–297.

Staal, G. B. Anti-juvenile hormone agents. *Annual Review of Entomology*. 1986, 31: 391–429.

Subramanian, S.; Shankarganesh, K. Insect hormones (as pesticides). In: *Ecofriendly Pest Management for Food Security*, Omkar (Ed.). Elsevier Inc., London. 2016, pp. 613–650.

31

Management of the Pests of Crops through Insect Sex Pheromones

Rajeev Kumar

CONTENTS

31.1 Introduction

Ecogen Incorporation and Mycogen Corporation have used two points and engineered two products combining more than one toxin of bacterium *Bacillus thuringiensis* when one of these products is effective in controlling gypsy moth and spruce budworm in forest and others against Colorado potato beetle and European corn borer on potatoes (Benhamou et al. 1994; Maddox 1994; Pedigo 1996; Lacey and Kaya 2000; Bailey et al. 2010). The efficacy of biopesticide management to enable the conservation of natural enemies and saving them from destruction was studied (Sheck 1991; Kalra and Khurana 2007). Maintaining cover crop, periodically moving it, and leaving the mulch reduces the number of aphids, leafhoppers, and codling moths (Gould 1991; Copping 2009; Kalra and Khurana 2007).

31.2 Pheromones

Use of pest control or suppressing their population dates back to the early 1900s in the form of attracting baits treated with insecticides or on food and killing them outright, but the current paradigm remains largely confined to improving the performance of toxic baits (Billen and Morgan 1998; Rust et al. 2004). In fact, characterization of ant trail pheromones over the past several years and little use of these compounds has been implemented in pest management (El- Sayed 2012).

To date, approximately 50 companies are involved in commercially producing synthetic pheromones (parapheromones) for more than 50 species of dangerous pests (Ridgway et al. 1990). Pheromones fit neatly into the *virtuous* spiral in greenhouse IPM where the use of one biological control agent such as a predatory spider mite encourages moving away from conventional pesticides for other pests (van Lenteren and Woets 1988).More than 670 genera from nearly 50 families of Lepidoptera in which female sex pheromones have been identified (Tumilsom et al. 1977; Arn et al. 1995; Fónagy et al. 2011). These pheromones are commonly used as attractants to facilitate contact with and dispersal of pathogens in pest populations (Pell et al. 1993).

Trail-marking pheromones: Pheromones are the specialty of foraging ants and mites and used to communicate the food requisites to the other members of the community, and it can be used to deceive and deprive them of the food and suppress their population. (Mashaly 2010, 2011; Mashaly et al. 2008; Sillam et al. 2007, Hölldobler and Wilson 1990; Jackson et al. 2006; Mashaly et al. 2008, 2010, 2011; Mashaly and Al-Khalifa 2012). These trails are then used by other foragers (Wilson and Pavan 1959). In other cases, however, trails may be laid to recruit workers for slave raids, colony emigration, or the repair of a breach in the nest wall (Wilson and Pavan 1959). Wilson (1962) had shown that the fire ant (*Solenopsis saevissima*) drags its sting and lays a trail in a manner similar to a pen inking a line.

Aggregation pheromones: Some of the coleopterans are known to attract their community members on food sites, reproductive habitats, and also hibernation sites, an example of which is bark beetle. Aggregation pheromones lead to the formation of animal groups near the pheromone source either by attracting animals from a distance or stopping ("arresting") passing conspecifics (Wyatt 2003). Nymphs of the German cockroach *Blattella germanica* (Sakuma and Fukami 1990). Their ability to attract females makes these pheromones well suited for the attract-annihilate method (Lanier 1990). Aggregation pheromones have been used successfully for controlling various Coleoptera and cotton boll weevil *Anthonomus grandis* in the USA (Hardee 1982) and bark beetles in North America and Europe (Lanier 1990). Innocenzi et al. (2001) characterized a male-produced aggregation pheromone of *Anthonomus rubi* as a 1:4:1 blend of grandlure I, grandlure II, and lavandulol (note that "grandlure" is the name given to four components in the aggregation pheromone lure of the cotton boll weevil *Anthonomus grandis* Boh).

The pheromone is to induce alert in conspecifics to raise a defense response and/or to initiate avoidance (Rechcigl and Rechcigl 1998). Weston et al. (1997) showed a dose response of attraction and repellency for several pure volatiles from the venom of the common and German wasps *Vespula vulgaris*

TABLE 31.1

Chemical Nature of the Pheromones

S. No.	Pheromone	Name of Insect Species	Chemical Formula	References
Alcohol				
1	Seudenol	*Dendroctonus pseudotsugae*	$C_7H_{12}O$	McKnight (1978)
2	Sulcatol	*Gnathotrichus sulcatus*	$C_8H_{16}O$	Borden et al. (1976)
3	Grandisol	*Anthonomus grandis*	$C_{10}H_{18}O$	Mori et al. (1978)
Acetates and propionate				
1	1. Acetate of (Z)-6-isopropenyl-3-methyl-3 9- decadien-1-1	*Aonidiella aurantii*	$C_{16}H_{26}O_2$	Roelofs et al. (1978)
2	2. Acetate of (E)-6-isopropyl-3 9-dimethyl-58- decadien-1-1	*Aonidiella citrina*	$C_{17}H_{30}O_2$	Roelofs et al. (1982)
3	3. Propionate of (Z)-6-isopropenyl-3 9-dimethyl-3 9-decadien-1-$_0$1	*Pseudaulacaspis pentagona*	$C_{18}H_{30}O_2$	Heath et al. (1980)
Aldehydes and ketones				
1	Franal	*Monomorium pharaonis*	$C_{17}H_{30}O$	Kobayashi et al. (1980)
2	Trogodermal	*Trogoderma granarium*	$C_{17}H_{32}O$	Levinso and Mori (1980)
3	311-Dimethyl-2-nonacosanone	*Blattella germanica*	$C_{17}H_{32}O$	
Acids esters and lactones				
1	4-Hexanolide	*Trogoderma granarium*	$C_6H_{10}O_2$	Ravid et al. (1978)
2	Callosobruchusic acid	*Callosobruchus chinensis*	$C_{10}H_{16}O_4$	Mori et al. (1983)
3	Dominicalure 1	*Rhizoperta dominica*	$C_{11}H_{20}O_2$	Williams et al. (1981)
Acetals and epoxides				
1	Dominicalure 2	*Rhyzopertha dominica*	$C_{12}H_{22}O_2$	Williams et al. (1981)
2	(Z)-5-Tetradecen-4-olide	*Popillia japonica*	$C_{14}H_{24}O_2$	Tumlinson et al. (1977)
3	Frontalin	*Dendroctonus frontalis*	$C_8H_{14}O_2$	Wood et al. (1976)
4	exo-Brevicomin	*Dendroctonus brevicomis*	$C_9H_{16}O_2$	Wood et al. (1976)
5	Lineatin	*Trypodendron lineatum*	$C_{10}H_{16}O_2$	Slessor et al. (1980)
6	α-Multistriatin	*Scolytus multistriatus*	$C_{10}H_{18}O_2$	Elliott et al. (1979)
7	Disparlure	*Porthetria dispar*	$C_{19}H_{38}O$	Iwaki et al. (1974), Vitae et al. (1976)

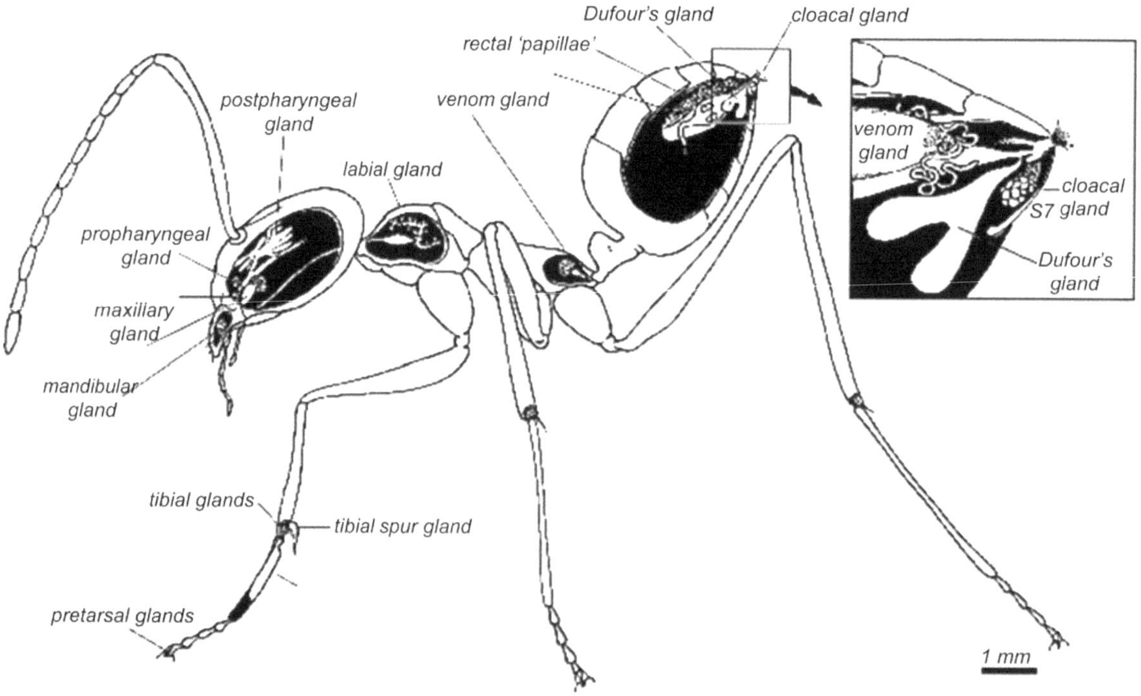

FIGURE 31.1 Diagrammatic sketch of the worker ant showing the distribution of different exocrine glands. Inset is the enlarged abdominal tip showing cloacal gland, Dufour's gland and venom gland (Wenseleers et al. 1998)

TABLE 31.2

Commercially Available Brands of Pheromones

S. No.	Pheromone	Brand Name (Pheromone Lure)	Categories	Manufacturer
1	*Helicoverpa armigera*	Heli Lure	Biopesticides/ biological pesticides	PCI, India
2	*Spodoptera litura*	Spodo Lure	Biopesticides/ biological pesticides	PCI, India
3	*Spodoptera frugiperda*	FAW Lure	Biopesticides/ biological pesticides	PCI, India
4	Fruit and shoot borer	Leucin Lure	Biopesticides/ biological pesticides	PCI, India
5	*Earias vitela* (spotted bollworm)	Ervit Lure	Biopesticides/ biological pesticides	PCI, India
6	Early shoot borer	ESB Lure	Biopesticides/ biological pesticides	PCI, India
7	Interrmode borer	INB Lure	Biopesticides/ biological pesticides	PCI, India
8	Top shoot borer	TSB Lure	Biopesticides/ biological pesticides	PCI, India
9	Fruit fly	Methyl Euginol Lure	Biopesticides/ biological pesticides	PCI, India
10	Cucurbits fruit fly	Cue Lure	Biopesticides/ biological pesticides	PCI, India
11	Red Palm weevil	R.P.W. Lure	Biopesticides/ biological pesticides	PCI, India
12	Diamond back moth (DBM)	D.B.M. Lure	Biopesticides/ biological pesticides	PCI, India
13	Rhinocerous beetle	Rhino Lure	Biopesticides/ biological pesticides	PCI, India
14	Coffee white stem borer	C.W.S.B. Lure	Biopesticides/ biological pesticides	PCI, India
15	Paddy leaf folder	L.F Lure	Biopesticides/ biological pesticides	PCI, India
16	Coco pod borer	C.P.B Lure	Biopesticides/ biological pesticides	PCI, India
17	Pink bollworm	Pectino Lure	Biopesticides/ biological pesticides	PCI, India

and *V. germanica*. The compounds are usually highly volatile (Francke et al. 1979). Alarm pheromones of aphids have been used commercially to increase the effectiveness of conventional pesticides (Howse et al. 1998). The synthetic alarm pheromones and the increased activity of the aphids in response to their alarm pheromone increase mortality because they contact more insecticide or fungal spores (Pickett et al. 1992). The trail pheromone is secreted by the poison gland, and Cross et al. (1982), Sillam-Dusse`s et al. (2007) have reported it in the genus *Atta* such as

31.3 Chemical Composition of Pheromones

Pheromones are chemically classified into different categories to serve different functions, and this range includes small hydrophobic molecules to water-soluble peptides (Sunamura et al. 2011). "lure and kill" or mass trapping, and other manipulations of pest behavior (Billen and Morgan 1998; Blum and Ross 1965; Fónagy et al. 2011). Pheromones fit neatly into the *virtuous* spiral, for example, in greenhouse IPM where the use of one biological control agent such as a predatory spider mite encourages (or requires) moving away from conventional pesticides for other pests (van Lenteren and Woets 1988; Tangchitphinitkan et al. 2007).

The key to all of these behavioral chemicals is that they leave the body of the first organism, pass through the air (or water), and reach the second organism where they are detected by the receiver (Free 1987). Some of the chemicals secreted by respective insects of economically important pests are listed with their finder's references in Table 31.1. These pheromones are released by unreceptive females to deter males from continuing mating attempts (Rechcigl and Rechcigl 1998; Taymour 2012).

Source of trail pheromones: Generally the source of trail pheromones are hind tibia,Dufour's gland, and venom glands in Myrmicinae; Pavan's gland in Dolichoderinae;pygidial gland in Ponerinae; post pygidial gland in Aenictinae; and the hind gut in Fig. 31.1. Dufour's gland of at least a portion in Myrmicinae, Dolichoderinae, Ponerinae, and Formicinae secretes and contains mixture of straight chain of hydrocarbons ranging between C9 and C27 (Morgan 2008).

31.3.1 Commercial Pheromones

Basically sex pheromone are microchemical matter secreted outside their bodies by female insects to be used for luring male insect mating; it is also called the sex information hormone or sex luring agent (Tables 31.1 and 31.2).

31.4 Role of Biopesticides in Integrated Pest Control

In a health-conscious society, the use of biopesticides could reach 10% (Burges 1998; Thakore 2006; Taymour 2012). Today, communities are more health-conscious than ever (Gould 1991; Sheck 1991; Rechcigl and Rechcigl 2000; Tangchitphinitkan et al. 2007; Sinha and Biswas 2008; EPA 2013).

REFERENCES

Francke W, Hindorf G, Reith W. (1979). Alkyl-1,6-dioxaspiro[4,5]-decanes, a new class of pheromones. *Naturwissenschaften* 66:618.

Free JB. (1987). *Pheromones of Social Bees.* Comstock Publishing, Ithaca.

Gould F. (1991). Arthropod behavior and the efficacy of plant protectants. *Annu Rev Entomol* 36:305–330.

Hölldobler B, Wilson EO. (1990). *The Ants.* Harvard University Press, Cambridge.

Howse PE, Stevens IDR, Jones OT. (1998). *Insect Pheromones and Their Use in Pest Management.* Chapman &Hall, London.

Innocenzi PJ, Hall DR, Cross JV. (2001). Components of male aggregation pheromone of strawberry blossom weevil Anthonomus rubi Herbst (Coleoptera: Curculionidae). *J Chem Ecol* 27:1203–1218.

Jackson DE, Martin SJ, Holcombe M, Ratnieks F. (2006). Longevity and detection of persistent foraging trails in Pharaoh's ant Monomorium pharaonis. *Anim Behav* 71:351–359.

Kalra A, Khurana SPS. (2007). Research and development priorities for biopesticides and biofertiliser product for sustainable agriculture in India. In: Teng PS (ed) *Business Potential for Agricultural Biotechnology.* Asian Productivity Organization, Tokyo.

Lacey LA, Kaya HK. (2000). *Field Manual of Techniques in Invertebrate Pathology: Application and Evaluation of Pathogens for Control of Insects and Other Invertebrate Pests.* Kluwer Academic, Dordrecht.

Lanier GN. (1990). Principles of attraction annihilation: mass trapping and other means. *Canad. F. S. Publication* 128:25–45.

Mashaly AMA. (2011). Trail communication in two Messor species Messor meridionalis and Messor foreli (Hymenoptera: Formicidae). *Ital J Zool* 78(4):524–531.

Mashaly AMA, Ali AS, Ali MF. (2008). Trail communication in Paratrechina longicornis (Latreille) and Paratrechina vividula (Nylander) (Hymenoptera: Formicidae). *Egypt J Zool* 51:449–459.

Mashaly AMA, Ali AS, Ali MF. (2010). Source optimal dose concentration and longevity of trail pheromone in two Monomorium ants (Formicidae: Hymenoptera). *J King Saud Univ Sci* 22(2):57–60.

Mashaly AMA, Al-Khalifa MS. (2012). The trail pheromone in pest control. In: Bandani AR (ed) *New Perspectives in Plant Protection*, vol 1. InTech, Rijeka, Croacia.

Morgan ED. (2008). Chemical sorcery for sociality: exocrine secretions of ants (Hymenoptera: Formicidae). *Myrmicl News* 11:79–90.

Pedigo LP. (1996). *Entomology and Pest Management*, 2nd edn. Prentice-Hall, Englewood Cliffs.

Pickett JA, Wadhams LJ, Woodcock CM, Hardie J. (1992). The chemical ecology of aphids. *Annu Rev Entomol* 37:67–90.

Rechcigl JE, Rechcigl NA. (1998). *Biological and Biotechnological Control of Insect Pests Insect Growth Regulators.* CRC Press, Boca Raton.

Ridgway RL, Silverstein RM, Inscoe MN. (eds.) (1990). Behavior – modifying chemicals for insect antijuvenile hormones: allyic alcohol derivatives. *J Agric Food Chem* 33:47–50.

Rust MK, Reierson DA, Klotz JH. (2004). Delayed toxicity as a critical factor in the efficacy of aqueous baits for controlling Argentine ants (Hymenoptera: Formicidae). *J Econ Entomol* 97:1017–1024.

Sakuma M, Fukami H. (1990). Dose/response relations in taxes of nymphs of the German cockroach Blattella germanica (L) (Dictyoptera: Blattellidae) to their aggregation pheromone. *Jpn J Appl Entomol Zool* 25:9–16.

Sheck AL. (1991). Biotechnology and plant protection. In: Pimentel D (ed). *Hand Book of Pest Management in Agriculture*, vol 2, 2nd edn. CRC Press, Boca Raton.

Sillam-Dussès D, Semon E, Lacey MJ, Robert A, Lenz M, Bordereau C. (2007). Trail-following pheromone in basal termites with special reference to Mastotermes darwiniensis. *J Chem Ecol* 33:1960–1977.

Sinha B, Biswas I. (2008). Potentials of biopesticides in Indian agriculture vis-a-vis rural development. *Ind Sci Technol.* NISTADS (CSIR), New Delhi.

Tangchitphinitkan P, Visetson S, Maketon M, Milne M. (2007). Effects of some herbal plant extracts against pharaoh ant *Monomorium pharaonis* (Linnaeus). *KMITL Sci Tech* J 7(2):155–159.

Taymour M. (2012). Pheromones in sex and reproduction: do they have a role in humans? *J Adv Res* 3:1–9.

Thakore Y. (2006). The biopesticide market for global agricultural use. *Ind Biotechnol* 2:194–208.

Tumlinson JH, Klein MG, Doolittle RE, Ladd TL, Proveaux AT. (1977). Identification of the female Japanese beetle sex pheromone: inhibition of the male response by an enantiomer. *Science* 197:789–792.

van Lenteren JC, Woets J. (1988). Biological and integrated pest control in green house. *Annu Rev Entomol* 33:239–269.

Weston RJ,Woolhouse AD, Spurr EB, Harris RJ, Suckling DM. (1997). Synthesis and use of spirochetals and other venom constituents as potential wasp (Hymenoptera: Vespidae) attractants. *J Chem Ecol* 23:553–568.

Wilson EO. (1962). Chemical communication among workers of fire ant Solenopsis saevissima (Fr Smith) I-the organization of mass foraging II – an information analysis of the odour trail III – the experimental induction of social responses. *Anim Behav* 10:134–164.

Wyatt TD. (2003). *Pheromones and Animal Behaviour.* Cambridge University Press, Cambridge.

Section XV

Management of the Diseases of Crops through Biotic Biopesticides (Natural Enemies) Biological Control

32

Biological Control of Harmful Forest Insects: The Biological Efficiency of Microbial Insecticides and the Possibility of Their Application in Forest Ecosystems

Mara Tabakovic-Tosic

CONTENTS

32.1 Introduction: Integrated Forest Protection and Biological Control of Forest Pests

Forests, as the natural resources, are significant productive-natural and ecological potentials. We are witnessing a dynamic era of a disturbed equilibrium, marked by recent climate imbalances, in synergy with other abiotic factors, which have initiated the "population explosion" of some economically harmful forests species, resulting in the mortality of trees as well as of entire forest complexes (Tabakovic-Tosic et al. 2011; Tabakovic-Tosic 2019).

The intensive application of chemical insecticides in forests has led to the disturbance of the bio-cenotic balance, to mass reproduction of some previously less dangerous pests and parasites, and to resistant strains. The effective substitution of classical poisonous substances by bio-preparations is a very significant improvement in the protection of forest plants from insect pest species.

32.2 Bacterium *Bacillus thuringiensis* ssp. *Kurstaki* and Baculoviruses as Active Ingredients of Insecticides Intended for Use in Forest Ecosystems and Their Biological Efficiency

Bacillus thuringiensis ssp. *kurstaki* (*Btk*) is currently the most efficient entomopathogen, which, as the active ingredient of microbial insecticides, has an important role in the system of the integral protection of forests.

The creation of high-potency formulation and application in the undiluted form, i.e., in ultra-low volume (ULV), combined with a shift in the public awareness of the necessity of the reduction of environmental pollution, has popularized the use of the *Btk* insecticide for the suppression of some defoliating species in the forests, and it was applied for the control of more than 180 Lepidoptera species. However, their sensitivity to this microbiological agent is not the same, and ranges from most sensitive to resistant. The forest insect pests, such as *Hyphantria cunea* (Drury), *Malacosoma* spp., *Stilpnotia salicis* L., *Hibernia defoliaria* L., etc., belong to the group of sensitive. The most harmful outbreaking insects, such as *Lymantria dispar* L., *Choristoneura fumiferana* Clemens, *Operophtera brumata* L., etc., belong to the mid-sensitive group. Almost total production of *Btk* insecticides has been aimed for decades at the suppression of the economically most harmful defoliating insect species that have been stated first, *C. fumiferana*in conifer forest stands, and *L. dispar* in broadleaf forest stands. The harmful forest insects from Family Noctuidae (Martouret 1975), belong to the most resistant group.

This type of insecticides is widely used, as a result of their efficacy and safety to other living organisms. A small number of the economically most important harmful insect species in the suppression of which *Btk* preparations were efficiently used mainly belong to Lepidoptera order [*Yponomeuta* spp. (Yponomeutidae); *Tortrix viridana* L., *Pandemis heparana* Den. et Schiff., *Archips podana* Scop., *Spilonota ocellana* Fabricius, *C. fumiferana*, *Choristoneura* spp. (Tortricidae); *Malacosoma neustria* L. (Lasiocampidae); *O. brumata*, *Erannis* spp. (Geometridae); *L. dispar* (Erebidae); *Thaumatopoea processionea* L. (Thaumatopoeidae); *H. cunea* (Arctiidae)] (Asano et al. 1976; Morris 1982; Morris and Moore 1983; Smirnoff and Valéro 1983; Fast and Dimond 1984; Marovic et al. 1997; Lacey et al. 2001; Tabakovic-Tosic 2008).

New studies on insecticidal activity are aimed at the identification of the isolate of *B. thuringiensis* ssp. *kurstaki*, which will be highly pathogenic to the economic harmful lepidopterous insects (Kati et al. 2007).

Baculoviruses are divided into three subgroups, but insecticide research has focused on subgroups A [nuclear polyhedrosis viruses (NPV)] and B [granulosis viruses (GV)] which have the greatest biological potential in the suppression of the forest harmful insect species (Granados and Federici 1986; Adams and McClintock 1991; Tanada and Hess 1991; Vail 1993; Cunningham 1995; Hunter-Fujita et al. 1998; Lacey et al. 2001; Moreau and Lucarotti 2007).

The role of baculoviruses in the population dynamics of their respective hosts varies extremely from species to species. In some species, such as *C. fumiferana*, naturally occurring baculovirus epizootics have never been observed in the field to terminate an outbreak (Coppel and Mertins 1977; Cunningham and Kaupp 1995; Lucarotti et al. 2004). In contrast, under some conditions, baculoviruses have the potential to initiate insect population fluctuations (Anderson and May 1980; Myers 1988; Tanada 1976; Olafson 1987; Moreau and Lucarotti 2007); therefore, they are highly appreciated in biological control programs, at least in the systems where they dominate the dynamics of their hosts.

The number of registered pesticides based on baculoviruses, although slowly, increases steadily. At present, there are more than 50 virus formulations, some of them being the same baculovirus preparations distributed under different trade names in different countries.

In the world market, the insecticides, the active ingredients of are *Neodiprion lecontei NPV, N. abietis NPV, N. sertifer NPV, Orgyia pseudotsugata NPV and Lymantria dispar NPV*, can be found.

32.3 Main Factors of the Efficient Application of Commercial Microbiological Insecticides in Forest Stands

Today. microbial insecticides are mainly prepared to be used in ultra-low volumes, which is very significant for their application in large forest ecosystems.

Starting from the facts that the microbial insecticides are of the highest quality, that their potency is high, and that they can be applied in the suppression of Lepidoptera species in forests (aerial application, ULV method), attention should be directed to the following factors which affect its biological effectiveness to the extent that it can be annulled:

Population level of the target insect: The selection of insecticide to be applied in the suppression of, for instance, gypsy moth larvae, primarily depends on the intensity of the attack, which is measured by the number of egg masses per forest unit area: 1–10 egg masses/ha—low, 11–100—medium, 101–500—strong, and above 500—severe. This classification is in correlation with the damages to the foliage volume, caused by larval nutrition. If we assume that *Btk* or NPV insecticide was 90% effective (which is very rare in natural conditions), 10% of the larvae from the target population remained alive. When it is transposed (only hypothetically) to the number of egg masses per hectare of forest area and percentage of the damage to the foliage at different intensities of the attack, it is concluded that the critical value of the number of egg masses for the application of *Btk* insecticide is 2,500, because the larvae from 250 egg masses (10%), stay alive, and they cause the foliage damage up to 40%, which corresponds to the positive value of biological effectiveness. In spite of the good quality of the bioinsecticide, any application of this preparation above this critical value of the intensity of the attack involves the risk of unsatisfactory results, i.e. the damage is the same as if there was no suppression at all (Tabakovic-Tosic 2008).

The degree of foliage expansion and larval instars: The determination of the right moment of the application requires a subjective judgment, based on the measuring of the foliage expansion (typically 45–60%), larval instar of the target insect (typically 30% first, 50% second, 20% third instar), the population density, and the predicted level of the defoliation prior to the treatment (Lewis et al. 1979; Reardon et al. 1994; Tabakovic-Tosic 2008). This percentage ratio of the larval instars is the result of the fact that for the older insects the higher lethal dosage is needed, and of the fact that the ingested quantity of the preparation often causes sublethal effects. The larvae which absorb a sublethal dose stop feeding temporarily, because of the damage to the midgut epithelium, they seem sick, but later the epithelium becomes revitalized, their health condition improves, they continue with the intensive diet, and their development continues. The damage caused by them is greater than the damage caused by the larvae from the untreated areas, as one of the results of the sublethal dose is the prolonged larval development.

Forest stand structure: The effectiveness of the suppression is greatly affected also by the stand elements, such as the age, percentage of the tree species, mixture proportion, the canopy cover (proportion of the ground covered with tree crowns), and form (vertical layers), crown development, which directly affects the insecticide activity and the application method. The complete canopy and vertical layers can have an adverse effect on the effectiveness, because of the greater possibility that the foliage in the lower stores is insufficiently, or not at all, covered, so the larvae which consume such leaves will not be affected by the active ingredients of the insecticides which have been applied. The tree species and the proportion of them in the mixture can affect the effectiveness of the preparation, i.e. in mixed stands, the leafing phenophases usually do not overlap (Tabakovic-Tosic 2008).

Meteorological conditions: Unfavorable meteorological conditions (precipitation, wind, low temperatures) in the period optimal for the aerial suppression often lead to the postponing of it. As a result, a number of larvae come of age when the biological effectiveness of the preparations is considerably lower or insufficient. Due to cold and rainy weather, immediately after the aerial application, larvae feeding intensity is lower, so the dose of the consumed bioinsecticide is sublethal. The effect of the sublethal dose is reflected also in a considerably lower size and weight of pupae and adults, reduced fecundity, as well as in the impossibility of chrysalidation of a number of larvae. Therefore, the effectiveness of the preparation is reflected in the reduced population density over the subsequent period, but the effects of the foliage protection from damage during the current season are insufficient or absent (Rothman and Myers 1996; Tabakovic-Tosic, 2008).

Sunlight: Natural sunlight (the active spectrum is between 290 and 400 nm) is the most destructive environmental factor affecting the persistence of entomopathogens. Salama et al. (1983) found that one day of direct sunlight could inactivate over 90% of *Btk* spores on the potted white spruce. Pozsgay et al. (1987) concluded that the wavelengths in the 300–380 nm range of the solar spectrum are largely responsible for the loss of toxicity in purified *Btk* crystals. Considering the above facts, the producers of the preparations recommend application early in the morning. In that case, in the part of the day following the application, *Btk* spores and protein-crystals are exposed to the destructive effect of the direct solar radiation. Based on our research, greater success and effectiveness would be achieved by the application in the late afternoon hours, when the application is followed by at least 10 hours without the direct effect of UV radiation (Tabakovic-Tosic 2008).

Orographic characteristics: Non-uniform distribution, which implies lower effectiveness of the applied preparation, is mainly the result of land fragmentation, i.e., the irregular accidented terrain, and the slopes.

Technology: The application equipment should be cleaned, calibrated, and characterized before each application (Reardon et al. 1994).

32.4 Conclusion and Future Prospects

Finally, in many countries, and particularly in the 28 European Union member states, the aerial application of pesticides in forests is prohibited by law (European Parliament and the Council of the European Union, Directive 2009/128/EC). Bearing in mind the structure of forest ecosystems, morphometric characteristics of the forest tree species, spatial characteristics, land fragmentation, outbreak of the biological harmful agents, on one hand, and the prohibition of the only possible method of application of the appropriate pesticides, on the other hand, the following conclusion can be made—the lawmakers in the domain of the forest protection from the harmful biotic factors have regulated by law the impossibility of the application of repressive methods of control by using suitable microbial pesticides, no matter whether they are proved to be safe by the ecotoxicological risk assessment (Tabakovic-Tosic 2019).

REFERENCES

Adams, J. R.; McClintock. J. T. Baculoviridae. Nuclear polyhedrosis viruses. Part 1. Nuclear polyhedrosis viruses of insects. In *Atlas of Invertebrate Viruses*; Adams, J. R., Bonami, J. R., eds. Boca Raton, FL: CRC Press, 1991; pp. 87–204.

Anderson, R. M.; May, R. M. Infection diseases and population cycles of forest insects. *Science*. 1980, *210*, 658–661.

Asano, S.; Sakakibara, H.; Nakamura, K. Susceptibility of the pine caterpillar, *Dendrolimus spectabilis* Butler (Lepidoptera, Lasiocampidae) to *Bacillus thuringiensis*. *Kontyu*. 1976, *44*, 217–227.

Coppel, H. C.; Mertins, J. W. *Biological Insect Pest Suppression*. Berlin: Springer, 1977.

Cunningham, J. C. Baculoviruses as microbial insecticides. In *Novel Approaches to Integrated Pest Management*; Reuveni, R., eds. Boca Raton, FL: Lewis Publishers, 1995; pp. 261–292.

Cunningham, J. C.; Kaupp, W. J. Insect viruses. In *Forest Pest Insects in Canada*; Armstrong, J. A., Ives, W. G. H., eds. Ottawa, ON: Natural Resources Canada, 1995; pp. 327–340.

European Parliament and the Council of the European Union. Directive 2009/128/EC, establishing a framework for community action to achieve the sustainable use of pesticides. *Official Journal of the European Union*. 2009, L309/71-86.

Fast, P. G.; Dimond, J. B. Susceptibility of larval instars of spruce budworm, *Choristoneura fumiferana* (Lepidoptera: Tortricidae), to *Bacillus thuringiensis*. *The Canadian Entomologist*. 1984, *116*, 131–137.

Granados, R. R.; Federici, B. A. *The Biology of Baculoviruses*, Vols. I and II. Boca Raton, FL: CRC Press, 1986.

Hunter-Fujita, F. R.; Entwistle, P. F.; Evans, H. F.; Crook, N. E. *Insect Viruses and Pest Management*. New York: Wiley, 1988; 620 pp.

Kati, H.; Sezen, K.; Nalcacioglu, R.; Demirbag, Z. A highly pathogenic strain of *Bacillus thuringiensis* serovar *kurstaki* in lepidopteran pests. *The Journal of Microbiology*. 2007,*45*(6), 553–557.

Lacey, L. A.; Frutos, R.; Kaya, K. H.; Vails, P. Insect pathogens as biological control agents: Do they have a future? *Biological Control*. 2001, *21*, 230–248.

Lewis, F. B.; McManus, M. L.; Schneeberger, N. F. *Guidelines for the Use of GYPCHEK to Control* the Gypsy Moth. USDA Forest Service Research Paper NE-441. Broomall, PA: U.S. Department of Agriculture, Forest Service, Northeastern Forest Experiment Station, 1979; 9 pp.

Lucarotti, J. C.; Eveleigh, S. E.; Royama, T.; Morin, B.; McCarthy, P.; Ebling, M. P.; Kaupp, J. W.; Guertin, C.; Arella, M. Prevalence of baculoviruses in spruce budworm (Lepidoptera: Tortricidae) populations in New Brunswick. *The Canadian Entomologist*. 2004, *136*, 255–264.

Marovic, R.; Minic, D.; Jancic, G.; Sikanja, R.; Maravic, M.; Nikolic, A. Control of *Euproctis chrisorrhoea* L. in the Lipovica preserve by domestic bioinsecticides. In *Book of Articles, International Scientific Symposium "50 years—Faculty of Forestry, Skopje, Macedonia."* 1997; pp 99–102.

Martouret, D. Remarques sur le titre des formules commerciales a base de *Bacillus thuringiensis*. *Entomophaga*. 1975, *20*(1), 3–9.

Moreau, G.; Lucarotti, C. A brief review of the past use of baculoviruses for the management of eruptive forest defoliators and recent developments on a sawfly virus in Canada. *The Forestry Chronicle*. 2007, *83*(1), 105–112.

Morris, O. N. Bacteria as pesticides: Forest applications. In *Microbial and Viral Pesticides*; Kurstak, E., ed. New York: Marcel Dekker, 1982; pp. 239–287.

Morris, O. N.; Moore, A. Relative potencies of 50 isolates of *Bacillus thuringiensis* for larvae of the spruce budworm, *Choristoneura fumiferana* (Lepidoptera: Tortricidae). *The Canadian Entomologist*. 1983, *115*, 815–822.

Myers, J. H. Can a general hypothesis explain population cycles of forest Lepidoptera? *Advances in Ecological Research*. 1988, *18*, 179–242.

Olofsson, E. Mortality factors in a population of *Neodiprion sertifer* (Hymenoptera: Diprionidae). *Oikos*. 1987, *48*, 297–303.

Pozsgay, M.; Fast, P.; Kaplan, H.; Carey, R. P. The effect of sunlight on the protein crystals from *Btk* HD-1 and NRD-12: A Roman spectroscopic study. *Journal of Invertebrate Pathology*. 1987, *50*, 246–253.

Reardon, R.; Dubois, N.; McLane, W. *Bacillus thuringiensis for Managing Gypsy Moth: A Review*. Morgantown, W. VA: USDA Forest Service—National Center of Forest Health Management (Technology Transfer). 1994; pp. 1–32.

Rothman, L. D.; Myers, J. H. Debilitating effects of viral diseases on host Lepidoptera. *Journal of Invertebrate Pathology*. 1996, *61*, 1–10.

Salama, H.; Foda, S. M.; Zaki, N. F.; Khalafallah, A. Persistence of *Bacillus thuringiensis* Berliner spores in cotton cultivations. *Zeitschrift für Angewandte Entomologie*. 1983, *95*, 321–326.

Smirnoff, W. A.; Valéro, J. R. Characteristics of a highly concentrated *Bacillus thuringiensis* formulation against spruce budworm, *Choristoneura fumiferana* (Lepidoptera: Tortricidae). *The Canadian Entomologist*. 1983, *115*, 443–444.

Tabakovic-Tosic, M. *Entomopathogenic Bacterium Bacillus thuringiensis ssp. kurstaki the Important Component of the Integral Protection of Forest Ecosystems*; Tabakovic-Tosic, M., ed. Belgrade: Institute of Forestry, special editions, 2008; 148 pp.

Tabakovic-Tosic, M. The application of pesticides in forestry. In *Use of Pesticides in Plant Production and environmental Protection*; Skoric, D., Andjelkovic, M., eds. Belgrade: Serbian Academy of Sciences and Arts—Department of Chemical and Biological Sciences, 2019; *Book16*, pp. 85–97. (in Serbian, Abstract and Resume in English).

Tabakovic-Tosic, M.; Golubovic-Curguz, V.; Tosic, D. New technological methods in the integrated forest protection in the Republic of Serbia. In Proceedings of the International Scientific Conference on Integrated Plant Protection: Strategy and Tactics, Minsk. 2011; pp. 49–55.

Tanada, Y. Ecology of insect viruses. In *Perspectives in Forest Entomology*; Anderson, J. F., Kaya, K. H., eds. New York: Academic Press, 1976; pp. 265–283.

Tanada, Y.; Hess, R. T. Baculoviridae. Granulosis virus. In *Atlas of Invertebrate Viruses*; Adams, J. R., Bonami, J. R., eds. Boca Raton, FL: CRC Press, 1991; pp. 227–257.

Vail, P. V. Viruses for control of arthropod pests. In *ACS* Proceedings *Series, Pest Management: Biologically Based Technologies*; Lumsden, R. D., Vaughn, J. L., eds. Washington, DC: American Chemical Society. 1993; pp. 30–39.

33

The Role of Natural Enemies and Biopesticides for Sustainable Management of Major Insect Pests of Legumes

Sathish Kota, Vinod Kukanur, Reddy Ramya Sree, Naveen Arora, Jagdish Jaba, and Rana DK

CONTENTS

33.1 Introduction

Pulses are the important components of a healthy diet and take an important place in the traditional diets throughout the World (Malaguti et al. 2014). pulses are damaged by a large number of insect species, both under field conditions and in storage (Clement et al. 2000).

Among legume insect pests, *Helicoverpa armigera* is the single largest yield shrinking factor in food legumes, causes an estimated loss of US$317 million in pigeonpea and $328 million in chickpea (ICRISAT 1992). Worldwide, it causes an estimated loss of over $2 billion annually, despite over $1 billion value of insecticides used to control *H. armigera* (Sharma 2005). Another pod borer *Maruca vitrata* causes loss to the tune of US$30 million annually (Saxena et al. 2002). Pigeonpea yield losses due to pod borer are 25–70%; Pod fly is second most important pest of pigeonpea in northern and central India, and cause 10 - 50 % yield loss. *Maruca* is reported to cause 5 - 25% yield loss in pigeonpea, pod bug can cause yield loss upto 30%. Soybean aphid, (*Aphis glycines*) can induce up to 58% yield losses in soybean crop (Wang et al. 1994) and annually $2.4 billion estimated losses in yield (Song et al. 2006, Tilmon et al. 2011). Legume flower thrips (LFT), *Mylothris sjostedti* Trybom in cowpea *V. unguiculata* in tropical Africa causes yield losses ranging from 20% to 100% (Karungi et al. 2000).

33.2 Management of Legume Insect Pest through Natural Enemies

33.2.1 *H. armigera* and *H. punctigera*

These are the major insect pests of legume crops; both these pests are managed by natural enemies such as parasitoids, *Trichogramma* and *Telenomus*, which are minute parasitoid wasps (shorter than 1 mm) which parasitize *Helicoverpa* eggs, but their activity levels are too low in chickpea and pigeonpea

FIGURE 33.1 *Campoletis chlorideae* (http://www.nbair.res.in).

because of trichome exudates. There are a numerous number of wasps that parasitize pod borer *Helicoverpa* larvae. Only the species considered to be most significant for pest management are described here. The Ichneumonid, *Campoletis chlorideae* Uchida (Figure 33.1), is most likely the main important larval parasitoid of pod borer, *H. armigera*, on pigeonpea and chickpea in India (Pawar et al. 1986). Late-instar *H. armigera* larvae are parasitized by Tachinids, but result in little reduction in larval density. In India, *Carcelia illota* (Curran), and to a lesser extent, *Goniophthalmus halli* Mesnil, and *Palexorista laxa* (Curran) parasitize up to 22% of *H. armigera* larvae on pigeonpea (Bhatnagar et al. 1983), and up to 54% larvae in chickpea. Prepupal and pupal mortality of *H. armigera* is reliable (Table 33.1).

Netelia producta parasitoid of *H. armigera* (Figure 33.2) is a pupal parasitoid of *Helicoverpa*, some armyworms, and is very active in chickpea crops. A female *Ichneumon promissorius* has a white band in the middle part of the antennae (Figure 33.3), whereas a male *Ichneumon promissorius* has orange and black antennae.

33.2.2 The Spotted Pod Borer, *Maruca vitrata* (Geyer)

There are a numerous number of parasites recorded on larvae/pupae of spotted pod borer, *Maruca vitrata* include Tachinids, Braconids, Chalcidids, Eulophids, Ichneumonids, Pteromalids, and Scelionids (Table 33.2).

33.2.3 Pea Pod Borer and Pea Weevils

Pea pod borer, *Etiella zinckenella* (Treitschke), is attacked by several species of natural enemies, such as the hymenopteran egg parasitoids, *Trichogrammatoidea armigera*. Larval parasitoids are *Bracon hebetor*, *Phanerotoma* sp., *Tetrastichus* sp., and *Phanerotoma planifrons*. Pea weevils (*Bruchus pisorum*) are attacked by several species of natural enemies, such as the hymenopteran egg parasitoids *Uscana senex* Grese (Trichogrammatidae) (Hormazabal & Gerding 1998) and *U. chiliensis* sp. nov (Pintureau et al. 1999).

33.2.4 Cow Pea Aphid

Aphis craccivora Koch and pea aphid, *Acyrthosiphon pisum* Haris (Homoptera: Aphididae), are the major hemipteran insect pest of lentils in America, Europe, Africa, Australia, and Asia. predators such as Syrphid flies, Rove beetle, green lacewing, ladybird beetle, and parasitoids such as *Aphidius* sp. and *Aphelinus* sp. While aphids are attacked by a number of natural enemies, Coccinellids especially prevent their rapid reproduction rate and may reduce infestation levels sufficiently (Table 33.3).

33.2.5 Bean Stem Maggot (*Ophiomyia phaseoli*)

Bean stem maggots, also known as bean flies, are often considered to be the most important field pest of beans in Africa. They account for yield losses ranging from 80% to 100%. The most common species include *O. phaseoli*, *O. spencerella*, and *O. centrosematis*. They attack the crop wherever it is grown. *Ophiomyia phaseoli* is a widely distributed pest of seedling bean.

33.2.6 Red Gram Pod Fly

Melanagromyza obtusa Malloch (Diptera: Agromyzidae) causes serious damage to the pods and seeds of pigeonpea from November to February, resulting in poor germination and making them unfit for human consumption. Until now, more than 25 hymenopteran parasitoids were known to attack this notorious pest (Table 33.4). Among them, *Euderus lividus* and *Ormyrus orientalis* are reported as major biocontrol agents of *M. obtusa* (Yadav et al. 2012).

33.2.7 The Bean Aphids

Aphis fabae is the main aphid pest of common bean in Africa and Asia. Aphids have experienced some adaptation in relation to host plant so that many aphid taxa have a biologically complex life cycle. In some instances, both larvae and adults of predators belonging to the family Coccinellidae (Ladybird beetles) feed on aphids. All species in the braconid subfamily Aphidiinae develop as endoparasitoids (inside) of aphids with one larva completing development in each host.

33.2.8 Tobacco Caterpillar

Spodoptera litura (Fab.) is a serious and regular pest of soybean in India. *Trichogramma chilonis*, *Tetrastichus*, and *Telenomus* are egg parasitoids of *Spodoptera* and *Ichneumon promissorius*, *Carcelia* sp., and *Diglyphus isaea* are larval parasitoids of *Spodoptera*.

33.2.9 Stem Fly

Melanagromyza sojae Zehntner (Diptera: Agromyzidae) has emerged as a major pest in the soybean growing areas of India over the last two decades. Two species of hymenopteran parasitoids from two families, namely Eurytomidae (*Eurytoma melanagromyza* Narendran) and Pteromalidae (*Chlorocytus* sp.)

TABLE 33.1

Parasitoids of Pulse Pod Borer *Helicoverpa armigera*

Order	Family	Stage	Scientific Name
Hymenoptera	Scelionidae	Egg parasitoids	*Telenomus* sp.
	Trichogrammatidae		*Trichogramma* sp., *T. achaeae*, *T. brasiliensis*, *T. chilonis*, *T. chilotraeae*, *Trichogrammatoidea* sp., *Trichogrammatoidea armigera*, *T. bactrae*, *T. fumata*
	Braconidae	Egg larval parasitoids	*Chelonus* sp., *C. heliopae*, *C. narayani*, *Microchelonus curvimaculatus*
Diptera	Sarcophagidae	Larval parasitoids	*Sarcophaga orientoides*
	Tachinidae		*Compsilura concinnata*, *Drino unisetosa*, *Drino imberbis*, *Eucelatoria bryani*, *Exorista fallax*, *Exorista japonica*, *Exorista xanthaspis*, *Somera cinerascens*, *Pales coerulea-nigra*, *Palexorista* sp., *P. laxa*, *P. solennis*, *Sisyropa apicata*, *Spallanzania* sp., *Strobiomyia aegyptia*, *Sturmiopsis inferens*, *Suensonomyia* sp., *Thecocarcelia incedens*, *Voria edentata*, *V. ruralis*, *Winthemia diversoides*
	Chloropidae		*Mepachymerus ensifer*
Hymenoptera	Bethylidae		*Goniozus* sp.
	Braconidae		*Apanteles* sp., *Apanteles* sp. *glomeratus* group, *A. glmeratus*, *A. rujicrus*, *Bracon* sp., *B. brevicornis*, *B. cushmani*, *B. gelechiae*, *B. greeni*, *B. hebetor*, *Microplitis maculipennis*, *Microplitis palidipes*, *Odentepyris* sp., *Paraphylax* sp., *Rogas* sp.
	Eulophidae		*Euplectrus* spp.
	Ichneumonidae		*Agrypon nox*, *Attractodes* sp., *Banchopsis rujicornis*, *Barichneumon* sp., *Campoletis chlorideae*, *C. maculipes*, *Campoletis (Ecphoropsis)* sp., *Disophrys* sp., *Enicospilus* sp., *Enicospilus shinkanus*, *Eriborus argenteopilosus*, *Eriborus pilosellus*, *Eriborus trochanteratus*, *Ichneumon* sp., *Metopius rufus*, *Netelia* sp., *Pristomerus* sp., *Temelucha* sp.
Nematoda	Mermithidae		*Hexamermis* sp., *Ovomermis albicans*
Dlptera	Tachinidae	Larval Pupal parasitoids	*Carcelia* sp., *C. (Stenometopia) illota*, *C. kolkiana*, *C. raoi*, *C. peraequalis*, *Goniophthalmus halli*
Hymenoptera	Ichneumonidae	Pupal parasitioids	*Xanthopimpla stemmator*, *Compoletis chloriedae*
	Chalcididae		*Brachymeria responsator*, *Tetrastichus ayyari*

Source: Manjunath et al. 1999

FIGURE 33.2 *Netelia producta* (http://www.nbair.res.in).

FIGURE 33.3 *Ichneumon promissorius* (http://www.nbair.res.in).

with an average parasitism of 12.50% and 11.14%, respectively, were the most prevalent parasitoid species throughout the season (Gaur et al. 2015).

33.2.10 Lycaenid Pod Borer

Natural enemies of Lycaenid pod borer include parasitoids *Trichogramma* sp., *Tetrastichus* sp., *Telenomus* sp., *Chelonus* sp., *Campoletis* sp., *Bracon* sp., *Carcelia* sp., etc. Predators include Lacewing, ladybird beetle, spider, red ant, dragonfly, robber fly, reduviid bug, praying mantis, black drongo (King crow), wasp, common mynah, big-eyed bug (*Geocoris* sp), earwig, ground beetle, Pentatomid bug (*Eocanthecona furcellata*), etc.

33.2.11 Field Bean Pod Borer

Natural enemies of field bean pod borer include parasitoids *Trichogramma* sp., *Tetrastichus* sp., *Telenomus* sp., Chelonus sp., *Campoletis* sp., *Bracon* sp., *Carcelia* sp., etc. Predators include Lacewing, ladybird beetle, spider, red ant, dragonfly,

TABLE 33.2

List of Parasitoid Complex of *Maruca vitrata*

Parasitoid	Life Stage Parasitized	Reference
Diptera		
Tachinidae		
Aplomya metallica (Weid.)	Larva	Agyen-Sampong (1978)
Exorista xanthaspis (Wiedemann)	Larva	Barrion et al. (1987)
Palexorista solemnis (Walker)	Larva	Barrion et al. (1987)
Peirbaea orbata (Wiedemann)	Larva	Barrion et al. (1987)
Zygobothria atropivora (Rob.-Desv.)	Larva	Barrion et al. (1987)
Zygobothria ciliata (Wulp)	Larva	Barrion et al. (1987)
Thelairosoma sp.	Larva	Usua and Singh (1977)
Pseudopetichaeta laevis (Vill.)	Larva	Amen-Sampong (1978)
Pseudaporichaeta sp.	Larva	Usua and Singh (1977)
Thecocarcelia incedens (Rond.)	Larva	Agyen-Sampong (1978)
Hymenoptera		
Baraconidae		
Apanteles sp.	Larva	Okeyo-Owuor et al. (1991)
Bracon sp.	Larva	Okeyo-Owuor et al. (1991)
Braunsia sp.	Pupa	Okeyo-Owuor et al. (1991), Agyen-Sampong (1978)
Cardiochiles philippinensis Ashm.	Larva	Barrion et al. (1987)
Chelonus sp.	Larva	Barrion et al. (1987)
Cremnops sp.	Larva-Pupa	Barrion et al. (1987)
Snellenius manitae Ashm.	Larva	Barrion et al. (1987)
Phanertoma handecasisella Cam.	Larva	Subasinghe and Fellows (1978)
Phanertoma sp.	Larva	Usua and Singh (1977)
Chalcididae		
Antrocephalus sp., *nr subelongatus*	-	Subasinghe and Fellows (1978)
Antrocephalus sp.	Pupa	Okeyo-Owuor et al. (1991)
Bmchymetia sp. A.	Larva-Pupa	Barrion et al. (1987)
Brachymeria sp. B.	Larva-Pupa	Barrion et al. (1987)
Eulophidae		
Nesolynx thymus (Gir.)		Subasinghe and Fellows (1978)
Tetrastichus sesamiue Risbec	Pupa	Okeyo-Owuor et al. (1991)
Tetrasrichus sp.	Pupa	
Ichneumonidae		
Caenopimpla arealis (Cushman)	Larva	Barrion et al. (1987); Usua and Singh (1977)
Charops nigrita Gupta and Maheswary	Larva	Barrion et al. (1987); Usua and Singh (1977)
Meloboris sinicus (Holmgren)	Larva	Barrion et al. (1987); Usua and Singh (1977)
Metopius rufus hrowni Ashm.	Larva	Barrion et al. (1987); Usua and Singh (1977)
Pteromalidae		
Trichomalopsis sp.	Larva-Pupa	Barrion et al. (1987)
Scelionidae		
Telenomus sp.		Subasinghe and Fellows (1978)

robber fly, reduviid bug, praying mantis, King crow, wasp, common mynah, *Geocoris* sp., earwig, ground beetle, Pentatomid bug (*Eocanthecona furcellata*), etc. The percentages of parasitism of natural enemies on different legume crops are provided in Table 33.5.

33.3 Stored Legumes

Legume seeds are stored in godowns and warehouses in huge and minute quantities and are severely infested by bruchid beetles, *Callosobruchus chinensis* and *C. maculatus* (Tables 33.6 and 33.7).

33.4 Insect Growth Regulators (IGR)

Overlap of insect growth regulator at 5% cyromazine and dipping treatment at 30°C showed a noteworthy decline in the reproductive rate in *C. maculatus*. Moreover, escalating the cyromazine concentration led to a decrease in food burning up and an increase of generation's lifetime (Al-Mekhlafi et al. 2012).

TABLE 33.3

Diversity of Natural Enemy of Cowpea

S. No	Common Name	Scientific Name	Order/Family	Population Status	Host
1	Ladybird beetle	*Micraspis discolor* (Fab.)	Coleoptera: Coccinellidae	High	Egg, nymph, adult of jassid and white fly; Egg larva of Lepidoptera
2	Ladybird beetle	*Coccinella transversilis* (Fab.)	Coleoptera: Coccinellidae	Medium	Egg, nymph, adult of jassid and white fly; Egg larva of Lepidoptera
3	Ladybird beetle	*Brumoides* sp.	Coleoptera: Coccinellidae	Medium	Egg, nymph, adult of jassid and whitefly; Egg larva of Lepidoptera
4	Ladybird beetle	*Menochilus sexmaculatus* (Fab.)	Coleoptera: Coccinellidae	Low	Egg, nymph, adult of jassid and whitefly; Egg larva of Lepidoptera
5	Ladybird beetle	*Coccinella septempunctata* Linn.	Coleoptera: Coccinellidae	Low	Egg, nymph, adult of jassid and whitefly; Egg larva of Lepidoptera
6	Ladybird beetle	*Illeis indica* Timberlake	Coleoptera: Coccinellidae	Low	Egg, nymph, adult of jassid and whitefly; Egg larva of Lepidoptera
7	Rove beetle	*Paedurus* sp.	Coleoptera: Staphylinidae	Medium	Egg, nymph, adult of jassid and whitefly; Egg larva of Lepidoptera
8	Ground beetle	*Ophionea* sp.	Coleoptera: Carabidae	Medium	Egg, nymph, adult of jassid and whitefly; Egg larva of Lepidoptera
9	Yellow hornet	*Vespa* sp.	Hymenoptera: Vespidae	Low	Nymph, adult of jassid
10	Black ant	Unspecified	Hymenoptera: Formicidae	Low	Lepidopteran adults
11	Lynx spider	*Oxyopes* sp.	Araneae: Oxyopidae	High	Nymph, adult of jassid and whitefly; adult of Lepidoptera
12	Long jawed spider	*Tetragnatha* sp.	Araneae: Tetragnathidae	High	Do

Source: Singh and Singh 2014, Niba 2011.

TABLE 33.4

List of Hymenopteran Parasitoids on *M. obtuse*

Species	Super Family	Family	Reference
Euderus sp.	Chalcidoidea	Eulophidae	
Diglyphus funicularis Khan			Khan (1985)
D. mandibularis Khan			Khan (1985)
Tetrastichus atomella			Ipe (1987)
Euderus spp.			Sithantham et al. (1987)
Aprostocetus sp.			Narendran (2005)
O. fredreki			Peter (1992)
Ormyrus sp.			Ketipearachchi (2002)
Eurytoma sp.		Eurytomidae	Ipe (1987)
E. robusta Mayr			Sithanantham et al. (1987)
Plutarchia indefensa (Walker)			Sithanantham et al. (1987)
Microdontomerus (=*Antistrophoplex*) sp.			Sithanantham et al. (1983)
Pseudotorymus (=*Senegalella*) sp.			Singh et al. (1991)
Callitula sp.		Pteromalidae	Makinson et al. (2005)
Eupelmus sp.		Eupelmidae	Thakur and Odak (1982)
Bracon sp.	Ichneumonoidea	Braconidae	Sah and Mehra (1986)
Tricopria sp.	Proctotrupoidea	Diapriidae	Thakur and Odak (1982)
Omytes sp.	Unreported	Unreported	Rao and Babu (2009)

TABLE 33.5

Percentage of Parasitism of Natural Enemies on Different Legume Crops

Crop	Pest	Parasitoid/Predator	Percentage Parasitism	References
Soybean	*Helicoverpa* spp.	Nabidae	50% predation of eggs and larvae	Pfannenstiel et al. (2002)
Pigeonpea	*Helicoverpa armigera*	*T. chilonis*	1.2% to 8.3% on pods and on leaves parasitism varied from 5.0% to 29.0%	(Tandon and Bakthavatsalam, 2005)
Chickpea	*Helicoverpa armigera*	*C. chloridae*	65.0 to 75.0% on larval parasitism	Jagdish Jaba and Agnihotri M (2016)
Pigeonpea	*Maruca vitrata*	Apanteles sp.	Parasitism high as 63% on *M. vitrata* larvae	(Huang et al. 2003)
Pigeonpea	*Maruca vitrata*	Trichogrammatoidea eldanae	< 50% of eggs parasitism	Belmain, S.R. et al. (2013)
S. cannabina	*M. vitrata*	*Apanteles taragamae*	Parasitism reached as high as 63% of *M. vitrata* larvae found attacking	Chi-Chung Huang et al. (2003)
Soybean and pigeonpea	*Etiella zinckenella*	Braconid, *Iconella* (=Apanteles) *etiellae* (Vier.)	3.4%	Alejandro E et al. (1988)
Pigeonpea	*Melanagromyza obtusa*	Three hemenopteran parasitoids *viz., Systasis dasyneurae* Mani, *Torymus* sp. Dalman (ectolarval parasitoids) and *Epitranus* sp. Walker (endopupal parasitoid)	The natural parasitism of these parasitoids ranged from 1.69 to 7.32, 5.08 to 29.27 and 2.13 to 4.88%, respectively	Chiranjeevi, B et al. (2017)
Cowpea	*H. armigera*	*Trichogramma* spp.	37.0% level of parasitism	Sithanantham et al. (1983)
Green gram	*H. armigera*	*Trichogramma* spp.	17.4	Pawar et al. (1986)
Pigeonpea	*H. armigera*	*Trichogramma* spp.	0.2	Pawar et al. (1986)
Groundnut	*H. armigera*	*Trichogramma* spp.	32.3	Sithanantham et al. (1983)

TABLE 33.6

Natural Enemies of Pulse Beetles

Natural Enemy	Description
Dinarmus acutus, D. basalis, D. vagabundus (Pteromalidae: Hymenoptera)	It is a solitary parasitoid that attacks larval, pre-pupal, and pupal stages of *Callosobruchus* sp. (Islam and Kabir 1995)
Pteromalus cerealella, P. tritici	It was thought to be a monophagous parasitoid of Angoumois moth (Fulton 1933). Later it was highly successful in killing *Callosobruchus maculates* (F.) as to other storage insects
Anisopteromalus calandrae	It is an ectoparasitoid on the larvae and pupae of pulse beetles *C. maculates* (F.) and this parasitoid preferred younger larvae of the pest
Uscana lariophaga, U. mukherjii (Hymenoptera: Trichogrammatidae)	Egg parasitoids: *U. lariophaga* females were able to find hosts up to 75 cm horizontal distance from the host patch, which was the largest distance tested (Stolk et al. 2005)
Eupelmus orientalis, E. vuilleti	Parasite larvae/pupae (Anonymous 2014)
Lariophagus distinguendus	Parasite larva (Anonymous 2014)
Heterospilus prospidis	Parasite larva (Anonymous 2014)
Cheyletus eruditus	Predator (Anonymous 2014)
Entomopathogenic fungus	1. *Beauveria bassiana* (Vanmathi et al. 2011); Shaheen et al. 2016) 2. An isolate (CA-2) of *B. brongniartii* (E-9) of *Metarhizium anisopliae* (Rodrigues and Pratissoli 1999)
Entomopathogenic nematodes	1. *Heterorhabditis bacteriophora, Steinernema siamkayai*, and *S. pakistanense* were among those nematodes that showed the highest virulence 2. The last larval stage of the pulse beetle seems to be more susceptible than the adult (Shahina and Salma 2009)

33.5 Biopesticides

The term biopesticides defines compounds that are used to manage agricultural pests by means of specific biological effects rather than as broader chemical pesticides. Globally, though biopesticides cover only about 1% of the total plant protection products, their number and the growth rate have been showing an increasing trend in the past two decades (Ranga Rao et al. 2007). In India, they represent only 2.89% of the overall pesticide market and are expected to exhibit an annual growth rate of about 2.3% in the coming years (Mazid et al. 2011).

33.5.1 Baculovirus

Baculovirus is one of the microbial pesticides that have a very narrow host range and generally infests the larvae of crop pests.

TABLE 33.7
Active Ingredients of Plant Extracts of Pulse Beetles

Name of the Plant	Potential Active Ingredients	References
Neem (*Azadirachta indica*)	Tetranortriterpenoids (Azadirachtin, melantriol, salannin, nimbin, nimbidin	Reddy (2010)
Karanj (*Pongammia pinnata*)	Furanoflavonoid (karanjin, pongamol, pongapin, glabrin, karanja chromene, karanjone and pongaglabrone)	Bringi (1987)
Castor (*Ricinus communis*)	Ricin and ricinine	Ramos-López (2010)
Clove (*Syzygium aromaticum*)	Eugenol (4 allyl-2-methoxy phenol; C10H12O2), an organic phenol compound	Shapiro (2012)
Mahua (*Madhuca latifolia*)	Saponin, an alkaloid glucoside (leaves), Sapogenin and other basic acid (seeds)	Sunitha and Sarojini (2013)
Mustard (*Brassica* sp.) Horseradish (*Armoracia rusticana*) Wasabi (*Eutrema japonicum*)	Allyl isothiocyanate (AITC)	Yu et al. (2003)
Soybean (*Glycine max*)	Emulsifiable concentrate formulations of soybean oil	EPA (1993)

FIGURE 33.4 NPV infected pod borer *H. armigera* larva (Source: ICRISAT).

Nuclear polyhedrosis viruses (NPVs) is a nucleic acid (double standard, circular DNA) enclosed in protein matrix, hence it is called polyhedral occlusion body (POB). NPV infects the nucleus of the cell and multiplies within the nucleus. Nuclear polyhedrosis viruses like Ha NPV and Sl NPV are increasingly being used as alternatives to chemicals for the control of polyphagous pests of pulses like gram pod borer (*Helicoverpa armigera*) and tobacco caterpillar (*Spodoptera litura*), respectively (Figure 33.4).

The majority of the commercially formed fungi (*Beauveria*, *Metarhizium*, *Lecanicillium*, and *Verticillium*) have evolved focused mechanisms for the enzymatic deprivation of the integument and for overcoming insect guard compounds; among them *Beauveria* and *Metarhizium* are used for the control of pulse pests like *Helicoverpa armigera*, *Spodoptera litura*, thrips, whiteflies, aphids, and mites (Figure 33.5).

33.5.2 Bacteria

Bacillus thuringiensis (Bt) has been known to be a pool of numerous insecticidal proteins, such as δ-endotoxins, cytolytic proteins, vegetative insecticidal proteins, etc. Among these, δ-endotoxins have been more efficiently utilized for protection of a variety of crops from various insect pests.

33.5.3 Plant-Incorporated Protectants (PIPs)

Limited success of conventional breeding and use of hazardous chemicals have directed to the production of transgenic chickpea plants expressing Bt genes resistance against pod borer, *H. armigera* (Jalu and Panera 2018). A few studies also conclude that *B. thuringiensis* is more effective against pulse beetle, *C. maculatus*, and could be used as one of the potential biocontrol agents in the management of stored product insect pests in the future (Malaikozhundan and Vinodhini 2018).

33.5.4 Biochemicals

Herbal pesticides/biochemicals are concentrates of biologically effective plant items like leaves, stem, roots, and entire plants particularly for the control of soft-bodied insect pests, which feed on the leaves and tender plant parts as flowers and developing grains and so forth. It contains Azadirachtin, a tetranortritarpinoid, a major active ingredient, which is known to disrupt the metamorphosis of insects. Neem Seed Kernel Extract reduces the larval population of *H. armigera* in chickpea and pod damage. Similarly, aqueous leaf extract of *Catharanthus roseus* spray on blakgram also reduced its population (Rajasekaran et al. 1987). Karanja oil (2%) was reported to prolong its larval

FIGURE 33.5 *Helicoverpa armigera* larvae infected with *B. bassiana*. Source: ICRISAT.

development and growth-inhibiting activity (Bajpai and Sehgal 1994). Nicotine sulfate isolated from tobacco leaves was found to be highly toxic to this borer (Patel et al. 1990).

33.6 Pheromones

Sex pheromone-baited traps can be used in the fields to capture males of newly emerged moths and reduce the number of males for mating. Sex pheromone of tobacco caterpillar (*Spodoptera litura*) at 10–12 traps/ha have been recommended for the monitoring of the pests of legumes. Sex pheromone trap for gram pod borer (*Helicoverpa armigera*) at 10–12 traps/ha have been suggested for the monitoring of the pulse crop pests like chickpea, soybean, etc.

33.7 Conclusion and Future Prospects

Grain legumes continue to be important crops not only from a nutritional security perspective, but also for their involvement in the health and fertility of the soil. Cultivation of these crops faces a number of biotic and abiotic stresses, reflected in the form of deteriorated yield and quality. Insecticide misuse has led to increased levels of resistance to insecticides, emergence of secondary pests, loss of biodiversity, and a rise in human health hazards. Integrated pest management (IPM) can help to minimize insecticide use. In recent years, pest management strategies are mostly concentrated on integrating various options such as resistant varieties, biological, mechanical, biopesticides, insect pheromones technology, and chemical control. The role of natural enemies in the management of major insect pests has been underestimated until now. The importance of mixed crops, crop rotations, sowing dates, although very well understood, have not been fully exploited.

REFERENCES

Al-Mekhlafi, F.A.; Ali Mashaly, A.M.; Wadaan, M.A.; Al-Mallah, N.M. Overlap effects of cyromazine concentration, treatment method and rearing temperature on southern cowpea weevil *Callosobruchus maculatus* reared on mung bean. *Pakistan Journal of Zoology*. 2012, 44(1), 285–290.

Alejandro, E.; Segarra, C.; Pedro, B. Notes on the natural enemies of *Etiella zinckenella* in Puerto Rico. *The Journal of Agriculture of the University of Puerto Rico*. 1988, 72(1), 153–159.

Bajpai, N.K.; Sehgal, V.K. Effect of neem products, nicotine and karanj on survival and biology of pod borer, Helicoverpa armigera Hubn. of chickpea. Proc. II AZRA Conf. on Recent Trends in Plant,Animal and Human Pest Management: Impact on Environment. Madras Christian College, Madras, 27–29 Dec,1994 pp:48.

Barrion, A.T.; Bandong, J.P.; De la Cruz, C.G.; Apostol, R.F.; Litsinger, J.A. Natural enemies of the bean pod-borer, *Maruca testulalis* in the Philippines. Tropical Grain Legume Bulletin. 1987, 34, 21–22.

Belmain, S.R.; Haggar, J.; Holt, J.; Stevenson, P.C. *Managing Legume Pests in Sub-Saharan Africa: Challenges and Prospects for Improving Food Security and Nutrition Through Agro-Ecological Intensification*. Chatham Maritime, UK: Natural Resources Institute, University of Greenwich, 2013, 34p.

Bhatnagar, V.S.; Sithanantham, S.; Pawar, C.S.; Jadhav, D.S.; Rao, V.K.; Reed, W. Conservation and augmentation of natural enemies with reference to IPM in chickpea and pigeonpea. In Proceedings of the International Workshop on Integrated Pest Control in Grain Legumes, Goiania, Brazil, 4–9 April 1983, 157–180.

Bringi, N.V. *Non-traditional Oil Seeds and Oils in India*. New Delhi: Oxford and IBH, 1987.

Chiranjeevi, B.; Patange, N.R. New host record of three hymenopteran parasitoids associated with *Melanagromyza obtusa* (Malloch) and their level of natural parasitism. *Journal of Entomology and Zoology Studies*. 2017, 5(4), 277–283.

Clement, S.L.; Wightman, J.A.; Hardie, D.C.; Bailey, P.; Baker, G. McDonald, G. Opportunities for integrated management of insect pests of grain legumes. In: Knight R. (ed.), *Linking Research and Marketing Opportunities for Pulses in the 21st Century*. Dordrecht: Springer, 2000, pp. 467–480.

Environmental Protection Agency, Prevention, Pesticides and Toxic substances (7508W), EPA-738-F-93-027. December, 1993, 1–4.

Fulton, B.B. Notes on *Habrocytes cerealellae*, parasite of the Angoumois grain moth. *Annals of Entomological Society of America*. 1933, 26, 536–553.

Gaur, N.; Sharma, P.; Nautiyal, A. Seasonal incidence of major insect-pests of soybean and their correlation with abiotic factors. *Journal of Hill Agriculture*. 2015, 6, 75–78.

Hormazabal, R.L.; Gerding, P.M. Release density of *Uscana senex* Grese (Hymenoptera: Trichogrammatidae) for control of *Bruchus pisorum* L. (Coleoptera: Bruchidae). *Agro-Ciencia*, 1998, 14, 157–161.

Huang, C-C.; Peng, W-K.; Taleka, N.S. Parasitoids and other natural enemies of *Maruca vitrata* feeding on *Sesbania cannabina* in Taiwan. *BioControl*. 2003, 48(4), 407–416.

International Crops Research Institute for the Semi-Arid Tropics. *The Medium-term Plan*. Patancheru: ICRISAT, 1992.

Ipe, M. Biosystematic studies on Agromyzidae from India. *Proceedings of the* Indian *Academy of Sciences* (*Animal Sciences*). 1987, 96, 573–581.

Islam, W.; Kabir, S.M.H. Biological control potential of *Dinarmus basalis* (Rond.) (Hymenoptera: Pteromalidae) larval-pupal ectoparasitoid of the pulse beetle, *Callosobruchus Chinensis* (L.). *Crop Protection*. 1995, 14(6), 439–443.

Jaba, J.; Agnihotri, M.; Chakravarty, S. Biology and relative parasitization of larval endoparasitoid *Campoletis cholrideae* Uchida on Heliocoverpa armigera Hübner under sole and chickpea-coriander intercropping system. *Journal of Biological Control*. 2016, 30(2), 84–90.

Jalu, R.K.; Panera, A.V. Transgenic Bt. chickpea: India's most required GM pulse crop. *The Pharma Innovation Journal*, 2018, 7(8): 551–552.

Karungi, J.; Adipala, E.; Nampala, P. Pest management in cowpea. Part 3. Quantifying the effect of field pests on grain yields in eastern Uganda. *Crop Protection*. 2000, 19, 343–347.

Ketipearachichi, Y. Hymenoptera parasitoids and hyper parasitoids of crop pests at Aralaganurla in north central province of Sri Lanka. *Annals of the Sri Lanka Department of Agriculture*, 2002, 4, 293–306.

Khan, M.A. Eulophid parasites (Hymenoptera: Eulophidae) of Agromyzidae in India. *Journal of the Bombay Natural History Society*. 1985, 82(1), 149–159.

Makinson, J.; Goolsby, J.; Kirk, A.; Meyerdirk, S. A new record and host association for the pigeon pea podfly, *Melanagromyza obtusa* (Malloch) (Diptera: Agromyzidae) and notes on its parasitoids in the Northern Territory, Australia. *Australian Entomologist*. 2005, 32, 79–82.

Malaikozhundan, B.; Vinodhini, J. Biological control of the Pulse beetle, *Callosobruchus maculatus* in stored grains using the entomopathogenic bacteria, *Bacillus thuringiensis*. *Microbial Pathogenesis*. 2018, 114, 139–146.

Manjunath, T.M.; Bhatnagar, V.S.; Pawar, C.S.; Sithanantham, S. Economic importance of *Heliothis* sp. in India and an assessment of their natural enemies and host plants. In: King E.G.; Jackson R.D. (eds.), Proceedings of the Workshop on Biological Control of *Heliothis*: Increasing the Effectiveness of Natural Enemies. Far Eastern Regional Research Office, United States Department of Agriculture, New Delhi, 1999, pp. 197–228.

Mazid Salma.; Jogen Ch. Kalita.; Ratul Ch. Rajkhowa. A review on the use of biopesticide in the insect pest management. *International Journal of Science and Advanced Technology*. 2011, 1(7), 169–178.

Miller, P.R.; McConkey, B.G.; Clayton, G.W.; Brandt, S.A.; Staricka. J.A.; Johnston, A.M.; Lafond, G.P.; Schatz, B.G.; Baltensperger, D.D.; Neill, K.E. Pulse crop adaptation in the Northern Great Plains, *Agronomy Journal*. 2002, 94, 261–272.

Narendran, T.C. Two new species of *Tetrastichus haliday* (Hymenoptera: Eulophidae) displaying some characters of Aceratoneuromyia Girault. *Bulletin of Pure and Applied Sciences (A)*. 2005, 24(1), 37–42.

Niba, A.S. Arthropod assemblage dynamics on cowpea, *Vigna unguiculata* (L) Walp in subtropical agroecosystem, South Africa. *African Journal of Agricultural Research*. 2011, 6, 1009–1015.

Okeyo-Owuor, J.B.; Oloo, J.B.; Agwaro, P.O. Natural enemies of legume pod-borer, Murucu *testululis* (Geyer) (Lepidoptera: Pyralidae) in small farming systems of western Kenya. *International Journal of Tropical Insect Science*.1991, 12, 35–42.

Patel, H.R.; Patel, D.J.; Patel, C.C.; Thaker, N.A. Affectivity of *Clerodendron inerme* Linn. and Periwinkle, *Catharanthus roseus* (Linn.) G. Don. for management of root-knot nematodes in okra. Presented in National Symposium on Problems and Prospects of Botanical Pesticides. CTRI, Rajahmundry, India, 1990, pp. 40–41.

Pawar, C.S.; Bhatnagar, V.S.; Jadhav. D.R. *Heliothis* species and their natural enemies, with their potential for biological control. *Proceedings: Animal Sciences*. 1986, 95(6), 695–703.

Peter, C. A note on the parasitoid fauna associated with red gram pod fly, *Melanagromyza obtusa* Malloch. *Journal of Insect Science*. 1992, 5, 88.

Pfannenstiel, R.S.; Yeargan, K.V. Identification and diel activity patterns of predators attacking *Helicoverpa zea* (Lepidoptera: Noctuidae) eggs in soybean and sweet corn. *Environmental Entomology*. 2002, 31(2), 232–241.

Pintureau, B.; Gerding, M.; Cisternas, E. Description of three new species of Trichogrammatidae (Hymenoptera) from Chile. *The Canadian Entomologist*. 1999, 131, 53–63.

Rajasekaran, B.; Jayraj, S.; Raghuramman, S.; Narayan swamy, T. Use of neem products for the management of certain rice pests and diseases. In: *Mid Term Appraisal Works on Botanical Pest Control of Rice*, T.N.A.U, Coimbatore, India 1987. p 13.

Ramos-López, M.A.; Pérez, S.; Rodríguez Hernández, G.C.; Guevara Fefer, P.; Zavala Sánchez, M.A. Activity of *Ricinus communis* (Euphorbiaceae) against *Spodoptera frugiperda* (Lepidoptera: Noctuidae). *African Journal of Biotechnology*. 2010, 9(9), 1359–1365.

Ranga Rao, G.V.; Rupela, O.P.; Rameshwar Rao, V.; Reddy, Y.V.R. Role of Biopesticides in crop protection: Present status and future prospects. *Indian Journal of Plant Protection*. 2007, 35(1), 1–9.

Rao, A.G.; Babu, T.R. Biological control of Arthopod pest of pulse crops. In: Upadhyay, R.K.; Mukerji, K.G.; Chandola, B.P.; Dubey, O.P. (eds.), *Integrated Pest and Disease Management*. New Delhi: APH Publishing Corporation, 2009, p. 247.

Reddy DS. Chapter 17: Botanicals. In: *Applied entomology, Pusa Agricultural Book Service*, New Delhi 2010, 109.

Rodrigues, C.; Pratissoli, D. Pathogenicity of *Beauveria brongniartii* (Sacc.) petch and *Metarhizium anisopliae* (Mots.) sorok and their effect on the corn weevil and the bean beetle. *Anais da Sociedade Entomológica do Brasil*. 1999, 19(2), 301–306.

Sah, B.N.; Mehra, B. P. New record of the parasites of *Melanagromyza obtusa* Malloch in Ranchi, India. *Indian Journal of Entomology*. 1986, 348, 359.

Sampong, A. Pests of cowpea and their control in Ghana. In: Singh, S.R.; van Emden, H.F.; Taylor, J.A. (eds.), *Pests of Grain Legumes: Ecology and Control*. London: Academic Press, 1978, pp. 85–92.

Saxena, K.B.; Chandrasena, G.D.S.N.; Hettiarachchi, K.; Iqbal, Y.B.; Fonseka, H.H.D; Jayasekara, S.J.B.A. Evaluation of pigeon pea accessions and selected lines for reaction to Maruca. *Crop Science.* 2002, 42, pp. 615–618.

Shaheen, F.A.; Akram, M.W.; Rashid, M.A.; Nadeem, M.; Saeed, M.; Husain, M.; Khalid, M. Resistance of chickpea (Cicer arietinum L.) cultivars against pulse beetle. Pak. J. Bot. 2006, 38: 1237–1244.

Shahina, F.; Salma, J. Laboratory evaluation of seven Pakistani strains of entomopathogenic nematodes against a stored grain insect pest, Pulse beetle *C. chinensis* (L.). *Journal of Nematology.* 2009, 41, 255.

Shapiro, R. Prevention of vector transmitted diseases with clove oil insect repellent. *Journal of Pediatric Nursing.* 2012, 27(4), 346–349.

Sharma, H.C. Strategies for Heliothis/Helicoverpa management: emerging trends and strategies for future research. New Delhi: Oxford and IBH, 2005.

Sharma, O.P.; Gopali, J.B.; Yelshetty, S. *Pests of Pigeon Pea and Their Management.* New Delhi: National Centre for Integrated Pest Management, IARI Campus, 2010.

Singh, C.; Singh, N.N. Occurrence of insect-pests infesting cowpea (*Vigna unguiculata* walpers) and their natural enemy complex in associations with weather variables. *Legume Research,* 2014, 37(6), 658–664.

Singh, S.; Singh, S.P.; Babu, S.S.; Sebastian, P.C. Bioecology of Hymenoptera parasites of Agromyzidae (Diptera) pests species in India. Memorial School Of Entomology. 1991, 11, 1–238.

Sithanantham, S.; Rao, V.R.; Reed, W. Parasites of the pigeon pea pod fly, *Melanagromyza obtusa* in India. *Journal of Biological Control.* 1987, 1(1), 10–16.

Sithanantham, S.; Rao, V.R.; Reed, W. Survey of pigeon pea pod fly in India. *International Pigeon Pea News Letter.* 1983, 2, 66–68.

Song, F.; Swinton, S.M.; DiFonzo, C. *Profitability Analysis of Soybean Aphid Control Treatments in Three Northcentral States.* Staff Paper 24. Lansing, MI: Department of Agricultural Economics, Michigan State University, 2006.

Stolk, C.; Ghimire, M.N.; Souquie, S.; van der Werf.; van Huis A. Host finding by Uscana lariophaga (Hymenoptera: Trichommatidae) in stored cowpea: the effect of distance, time interval, host patch size and spatial orientation. Bulletin of Entomological Research. 2005, 95(3):231–241.

Subasinghe, S.M.C.; Fellows, R.W. Recent trends in grain legume pest research in Sri Lanka. In Singh, S.R.; van Emden, H.F.; Taylor, J.A. (eds.), *Pests of Grain Legumes: Ecology and Control.* London: Academic Press, 1978, pp. 37–41.

Sunitha, M.; Sarojini, P. *Madhuca Lonigfolia* (Sapotaceae): A review of its traditional uses and nutritional properties. *International Journal of Humanities and Social Science Invention.* 2013, 2(5), 30–36.

Tandon, P.L.; Bakthavatsalam, N. Electrophysiological and olfatometric responses of *Helicoverpa armigera* and *Trichogramma chilonis* Ishii, to volatiles of trap crops *tagetes erecta* L. and *Solanum viarum* Dunal. *Journal of Biological Control.* 2005, 19, 9–16.

Thakur, B.S.; Odak, S.C. New record of the parasites of *Melanagromyza obtusa* Malloch. *Science and Culture.* 1982, 48, 80.

Tilmon, K.J.; Hodgson, E.W.; O'Neal, M.E. Biology of the soybean aphid, Aphis glycines (Hemiptera: Aphididae) in the United States. *Journal of Integrated Pest Management.* 2011, 2(2), 1–7.

Usua, E.J.; Singh, S.R. Parasites and predators of the cowpea podborer, *Maruca testuhlis* (Lepidoptera: Pyralidae). *Nigerian Journal of Entomology.* 1977,1, 100–102.

Vanmathi Shifa, J.; Padma Latha, C.; Ranjit Singh, A.J.A. Impact of entomopathogenic fungus, *Beauveria bassiana* on stored grains pest, *Callosobruchus maculates* (F.) (Coleoptera: Bruchidae). *Journal of Biopesticides.* 2011, 4(2), 194–197.

Wang, X.B.; Fang, Y.H.; Lin, S.Z. A study on the damage and economic threshold of the soybean aphid at the seedling stage. *Plant Protection.* 1994, 20, 12–13.

Yadav, A.K.; Yadav, S.; Singh, M.K. Effect of temperature on the population of parasitoids and their impact on the pest Melanagromyza obtusa (Diptera:Agromyzidae). Journal of Environmental Science. 2012, 1: 45–50.

Yu, J.C.; Jiang, Z.T.; Li, R.; Chan, S.M. Chemical composition of the essential oils of *Brassica juncea* (L.) Coss. grown in different regions, Hebei, Shaanxi and Shandong, of China. *Journal of Food Drug Analysis.* 2003, 11, 22–26.

34

Conservation Biological Control: Definitions, Bases, and Practices

Simone Mundstock Jahnke and Gisele De Souza Da Silva

CONTENTS

34.1 Introduction

The diversification of agricultural environments enables an increase in associated communities, through the implementation of plant architectures that support populations of natural enemies (NE). Named "conservation biological control" (CBC), this type of management is commonly based on managing vegetation patterns at the local scale (e.g., flowering strips that promote parasitoids by supplying nectar) or at wider scale (e.g., woodland to serve as donor habitat for NE). Such vegetation management also offers scope to provide agriculture with additional ecosystem services as diverse as pollination and carbon sequestration. These aspects point to promising results from the use of conservation biological control as a tool for agriculture that, in addition to being socially and economically sustainable, can minimize environmental impacts and assist in the preservation of the natural environment.

34.2 Theoretical Ecological Bases of Conservation Biological Control

We will start by thinking about the "Island Biogeography Theory" (IBT), which was proposed by MacArthur and Wilson (1967) and addresses the "species-area relationship," between species richness and size of the areas of islands or habitat patches. In addition, a "colonization-extinction" model simulates a simple system in which inputs and outputs are balanced (Towsend et al., 2006). Some factors can increase extinction rates on an "island," quasi: reduction of habitat heterogeneity; intensification of specific inter and intra competitions due to scarcity of resources; secondary extinctions due to the disappearance of key species; increased risk of stochastic extinctions. In a general sense, IBT is applied to oceanic islands or to continental "islands," understood as fragments of native vegetation surrounded by agricultural land or urban centers (Laurance, 2008).

We will, however, take a different perspective. Imagine that a certain agricultural area is an "island." Regarding ecological processes, this "island" can be seen as a recent ecosystem, in the early stages of succession, where net primary productivity is maximum and diversity is minimal (Gliessman, 2001). This is truer in annual crops, of large monocultures. The concept of metapopulation (Levins, 1969) is defined as a network formed by populations spatially structured in clusters, whose individuals reproduce locally and where migration between populations can influence local dynamics (Hanski, 1994; With, 2004). Therefore, models of metapopulation dynamics become important tools for conservation biological control.

First it is necessary to reduce the agroecosystem isolation from wild areas or with more diversified vegetation and that can be repositories of NE (Ferreira et al., 2014; da Silva et al., 2016).

Within the agricultural system, numerous ecological interactions will occur and biodiversity has a fundamental role. For Gliessman (2001) and Altieri (2012), the components of biodiversity in an agroecosystem can be considered according to the role they play in their functioning.

Service provider organisms will colonize the agroecosystem from the surroundings and will develop in the system depending on its management and structure (Gliessman, 2001). One way used to assess the effect of habitat management on CBC is the analysis of comparative diversity, tools used in community ecology(Magurran, 2011). Measuring the biodiversity of NE and comparing between agricultural systems with different managements or between adjacent natural systems has been used to understand structural and functional complexities of agroecosystems. This is done using diversity indices, comparative diversity (beta), and richness estimators, among others. An example is the work that evaluated the predator diversity of an irrigated rice organically production agroecosystem and a preserved area (Figure 34.1) (Ferreira et al., 2014). The biggest richness of the preserved area can be an NE repository for the rice field.

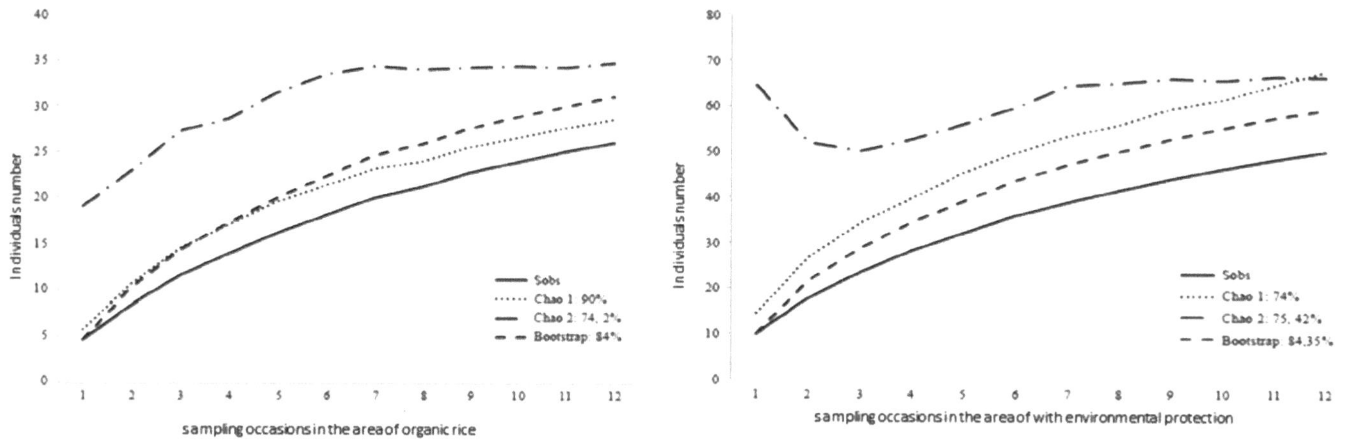

FIGURE 34.1 Morphospecies curve for three estimators (chaol, chao 2, and bootstrap 200 times randomized) and collector curve (Sobs): (A) in organic cultivation of irrigated rice and (B) protected area Refugio Banhado dos Pachecos, from May 2011 to April 2012.Viamão, RS. (Adapted from Ferreira et al., 2014.)

The study of ecological networks has emerged as a new tool to assess interactions between different trophic levels. Usually, networks are built by links and are important summaries of patterns of organization in biological communities (Morin, 2011). Pacheco et al. (2018) evaluated the interaction between Noctuidae larvae and their parasitoids in crops in Canada. The constructed network indicates the interaction between two lepidopterans' hosts present in different plants and the associated parasitoids (Figure 34.2).

34.3 CBC in Conventional Cultivation Systems

The conservation of natural enemies has been considered a challenge, especially for producers who follow the model of conventional industrialized agriculture. The challenge lies in the inability of NE to establish and perform well in fragmented landscapes, crops with high pesticide residue and/or in monocultures that lack alternative food or shelter. Even so, in conventional arming, the idea of Integrated Pest Management (IPM) advocates the use of different control methods (Norris et al., 2003), associating biological control. Developing reliable tactics to address this problem requires an interdisciplinary approach involving knowledge of conservation, trophic, community, and chemical ecology.

Soil management in conventional crops involves tillage, addition of chemical fertilizers, maintenance of Hydrogen Potential (pH), and other practices and can affect both pests and NE (Van Driesce & Bellows, 1996). A study evaluated short- and long-term effects of pesticides in golf turfs and recorded turfs that regularly received high rates of pesticides exhibited suppression in soil biological process while the beneficial biota seemed to tolerate only a low level of pesticide intake (Gan & Wickings, 2017).

Disturbances by machinery or environmental stress conditions (drought, flood, snow, among others), can raise dust and act on plants in nearby crops. Dusty leaves are often neglected by foraging parasitoids (Van Driesce & Bellows, 1996), possibly because they alter volatile organic compounds emitted by the plant or the host. In predatory mites of the Phytoseiidae family, on the other hand, the presence of dust in the phylloplane does not influence predation on Pacific spider mites (Oi & Barnes, 1989).

Water management also influences the presence of NE. Liu et al. (2014) showed that high relative humidity RH% may cause a decrease in the antioxidant system of *Spodoptera exigua* (Hübner) larvae, which may explain high RH% is always associated with high mortality in insects due to infection with pathogens.

The management of crop residues affects populations of decomposers and can impact generalist predators through trophic cascades (Halaj & Wise, 2002). Sereda et al. (2015) evaluated the presence of Collembola and predators in the soil in conventional and organic corn and wheat crops and found significantly more NE, (beetles and spiders), in plots with litter and under organic agriculture. Even so, in the conventional property, the presence of residues increased the presence of predators.

For natural enemies, pesticides can have direct effects, such as mortality or indirect, reducing longevity, fertility, causing repellency or loss of hosts due to changes in chemical cues or synchronization with prey or hosts (Van Driesce & Bellows, 1996).

At this point, IPM practices recommend the decrease in applications, and the selectivity of pesticides to NE (Crócomo, 1990). The decrease in the number of applications is related to the monitoring of pests by sampling for decision making according to the level of pest control (Pimentel & Peshin, 2014). As for selectivity, a working group linked to the International Organization for Biological Control (IOBC) has conducted studies to develop methodological standards for laboratory, semi-field, and field tests on the effect of insecticides on beneficial organisms, seeking to select selective products and resistant NE (Morales-Ramos et al., 2014).

Ecological selectivity is characterized by differences in niches and behaviors between pests, NE, and pollinators. This can be obtained through the time of application of the product, place of application, periods of less occurrence of NE (Foerster, 2002).

Physiological selectivity refers to the product's action being more effective on the pest and not on the NE when both come into contact with the insecticide or its residues, for example by reducing the absorption of the chemical through the integument or by increasing the degradation of the toxic substance by the enzymatic system (Pedigo, 1988). Heimpel and Mills (2017) warn of the fact that attention should also be paid to sublethal effects and that the importance of pesticide-induced disruption of biological control in conventional management systems cannot be underestimated.

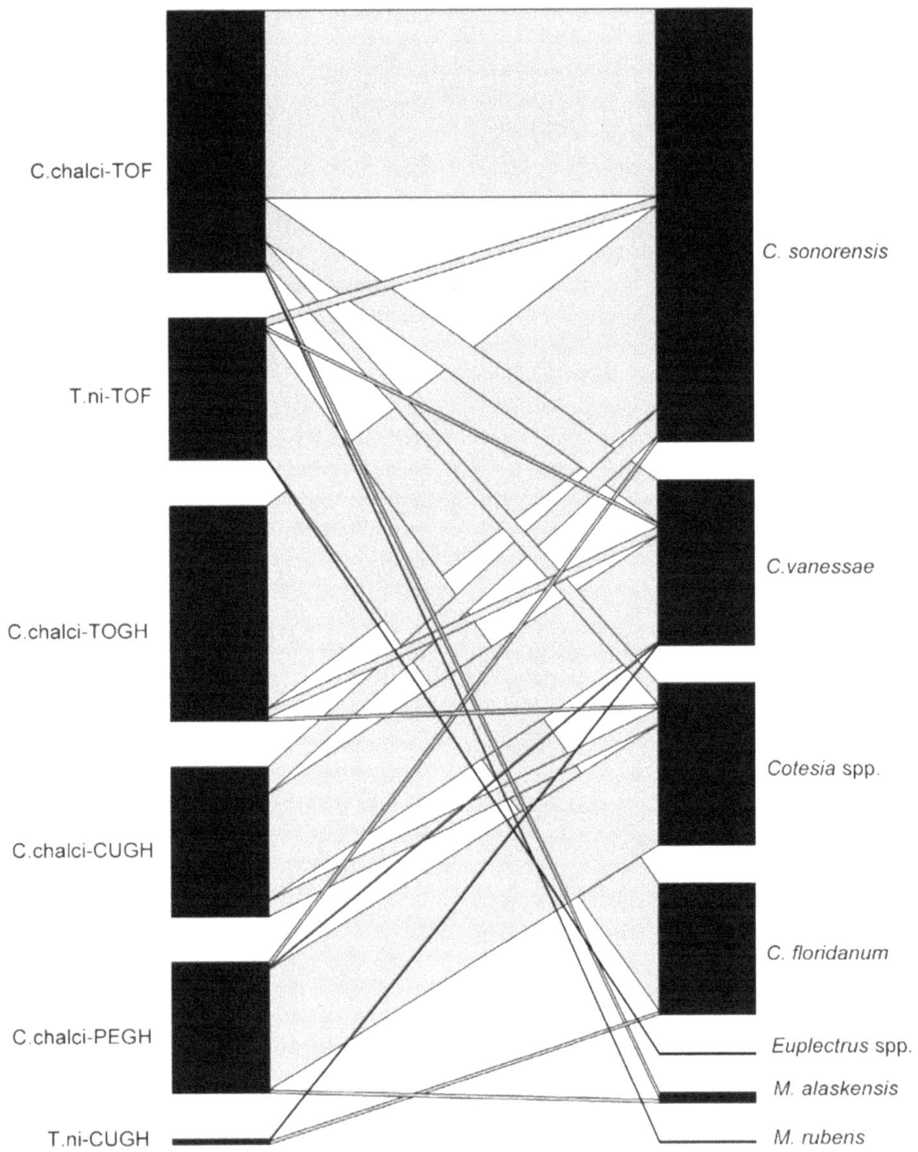

FIGURE 34.2 Food network (left) of phytophagous hosts *Chrysodeixis chalcites* (*C. chalci*) and *Trichoplusia ni* (*T. ni*), plants and agroecosystems (field or greenhouse crops) and (right) their parasitoids during the 2009–2010 seasons in tomato (TO), cucumber (CU) and pepper (PE) samples in the field (F) and greenhouse (GH). The width of the bars is a proportional representation of the abundance of host and parasitoid species in a specific host plant and agroecosystem. (Modified from Pacheco et al., 2018 – free access.)

34.4 The Role of Wild Vegetation in Biological Control

Through agricultural practices guided by the knowledge of ecological processes and looking to achieve characteristics similar to those of natural ecosystems, the sustainability of the food system can be promoted (Gliessman, 2001; Altieri, 2012). No other aspect of agricultural systems provides as many fundamental ecological services to ensure the protection of plants against insect pests as the diversity of vegetation (Altieri et al., 2003).

The ecological processes of a food web comprise a relationship between landscape structure and control of pest organisms within a multitrophic-level context (Pywell et al., 2011). Beyond the plants, herbivores, and natural enemies, we have microorganisms involved possibly respond to landscape heterogeneity at very different spatial scales (Tscharntke et al., 2005). Microorganisms in noncrop areas can be associated with both phytophagous and NE (Ekesi et al., 2005) exerting together with other factors for the biological control of local pests. Some aphid entomopathogenic fungi species, for example, are known to overwinter in hosts in hedges and forest borders (Nielsen et al., 2001) or in grass and legume field margins and woodlands (Pell et al., 2010), showing a complex interaction within NE, phytophagous insects, and microorganisms.

Natural enemies must have a rapid numerical response to herbivore density for effective pest control and so, they must be present in the crop habitat and nearby areas (Tscharntkeet al., 2007). Many species of NE benefit from frequent movement between crop and noncrop habitats, because they require herbivorous insects as hosts or prey, floral food resources for adults (Lee & Heimpel, 2008), and refuges from disturbance (Landis et al., 2000; Cronin & Reeve, 2005).

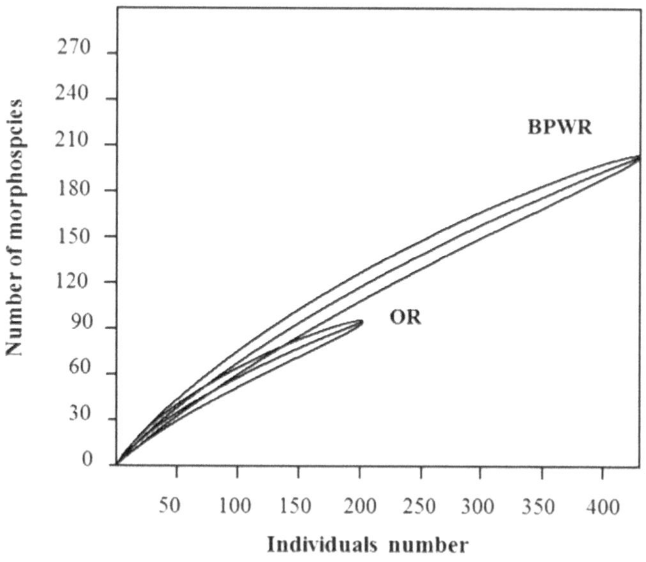

FIGURE 34.3 Rarefaction richness curves of Hymenopteran parasitoids collected at the organic rice site (OR) and at the Banhado dos Pachecos Wildlife Refuge (BPWR), from May 2011 to April 2012. Viamão, RS, Brazil. (Modified from da Silva et al., 2016.)

The refuges of native vegetation which, in general, have a greater richness of species (Figure 34.3) and can serve as a repository of NE, for both parasitoids (da Silva et al., 2016; da Silva et al., 2019) and predators (Ferreira et al., 2014; Medeiros et al., 2017) of insect crop pests.

Medeiros et al. (2017) in wheat crop found that species richness of Syrphidae was positively associated with the percentage of noncrop habitats. Moreover, Beta diversity (i.e., species composition variation between sites; Zellweger et al., 2017) was related to noncrop habitats, suggesting that the reduction in noncrop habitats implies in species loss. Da Silva et al. (2019) found a negative correlation between parasitoids abundance and the distance from the forest area for the rice crop in conventional management area, but this was different for organic management area. This is discussed in Chapter 40 and refers to the fact that the richness may not be so dependent on the proximity of the preserved area, because levees can serve as corridors extending the distances traveled by parasitoids (Simões-Pires et al., 2016).

A study with *Zizania caduciflora* (Turcz.) surrounding a rice crop in China evidenced that this vegetation supports *Sogatella* sp., a Delphacidae that is not considered a rice pest (Yu, 2001). However, this insect supports the parasitoid *Anagrusoptabilis* (Perkins), which is also an important parasitoid of *Nilopavarta lugens* (Stål), a rice pest (Zheng et al., 1999). In this way, the parasitoid can survive on the spot while the crop is not developing (Yu, 2001). It has also been demonstrated that several plant species spontaneously growing in borders of alfalfa agroecosystems act as sources of *Lysiphlebustes taceipes* (Cresson), important biocontrol agent of *Aphis craccivora* Koch (Zumoffen et al., 2015; Zumoffen et al., 2018).

In Asia, it appears that even with the harmful effects of insecticides, the level of parasitism in Delphacidae increases when the habitat is manipulated, emphasizing that adjacent areas that can support and allow the persistence of parasitoids (Gurr et al., 2011). Studies have investigated the impact of habitat diversification on the activities of *Trichogramma* spp. and most of these recorded an increase the parasitism level (Romeis et al., 2005). In many cases, invasive plants or other wild vegetation around cultivated areas shelter alternative hosts and prey, providing seasonal resources that contribute to maintaining the life cycles closing gaps in the temporal distribution of these species in the agricultural landscape (Altieri et al., 2003).

34.5 The effect of flower strips and intercropping

The landscape management at the local level, of the "push-pull" and "attract and reward" type, show efficiency in the manipulation of natural enemies, increasing their recruitment in the cultures and the rates of parasitism/predation. The push-pull system is a biological control strategy by using repellent "push" plants and trap "pull" plants (Cook et al., 2007). The "attract and reward" strategy deals with the activity of attracting NE by herbivore-induced plant volatiles (HIPVs) in the field. However, for success in CBC, it is necessary to know the structure and function of each trophic level of a food web in the environment (Eilenberg et al., 2001). This involves the manipulation of the environment through the use of some plants that can act as an alternative food source for NE and enabling adequate refuge for adverse conditions and other factors, providing increased fertility, reproductive success, and the efficiency in a reduction of pest populations (Wackers et al., 2007).

Favoring the habitat for NE has become the subject of much research and with that the term "habitat management" (Gurr et al., 2017). Plant diversification reduces herbivory in different ways. Root (1973) suggests that the diversity of olfactory and visual stimuli associated with various plant species, makes it difficult for herbivores to locate and colonize plants. In addition, there could be less supply of plants with good nutritional quality (Jonsson et al., 2015) which increases herbivore development time and leaves them more exposed to NE (Kagata et al., 2005). The second way in which diversity can decrease pest damage, is related to the "top-down" system, whose predator affects a basal trophic level through direct consumption of an intermediate trophic level (Kagata et al., 2005) and is associated with the attract and reward systems.

The strips of wild plants must be integrated into the culture in a practical way for producers and distributed in space and time so that they favor NE. Aguiar-Menezes & Silva (2011) present suggestions of flower arrangements between cultures in a technical bulletin (Figure 34.4).

The positive effect of flowering plants is shown in a lettuce crop where an increase in the number of *Eupeodes fumipennis* (Thompson) was observed in treatments with *Lobularia maritima* (L.), which resulted in fewer aphids (Hogg et al., 2011). Cahenzlia et al. (2019) showed that the strips of perennial flowers in alleyways between apple tree rows increases NE and reduces the main apple pests and the damage associated with fruits. In greenhouses, the floristic resources combined with the use of plants that act as a reservoir of aphids, of non-pest species, "insectary plants" act directly in the maintenance of sirphids and parasitoids over time (Pineda, 2008).

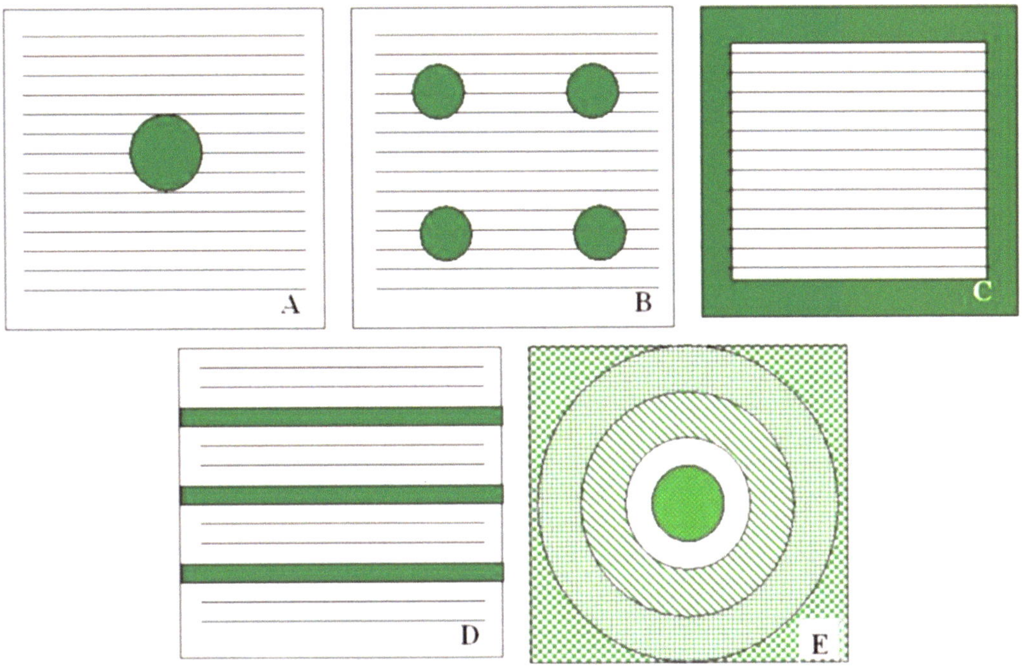

FIGURE 34.4 Distribution arrangements of attractive plants (green) for natural enemies: central island (A); several islands (B); border (C); strips (C); mandala (D). (Modified from Aguiar-Menezes & Silva, 2011.)

Intercropping, are practices that advocate the combined drawings of two or more plants in the same area (Altieri & Nicholls, 2010). Among the desired characteristics of intercropping plants, we highlight the presence of nectaries in floral or extrafloral whorls that also provide alternative food for NE (Pacini & Nicolson, 2007).

Ocimum basilicum L., for example, favors the appearance of *Oriuslaevigatus* (Fieber), a predator of thrips, aphids, white flies, and mites, intercropped with peppers (Montserrat et al., 2012). The advantage of these intercropped crops is that it can have a high productive potential, and is also commercialized (Kahn, 2014). However, Venzon et al. (2006) reported that only the diversification of vegetation will not result in a reduction in the population of herbivores, as they observed that they can also benefit from the resources offered by plants.

These volatile signals can warn other plants, used in intercropping systems, of impending attacks. Studies show that plants can be "activated" by synthetic HIPVs signals and therefore become sources of natural volatiles for recruiting NE (Kahn et al., 2008, Aartsma et al., 2017). Based on the system, "attract and reward" the combination of two aspects of applied insect ecology, HIPVs and CBC was tested, to improve immigration of beneficial taxa into crops and nectar plants to maintain their populations (Simpson et al., 2011).

34.6 Conclusions and Future Prospects

The action of hyperparasitoids (fourth trophic level organisms) that can disrupt biological pest control by suppressing the populations of their parasitoid hosts (Cusumano et al., 2020), also needs to be considered. Research suggests that a "push-pull" strategy could be a promising approach to "push" hyperparasitoids away from their parasitoid hosts and "pull" them into traps using synthetic info-chemicals.

REFERENCES

Aartsma Y.; Bianchi F. J.; van der Werf W.; Poelman E. H.; Dicke M. Herbivore-induced plant volatiles and tritrophic interactions across spatial scales. *New Phytol.* 2017, 216, 1054–1063.

Aguiar-Menezes, E. L.; Silva, A. C. *Plantas atrativas para inimigos naturais e sua contribuição no controle biológico de pragas agrícolas.* Embrapa Agrobiologia: Rio de Janeiro, 2011; p. 60.

Altieri, M. A. *Agroecologia: Bases científicas para uma agricultura sustentável.* 3rd ed. rev. ampl.; São Paulo. Expressão Popular: Rio de Janeiro: AS-PTA, 2012; p. 400.

Altieri, M. A; Nicholls, C. Diseños agroecológicos para aumentar la biodiversidad de entomofauna benefica en agroecosistemas. Medellin: SOCLA, 2010.

Altieri, M. A.; Silva, E. N.; Nicholls, C. I. O papel da biodiversidade no manejo de pragas. Holos: Ribeirão Preto, São Paulo, 2003; p. 223.

Cahenzlia, F.; Sigsgaard, L.; Daniel, C.; Herz, A.; Jamar, L.; Kelderere, M.; Jacobsenb, S. K.; Kruczyńskaf, D.; Matrayc, S.; Porcelg, M.; Sekreckaf, M.; Świergielg, W.; Tasing, M.; Telfsere, J.; Pfiffner, L. Perennial flower strips for pest control in organic apple orchards - A pan-European study. *Agric.Ecosyt.* Environ. 2019, 278, 43–53.

Cook, S. M.; Khan, Z. R.; Pickett, J.A.The use of push-pull strategies in integrated pest management. *Annu. Rev.Entomol.* 2007, 52(1), 375–400.

Crócomo, W. B. (org.). *ManejoIntegrado de Pragas.* UNESP: Botucatu, São Paulo, 1990; P. 358.

Cronin, J. T.; Reeve, J. D. Host-parasitoid spatial ecology a plea for a landscape-level synthesis. *Proc. Roy. Soc. London Ser. B.* 2005, 272(1578), 2225–2235.

Cusumano, A.; Harvey, J. A.; Bourne, M. E.; Poelman, E. H.; Jetske G.; de Boer, J. G. Exploiting chemical ecology to manage hyperparasitoids in biological control of arthropod pests. *Pest Manage. Sci.* 2020, 76(2), 432–443. doi: 10.1002/ps.5679.

De Souza Da Silva, G.; Jahnke, S. M.; Ferreira, M. L. G. Hymenoptera parasitoids in protected area of Atlantic Forest biomes and organic rice field: 2 compared assemblages. *Rev. Colomb. Entomol.* 2016, 42(2), 110–117.

De Souza Da Silva, G.; Jahnke, S. M.; Johnson, N. Riparian forest fragments in rice fields under different management: Differences on hymenopteran parasitoids diversity. *Braz. J. Biol.* 2019, 79(1), 1–11.

Eilenberg, J.; Hajek, A.; Lomer, C. Suggestions for unifying the terminology in biological control. *Biol. Control.* 2001, 46, 387–400.

Ekesi, S.; Shah, P. A.; Clark, S. J.; Pell, J. K. Conservation biological control with the fungal pathogen Pandora neoaphidis: Implications of aphid species, host plant and predator foraging. *Agric. Forest. Entomol.* 2005, 7(1), 21–30.

Ferreira, M. L. G.; Jahnke, S. M.; Morais, R. M.; Da silva, G. S. de. Diversidad de insectos depredadores en área orizícola orgánica y de conservación en Viamão, RS, Brasil. *Rev. Colomb. Entomol.* 2014, 40(1), 120–128.

Foerster, L. A. Seletividade de inseticidas a predadores e parasitoides. In Parra, J. R. P.; Botelho, P. S. M.; Corr~ea-Ferreira, B. S (eds.), *Controle Biológico no Brasil: parasitoides e predadores.* Manole: São Paulo, 2002; pp. 95–114.

Gan, H.; Wickings, K. Soil ecological responses to pest management in golf turf vary with management intensity, pesticide identity, and application program. *Agric. Ecosyst. Environ.* 2017, 246, 66–77.

Gliessman, S. R. *Agroecologia: Processos ecológicos em agricultura sustentável.* 2 ed. Editora UFRGS: Porto Alegre, RS, 2001; p. 653.

Gurr, G. M.; Liu, J.; Read, D. M. Y.; Catindig, J. L. A.; Cheng, J. A.; Lan, L. P.; Heong, K. L. Parasitoids of Asian rice planthopper (Hemiptera: Delphacidae) pests and prospects for enhancing biological control by ecological engineering. *Ann. Appl. Biol.* 2011, 158(2), 49–176.

Gurr, G. M.; Wratten, S. D.; Landis, D. A.; You, M. Habitat management to suppress pest populations: Progress and prospects. *Annu. Rev.Entomol.* 2017, 62(1), 91–109.

Halaj, J. D.; Wise, D. H. Impact of a detrital subsidy on trophic cascades in a terrestrial grazing food web. Ecology. 2002, 83(11), 3141–3151.

Hanski I. A practical model of metapopulation dynamics. *J Anim Ecol* 1994, 63,151–162.

Heimpel, G. M; Mills, N. J. *Biological Control: Ecology and Application.* Cambridge University Press: Cambridge, UK, 2017; p. 379.

Hogg, B. N.; Nelson, E. H.; Mills, N. J.; Daane, K. M. Floral resources enhance aphid suppression by a hoverfly. Entomol. Exp. Appl. 2011, 141(2),138–144.

Jonsson, M.; Straub, C. S.; Didham, R. K.; Buckley, H. L.; Case, B. S.; Hale, R. J.; Gratton, C. Wratten, S. D. Experimental evidence that the effectiveness of conservation biological control depends on landscape complexity. *J. Appl. Ecol.* 2015, 52(5), 1274–1282.

Kagata, H.; Nakamura, M.; Ohgushi, T. Bottom-up cascade in a tritrophic system: Different impacts of host-plant regeneration on performance of a willow leaf beetle and its natural enemy. *Ecol. Entomol.* 2005, 30(1), 58–62.

Khan, Z. R. Push–pull farming systems. *Curr. Opin. Biotech.* 2014, 26, 125–132.

Kahn, Z. R.; James, D. G.; Midega, C.A.O.; John A.Pickett, J.A. Chemical ecology and conservation biological control. *Biol. Control.* 2008, 45 (2), 210–224. https://doi.org/10.1016/j.biocontrol.2007.11.009.

Landis, D. A.; Wratten, S. D.; Gurr, G. M. Habitat management to conserve natural enemies of arthropod pests in agriculture. *Annu. Rev. Entomol.* 2000, 45(1), 175–201.

Laurance, W. F. Theory meets reality: How habitat fragmentation research has transcended island biogeographic theory. *Biol. Conserv.*2008, 141(7), 1731–1744.

Lee, J. C.; Heimpel, G. E. Floral resources impact longevity and oviposition rate of a parasitoid in the field. *J. Animal. Ecol.* 2008, 77(3), 565–572.

Levins R. Some demographic and genetic consequences ofenvironmental heterogeneity for biological control. *Bul. Entomol. Soc. Am.* 1969, 15, 237–240.

Liu, X.; Wang, Y.; Wang, X.; Zhou, L.; Si, S. Effect of relative humidity on entomopathogens infection and antioxidant responses of the beet armyworm, spodoptera exigua (Hübner). *Afr. Entomol.* 2014, 22(3), 651–659.

MacArthur, R. H.; Wilson, E. O. *The Theory of Island Biogeography.* Princeton University Press: Princeton, USA, 1967; p. 203.

Magurran, A. E. *Medindo a diversidade biológica.* Editora UFPR: Curitiba, 2011; p. 261.

Medeiros, H. R.; Hoshino, A. T.; Ribeiro, M. C.; Morales, M. N.; Martello, F.; Pereira Neto, O. C.; Carstensen, D. W.; Menezes Junior, A. O. Noncrop habitats modulate alpha and beta diversity of flower flies (Diptera, Syrphidae) in Brazilian agricultural landscapes. *Biodivers. Conserv.* 2017, 27(3), 1309–1326.

Montserrat, C.; Vila, E.; Salvador, E. Janssen, D. Utilizacion de Menthauaveolens y Ocimumbasilicum L. como plantas de refugio para adelantarlainstalación de Oriuslaevigatus F. (Hemiptera: Anthocoreidae) em cultivo de pimiento. *Bol. San. Veg. Plagas.* 2012, 8(2), 311–319.

Morales-Ramos, J. A.; Rojas, M. G.; Shapiro-Ilan, D. I. *Mass Production of Beneficial Organisms, Invertebrates and Entomopathogens.* Academic Press: Cambridge, MA. 2014; p. 764.

Morin P. J. *Community Ecology.* 2 ed. Blackwell Science: Malden, MA, 2011; p. 413.

Nielsen, C.; Eilenberg, J.; Dromph, K. M. *Entomophthorales on Cereal Aphids: Characterisation, Growth, Virulence, Epizootiology and Potential for Microbial Control.* Ministry of the Environment: Copenhagen, 2001; p. 58.

Norris R. F.; Caswell-Chen E. P.; Kogan, M. *Concepts in Integrated Pest Management.* Pearson Education Inc.: Upper Saddle River, NJ, 2003; p. 586.

Oi, D. H.; Barnes, M. M. Predation by the western predatory mite (Acari: Phytoseiidae) on the pacific spider mite (Acari: Tetranychidae) in the presence of road dust. *Environ. Entomol.* 1989, 18(5), 892–896.

Pacheco, M. H.; Vanlaerhoven, S. L.; Garcia, M. A.; Hunt, D. W. Food web associations and effect of trophic resources and environmental factors on parasitoids expanding their host range into non-native hosts. *Entomol. Exp. Appl.* 2018, 166(4), 277–288.

Pacini, E.; Nicolson, S. W. Nectar production and presentation. In Nicolson, S. W; Nepi, M.; Pacini, E. (eds.), *Nectaries and Nectar.* Springer: Dordrecht, 2007; pp. 167–214.

Pedigo, L. P. *Entomology and Pest Management*. Macmillan: New York, 1988; p. 646.

Pell, J. K.; Hannam, J. J.; Steinkraus, D. C. Conservation biological control using fungal Entomopathogens. *Biol. Control*. 2010, 55, 187–198.

Pimentel, D.; Peshin, R. (Eds.). *Integrated Pest Management: Pesticide Problems*, Vol. 3. Springer: Dordrecht, 2014; p. 474.

Pineda, A. Los sirfidos (Diptera, Syrphidae) enelcontrol integrado de plagas de pulgónemcultivos de pimiento de invernadero. TesedeDoutorado. Centro Iberoamericano de La Biodiversidad (CIBIO), Universidad de Alicante: Valência, 2008; p. 163.

Pywell, R. F. Meek, W. R.; Loxton, R.G.; Nowakowski, M.; Carvell, C.; Woodcock, B. A. Ecological restoration on farmland can drive beneficial functional responses in plant and invertebrate communities. *Agric. Ecosyst. Environ.*2011, 140(1–2), 62–67.

Romeis, J.; Babendreier, D.; Wackers, F.L.; Shanower, T. G. Habitat and plant specificity of Trichogramma egg parasitoids—Underlying mechanisms and implications. *Basic. Appl. Ecol.* 2005, 6(3), 215–236.

Root, R. B. Organization of a plant-arthropod association in simple and diverse habitats: The fauna of collards (Brassica oleracea). *Ecol. Monogr.* 1973, 43, 94–125.

Sereda, E.; Wolters, V.; Birkhofer, K. Addition of crop residues affects a detritus-based food chain depending on litter type and farming system. *Basic. Appl. Ecol.* 2015, 16(8), 746–754.

Simões-Pires, P. R.; Jahnke.; S. M.; Redaelli, L. R. Influence of the vegetation management of the leeves in irrigated rice organic in diversity of Hymenoptera parasitoids. *Braz. J. Biol.* 2016, 76(3), 774–781.

Simpson, M. Gurr, G. M.; Simmons, A. T.; Wratten, S. D.; James, D. G.; Leeson, G.; Nicol, H. I.; Orre, Gordon, S. G. U. Attract and reward: Combining chemical ecology and habitat manipulation to enhance biological control in field crops. *J. Appl. Ecol.* 2011, 48(3), 580–590.

Towsend, C. R.; Begon, M.; Harper, J. L. *Fundamentos em Ecologia*. EditoraArtMed: Oxford, 2006; p. 945.

Tscharntke, T.; Bommarco, R.; Clough, Y.; Crist, T. O.; Kleijn, D.; Rand, T. A.; Tylianakis, J. M.; Van Nouhuys, S.; Vidal, S. Conservation biological control and enemy diversity on a landscape scale. *Biol. Control*. 2007, 43(3), 294–309.

Tscharntke, T.; Klein, A. M.; Kruess, A.; Steffan-Dewenter, I.; Thies, C. Landscape perspectives on agricultural intensification and biodiversity—Ecosystem service management. *Ecol. Lett.* 2005, 8(8), 857–874.

Van Driesce, R. G.; Bellows, T. S. (eds.). *Biological Control*. Chapman & Hall: New York, 1996; p 447.

Venzon, M.; PaulaJúnior, T. L.; Pallini, A. Controle alternativo de Pragas e Doenças. In: Venzon, M.; Rosado, M. C.; Euzébio, D. E.; Pallini, A. (eds.), *Controle Biológico conservative*. Universidade Federal de Viçosa: Viçosa, MG, 2006; p. 360.

Wackers, F. L.; Romeis, J.; Van Rijn, P. Nectar and pollen feeding by insect herbivores and implications for multitrophic interactions. *Annu. Rev.Entomol.* 2007, 52, 301–323.

With, K. A. Metapopulation dynamics: Perspectives from landscape ecology. Chapter: 2. In: Hanski, I.; Gaggiotti, O. E. (eds.), *Ecology, Genetics, and Evolution of Metapopulations*. Elsevier Academic Press, 2004; pp. 23–44. https://doi.org/10.1016/B978-0-12-323448-3.X5000-4

Yu, X. P. Role of Saccharosydne procerus on Zizaniacaduciflora as an alternate host for Anagrusnilaparvatae the egg parasitoid of the brown planthopper Nilaparvatalugens, which attacks temperate rice. In: Mew, T. W.; Borromeo, E.; Hardy, B. (eds.), Proceedings of the Impact Symposium on Exploiting Biodiversity for Sustainable Pest Management, Kunming, China, International Rice Research Institute: Filipinas, 2001; pp. 15–23.

Zellweger, F.; Roth, T.; Bugmann, H.; Bollmann, K. Beta diversity of plants, birds and butterflies is closely associated with climate and habitat structure. *Global Ecol. Biogeogr.* 2017, 26(8), 898–906.

Zheng, X.; Yu, X.; Lu, Z.; Chen, J. Dispersal patterns of natural enemies of rice planthoppers between zizania and rice fields. *Acta. Agric. Zhejiang.* 1999, 11(6), 339–343.

Zumoffen, L.; Rodriguez, M.; Gerding, M.; Salto, C. E.; Salvo, A. Plants, aphids and parasitoids: trophic interactions in agroecosystems in the province of Santa Fe, Argentina and a key to the identification of Aphidiinae and Aphelinidae (Hymenoptera) known to the region. *Rev. Soc. Entomol. Argent.* 2015, 74(3–4), 133–144.

Zumoffen, L.; Signorini, M.; Salvo, A. Bidirectional movement of aphid parasitoids (raconidae: Aphidiinae) between crops and non-crop plants in agroecosystems of central Argentina. *Appl. Entomol. Zool.* 2018, 53(3), 1–9.

35

Management of Insect Pests by means of Entomopathogenic Nematodes

Muhammad Sarwar and Zahid Mukhtar

CONTENTS

35.1 Introduction

Entomogenous microorganisms that arise in insects have the power in bringing about a certain degree of natural or microbial control of insect pests. Several entomopathogenic agents, when are inundatively introduced into a variety of habitats, can provide effective long-term to short-term control (Sarwar, 2015a,d). Nematodes feeding on bacteria, plants, and others can be parasites of animals. Nematodes are usually considered pests because of the diseases they cause in humans and animals, and the economic impact they have on many agricultural products (Adams and Nguyen, 2008).

They live parasitically inside the infected insect host, hence they are also termed endoparasites nematodes (Sarwar et al., 2020). Entomopathogenic nematodes occur naturally in soil environments and locate their host in response to carbon dioxide, vibration, and other chemical cues (Kaya and Gaugler, 1993; Sarwar, 2015b,c).

Nematode attacks an insect pest, kills or hampers the development of the insect host, and is capable of mass production hence can be used as an effective biological-control agent. In addition to insects, nematodes can parasitize spiders, leeches, annelids, crustaceans, and mollusks (Siddiqi, 2000; Sarwar, 2016).

The families Steinernematidae, Neosteinernematidae, and Heterorhabditidae (Superfamily Rhabditoidea) are unique because they are the only nematodes, which have developed the ability to carry and introduce symbiotic bacteria into the body of insects (Poinar, 1990). Of all nematodes studied for biological control of insects, those in the families Steinemernatidae and Heterorhabditidae have aroused the most interest, and information about them is growing exponentially. These two families, mutualistically associated with bacteria in the genus Xenorhabdus, are similar in their actions. Xenorhabdus species, which are gram-negative, facultatively anaerobic rods, belong to the family Enterobacteriaceae. Both groups are in the order Rhabdita, the bacteria-feeding nematodes (Poinar, 1990; Sarwar and Sarwar, 2018). They have a broad host range, are safe to vertebrates, plants, and other non-target organisms, easily applied using standard spray equipment, compatible with many chemical pesticides, and are amenable to genetic selection (Georgis, 1990a; Zimmerman and Cranshaw, 1990).

The bacterial cells, voided from the nematode's intestine into the hemolymph, propagate and kill the host by bacterial septicemia within 48 hours. The nematodes feed on the bacterial cells and host tissues, produce two or three generations, and emerge from the host as infective juveniles to search for new hosts (Bednarek and Nowicki, 1991).

Mutualism occurs between the nematode and bacterium Xenorhabdus species. The bacterium requires the nematode for protection from the external environment, penetration into the host's hemocoel, and possibly inhibition of the host's antibacterial proteins (Akhurst et al., 1990). The nematodes provide

FIGURE 35.1 Entomopathogenic nematode assaulted insect.

shelter to the bacteria, which, in return, kill the insect host and provide nutrients to the nematode. Together, the nematodes and bacteria feed on the liquefying host, and reproduce for several generations (Figure 35.1) inside the cadaver maturing through the growth stages of J2-J4 into adults (Boemare, 2002; Gaugler et al., 1997).

Nematode parasites of insects (also called entomophilic nematodes) can be found in the orders Aphelenchida, Ascaridida, Mermithida, Oxyurida, Rhabditida, Spirurida, and Tylenchida (Smart and Nguyen, 1994). Several species of *Heterorhabditis* and *Steinernema* are available in multiple commercial formulations, primarily for managing soil insect pests (Table 35.1).

35.2 Population Distribution of Nematodes

Entomopathogenic nematodes have limited dispersal ability and are typically found in patchy distributions, The founding of new populations and movement between patches may depend on the movement of infective juveniles or the movement of infected hosts (Lewis et al., 1998). Some studies suggest that entomopathogenic nematodes may also use non-host animals, such as isopods and earthworms for transport (Eng et al., 2005) or can be scavengers (San-Blas and Gowen, 2008).

35.3 Life Cycle of Nematodes

Insect parasitic nematodes are mobile and move short distances in search of host insects. The active stage of the nematode that invades an insect is the juvenile (dauer larva) third stage. Nematodes use carbon dioxide, vibration, and other chemicals produced in waste products of insects as cues to find their hosts. Steinernema nematodes enter the insect through natural openings, such as the mouth, spiracles, and anus, then penetrate into the body cavity and enter into the hemocoel. Heterorhabditis nematodes use natural openings, but also can enter by piercing the body wall (Bedding and Molyneux, 1982).

Both *Heterorhabditis* and *Steinernema* are mutualistically associated with bacteria of the genera *Photorhabdus* (Figure 35.2) and *Xenorhabdus* (Figure 35.3), respectively (Ferreira and Malan, 2014). The insect cadaver becomes red if the insects are killed by Heterorhabditids and brown or tan if killed by Steinernematids (Kaya and Gaugler, 1993). Associated color changes may occur, for instance, caterpillars parasitized by Heterorhabditis may have a reddish-brown color.

Reproduction differs in Heterorhabditid and Steinernematid nematodes. Infective juveniles of Heterorhabditid nematodes become hermaphroditic adults, but individuals of the next generation produce both male and females, whereas in Steinernematid nematodes all generations are produced by males and females (gonochorisism) (Grewal et al., 2005). Insect parasitic nematodes typically kill their host insect within two to three days after invading the body cavity and a large number of infective juveniles are eventually released into the environment to infect other hosts, and continue their life cycle (Kaya and Gaugler, 1993).

TABLE 35.1
Commercial Use of Entomopathogenic Nematodes *Steinernema* and *Heterorhabditis* as Bioinsecticides

Pathogen Species	Target Pests
Steinernema	
Steinernema glaseri	White grubs (scarabs, especially Japanese beetle, *Popillia* sp.), banana root borers
Steinernema kraussei	Black vine weevil, *Otiorhynchus sulcatus*
Steinernema carpocapsae	Turfgrass pests: billbugs, cutworms, armyworms, sod webworms, chinch bugs, crane flies. Orchard, ornamental and vegetable pests banana moths, codling moths, cranberry girdlers, dogwood borers and other clearwing borer species, black vine weevils, peachtree borers, shore flies (*Scatella* spp.)
Steinernema feltiae	Fungus gnats (*Bradysia* spp.), shore flies, western flower thrips, leafminers
Steinernema scapterisci	Mole crickets (*Scapteriscus* spp.)
Steinernema riobrave	Citrus root weevils (*Diaprepes* spp.), mole crickets
Heterorhabditis	
Heterorhabditis bacteriophora	White grubs (scarabs), cutworms, black vine weevils, flea beetles, corn root worms, citrus root weevils
Heterorhabditis megidis	Weevils
Heterorhabditis indica	Fungus gnats, root mealy bugs, grubs
Heterorhabditis marelatus	White grubs (scarabs), cutworms, black vine weevils
Heterorhabditis zealandica	Scarab grubs
Heterorhabditis heliothidis	Wide variety of soil-dwelling and boring insect larvae

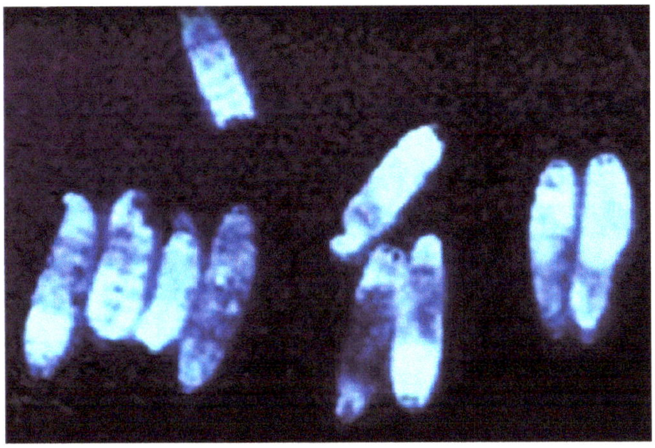

FIGURE 35.2 *Photorhabdus.*

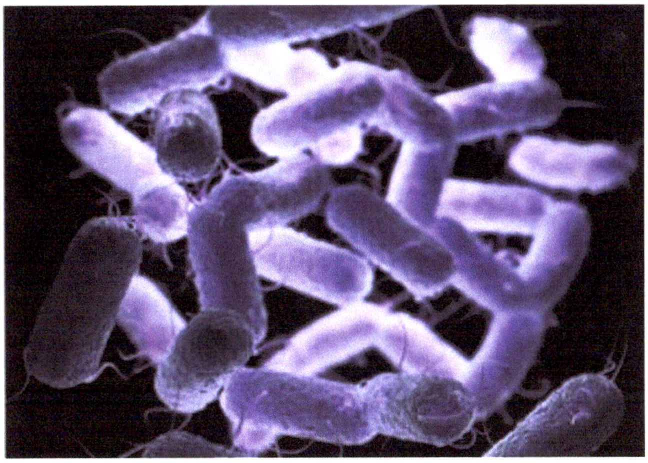

FIGURE 35.3 *Xenorhabdus.*

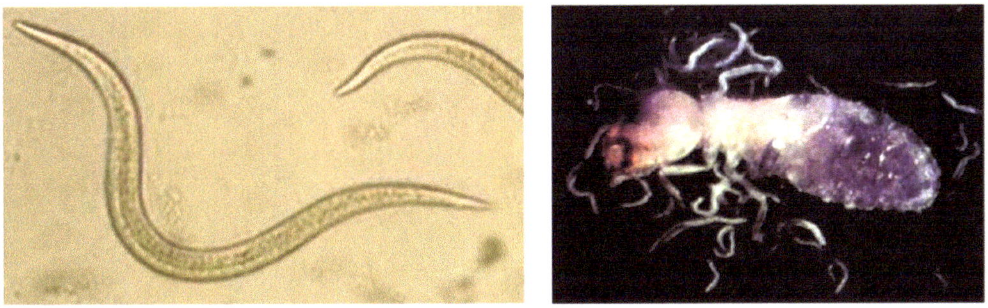

FIGURE 35.4 *Steinernema carpocapsae* and invaded host.

35.4 Parasitic Nematodes for Insects Control

Entomopathogenic nematodes fit nicely into integrated pest management programs because they are considered non-toxic to humans, relatively specific to their target pest (s), and can be applied with standard pesticide equipment (Shapiro-Ilan et al., 2006). Because the nematodes are susceptible to drying and ultraviolet light, they are most effective against insects that occur in moist dark locations (Nickle, 1991).

35.4.1 Steinernema Travassos

Steinernema is the most widely researched species for insect controls. In the insect cadaver, the fertilized eggs hatch as first larval juvenile (J1), and these animals molt into J2, J3, J4, and adults. It is the most readily available for yard and garden use because it is easier to rear and handle. In field applications, *Steinernema carpocapsae* tends to be most effective against caterpillar larvae. In laboratory and field trials, it has controlled sod webworms, cutworms, and certain borers (raspberry crown borer, carpenter worm).

35.4.1.1 Steinernema carpocapsae (Weiser) *Nematode*

Nematode *S. carpocapsae* (Figure 35.4) is a classic sit-and-wait ambush attacker near the soil surface and will attach to passing pest insects. This nematode is most effective against flea larvae and caterpillars in lawns, garden soil, and under trees where larvae pupate. It tends to be most effective when applied against highly mobile surface-adapted insects. For managing the branch and twig borer (*Melagus confertus*) in grapes, *S. carpocapsae* is one of the recommended options (Varela et al., 2015).

Essentially the nematodes serve as mobile vectors for their insect-pathogenic bacteria cargo. Symbiotic bacteria observed in the intestinal tract (Figure 35.5) of *S. carpocapsae* are mutualistically grown in the host. Infected insect hosts die quickly, the bacteria proliferate, the nematodes feed on bacteria and insect tissues, and reproduce.

35.4.1.2 Steinernema feltiae *(Filipjev) Nematode*

They are the most effective against larval control of several fly species (Sciaridae, Phoridae, leaf miners, domestic fly) and also of some moth larvae (Sarwar, 2020). The entomopathogenic nematode, *S. feltiae*, has reduced raspberry crown borer (*Pennisetia marginata*) populations by 33–67% (Capinera et al., 1986).

35.4.2 Heterorhabditis Poinar

The *Heterorhabditis* nematodes are most effective against Japanese beetles, weevils, and many other target pests in lawn and garden. Several species are sold, including *Heterorhabditis bacteriophora* Poinar, *H. megadis*, and *H. indica*. *Heterorhabditis* also effectively controls many nursery pests that feed in the root

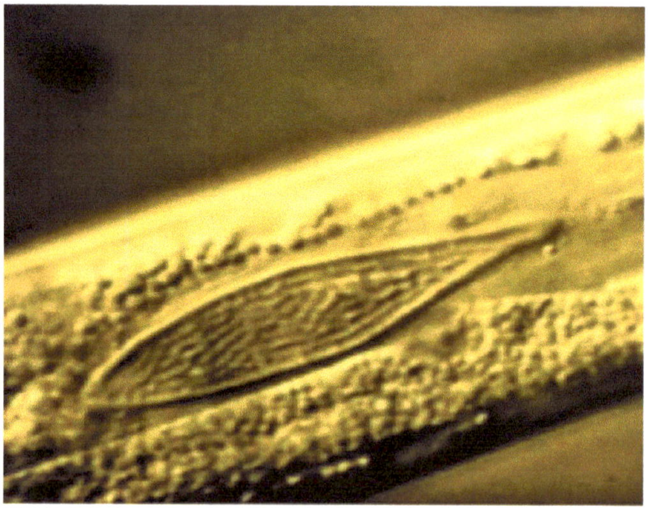

FIGURE 35.5 Symbiotic bacteria in *Steinernema carpocapsae* intestine.

zone, such as black vine weevils and citrus-infesting root weevils (Rasmann et al., 2005). (Figure 35.6)

35.4.2.1 Heterorhabditis bacteriophora Poinar *Nematode*

Nematode *Heterorhabditis bacteriophora* (Figure 35.7) can be as effective against white grubs as chemical insecticides, but is less effective against mole crickets. Interactions with other organisms. *H. bacteriophora* is an insect parasite used for the biological control of insects and an obligate host for symbiotic *Photorhabdus luminescens* bacteria (Figure 35.8) and reproduces for 2–3 generations on the host (O'Halloran and Burnell, 2002).

They are highly efficient when the pest is more widely dispersed in the soil because they have a "tooth" to rupture the insect's skin. The nematodes are effective against grubs and the larva or grub stage of Japanese beetles, northern masked chafer, European chafer, rose chafer, fly larvae, oriental beetles, June beetles, flea beetles, termites, ants, bill-bugs, cut-worms, army worms, black vine weevils, strawberry root weevils, fungus gnats, sciarid larvae, sod web-worms, girdler, citrus weevils, maggots and other Diptera, mole crickets, iris borer, root maggot, cabbage root maggot and carrot weevils.

35.5 Commercialization

Nematodes are open to mass production and do not require specialized application equipment since they are compatible with standard agrochemical equipment, including various sprayers (backpack, pressurized, mist, electrostatic, fan, and aerial) and irrigation systems (Cranshaw and Zimmerman, 2013).

35.5.1 Collection

Steinernema is collected mostly in loamy-sand and sandy-loam soils with a pH below 6, whereas Heterorhabditis can be commonly collected in sandy and loamy-sand soils with pH higher than 6. After incubation the insects are removed and examined. Dead insects from each sample are rinsed and placed on a moist filter paper in a petri dish for 3 to 4 days. The *G. mellonella* larvae showing signs of parasitism by Steinernematids or Heterorhabditids (Poinar, 1979) are transferred to a modified White trap (White, 1929) to collect the infective juveniles (IJ). Living insects from each site are also rinsed and placed on a filter paper in a petri dish for 3 to 4 days at 23°C and then observed for

FIGURE 35.6 *Steinernema feltiae* and host incursion.

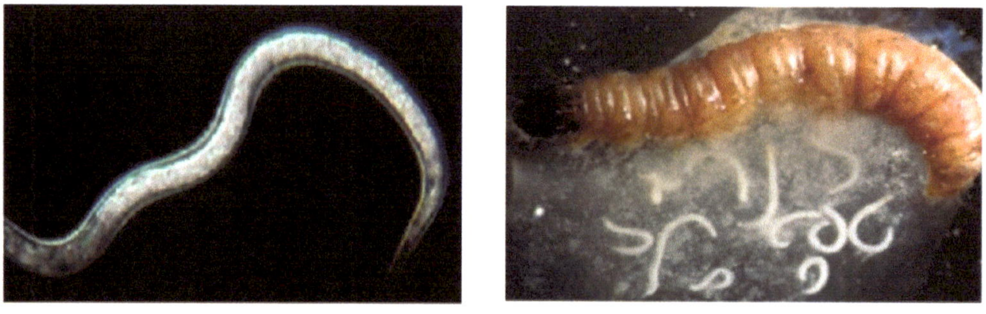

FIGURE 35.7 *Heterorhabditis bacteriophora* and attacked host.

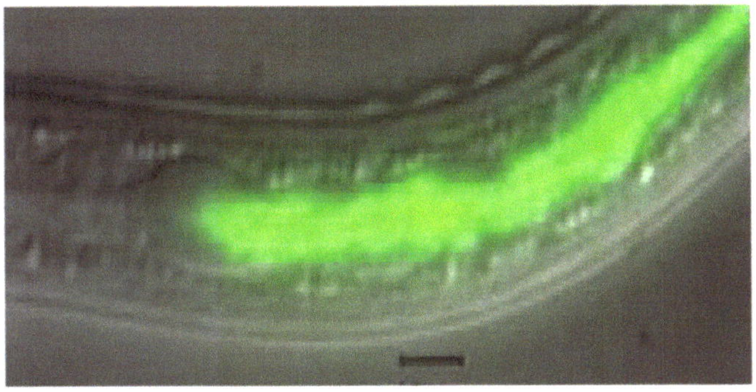

FIGURE 35.8. *Photorhabdus luminescens* in intestine of *H. bacteriophora*.

parasitism. When IJs are recovered, they are rinsed three times with sterile Ringer solution and finally added into a *G. mellonella* larvae (Rosa et al., 2000).

35.5.2 Mass Production

Mass production of nematodes is influenced by the amount of progeny required, time, resources, the costs of production, as well as the level of expertise available. The differences in nematode life cycle and bacterial symbiosis play major role in final nematodes yields.

Large-scale commercial yields of entomopathogenic nematodes require economies of production (decreasing costs in yields with an increasing scale of operation). Without such economies of scale, commercial production of nematodes will be confined to cottage industries. Mass production of entomopathogenic nematodes has evolved from the first large-scale *in vitro* solid media production by Glaser et al. (1940) to the *in vivo* production by Dutky et al. (1964), to the three-dimensional solid media *in vitro* process (Bedding, 1981), and to the *in vitro* liquid-fermentation production method (Friedman, 1990). Once upon a time, commercial nematodes have been produced monoxenically using the solid-media process developed by Bedding (1984) or the liquid-fermentation method.

Currently, entomopathogenic nematodes are produced by different methods either *in vivo* or *in vitro* (solid and liquid culture) (Shapiro-Ilan and Gaugler, 2012). *In vivo* production is a simple process of culturing a specific entomopathogenic nematode in live insect hosts, which requires minimal technology and involves the use of a surrogate host (typically larvae of wax moth (*Galleria mellonella*), trays, and shelves. However, this method is not cost-effective for scaled-up productions and may be ideal only for small markets or laboratory studies (Shapiro-Ilan et al., 2002).

The quality of nematodes produced in vitro solid culture is similar to that produced in vivo. A high quality of nematodes can be produced using liquid culture provided good media as well suitable environmental conditions in the bioreactor.

35.5.3 Formulation and Storage

Formulation of nematodes into a stable product has played a significant role in commercialization of these biological-control agents. Active nematodes must be immobilized to prevent depletion of their lipid and glycogen reserves. Immobilization has been accomplished by maintenance in aqueous suspensions at low temperatures (5–15°C), but this approach is not commercially desirable. Further advancements have been made with the development of a flowable concentrate that will revolutionize nematode applications and eliminate many of the constraints associated with current formulations, namely the dissolution of the carrier or separation of the nematodes from the carrier. Steinernematids, especially *S. carpocapsae*, can be maintained for up to 5 months at room temperatures or up to 12 months under refrigeration (Georgis, 1990b).

Formulated entomopathogenic nematodes can be stored for 2 to 5 months depending on the nematode species, and storage media and conditions. The quality of the nematode product can be determined by nematode virulence and viability assays, age, and the ratio of viable to non-viable nematodes (Grewal et al., 2005).

35.5.4 Application Considerations

Entomopathogenic nematodes can be applied with most horticultural equipment including pressurized sprayers, mist blowers, and electrostatic sprayers. In general, large diameter nozzles (orifices) and high volumes (up to 400 gallons per acre) are recommended. It is also important to ensure an adequate agitation during application because nematodes settle quickly in suspension, high pressures (> 300 psi) should also be avoided and nematodes can be kept cool by adding ice packs to the spray suspension. Entomopathogenic nematodes are compatible with many (but not all) insecticides, fungicides, and herbicides, but fresh manure or high rates of chemical fertilizers (urea) can be detrimental (Shapiro-Ilan et al., 2010).

Insect parasitic nematodes are sold in the infective juvenile dauer larva stage and are barely visible to the unaided eye. They typically are used at rates of 250 million to 2 billion per acre (approximately 6,000 to 46,000 per square foot). To apply, dilute the nematodes in water and drench the soil or inject them into plants. Regular spray equipment can be used and nematodes are quite tolerant of pressures found in many sprayers.

Innovative research has yielded insights into the fundamentals of the genera *Steinernema* and *Heterorhabditis*. Substantial progress in research on biology including major advances in genomics, nematode-bacterial symbiont interactions, ecological relationships, and foraging behavior and application has been made in the past decade, and the numbers of target pests shown to be susceptible to nematodes have continued to increase (Lacey et al., 2015).

35.6 Conclusion and Future Prospects

Entomopathogenic nematodes are a group of nematodes (thread-worms), causing death to insects. While most parasitic nematodes might be seen as harmful, entomopathogenic nematodes are beneficial to humans. Entomopathogenic nematodes are a welcome addition to the natural enemy pool of insects and can be integrated with various control measures for management of those target pests where individual tactics alone are inadequate. Entomopathogenic nematodes play a role underground reminiscent of that played by insect parasitoids. Like parasites or predators they have chemoreceptors and are motile. Like pathogens they are highly virulent, killing their hosts quickly, and can be cultured easily in vivo or in vitro.

Their field efficacy could be improved simply by matching nematode species and strains against those insects for which they are best adapted. Further understanding of their biology, behavior, ecology, and genetics will enhance the use and production of the most adapted species for insect control in the field. Recent advances made in nematode behavior and ecology clearly demonstrate that they are not generalist pathogens and their behavior restricts much of activity to a certain soil stratum, thus eliminating many insects from infection. The occurrence of quiescent nematodes suggests that they have evolved for effective survival strategies, and the J2 cuticle on some nematode species seem to play a significant role in desiccation tolerance and protection against fungal antagonists. They can be used effectively and selectively as inundative agents against numerous insect pests and have the potential to be used as inoculative agents for classical biological control. Finally, conservation and augmentation of natural nematode populations through proper management practices and periodic nematode releases offer exciting possibilities for insect pests suppression.

REFERENCES

Adams, B. J.; Nguyen, K. B. Nematode parasites of insects. In: Capinera J. L. (Ed.), *Encyclopedia of Entomology*. Springer, Dordrecht. 2008, Vol. 1, p. 12.

Akhurst, R. J.; Boemare, N. E. Biology and taxonomy of Xenorhabdus. In: Gaugler, R.; Kaya, H. K. (Eds.), *Entomopathogenic Nematodes in Biological Control*. Boca Raton, FL: CRC Press. 1990, pp. 75–90.

Bedding, R. A. Large scale production, storage and transport of the insect-parasitic nematodes Neoaplectana spp. And Heterorhabditis spp. *Ann. Appl. Biol.* 1984, 104: 117–120.

Bedding, R. A. Low cost in vitro mass production of Neoaplectana and Heterorhabditis species (Nematoda) for field control of insect pests. *Nematologica*. 1981, 27: 109–114.

Bedding, R.; Molyneux, A. Penetration of insect cuticle by infective juveniles of *Heterorhabditis* spp. (Heterorhabditidae: Nematoda). *Nematologica*. 1982, 28: 354–359.

Bednarek, A.; Nowicki, T. New estimation method for the density of entomogenous nematodes (Rhabditida: Steinernematidae) in the soil. *Rev. Nematol.* 1991, 14: 638–639.

Boemare, N. Biology, taxonomy, and systematics of photorabdus and xenorhabdus. In: Gaugler I, (ed.), *Entomopathogenic Nematology*. New Jersey: CABI. 2002, pp. 57–78.

Capinera, J. L.; Cranshaw, W. S.; Hughes, H. G. Suppression of raspberry crown borer *Pennisetia marginata* (Harris) (Lepidoptera: Sesiidae) with soil applications of *Steinernema feltiae* (Rhabditida: Steinernematidae). *J. Invertebr. Pathol.* 1986, 48: 257–258.

Cranshaw, W. S.; Zimmerman, R. Insect parasitic nematodes. In: *Insect Parasitic Nematodes*. Denver, CO: Colorado State University Extension. 2013, p. 3.

Dutky, S. R.; Thompson, J. V.; Can-twell, G. E. A technique for the mass propagation of the DD-136 nematode. *J. Insect Pathol.* 1964, 6: 417–422.

Eng, M. S.; Preisser, E. L.; Strong, D. R. Phoresy of the entomopathogenic nematode *Heterorhabditis marelatus* by a non-host organism, the isopod *Porcellio scaber*. *J. Invertebr. Pathol.* 2005, 88(2): 173–176.

Ferreira, T.; Malan, A. P. *Xenorhabdus* and *Photorhabdus*, bacterial symbionts of the entomopathogenic nematodes *Steinernema* and *Heterorhabditis* and their in vitro liquid mass culture: A review. *Afr. Entomol.* 2014, 22: 1–14.

Friedman, M. J. Commercial production and development. In: Gaugler, R.; Kaya, H. K. (Eds.), *Entomopathogenic Nematodes in Biological Control*. Boca Raton, FL: CRC Press. 1990, pp. 153–172.

Gaugler, R.; Lewis, E.; Stuart, R. J. Ecology in the service of biological control: The case of entomopathogenic nematodes. *Oecologia*. 1997, 109: 483–489.

Georgis, R. Commercialization of steinernematid and heterorhabditid entomopathogenic nematodes. In: Brighton Crop Protection Conference, Pests and Diseases, November 19–22, 1990, Brighton, UK. 1990a, Vol. 1, pp. 275–280.

Georgis, R. Formulation and application technology. In: Gaugler, R.; Kaya, H. K. (Eds.), *Entomopathogenic Nematodes in Biological Control*. Boca Raton, FL: CRC Press. 1990b, pp. 173–191.

Glaser, R. W.; McCoy, E. E.; Girth, H. B. The biology and economic importance of a nematode parasitic in insects. *J. Parasitol.* 1940, 26: 479–495.

Grewal, P. S.; Ehlers, R. U.; Shapiro-Ilan, D. I. *Nematodes as Biocontrol Agents*. New York, NY: CABI. 2005, p. 528.

Kaya, H. K.; Gaugler, R. Entomopathogenic Nematodes. *Annu. Rev. Entomol.* 1993, 38: 181–206.

Lacey, L. A.; Grzywacz, D.; Shapiro-Ilan, D. I.; Frutos, R.; Brownbridge; M.; Goettel, M. S.Insect pathogens as biological control agents: Back to the future. *J. Invertebr. Pathol.* 2015, 132: 1–41.

Lewis, E. E.; Campbell, J. F.; Gaugler, R. A conservation approach to using entomopathogenic nematodes in turf and landscapes. In: P Barbosa (Ed.), *Conservation Biological Control*. San Diego: Academic Press. 1998, pp. 235–254.

Nickle, W. R. *Manual of Agricultural Nematology*. New York: Marcel Dekker. 1991, p. 1035.

O'Halloran, D. M.; Burnell, A. M. Olfaction and odour discrimination in the insect parasitic nematode *Heterorhabditis bacteriophora*. *Nematology*. 2002, 4, 206.

Poinar, G. O. J. *Nematodes for Biological Control of Insects*. Boca Raton, FL: CRC Press. 1979, p. 277.

Poinar, G. O. J. Taxonomy and biology of Steinernematidae and Heterorhabditidae. In: *Entomopathogenic Nematodes in Biological Control*, Boca Raton, FL: CRC Press. 1990, pp. 23–61.

Rasmann, S.; Kollner, T. G.; Degenhardt, J.; Hiltpold, I.; Toepfer, S.; Kuhlmann, U.; Gershenzon, J.; Turlings, T. C. Recruitment of entomopathogenic nematodes by insect-damaged maize roots. *Nature*. 2005, 434, 732–737.

Rosa, J. S.; Bonifassi, E.; Amaral, J.; Lacey, L. A.; Simoes, N.; Laumond, C. Natural occurrence of entomopathogenic nematodes (Rhabditida: Steinernema, Heterorhabditis) in the Azores. *J. Nematol.* 2000, 32 (2): 215–222.

San-Blas, E.; Gowen, S. R. Facultative scavenging as a survival strategy of entomopathogenic nematodes. *Int. J. Parasitol.* 2008, 38: 85–91.

Sarwar, M. Biopesticides: An effective and environmental friendly insect-pests inhibitor line of action. *Int. J. Eng. Adv. Res. Technol.* 2015a, 1(2): 10–15.

Sarwar, M. Foodstuff contaminations with foodborne pathogens vehicled by insect vectors. *Int. J. Bioinf. Biomed. Eng.* 2015b, 1(3): 352–358.

Sarwar, M. Insect vectors involving in mechanical transmission of human pathogens for serious diseases. *Int. J. Bioinf. Biomed. Eng.* 2015c, 1(3): 300–306.

Sarwar, M. Life history of house fly *Musca domestica* Linnaeus (Diptera: Muscidae), its involvement in diseases spread and prevention of vector. *Int. J. Res. Appl. Chem.* 2016, 1(7): 23–34.

Sarwar, M. Microbial insecticides- An ecofriendly effective line of attack for insect pests management. *Int. J. Eng. Adv. Res. Technol.* 2015d, 1(2): 4–9.

Sarwar M. Typical flies: Natural history, lifestyle and diversity of Diptera, In: Sarwar, M. (Ed.), *Typical Flies: Natural History, Lifestyle and Diversity of Diptera.* London, UK: InTech. 2020, p. 50.

Sarwar, S.; Sarwar, M. Involvement of insects (Insecta: Artropoda) in spreading of plant pathogens and approaches for pests management. *Am. J. Microbiol. Immunol.* 2018, 3(1): 1–8.

Sarwar, M.; Mariya, A.; Sidra, S.; Romana, I. Different nematodes and plasmodiophorids as vectors of plant viruses. In: Awasthi, L. P. (ed.), *Applied Plant Virology: Advances, Detection, and Antiviral Strategies.* London: Elsevier Inc. 2020, pp. 275–290.

Shapiro-Ilan, D. I.; Cottrell, T. E.; Mizell, R. F.; Horton, D. L.; Behle, B.; Dunlap, C. Efficacy of *Steinernema carpocapsae* for control of the lesser peach tree borer, *Synanthedon pictipes*: Improved aboveground suppression with a novel gel application. *Biol. Control.* 2010, 54: 23–28.

Shapiro-Ilan, D. I.; Gouge, D. H.; Koppenhofer, A. M. Factors affecting commercial success: Case studies in cotton, turf and citrus. In: Gaugler R. (ed.), *Entomopathogenic Nematology.* New York, NY: CABI. 2002, pp. 333–356.

Shapiro-Ilan, D. I.; Gough, D. H.; Piggott, S.J.; Patterson, F. J. Application technology and environmental considerations for use of entomopathogenic nematodes in biological control. *Biol. Control.* 2006, 38: 124–133.

Shapiro-Ilan, D. I.; Han, R.; Dolinksi, C. Entomopathogenic nematode production and application technology. *J. Nematol.* 2012, 44: 206–217.

Siddiqi, M. R. *Tylenchida: Parasites of Plants and Insects.* Wallingford: CABI. 2000, p. 848.

Smart, G. C.; Nguyen, K. B. Role of entomopathogenic nematodes in biological control. In: Rosen, D.; Bennett, F. D.; Capinera, J.L. (Eds.), *Pest Management in the Subtropics: Biological Control—A Florida Perspective.* Andover, UK: Intercept. 1994, pp. 231–252.

Varela, L. G.; Haviland, D. R.; Bentley, W. J.; Zalom, F. G.; Bettiga, L. J.; Smith, R. J.; Daane, K. M. *UC IPM Pest Management Guidelines: Grape.* Davis, CA: UC ANR. 2015, p. 3448.

White, G. F. A method for obtaining infective nematode larvae from cultures. *Science.*1929, 66: 302–303.

Zimmerman, R. J.; Cranshaw, W. S. Compatibility of three entomogenous nematodes (Rhabditida) in aqueous solutions of pesticides used in turfgrass maintenance. *J. Econ. Entomol.* 1990, 83: 97–100.

36

Biological Pest Control in Different Crops Using Trichogramma Pretiosum

Cassia Renata Pinheiro, Jéssica Batista Torres Araújo Oliveira, and Glacy Jaqueline Da Silva

CONTENTS

36.1 Introduction

Wasps of the genus *Trichogramma* (Hymenoptera: Trichogrammatidae) are distributed in more than 180 recognized species. They are tiny insects, with lengths varying from 0.2 to 1.5 mm, exclusively egg parasitoids, used worldwide to combat various agricultural pests, mainly from the Lepidoptera Ordem. As a result, a decrease in the population of the pest is observed, without damage to other beneficial insects, human health, and the environment, mainly because they are naturally occurring insects in almost all terrestrial habitats (Smith, 1996; Knutson, 1998).

The interest in using *Trichogramma* for biological control of lepidopteran pests has been considered for more than 100 years, driven mainly by the fact that it controls important agricultural pests, which are caterpillars, although mass farming was the greatest barrier to success in the production process. In 1930 Flanders demonstrated the possibility of using *Sitotroga cerealella* (Olivier) eggs, as an alternative host for parasitoid multiplication, facilitating rearing systems and multiplication (Flanders, 1930). It was already possible to efficiently mass-produce wasps (Bigler, 1986; Hassan, 1993; Hawlitzky, 1986; Morrison et al., 1978; Newton, 1993; van Lenteren et al., 1982).

Trichogramma is the most studied egg parasitoid worldwide. Among these species, we highlight *T. pretiosum*, *T. evanescens* Westwood, *T. dendrolimi* Matsumura, *T. brassicae* Bezdenko, *T. ostriniae* Pan & Chen, *T. chilonis* Ishii, *T. exiguum* Pinto & Platner, and *T. euproctidis* Girault, being the most used for pest control in different cultures and countries (Li, 1994; Hassan, 1996).

Trichogramma pretiosum (Riley, 1879) is the most widespread species in the Americas and one of the most common wasps found in agricultural areas, due to its unique importance as an egg parasite of dozens of species, such as *Chrysoperla* sp. (Neuroptera: Chrysopidae) (Zucchi et al., 2010), *Heraclides astyalus* (Lepidoptera: Papilionidae), *Spodoptera* spp., *Tuta absoluta* (Lepidoptera: Gelechiidae), *A. argillacea* eggs (Querino et al., 2016), *Neoleucinodes elegantalis* (Lepidoptera: Crambidae) (Oliveira et al., 2019), *Helicoverpa armígera* (Lepidoptera: Noctuidae) (Laurentis et al., 2019), and others.

36.2 Rearing Systems

Insects of the genus *Trichogramma* are holometabolic, that is, in the development process, they pass through the egg, larva, pre-pupa phase, which is characterized by the presence of urate salts, and the pupa, where the insect's coloring making it darker as it progresses, due to sclerotization of the cuticle. The urate salts give a dark color to the parasitized eggs, which is a striking characteristic of *Trichogramma* parasitism (Masry and Wakeil, 2019).

A big step toward making this possible was the possibility of creating alternative hosts. This methodology is still in use in countries as Colombia, Peru, and Germany (Parra et al., 2015). However, the choice of alternative host for the creation of *Trichogramma* changed after Lewis et al. (1976) demonstrated that *Anagasta kuehniella* Zeller eggs are more suitable than *S. cerealella*, allowing the development of more fertile wasps with greater longevity. The creation system developed in Brazil for mass production of the flour moth follows the French system (Daumal et al., 1975), requiring a series of modifications to be able to meet the conditions available in the country (Parra, 2010; Parra et al., 2014). Flour moth larvae are raised on artificial diets based on wholewheat flour (97%) and yeast (3%) or a mixture of wheat (40%) and cornflour (60%) (Parra et al., 2014). The sequence used in the *A. kuehniella* production system was described by Parra (2010).

Several factors influence the mass creation of *Trichogramma*, such as the larval density that can culminate in an increase in temperature in the creation unit, which can increase the production of smaller adults, with less fertility and low longevity (Coelho Jr et al., 2016a; Coelho and Parra, 2013a). The concentration of CO_2 in the rearing rooms is another parameter that must be considered as it can reduce egg production (Coelho and Parra, 2013b).

The quality control is fundamental to ensure that parasitoids created in the laboratory are competitive (van Lenteren, 2003; Coelho et al., 2016b). In Brazil, the recommendations of the International Organization for Biological Control (IOBC) for the evaluation of the quality control of mass-reared parasitoids were implemented (Prezotti et al., 2002).

The low cost to produce the parasitoid is extremely important to make it a viable commercial product (Parra, 2002), The selection of wasps resistant to certain chemical molecules is also a promising line of research (Whitten and Hoy, 1999). Haplodiploid wasps suffer much lower levels of inbreeding depression when compared to diploid species (Legner, 1979; Antolin, 1999; Luna and Hawkins, 2004; Prezotti et al., 2004).

36.3 Biological Control in Major Crops

Although the use of *Trichogramma* may at first seem simple, effective pest control is determined by many factors, including: the species of *Trichogramma* used; the quality and suitability of the parasitic product; the number of parasitoids per area; the release method and other interactions between the parasitoid, the targets, the culture and the environment (Bourchier and Smith, 1996).

An important pest for the bean crop is *Chrysodeixis includens* (Walker) (Lepidoptera: Noctuidae), commonly called the soybean looper (Eichlin & Cunningham, 1978, Lafontaine & Poole, 1991). For bean plants, the insect has grown in importance because of defoliation, especially in the latest cropping seasons, which significantly reduces the productivity of the crops (Baldin et al., 2014). Synthetic insecticides have been adopted as the main control method used to combat soybean looper in common bean crops (Ramos et al., 2017; McPherson & MacRae, 2009). However, the soybean looper has shown a higher tolerance to insecticide dosages when compared to other caterpillars that affect beans (Bernardi et al., 2012). Faced with this scenario, bean producers have been looking for efficient and viable alternatives, such as the egg parasitoid, *Trichogramma pretiosum* Riley (Trichogrammatidae), which besides controlling *C. includens* is also effective in *H. armigera*.

Spodoptera frugiperda (Smith) (Lepidoptera: Noctuidae) is one of the main pests in the maize crop, (Querino and Zucchi, 2019; Kuate et al., 2019; Goergen et al., 2016). It is capable of reducing the production of this grain by 45% (Blanco et al., 2014), with reported yield losses of up to 72% (Murúa et al., 2006).

The inclusion of biological control with microorganisms such as *Bacillus thuringiensis* has been very efficient, however, even for these cases it is already possible to find reports of *S. frugiperda* resistance. To assist in the integrated pest management program, *Trichogramma pretiosum* has gained increasing prominence, mainly in Latin American countries (Van Lenteren et al., 2003).

Figueiredo et al. (2015) tested the application of *T. pretiosum* in organic corn. The results showed that 79.2% of the egg masses of *S. frugiperda* were parasitized resulting in a productivity gain of 19.4%. However, more recent studies suggest that the use of *Trichogramma pretiosum* should be reevaluated in biological control programs for *S. frugiperda*, since field cage trials showed a parasitism rate of less than 10% (Jaraleño-Teniente et al., 2020).

Many horticultural pests have been controlled with egg parasitoids, especially *Trichogramma pretiosum* Riley, which has had excellent results in the control of *Plutella xylostella* L. (Plutellidae), reaching parasitism rates of 80% (Meira et al., 2011). Other pests that have efficient control with the use of *T. pretiosum* include *Trichoplusia ni* (Hübner) (Noctuidae) (Krechemer and Foerster, 2015).

In tomato, *T. pretiosum* has been effective in controlling *T. absoluta*, *Neoleucinodes elegantalis* (Guenée) (Crambidae), and *Helicoverpa zea*, with applications associated with the biological insecticide *Bacillus thuringiensis* (Berliner) or chemical insecticides with high selectivity (Silva et al., 2006).

In the soy plantations, the efficiency of *T. pretiosum* has been reported in *Anticarsia gemmatalis* (Hohmann et al., 1989; Frías et al., 1993; Foerster and Avanci, 1999), Chloridea spp. (Martin et al., 1976), *C. virescens* Andrade et al., 2011), *C. includens* (Martin et al., 1976), *E. lignosellus* (Xavier et al., 2011), *H. armigera* (Carvalho et al., 2017), *R. nu* (Frías et al., 1993), species of the spodoptera complex (Siqueira et al., 2012), and *T. ni* (Martin et al., 1976; Milanez et al., 2009).

The timing for the release of *T. pretiosum* can be determined by the beginning of the capture of males in pheromone traps for the main lepidopteran pests. In this phase, two or three releases of 50.000-100.000 parasitoids per hectare are carried out. At the beginning of the reproductive stage, it is recommended to carry out two more releases containing 100.000-200.000 parasitoids per hectare, since the pressure of the pest is greater at this stage (Bueno et al., 2012a, b).

The biological management of soybean pests using *T. pretiosum* has shown good results, being as efficient as chemical insecticides (Corrêa-Ferreira et al., 2010) or even more, probably depending on ideal conditions for insect release.

36.4 Conclusion and Future Prospects

Management strategies that involve efficient, specific, and sustainable control are interesting and should be included in integrated pest management programs. The advances related to the creation of wasps, egg transport, and application strategies in the field have enabled an increase in their area of operation and can reach the most remote places with expected viability and efficiency. Investigations are still needed to select wasps that are more resistant to abiotic stresses and compatible with the applications of the most common pesticides in crops of agricultural importance. However, with the growing demand for sustainable and efficient alternatives, the lines of research with studies related to the use of *Trichogramma* are embracing and promising around the world.

REFERENCES

Andrade, G.S.; Pratissoli, D.; Dalvi, L.P.; Desneux, N.; dos Santos Jr, H.J.G. Performance of four *Trichogramma* species (Hymenoptera: Trichogrammatidae) as biocontrol agents of *Heliothis virescens* (Lepidoptera: Noctuidae) under various temperature regimes. *J Pest Sci.* 2011, 84, 313–320.

Antolin, M.F. A genetic perspective on mating systems and sex ratios of parasitoid wasps. *Popul Ecol.* 1999, 41(1), 29–37.

Baldin, E.; Lourenção, A.; Schlick-Souza, E. Outbreaks of *Chrysodeixis includens* (Walker) (Lepidoptera: Noctuidae) in common bean and castor bean in São Paulo State, Brazil. Bragantia. 2014, 73, 458–461.

Bernardi, O.; Malvestiti, G.S.; Dourado, P.M.; Oliveira, W.S.; Martinelli, S.; Berger, G.U.; Head, P.G.; Omoto, C. Assessment of the high-dose concept and level of control provided by MON 87701 × MON 89788 soybean against Anticarsia gemmatalis and Pseudoplusia includens (Lepidoptera: Noctuidae) in Brazil. Pest Management Science. 2012, 68, 1083–1091.

Bigler, F. Mass production of *Trichogmmma maidis* Pint. et Voeg. and its field application against *Ostrinia nubilalis* Hbn. in Switzerland. *J Appl Entomol*. 1986, 10(1), 23–29.

Blanco, C.A.; Pellegaud, J.G.; Nava-Camberos, U.; Lugo-Barrera, D.; Vega-Aquino, P.; Coello, J.; Terán-Vargas, A.P.; Vargas-Camplis, J. Maize pests in Mexico and challenges for the adoption of integrated pest management programs. J Integr Pest Manage. 2014. 5, 1–9.

Bourchier, R.S.; Smith, S.M. Influence of environmental conditions and parasitoid quality on field performance of *Trichogramma minutum*. *Entomol Exp Appl*. 1996, 80, 461–468.

Bueno, A.F.; Sosa-Gómez, D.R.; Corrêa-Ferreira, B.S.; Moscardi, F.; Bueno, R.C.O.F. Inimigos naturais das pragas da soja. In: Hoffmann-Campo, C.B.; Corrêa-Ferreira, B.S.; Moscardi, F. (eds.) *Soja: manejo integrado de insetos e outros artrópodes-praga*. Embrapa, Londrina, Brasília. 2012a, 493–629.

Bueno, R.C.O.F.; Parra, J.R.P.; Bueno, A.F. *Trichogramma pretiosum* parasitism and dispersal capacity: A basis for developing biological control programs for soybean caterpillars. *Bull Entomol Res*. 2012b, 102, 1–8.

Carvalho, G.S.; Silva, L.B.; Veras, M.S.; dos Santos, E.C.B.; Almeida, M.S.; Reis, S.S. Biological parameters of three *Trichogramma pretiosum* strains (Riley, 1879) (Hymenoptera: Trichogrammatidae) on eggs *Helicoverpa armigera* (Hübner, 1805) (Lepidoptera: Noctuidae). *Acta Sci Biol Sci*. 2017, 39, 349–355.

Coelho Jr, A.; Parra, J.R.P. Effect of carbon dioxide (CO_2) on mortality and reproduction of *Anagasta kuehniella* (Zeller, 1879), in mass rearing, aiming at the production of *Trichogramma* spp. *An Acad Bras Ciênc*. 2013a, 85(2), 823–831.

Coelho Jr, A.; Parra, J.R.P. The effect of rearing in different temperature regimes on the reproduction of *Anagasta kuehniella* (Lepidoptera: Pyralidae). *Environ Entomol*. 2013b, 42(4), 799–804.

Coelho Jr, A.; Reigada, C.; Haddad, M.L.; Parra, J.R.P. Effects of temperature increase caused by larval metabolism on the reproduction of *Anagasta kuehniella* (Lepidoptera: Pyralidae), a factitious host for *Trichogramma*. *Biocontrol Sci Technol*. 2016a, 26(5), 630–639.

Coelho Jr, A.; Rugman-Jones, P.F.; Reigada, C. et al. Laboratory performance predicts the success of field releases in inbred lines of the egg parasitoid *Trichogramma pretiosum* (Hymenoptera: Trichogrammatidae). *PLoS One*. 2016b, 11(1), 146–153.

Corrêa-Ferreira, B.S.; Alexandre, T.M.; Pelizzaro, E.C.; Moscardi, F.; Bueno, A.F. *Práticas de manejo de pragas utilizadas na soja e seu impacto sobre a cultura*. Embrapa Soja, Londrina, Brasília. 2010, Vol. 78, 1–16.

Daumal, J.; Voegelé, J.; Brun, P. Les Trichogrammes: II. Unité de production massive etquotidienne d'un hôte de substitution *Ephestia kuehniella* Zell. (Lepidoptera, Pyralidae). *Ann Zool Ecol Anim*. 1975, 7, 45–49.

Eichlin, T.D.; Cunningham, H.B. The Plusiinae (Lepidoptera: Noctuidae) of America North of México, emphasizing genitalic and larval morphology. Technical Bulletin number 1567. USDA Agricultural Research Service, Washington, DC. 1978.

Figueiredo, M.L.C.; Cruz, I.; Silva, R.B. et al. Biological control with *Trichogramma pretiosum* increases organic maize productivity by 19.4%. *Agron Sustainable Dev*. 2015, 35(3), 1175–1183.

Flanders, S. Mass production of egg parasites of the genus *Trichogramma*. *Hilgardia*. 1930, 4(16), 465–501.

Foerster, F.A.; Avanci, M.R.F. Parasitóides de ovos de *Anticarsia gemmatalis* Hübner (Lepidoptera: Noctuidae) na cultura da soja. *An Soc Entomol Bras*. 1999, 28, 545–548.

Frías, E.A.; Ovruski, S.M.; Popich, S.B. Parasitoides de huevos de lepidópteros noctuidos encontrados en cultivos de soja y su evaluación como agentes de control. *Cirpon*. 1993, 9, 29–35.

Goergen, G.; Kumar, P.L.; Sankung, S.B.; Togola, A.; Tamò, M. First report of outbreaks of the fall armyworm *Spodoptera frugiperda* (J E Smith) (Lepidoptera, Noctuidae), a new alien invasive pest in west and central Africa. *PLoS One*. 2016, 11, e0165632.

Hassan, S.A. Mass production of *Trichogramma*: breeding of the angoumois grain moth *Sitotroga cerealella* (Oliv.) as an alternative host. In: Parra, J.R.P.; Zucchi, R.O. (Org.) *Curso de controle biológico com Trichogramma*. ESALQ, Piracicaba. 1996, 45–64.

Hassan, S.A. The mass rearing and utilization of *Trichogramma* to control lepidopterous pests: Achievements and outlook. *Pestic Sci*. 1993, 37, 387–391.

Hawlitzky, N. Etude de la biologie dc la pyrale du mais, Osrrinia nubilalis Hbn. en rtgion parisienne durant quatre ann& et recherche d'6ltments prtvisionnels du debut de ponte. *Acta Oecol Oecol Appl*. 1986, 1, 47–68.

Hohmann, C.L.; Silva, S.M.T.; Santos, W.J. Lista preliminar de Trichogrammatidae encontrados no Paraná. *An Soc Entomol Bras*. 1989, 18, 203–206.

Jaraleño-Teniente, J.; Lomeli-Flores, J.R.; Rodríguez-Leyva, E.; Bujanos-Muñiz, R.; Rodríguez-Rodríguez, S.E. Egg parasitoids survey of *Spodoptera frugiperda* (Smith) (Lepidoptera: Noctuidae) in maize and sorghum in central Mexico. *Insects*. 2020, 11, 157.

Knutson, A. *The Trichogramma Manual: A Guide to the Use of Trichogramma for Biological Control with Special Reference to Augmentative Releases for Control of Bollworm and Budworm in Cotton: Texas Agricultural Extension Service*. The Texas A&M University System, College Station, TX, 1998.

Krechemer, F.S.; Foerster, L.A. Temperature effects on the development and reproduction of three *Trichogramma* (Hymenoptera: Trichogrammatidae) species reared on *Trichoplusia ni* (Lepidoptera: Noctuidae) eggs. *J Insect Sci*. 2015, 15(1), 1–6.

Kuate, A.F.; Hanna, R.; Doumtsop Fotio, A.R.P. et al. *Spodoptera frugiperda* Smith (Lepidoptera: Noctuidae) in Cameroon: Case study on its distribution, damage, pesticide use, genetic differentiation and host plants. *PLoS ONE*. 2019, 14, 4.

Lafontaine, J.D.; Poole, R.W. Noctuoidea (part): Plusiinae. In Hodges, R.W.; Davis, D.R.; Dominic, T.; Ferguson, D.C.; Franclemont, J.G.; Munroe, E.G.; Powell, J.A. (eds.) *The Moths of America North of Mexico, fasc.25.1*. Allen Press Lawrence, KS. 1991, 1–182.

Laurentis, V.L.; Ramalho, D.G.; Santos, N.A. et al. Performance of *Trichogramma pretiosum* Riley (Hymenoptera: Trichogrammatidae) on eggs of *Helicoverpa armigera* (Hübner) (Lepidoptera: Noctuidae). *Sci Rep*. 2019, 9, 1156.

Legner, E.F. Prolonged culture and inbreeding effects on reproductive rates of 2 pteromalid parasites of muscoid flies. *Ann Entomol Soc Am.* 1979, 72(1), 114–118.

Lewis, W.J.; Nordlund, D.A.; Gross Jr, H.R. et al Production and performance of *Trichogramma* reared in eggs of Heliothis zea and other hosts. *Environ Entomol.* 1976, 5(3), 449–452.

Li, L.Y. Worldwide use of *Trichogramma* for biological control on different crops: A survey. In: Wajnberg, E.; Hassan, S.A. (eds.), *Biological Control with Egg Parasitoids.* CABI, Oxfordshire, UK. 1994, 37–53.

Luna, M.G.; Hawkins, B.A. Effects of inbreeding versus outbreeding in *Nasonia vitripennis* (Hymenoptera: Pteromalidae). *Environ Entomol.* 2004, 33(3), 765–775.

Martin, P.B.; Lingren, P.D.; Greene, G.L.; Ridgway, R.L. Parasitization of two species of Plusiinae and Heliothis spp. after releases of *Trichogramma pretiosum* in seven crops. *Environ Entomol.* 1976, 5, 991–995.

Masry, S.H.D.; El-Wakeil, N. Egg parasitoid production and their role in controlling insect pests. In: El-Wakeil, N.; Saleh, M. (eds.), *Cottage Industry of Biocontrol Agents and Their Applications.* Springer, Berlin. 2019, 3–47.

McPherson, R.M.; Macrae, T.C. Evaluation of transgenic soybean exhibiting high expression of a synthetic *Bacillus thuringiensis* cry1A transgene for suppressing lepidopteran population densities and crop injury. *J Econ Entomol.* 2009, 1640–1648.

Meira, A.L.; Pratissoli, D.; Souza, L.P.; Strum, G. Seleção de espécies de *Trichogramma* sp. em ovos da traça-das-crucíferas, *Plutella xylostella. Rev Caatinga.* 2011, 24(2), 1–8.

Milanez, A.M.; Pratissoli, D.; Polanczyk, R.A. et al. Avaliação de *Trichogramma* spp. para o controle de *Trichoplusia ni. Pesqui Agropecu Bras.* 2009, 44, 1219–1224.

Morrison, R.K.; Jones, S.L.; Lopez, J.D. A unified system for the production and preparation of *Trichogramma pretiosum* for field release. *Southwest Entomol.* 1978, 3, 62–68.

Murúa, G.; Molina-Ochoa, J.; Coviella, C. Population dynamics of the fall armyworm, *Spodoptera frugiperda* (Lepidoptera: Noctuidae) and its parasitoids in northwestern Argentina. *Fla. Entomol.* 2006, 89, 175–182.

Newton, P.J. Increasing the use of trichogrammatids in insect pest management: A case study from the forests of Canada. *Pestic Sci.* 1993, 37, 381–386.

Oliveira, R.C.M., Pastori, P.L., Coutinho, C.R., Juvenal, S.O., Aguiar, C.V.S. Natural parasitism of *Trichogramma pretiosum* (Hymenoptera: Trichogrammatidae) in *Neoleucinodes elegantalis* (Lepidoptera: Crambidae) eggs on tomato (Solanales: Solanaceae) in the Northeast region, Brazil. *Braz J Biol.* 2019, 80, 474–475.

Parra, J.R.P. Criação massal de inimigos naturais. In: Parra, J.R.P., Botelho, P.S.M.; Corrêa-Ferreira, B.S. et al. (eds.), *Controle biológico no Brasil: parasitóides e predadores.* Manole. 2002, 143–164, Barueri, SP, Brazil.

Parra, J.R.P. Mass rearing of egg parasitoids for biological control programs. In: Cônsoli, F.L.; Parra, J.R.P.; Zucchi, R.A. (eds.), *Egg parasitoids in agroecosystems with emphasis on Trichogramma.* Springer, Berlin. 2010, 267–292.

Parra, J.R.P.; Coelho Jr, A.; Geremias, L.D. et al. Criação de *Anagasta kuehniella*, em pequena escala, para produção de *Trichogramma. Occasio.* 2014. 1–32.

Parra, J.R.P., Zucchi, R.A.; Coelho Jr, A. et al. *Trichogramma* as a tool for IPM in Brazil. In: Vinson, B.; Greenberg, S.M.; Liu, T. et al. (eds.), *Augmentative Biological Control Using Trichogramma spp.: Current Status and Perspectives.* Northwest A&F University Press, Yangling, China. 2015, 472–496.

Prezotti, L.; Parra, J.R.P., Vencovsky, R. et al. Effect of the size of the founder population on the quality of sexual populations of *Trichogramma pretiosum*, in laboratory. *Biol Control.* 2004, 30(2), 174–180.

Prezotti, L.; Parra, J.R.P., Vencovsky, R. et al. Teste de vôo como critério de avaliação da qualidade de *Trichogramma pretiosum* Riley (Hymenoptera: Trichogrammatidae): adaptação de metodologia. *Neotrop Entomol.* 2002, 31(3), 411–417.

Querino, R.B.; Zucchi, A.R. Annotated checklist and illustrated key to the species of *Trichogramma Westwood* (Hymenoptera: Trichogrammatidae) from South America. *Zootaxa.* 2019, 4656(2), 201–231.

Querino, R.B.; Silva, N.N.P.; Zucchi, R.A. *Trichogramma* na Amazônia: visão geral e potencialidades. In: Silva, N.M.; Adaime, R.; Zucchi, R.A. (eds.) *Pragas agrícolas e florestais na Amazônia.* Embrapa Amapá, Macapá. 2016, 597–606.

Ramos, R.; Sediyama, C.; Queiroz, E. et al. Toxicity of insecticides to *Chrysodeixis includens* and their direct and indirect effects on the predator Blaptostethus pallescens. *J Appl Entomol.* 2017, 141, 677–689.

Silva, J.B.C.; Giordano, L.B.; Furumoto, O. et al. *Cultivo de tomate para industrialização.* Embrapa Hortaliças. 2006, Brasília, DF - Brasil.

Siqueira, J.R.; Bueno, R.C.O.F.; Bueno, A.F.; Vieira, S.S. Preferência hospedeira do parasitoide de ovos *Trichogramma pretiosum. Ciênc Rural.* 2012, 42, 1–5.

Smith, S.M. Biological control with *Trichogramma*: advances, successes, and potential of their use. *Annu Rev Entomol.* 1996, 41, 375–406.

Van Lenteren, J.C. Need for quality control of mass produced biological control agents. In: Van Lenteren, J.C. (ed.) *Quality Control and Production of Biological Control Agents: Theory and Testing Procedures.* CABI, Oxfordshire, UK. 2003, 1–18.

Van Lenteren, J.C.; Bueno, V.H. Augmentative biological control of arthropods in Latin America. *BioControl.* 2003, 48, 123–139.

Van Lenteren, J.C.; Glas, R.C.G.; Smits, P.H. Evaluation of control capabilities of *Trichogramma* and results of laboratory and field research on *Trichogramma* in the Netherlands. In: Les trichogrammes. Ier Symposium International, Antibes, 20–23 avril 1982. 1982, 257–269.

Whitten, M.J.; Hoy, M.A. Genetic improvement and other genetic considerations for improving the efficacy and success rate of biological control. In: Bellows, T.S.; Fisher, T.W. (eds.) *Handbook of Biological Control: Principles and Applications of Biological Control.* Academic Press, New York. 1999, 271–296.

Xavier, L.M.S.; Laumann, R.A.; Borges, M. et al. *Trichogramma pretiosum* attraction due to the *Elasmopalpus lignosellus* damage in maize. *Pesqui Agropecu Bras.* 2011, 46(6), 578–585.

Zucchi, R.A.; Querino, R.B.; Monteiro, R.C. Diversity and hosts of *Trichogramma* in the New World, with emphasis in South America. In: Cônsoli, F.L.; Parra, J.R.P.; Zucchi, R.A. (eds) *Egg Parasitoids in Agroecosystems with Emphasis on Trichogramma.* Springer, Dordrecht. 2010, 219–236.

37

Biological Control of Mite Pests in Organic Farming

Hany Mohamad Galal El-Kawas and Mohamad Mohamad Ahamad Khedr

CONTENTS

37.1 Introduction

Spider mites and other phytophagous mites were generally minor pests of agricultural crops prior to World War II. This changed rapidly after the war, with the extensive use of pesticides and fertilizers (Huffaker et al., 1970). Spider mites developed resistance to multiple pesticides, especially in greenhouses and vegetable production systems where extensive use of pesticides (Georghiou & Saito, 1983). Pesticides are harmful to natural enemies, soil, water supply, and agricultural workers, and are expensive. Pesticides are simpler to deploy when compared to integrated pest management (IPM), which requires detailed information about the crop, pests, and their natural enemies. The experienced growers act before problems reach devastating levels because they know that pests are more effectively controlled when their numbers are low (Dhooria, 2006).

37.2 Biological Control

To save on the high cost of importing acaricides and to replace those that are made useless by mite resistance, the emphasis was on using biological control. According to van Lenteren (2012), Acari is the second biggest group (after Hymenoptera) of natural enemies used commercially in the biological control of arthropod pests.

Predatory mites are distinguished by certain characteristics such as their simple mass rearing, they can control different species of pests, the ability to release them by mechanical means; they are relatively small in size and do not spread across a wide area.

Mesostigmata is a large mite group containing about 11,500 species corresponding to about 20% of all known mite species (Beaulieu et al., 2011; Zhang, 2013). These are found in different habitats, but are most abundant and diverse in soil (El-Kawas, 2011; El-Kawas & Negm, 2018), where they feed on small insects, nematodes, collembola, or on phytophagous and mycophagous mites (Krantz, 1978). Prostigmata predators are also found in diverse habitats, including soils and overlaying litter layers as well as aerial parts of plants such as species of genus *Agistemus*, and they play an important role in biological control of economic pests (El-Sawi & Momen, 2006; Momen, 2012).

37.3 Mesostigmata as Biological Control Agents

Mesostigmata have been studied because of their potential to be used as biological control agents of agricultural pests or of parasites. Few studies have been conducted to validate the potential of the Laelapidae species in the open field conditions. Most of these studies refer to the use of *Gaeolaelaps aculeifer* (Momen et al., 2020). They are widespread in soil, litter, manure, decaying woods, and galleries of bark beetles (Lindquist et al., 2009b; Castilho et al., 2012).

Predatory mites of the family Phytoseiidae are successfully used as biocontrol agents in controlling phytophagous mites, thrips, and whitefly (Helle & Sabelis, 1985). Several (researchers) have studied the biology of these predators and concluded that there were many food types of phytoseiids, including small arthropods, nematodes, fungi, and pollen (McMurtry et al., 2013). The family Ascidae has many predators such as *Gamasellodes adrianae*, *G. bicolor*, and *G. claudiae*, which were confirmed experimentally under laboratory conditions to predate on nematodes (Walter & Ikonen, 1989; Walter, 2003). The family Melicharidae was represented by five species in Egypt, while the most dominant genus was *Proctolaelaps* which have been shown to develop and reproduce on nematodes. *Agistemus exsertus* is one of the most important species of the family Stigmaeidae (Acari: Prostigmata) in Egypt, since it was associated with many economic pests and able to attack tetranychoid and eriophyoid mites as well as scale insects and whitefly in field crops and orchards.

37.3.1 Family Phytoseiidae

Generally, predatory phytoseiid mites are related to the phytophagous mites and/or insect pests in the open field conditions. Therefore, field applications conducted to evaluate the efficacy of these predators as successful biological control agents against several agricultural pests infested many crops growing in the greenhouses/field. The mite predator *Phytoseiulus persimilis* (Athias-Henriot), *A. swirskii* (Athias-Henriot), *Neoseiulus californicus* (McGregor), *P. macropilis* (Banks), *N. cucumeris* (Oudemans), and *Neoseiulus zaheri* are the most phytoseiid species used in releasing program to suppress *T. urticae* and some insect pests. The mite predators *N. californicus* and *P. persimilis* were used successfully for controlling *T. urticae* on strawberry, where their release resulting in reduction percentages of *T. urticae* infestation (71.78 and 97.20%, respectively) (El-Saiedy, 2003). The predatory mite species *N. californicus* and *P. macropilis* were released against *T. urticae* infesting two cultivars of cucumber that grown in plastic houses. Both predators showed maximum percentages of reduction of *T. urticae* (97.36 and 97.51%) and (87.14 and 92.50%) in both cultivars, respectively (Hassan et al., 2007).

Neoseiulus californicus has been used effectively in the management of *T. urticae* in addition to *Panonychus ulmi* and *Oligonychus perseae* in several countries (Jolly, 2000; Elmoghazy et al., 2011; Greco et al., 2011).

The predatory mite *N. cucumeris* caused a greater reduction in *Aphis gossypii* and *T. urticae* infestations than *N. californicus* released in open field on two eggplant cultivars (El-Kholy & El-Saiedy, 2009). The mites *N. californicus* and *A. swirskii* gave a high performance to control *T. urticae* on two cultivars of *Vicia faba* in open field at Beheira Governorate (Elmoghazy et al., 2011). *N. Californicus* significantly reduces population density of *T. urticae* (87.22 and 74.22% mite/leaf reduction) compared to *T. swirskii* (57.49 and 41.5% mite/leaf reduction) for both cultivars, respectively. Releasing of *N. californicus* in addition to *A. swirskii* can give good results in controlling *T. urticae* (Hassan, 2013). Furthermore, the release of *P. persimilis* and *N. californicus* resulted in reduction of *T. urticae* population on sweet pepper growing in greenhouses during two successive seasons (Hassan, 2013). Under both open field and plastic low tunnels conditions, releasing *P. persimilis* and *A. swirskii* to control *T. urticae* on two watermelon and muskmelon cultivars revealed excellent results (Momen et al., 2020). The highest percentage of reduction was attained by *P. persimilis*, followed by *T. swirskii* on all tested plants (Abou-Awad et al., 2017). Kame et al. (2018) examined two sweet pea cultivars, releasing *P. persimilis* and *N. californicus*, under high plastic-net condition tunnels, releasing both species gave the best results in reducing the population of *T. urticae* as compared to *Typhlodromus negevi* and *E. scutalis*, probably because both species are pollen feeder.

37.3.2 Family Laelapidae

Laelapidae is a large family, with over 50 genera and several hundred species (Gerson et al., 2003). Many laelapids are ectoparasitic on small mammals or associated with various arthropods, other species are free-living soil-borne predators. This family includes about 1,300 species classified in over 90 genera (Beaulieu et al., 2011).

Four predatory mites of the Laelapidae family were also used for the control of pests that spend part of their life in the soil. These species are *Androlaelaps casalis*, *Gaeolaelaps aculeifer* (Canestrini), *Stratiolaelaps miles*, and *Stratiolaelaps scimitus*. *Cosmolaelaps barbatus* (Moreira, Klompen & Moraes) and *C. jaboticabalensis* (Moreira, Klompen & Moraes) were observed to feed and oviposit on *Rhizoglyphus echinopus* (Fumouze & Robin) (Astigmatina) (unpublished observation for Moreira & de Moraes, 2015). Complete development on the same prey was reported for *C. claviger* (Berlese) (Afifi & van der Geest, 1984) and *C. vacuus* (Michael) (Abou-Awad et al., 1989), and on *Tyrophagus putrescentiae* Schrank (Astigmatina) for *Cosmolaelaps paulista* Freire and Moraes (Freire & Moraes, 2007), *C. jaboticabalensis* (Moreira & de Moraes, 2015) and *Cosmolaelaps qassimensis* Fouly & Al-Rehyiani (Fouly & Abdel-Baky, 2015). Afifi & van der Geest (1984) reported *C. claviger* to oviposit on immature stages of *Tetranychus urticae* Koch (Prostigmata). Walter (1987) observed complete development and oviposition of *C. vacuus* on *Oppiella nova* (Oudemans) (Oribatida) and *T. putrescentiae*. *Cosmolaelaps simplex* Berlese was able to complete development on *Caloglyphus rodriguezi* Samsinak (Astigmatina), while its larval stage was observed to be completed without feeding (Al Rehiayani & Fouly, 2005) and *C. qassimensis* on *Caloglyphus rodriguez* Samsinak (Fouly & Abdel- Baky, 2015).

In Egypt, six laelapids were recorded with crops and weeds during a survey in some governorates in Egypt (Moment et al., 2020) namely, *Stratiolaelaps marianum*, *Hypoaspis reticulatus* (Xu & Liang), *Ololaelaps chanti*, *Cosmolaelaps keni*, *Geolaelaps aculeifer*, and *Androlaelaps aegypticus*. In a recent study, Joharchi and Negm (2020) revised the taxonomic status of ten laelapid species from Egypt including six new records.

37.3.3 Family Ascidae

Ascids are free-living predators inhabiting upper soil layers, plant parts, stored products, grasses, and in organic matter where they feed on nematodes and arthropods. *Arctoseius cetratus* (Sellinck) observed the predation of *Tarsonemus myceliophagus* (Hussey) and preyed on eggs and first instar larvae of *Drosophila melanogaster* (Drosophilidae) and of *Lycoriella auripila* (Sciaridae) Binns (1972). *Lasioseius parberlesei* Tseng is a natural enemy of *Steneotarsonemus spinki* Smiley (Tarsonemidae), a main acarine pest of rice (Tseng, 1984). Walter (1987) reported that *Gamasellodes vermivorax* Walter feed on mites of very different families (including predatory mites): namely of the families Alicorhagiidae and Tydeidae, as well as on *Asca nesoica* Athias-Henriot (Ascidae), *Cosmolaelaps vacua* (Michael) (Laelapidae), *Pilogalumna* sp. (Oribatida) and *T. putrescentiae* (Schrank). The mite *Asca macromela* (Walter, Halliday, and Lindquist) was supposed by Walter et al. (1993) to be a predator of tenuipalpid species *Brevipalpus* (Donnadieu).

In Egypt, *Protogamasellus minutus* (Nasr) developed on the mite *Rhizoglyphus robini* Claparède (Acaridae) and *T. putrescentiae* (Afifi et al., 1986) and the mite *P. primitivus similis* (Genis) was observed on the same previous mite prey by Nawar and Nasr (1989). Nine ascid species were surveyed in some Egyptian governorates: *Lasioseius africanus* Nasr; *L. aegypticus* Afifi; *L. zahri*; *L. athiasae* Nawar & Nasr; *L. lindquisti* Nawar & Nasr; *Blattisocius tarsalis* (Berlese); *Cheiroseius egypticus* Reda, Nawar & El -Borolossy; *C. nepalensis* (Evans &. Hyatt) and *C. crassipes* Ramadan (Momen et al., 2020).

37.3.4 Family Rhodacaridae

Three genera of the family Rhodacaridae were reported by various authors, included *Rhodacarellus* (Willmann), *Multidentorhodacarus* (Karg), and *Protogamasellopsis* sp. Nematodes have been reported many times to be preyed by rhodacarids (Castilho et al., 2012). Only one species was recorded from Egypt namely *Protogamasellopsis denticus* Nasr. This species was able to feed on insect and mite foods. Insects prey included: *Bactrocera zonata*, *Phothorimaea operculella*, *Agrotis ipsilon*, and *Agrotis ipsilon* while the mites' prey included *Rhizoglyphus robini* (Claparède) and *T. putrescentiae* (Momen et al., 2020).

37.3.5 Family Melicharidae

The most important genus in the Melicharidae family is *Proctolaelaps*, and can play an important role in biological control. *Proctolaelaps bickleyi* (Bram), *P. deleoni* have been formed and reproduced on different food types, such as nematodes, mites, pollen, and fungi. *Proctolaelaps cyinchuanensis* was observed to feed on *Suidasia pontifi* (Navasero et al., 2004). Most *Proctolaelaps* species studied can develop and reproduce on several food items, indicating the polyphagous habit of members of this genus. In Egypt, five species were recorded during a survey at some governorate in Egypt as follow: *Proctolaelaps bickleyi* (Bram); *P. pygmaeus* (Müller); *P. orientalis* Bhattacharyya; *P. naggarii* (Taha & Mohmoud) and *P. aegyptiaca* (Nasr) (Momen et al., 2020).

37.4 Prostigmata as Biological Control Agents

Prostigmata is the larger of the two suborders of the order Trombidiformes with the mite superorder Acariformes (Walter et al., 2009). It consists of 36 superfamilies, some of which, are predatory or parasitic.

These mites occupy a wide range of habitats, many being edaphic, others living in the water, on plants or on animals. Feeding habits include predation, parasitism, fungivory, pollenophagy, etc. Several major groups contain pest species of agricultural importance.

37.4.1 Family Stigmaeidae

The species of family Stigmaeidae belongs to superorder Acariformes, order Trombidiformes, suborder Prostigmata, supercohort Eleutherengoides, cohort Raphignathina, Superfamily Raphignathoidea (Lindquist et al., 2009a). The species of this family have a worldwide range of distribution in all biogeographical regions though most of them were discovered from the Palaearctic, Oriental, Nearctic, Afrotropical and Australian regions (Fan & Zhang, 2005).

The species of Stigmaeidae are generalist main predators feeding on gall and rust mites (Eriophyoidea), spider mites (Tetranychidae), false spider mites (Tenuipalpidae), broad mites (Tarsonemidae), and other mites in the field (Gerson et al., 2003) and greenhouses (Zhang, 2003). They are also known to feed on pollen of various plants, eggs and crawlers of small insects such as Hemiptera and eggs of Lepidoptera.

Momen et al. (2020) mentioned that three important predatory stigmaeid species were recorded in Egypt; namely, *Agistemus exsertus* (Gonzalez), *A. olive* (Romeih), and *A. vulgaris* (Soliman & Gomaa) were recorded during a survey conducted at some governorate in Egypt. The functional responses have been studied for *A. exsertus* (Abou-Awad & Elsawi, 1993; El Sawi & Momen, 2006) and *Zetzellia mali* (Ewing) (Santos, 1982; Lawson & Walde, 1993).

The main problems reducing the efficacy of stigmaeids are artificial food, the techniques reducing the cannibalistic behavior and rearing of the pesticide resistance strains are probably the three key factors for the commercial application of Stigmaeidae (Momen et al., 2020).

REFERENCES

Abou-Awad, B. A.; Afia, S. I.; El-Saiedy, E. Population dynamics of tetranychid mite and its predator on watermelon and muskmelon and effect of mite feeding on the phytochemical components of the host plants. *Biosci. Res.* 2017, *14*, 879–886.

Abou-Awad, B. A.; Elsawi, S. A. Biology and life table of the predacious mite, *Agistemus exsertus* Gonz. (Acari: Stigmaeidae). *Anz. Schaedlingskd. Pflanzenschutz Umweltschutz.* 1993, 66, 101–103.

Abou-Awad, B. A.; Nasr, A. K.; Gomaa, E. A.; Abou-Elela, M. Feeding, development and reproduction of the predatory mite, *Hypoaspis vacua* on various kinds of food substances (Acari: Laelapidae). *Int. J. Trop. Insect. Sci.* 1989, *10*, 503–506.

Afifi, A.; van der Geest, L. Notes on the development and biology of the predaceous soil mite *Cosmolaelaps claviger* (Berlese 1883) (Gamasida: Laelapidae). In: Griffiths, D.; Bowman, C. Eds.; *Acarology IV, Vol I.* Ellis Horwood Limited, Chichester, 1984, pp. 585–590.

Afifi, A. M.; Hassan, M. F.; Nawar, M. S. Notes on the biology feeding habits of *Protogamasellus minutus* Hafez, El-Badry & Nasr (Acari: Gamasida: Ascidae). *Bull. Entomol. Soc. Egypt.* 1986, *66*, 251–259.

Al Rehiayani, S.M.; Fouly, A. H. *Cosmolaelaps simplex* (Berlese), a polyphagous predatory mite feeding on root-knot nematode *Meloidogyne javanica* and citrus nematode *Tylenchulus semipenetrans. Pak. J. Biol. Sci.* 2005, *8*, 168–174.

Beaulieu, F.; Dowling, A. P. G.; Klompen, H.; Moraes, G. J.; Walter, D. Superorder parasitiformes reuter, 1909. In: Zhang, Z.-Q. Ed.; Animal biodiversity: An outline of higher-level classification and survey of taxonomic richness (Addenda 2013). *Zootaxa.* 2011, 3703, 123–128.

Binns, E. S. *Arctoseius cetratus* (Sellnick) (Acarina: Ascidae) phoretic on mushroom sciarid flies. *Acarologia.* 1972, 14, 350–356.

Castilho, R. C.; de Moraes, G. J.; Halliday, B. Catalogue of the mite family Rhodacaridae Oudemans, with notes on the classification of the Rhodacaroidea. *Zootaxa.* 2012, *3471*, 1–69.

Dhooria M. S. Management of mites under organic farming. In: *Fundamentals of Applied Acarology.* Springer, Singapore, 2006.

El-Kawas, H. M. G. *Acarines As Biological Control Agents: An Overview of Bio-Relationships Between Mites and Insects in Egypt.* Lap Lambert Academic. 2011, 128pp.

El-Kawas, H. M. G.; Negm, M. W. Parasitic and phoretic mites (Arachnida, Acari) reported from insects (Arthropoda: Insecta) in Egypt. *J. Insect Biodivers. Syst.* 2018, *4*(1), 57–71.

El-Kholy, M. Y.; El-Saiedy, E. M. A. K. Biological control of *Thrips tabaci* (Lind.) and *Aphis gossypii* (Glover) using different predatory Phytoseiid mites and the biocide Vertimec on eggplant at Behaira Governorate. *Egypt. Acad. J. Biol. Sci.* 2009, *2*, 13–22.

El-Saiedy, E. M. A. Integrated control of red spider mite *Tetranychus urticae* Koch on strawberry plants. Ph.D. Dissertation, Faculty of Agriculture, Cairo University, Egypt. 2003, p. 170.

El-Sawi, S. A.; Momen, F. M. *Agistemus exsertus* Gonzalez (Acari: Stigmaeidae) as a predator of two scale insects of the family Diaspididae (Homoptera: Diaspididae). *Arch. Phytopathol. Plant. Prot.* 2006, *39*, 421–427.

Elmoghazy, M. M. E.; El-Saiedy, E. M. A.; Romeih, A. H. M. Integrated control of the two spotted spider mite *Tetranychu surticae* Koch (Tetranychidae) on faba bean *Vicia faba* (L.) in an open field at Behaira Governorate, Egypt. *Int. J. Environ. Sci. Eng.* 2011, *2*, 93–100.

Fan, Q.-H.; Zhang, Z.-Q. *Raphignathoidea (Acari: Prostigmata).* 52, *Fauna of New Zealand.* Manaaki Whenua Press, Lincoln/Canterbury, 2005, p. 396.

Fouly, A. H.; Abdel-Baky, N. F. Influence of prey types on the biological characteristics of *Cosmolaelaps qassimensis* (Acari: Laelapidae). *J. Entomol.* 2015, *12*, 21–29.

Freire, R.; Moraes, G. Description of a new species of *Cosmolaelaps* Berlese (Acari: Laelapidae: Hypoaspidinae) from Brazil and its biological cycle. *Intl. J. Acarol.* 2007, *33*, 353–358.

Georghiou, G. P.; Saito, T. *Pest Resistance to Pesticides.* Plenum Press, New York. 1983.

Gerson, U.; Smiley, R. L.; Ochoa, R. *Mites (Acari) for Pest Control.* Blackwell Science, Oxford, UK. 2003, p. 539.

Greco, N. M.; Liljesthröm, G. G.; Ottaviano, M. F. G.; Cluigt, N.; Cingolani, M. F.; Zembo, J. C.; Sánchez, N. E. Pest management plan for the two-spotted spider mite, *Tetranychus urticae*, based on the natural occurrence of the predatory mite *Neoseiulus californicus* in strawberries. Int. J. Pest. Manage. 2011, *57*, 299–308.

Hassan, A. S. Biological and chemical control of two spotted spider mite and important insects infesting sweet pepper in green houses in Egypt. Ph.D. dissertation, Faculty of Agriculture, Cairo University, Egypt. 2013.

Hassan, M. F.; Ali, F. S.; Hussein, A. M.; Mahgob, M. H. Control measures of *Tetranychs urticae* Koch on two cucumber cultivars in plastic houses. *Acarines.* 2007, *1*, 11–15.

Helle, W.; Sabelis, M. W. *Spider Mites: Their Biology*, Natural Enemies and Control. Elsevier, Amsterdam. 1985, p. 458.

Huffaker, C. B.; van de Vrie, M.; McMurtry, J. A. Ecology of tetranychid mites and their natural enemies: A review. II. Tetranychid populations and their possible control by predators: An evaluation. *Hilgardia.* 1970, *40*(11), 391–458.

Joharchi, O. and Negm, M.W. Soil-inhabiting mites of the family Laelapidae from Assiut governorate, Egypt. *Zootaxa.* 2020, 4759(4), 488–510.

Jolly, R. L. The predatory mite *Neoseiulus californicus*: Its potential as biological controlagent for the fruit tree red spider mite, Panonychus ulmi. In: The BCPC Conference: Pests and Diseases, Brighton Hilton Metropole Hotel, Brighton, UK. 2000, 1, 487–490.

Kame, M. S.; Afia, S. I., El-Saiedy, E. M. A. Biological control of *Tetranychus urticae* (Acari: Tetranychidae) using four predatory mites (Acari: Phytoseiidae) on two sweet pea cultivars. *Biosci. Res.* 2018, *15*, 185–191.

Krantz, G. W. *A Manual of Acarology*, 2nd Edn. Oregon State University Bookstores, Corvallis. 1978, p. 509.

Lawson, A. B.; Walde, S. J. Comparison of the responses of two predaceous mites, *Typhlodromus pyri* and *Zetzellia mali*, to variation in prey density. *Exp. Appl. Acarol.* 1993, *17*, 811–821.

Lindquist, E. E.; Krantz, G. W.; Walter, D. E. Chapter 8: Classification. In: Krantz, G. W.; Walter, D. E. Eds.; *A Manual of Acarology*, 3rd Edn. Texas Tech University Press, Lubbock. 2009a, pp. 97–103.

Lindquist, E. E.; Krantz, G. W.; Walter, D. E. Order Mesostigmata. In: Krantz, G. W.; Walter, D. E. Eds.; *A Manual of Acarology*, 3rd Edn. Texas Tech University Press, Lubbock. 2009b, pp. 124–232.

McMurtry, J. A.; Moraes, J. G.; Sourassou, F. N. Revision of the lifestyles of phytosiid mites and implication for biological control strategies. *Syst. Appl. Acarol.* 2013, *4*, 297–320.

Momen, F. M. Influence of life diet on the biology and demographic parameters of *Agistemus olive* Romeih, a specific predator of eriophyid mites (Acari: Stigmaeidae and Eriophyidae). *Trop. Life. Sci. Res.* 2012, *23*, 25–34.

Momen, F.; Fahim, S.; Barghout, M. Mass production of predatory mites and their efficacy for controlling pests. In: El-Wakeil, N.; Saleh, M.; Abu-Hashim, M. Eds.; *Cottage Industry of Biocontrol Agents and Their Applications.* Springer Nature Switzerland AG, Cham, Switzerland. 2020, pp. 157–200.

Moreira, G. F.; de Moraes, G. J. The potential of free-living laelapid mites (Mesostigmata: Laelapidae) as biological control agents. In: Carrillo, D.; de Moraes, G.; Peña, J. Eds.; *Prospects for Biological Control of Plant Feeding Mites and Other Harmful Organisms. Progress in Biological Control, 19.* Springer, Cham. 2015.

Navasero, M. M.; Hirao, G. A.; Santiago, D. R.; Navasero, M. V.; Raros, L. C. Laboratory rearing technique for the predatory *Proctolaelaps yinchuanensis* Xue, Sui & Yi (Ascidae: Gamasida: Acari) using *Suidasia pontifica* Oudemans (Suidasiidae: Acaridae: Acari). *Philip. Entomol.* 2004, *18*,180.

Nawar, M. S.; Nasr, A. K. Biology of the Ascid mite *Protogamasellus primitivus similis* Genis, Loots and Ryke with description of immature stages (Acari: Mesostigmata: Ascidae). *Bull. Fac. Agr.* Cairo Univ. 1989, *68*, 85–94.

Santos, M. A. Effects of low prey densities on the predation and oviposition of *Zetzellia mali* (Acarina: Stigmaeidae). *Environ. Entomol.* 1982, *11*, 972–974.

Tseng, Y. H. Mites associated with weeds, paddy rice and upland rice fields in Taiwan. In: Griffiths, D. A.; Bowman; C. E. Eds.; *Acarology VI.* Ellis Horwood, Chichester. 1984, Vol. 2, pp. 770–780.

van Lenteren, J. C. The state of commercial augmentative biological control: Plenty of natural enemies, but a frustrating lack of uptake. *Biol. Control.* 2012, *57*, 1–20.

Walter, D. E. The genus *Gamasellodes* (Acari: Mesostigmata: Ascidae) New Australian and North American species. *Syst. Appl. Acarol.* (Spec. Publ. Lond.). 2003, *15*, 1–10.

Walter, D. E. The influence of prey type on the development and reproduction of some predatory soil mites. *Pedobiologia.* 1987, *30*, 419–424.

Walter, D. E.; Halliday, R. B.; Lindquist, E. E. A review of the genus *Asca* (Acarina: Ascidae) in Australia, with descriptions of three new leaf-inhabiting species. *Invert. Taxon.* 1993, *7*, 1327–1347.

Walter, D. E.; Ikonen, E. K. Species, guilds, and functional groups: Taxonomy and behavior in nematophagous arthropods. *J. Nematol.* 1989, *21*(3), 315–327.

Walter, D. E.; Lindquist, E. E.; Smith, I. M.; Cook, D. R.; Krantz, G. W. Order Trombidiformes. In: Krantz, G. W.; Walter, D. E. Eds.; *A Manual of Acarology*, 3rd Edn. Texas Tech University Press, Lubbock. 2009, pp. 233–420.

Zhang, Z.-Q. *Mites of Greenhouses: Identification, Biology and Control.* CABI, Wallingford. 2003, 244pp.

Zhang, Z.-Q. Phylum Arthropoda. *Zootaxa.* 2013, *3703*, 17–26.

Section XVI

Management of the Diseases of Crops through Parasitoids Biopesticides

38

Management of the Diseases of Crops through Parasitoids Biopesticides: Family Eulophidae

Aparna Shree Singh

CONTENTS

38.1 Introduction

At present, about two million species of organisms are scientifically identified and named, while an estimated 5–100 million or more await discovery (Wilson, 2003). Every species is thus a mosaic of physical variety and genetic constitution that can lead to both taxonomic and ecological ambiguity in interpreting its integrity and the ways in which it may evolve and persist.

38.2 Insects: The Most Successful Living Organisms

Considerable debate continues over how many species of insects are in the world. Estimates range from 2 to 50 million (Stork, 1993). The lower figure is given by Hodkinson and Casson (1991). The higher figure of up to 50 million is provided by Erwin (1988, 1993). The number of insects described at present is estimated to be 9,25,000 (Grimaldi and Engel, 2005) in a total biota described to date of 1.4 to 1.8 million (Stork, 1988, 1993; May, 1990; Hammond, 1992). Using the 9,25,000 species described, versus the estimate of 5 million total, Grimaldi and Engel (2005) suggested that only about 20% of the insects are named. They have invaded every niche except the oceanic benthic zone (Grimaldi and Engel, 2005). They control populations of other organisms, and provide a major food source for other taxa (Majer, 1987), (Shurin et al., 2005).

38.3 Parasitoids: The Influential Microscopic Insects

Amid insects, the most widely recognized Hymenopterans—bees, ants, and wasps—have long been part of art, ritual, and folklore worldwide (Hanson and Gauld, 1995). The primitive lineages were plant feeding and parasitoid mode of life did not appear until about 210 million years ago (Grimaldi and Engel, 2005). An estimation of total number of insect parasitoids suggests being 87,000 species or roughly 10% of all described species (Eggleton and Belshaw, 1992). More than 80% of all parasitoid species belong to the Hymenoptera (Eggleton and Belshaw, 1992; Quicke, 1997).

Chalcids are of great importance in the regulation of populations of pest species successfully used in the biological control of insect pests (Noyes, 1985). Successful biological control projects of insect pests have used chalcidoids (Greathead, 1986). More than 800 species of Chalcidoidea have been used in biological control programs and overall have had the greatest rate of establishment and success (Noyes, 2019). Chalcids are grouped into 19 subfamilies and are known to attack hosts from 13 orders of insects (Boucek, 1988).

38.4 Eulophids: Minute Parasitoids with Gigantic Impacts

Within the superfamily Chalcidoidea, the Eulophidae (Westwood, 1828) is one of the largest families. The family is present and common in all geographic regions (Noyes, 2002). The species of these phytophagous genera again display various lifestyles (Gauthier et al., 2000; Noyes, 2002). The family is comprised of four subfamilies, i.e., Tetrastichinae, Eulophinae, Entedoninae, and Euderinae (Graham, 1975; Boucek and Graham, 1978).

38.5 Biology of Family Eulophidae

Eulophids attack various gall-forming species of insects, eriophyid mites (Boucek and Askew, 1968), and also gall-forming nematodes (Berg et al., 1990). The European species *Dahlbominus fuscipennis* is an ectoparasitoid of prepupae or pupae of diprionid sawflies in cocoons (Morris and Cameron, 1935). *Eulophus* and *Euplectrus* species are gregarious ectoparasitoids of exposed, leaf-feeding lepidopterous larvae often pupate on leaves around their dead host (Gradwell, 1957). Among these, *Mestocharis bimacularis* is unusual in attacking dytiscid beetle eggs when they are exposed by fluctuating water levels (Jackson, 1964). *Entedon ergias*, which oviposits in the eggs of various beetles, especially those of scolytids, is unusual in being an egg/larval parasitoid (Beaver, 1966). *Tetrastichus mandanis* also starts its larval life as an endoparasitoid in an egg (of a delphacid) but emerges from the egg during its second instar, and becomes a predator that actively searches out other homopteran eggs (Rothschild, 1966). Another British species, *Tetrastichus*, is recorded as having a completely predacious larva that develops by consuming a number of cecidomyiid larvae in a single pod of *Sarothamnus* (Parnell, 1963).

Many endoparasitic species are idiobionts (e.g., most *Chrysocharis* spp.), but some (e.g., *Chrysocharis phryne* and *Achrysocharoides* spp.) are koinobionts (Askew and Shaw, 1979). Many idiobiont species can behave as facultative hyperparasitoids (Askew, 1968; Askew and Shaw, 1979) and hyperparasitism is usual or even obligatory in some species. *Pediobius brachycerus*, for example, is an obligatory hyperparasitoid in spider's egg sacs (Viggiani, 1964; Askew and Ruse, 1974; Bryan, 1983).

The egg is normally elongate oval or kidney-shaped (Cameron, 1939; Clancy, 1946; Askew and Ruse, 1974), or occasionally with a long anterior filament (Viggiani, 1975) which probably serves to anchor the egg to the integument of the host (Silvestri, 1911). The mature larva is generally not hairy (Askew, 1968). The larvae of some species (e.g., *Diglyphus, Chrysocharis*) construct a circle of little fecal pillars about themselves (Viggiani, 1964). The larva pupates within this circle and the pillars harden to serve as "pit props," preventing the collapse of the host mine as the plant tissue dries out (see Askew, 1971). The larvae of *Euplectrus bicolor* pupate near or beneath the host and are enclosed in flimsy cocoons (Swezey, 1924). A few species (e.g., *Necremnus* spp.) may overwinter as adults (Cameron, 1939; Askew, 1968). A few species of, for example, *Eulophus* and *Sympiesis*, have different seasonal color forms, and this condition may occur in the pupae as well as the adults (Gradwell, 1958).

The mating behavior of some eulophids is very complex and has been described in detail for Melittobia species by Assem and coworkers (1975, 1978, 1982). Mating behavior can be used to separate closely related species (van dem Assem and Maeta, 1980; Dahms, 1984a) and the evolution of different patterns of behavior may help in understanding taxonomic relationships (in den Bosch and van den Assem, 1986).

An interesting phenomenon has been reported for *Melittobia Acasta* (Balfour-Browne, 1922; van den Assem, 1975). The unmated female will mate with one of her emerging sons.

38.6 Identification and Classification of Eulophidae

38.6.1 Entedoninae

Members are characterized by scutellum with one pair of setae; submarginal vein with two setae and frontofacial sutures when present separated from the anterior ocellus by at least one-third the distance from ocellus to torulus. In some genera (i.e., *Entedon, Pediobius, Horismenus*), the body is rather strongly sclerotized, but in other genera (i.e., *Omphale*) the body is weakly sclerotized and specimens often shrivel badly after death.

38.6.2 Euderinae

Females are recognized by Gt7 and Gt8 being separated from each other anterior to the cerci so that eight distinct gastral terga are visible. All other eulophids have Gt7 and Gt8 fused into a syntergum, so only seven gastral terga are visible. Males are sometimes difficult to identify as Euderinae, but in both the sexes of all genera the scutellum lacks submedian grooves, and the mesoscutum has notauli that are usually straight, deep, and complete.

38.6.3 Eulophinae

The Eulophinae appear to be the most primitive subfamily because they have less specialized features.

38.6.4 Tetrastichinae

The majority of species are recognized by the following combination of features: Scutellum with submedian and sublateral grooves and postmarginal vein greatly reduced (less than half the length of the stigma vein) or absent; small body size, uniform morphology, and a tendency for specimens to shrivel upon death make this an extremely difficult group to study.

38.7 Evidence from Globe

Parasitoids from Superfamily from order hymenoptera have been introduced all over the world (Noyes, 2020). Following are the examples of successful incorporation of parasitoids in various crops in different geographical regions

38.7.1 *Pediobius foveolatus* (Crawford) (Eulophidae: Hymenoptera)

This parasite was recorded as a larval pupal parasitoid of Epilachna beetle causing 36.6% mortality (Noyes, 2020).

38.7.2 *Tamarixia radiata* (Hymenoptera: Eulophidae)

Diaphorina citri (Kuwayama) is a global pest of citrus that transmits the bacteria Candidatus liberibacter spp. associated with the disease Huanglongbing (HLB) or greening. HLB is widespread in almost all citrus-producing regions except the (Parra et al., 2016). Infection of groves results in complete loss of productive capacity within four years, and young citrus trees never produce (Belasque and Yamamoto, 2010). Natural enemies play an essential role in regulating the population of *D. citri* in the field, and the elimination of these control agents by intensive use of insecticides can increase the spread of the disease (Quershi et al., 2009). *Tamarixia radiata* (Hymenoptera: Eulophidae), an idiobiont ectoparasitoid, is associated with the citrus psyllid and found to be more effective than other parasitoids belonging to Encyrtidae. *T. radiata* is believed to be more efficient than *D. aligarhensis* in controlling *D.citri* (Parra et al., 2016).

38.7.3 *Ageniaspis citricola* Logviniskaya (Hymenoptera: Eulophidae)

The citrus leafminer (CLM), *Phyllocnistis citrella* Stainton (Lepidoptera: Gracillariidae), is a pest native of Eastern and Southern Asia that since 1993 invaded all citrus-growing regions in America and the Mediterranean basin. The CLM attacks and causes problems mainly in young trees, nurseries, and overgraftings (Vercher et al., 2004). The first attempt to control CLM in the tropics was using broad spectrum insecticides, but chemical control appeared to be a costly and short-term solution (Knap et al., 1995; Argov et al., 1996)] in (Vercher et al., 2004). In the USA, an endoparasitic wasp of Asian origin, *Ageniaspis citricola* Logviniskaya (Hymenoptera: Eulophidae) was imported from Australia and released in Florida during 1994 and 1995 (Hoy et al., 1997). The population of *Ageniaspis citricola* quickly established and dispersed throughout the state, reaching parasitism levels near 100% in some areas (Walker and Hoy, 2003).

38.7.4 *Tetrastichus howardi* Olliff (Hymenoptera: Eulophidae)

Diatraea saccharalis Fabricius (Lepidoptera: Crambidae) is a major pest of sugarcane in the Americas (White and Wilson, 2012; Dinardo-Miranda et al., 2012; Svedese et al., 2013) and is capable of causing loss in biomass, death of the apical meristem and reduction in sugar and alcohol production (Simões et al., 2012; Rossato et al., 2013). *Diatraea saccharalis* caterpillars develop inside the sugarcane stalks, which diminishes the efficacy of the insecticides used to control them (Pinto et al., 2009); therefore the attention paid to the use of biological control agent to suppress *D. saccharalis* infestations (Cruz et al., 2011; Rodrigues et al., 2013). Hymenopteran parasitoids are frequently used as natural enemies in numerous applied biological control programs targeting *D. saccharalis* (Zanuncio et al., 2008; Mafi and Ohbayashi, 2010; Tavares et al., 2011, 2013). The parasitoid *Tetrastichus howardi* Olliff (Hymenoptera: Eulophidae) has been recorded from the pupae of several lepidopteran families, including Crambidae, Noctuidae, and Plutellidae (Kfir et al., 1993; Moore and Kfir, 1995; Kfir, 1997; Baitha et al., 2004; Hayat and Shahi, 2004; Prutz et al., 2004; Silva-Torres et al., 2010; Duong et al., 2011; Costa et al., 2014), and it has been extensively used to control several lepidopteran pests. *Tetrastichus howardi* was also found parasitizing *D. saccharalis* larvae (Vargas et al., 2011). *Tetrastichus howardi* Olliff (Hymenoptera: Eulophidae) parasitizes the larvae, pupae, and adults of *Diatraea saccharalis*, hence a competent parasitoid for the biological control of *D. saccharalis* in commercial sugarcane production in Brazil and other industries where this stem borer causes significant damage.

38.8 Eulophidae Species Discovered and Reported in India

The family currently includes 297 genera and 4,472 species placed in four subfamilies as follows: Euderinae (17/148), Eulophinae (97/1316), Entedoninae (87/1307), Tetrastichinae (91/1644), and (5/57) are unplaced (Universal Chalcidoidea database).

The details of species from family Eulophidae available in India are provided in Table 38.1.

38.9 Conclusion and Future Prospects

Mother Nature has its own intricacies to keep the biosphere stress-free and healthy, whenever a component exceeds the provided perimeter, it eventually transformed into catastrophic extinction events in different eras. There could be many reasons and causes for the previous extinction of species, but in the present age, the population of *Homo sapiens* has increased, leading to receding resources, habitat destruction, conflicts between man and wild populations, decreasing tree cover, incompatible agricultural practices, declining of water reservoirs, and so on, all leading to species extinction. Species-A Resource, which could have harbored the competencies to resolve environmental issues.

Eulophid parasitoids could prove to be a significant tool for sustainable management in organic farming and integrated pest management. Therefore, it is obligatory first to determine the species taxonomically and then to study whether or not it can be instrumental in various possible areas to reduce the stress on the environment.

TABLE 38.1
Genera & Species of Family Eulophidae in India

S.No	Name of Genus	Number of Species in India	Name of Species
1	Aceratoneuromyia	2	*Aceratoneuromyia indica, A. wayanadensis*
2	Alophomorphella	2	*Alophomorphella bharathica, A.viola*
3	Anaprostocetus	5	*Anaprostocetus acuminatus, A.areos A.dehraensis, A.keralicus, A.sringeriensis*
4	Anselmella	1	*Anselmella kerrichi*
5	Anumanniola	2	*Anumanniola lasallei,A.narendrani*
6	Apleurotropis	2	*Apleurotropis sivadasani, A.viswanathan*
7	Aprostocetus	81	*Aprostocetus ajmerensis, A. amalius, A. angus, A. annulicornis, A. artemesiae,A. arunachalicus, A. ashotus A. asphondyliae, A. asthenogmus, A. atomell, A. bangaloricus, A. barnallicus, A. benazeer A. calasmus, A. calius, A. calycopteridis A. cauperdatus, A. coimbatorensis, A. dala, A. decatus A. dellus, A. dhireni, A. diclus, A. diplosidis A. doonensis, A. fasciatus, A. ficicolae A. flavidus, A. gala, A. hagenowii, A. harithus A. hayati, A. ille, A. indicus, A. jaipurensis A. karicus, A. kaya, A. kodei, A. kuriani, A. lasallei A. licalis, A. maculatus, A. malabarensis, A. malcis A. malloticolae, A. manja, A. microcosmus, A. nadicus, A. nagri, A. nainitalensis, A. namdaphicus A. narius, A. neglectus, A. neogala, A. neovanilae A. niger, A. nigricornis, A. obtusae, A. percaudatus A. psyllidis, A. purpureus, A. rajai, A.* retura *A. ricosus, A. sankarani, A. santalinus, A. satpurensis A. serycus, A. shimlicus, A. silent, A. sringeriensis A. taruna, A. travancorensis, A. tritrichia, A. uthonia A. vanilla, A. vatiata, A. versicolor, A. vimania A. yesica, A. yoshimotoi*
8	Aprostoporoides	2	*Aprostoporoides curiosus, A. manjericus*
9	Aroplectrus	2	*Aroplectrus contheylae, A. dimerus*
10	Asecodes	1	*Asecodes turcicum*
11	Astichus	1	*Astichus santhoshi*
12	Aulogymnus	2	*Aulogymnus arifaokhani,A.indicus*
13	Baeoentedon	1	*Baeoentedon farazi*
14	Baryscapus	6	*Baryscapus barica*
15	Ceranisus	3	*Ceranisus femoratus, C.javae, C.menes*
16	Ceratoneura	1	*Ceratoneura indi*
17	Chrysocharis	11	*Chrysocharis ajugus, C. echinata, C. funicularis C. horticola, C. imphalensis, C. indicus, C. johnsoni C. longiclavatus, C. longiscapus, C. nitida, C. pubicornis*
18	Chrysonotomyia	12	*Chrysonotomyia appannai, C.elongata, C.hyalinipennis, C.locustivora, C.longiclava C.malabarica, C.obesula, C.postmarginaloides C.pubipennis, C.rexia, C.ricini, C.younusi*
19	Cirrospilus	8	*Cirrospilus abalus,C. acadius, C. brevicorpus C. denitus, C. ingenuus, C. kanpurensis C. ornatus, C. pondicheri*
20	Citrostichus	1	*Citrostichus phyllocnistoides*
21	Closterocerus	7	*Closterocerus agromyzae, C. cinctiventris C. insignis, C. longiscapus, C. phytomyzae C. pulcherrimus, C. scapiatus*
22	Deutereulophus	2	*Deutereulophus malabarensis, D. tunctatus*
23	Dicladocerus	4	*Dicladocerus antennalis, D. indicus, D. liriomyza D. viggianii*

(Continued)

S.No	Name of Genus	Number of Species in India	Name of Species
24	Diglyphomorphomyia	9	*Diglyphomorphomyia ebifurcata, D. kairali D. nexius, D. nigraella, D. palodica, D. rufescens D. scolofronta, D. sholayarica, D. sringeriensis*
25	Diglyphus	8	*Diglyphus frontolatus, D. funicularis, D. guptai D. horticola, D. indicus, D. isaea, D. kashmiricus* D. mandibularis_
26	Dubeyiella	1	*Dubeyiella indica*
27	Elachertus	20	*Elachertus adimalicus, E. aenigmatis, E. asokai E. asycritus, E. breviclavus, E. dericus, E. indicus E. kainophanestus, E. kashmiricus, E. lanotus E. malabaricus, E. nedumbassericus, E. nigrithorax E. nuperus, E. piperis, E. propodiatus, E. radhakrishnai E. sinui, E. zandanicus, E. zellus*
28	Elasmus	60	*Elasmus alami, E. alaris, E. albopictus, E. anamalaianus E. anticles, E. bathyskius, E. binocellatus, E. brevicornis E. caligneus, E. cameroni, E. cavicolous, E. ceylonicus E. claripennis, E. colemani, E. cyanomontanus E. dalhousieanus, E. deccanus, E. dexotus, E. dubiatus E. flavescens, E. flavocorpus, E. fulvicornis, E. grimmi E. homonae, E. hutsoni, E. hyblaeae, E. indicoides E. indicus,E. johnstoni, E. khandalus, E. kodaianus E. kollimalaianus, E. krishnagiriensis, E. longicornis E. longiscapus, E. longiventris, E. lutens, E. mahabalii E. munnarus, E. neoflavescens, E. neoflavocorpus E. neopunctulatus, E. nephantidis, E. nigritus, E. noyesi E. nuperus, E. philippinensis, E. platyedrae, E. polistis E. pulchellus, E. punctulatus, E. punensis, E. queenslandicus, E. rajasthanicus, E. scutellofurvus E. sigillatus, E. striptogasteri, E. syleptae, E. viridiscutellatus, E. zatonus, E. zehntneri*
30	Entedon	4	*Entedon gunturensis, E. longicorpus, E. pempheridis E. thoubalensis*
31	Entedonomphale	1	*Entedonomphale postmarginalis*
32	Euderomphale	3	*Euderomphale longipedicelus, E. secunda, E. stomale*
34	Euderus	9	*Euderus agromyzae, E. agromyzae, E. albitarsis E. alcidodes, E. carpomyiae, E. gossypii, E. lividus E. pempheriphila, E. torymoides*
35	Eulophinusia	2	*Eulophinusia indica, E. keralensis*
36	Eulophomorpha	1	*Eulophomorpha flavicornis*
37	Eulophus	3	*Eulophus chennaicus, Eulophus razaki, Eulophus velosus*
38	Euplectrophelinus	3	*Euplectrophelinus calicuticus, E. saintpierrei, E. sureshani*
40	Euplectrus	34	*Euplectrus asotus, E. bristlis, E. ceylonensis E. coimbatorensis, E. dubeyi, E. euplexiae, E. gopimohani E. himalayaensis, E. indicus, E. latifrons, E. lenocos E. leucostomus, E. longiscapus, E. longiscapus, E. manii E. manjericus, E. maternus, E. mathuri, E. narariae E. nuperus, E. nyctemerae, E. pantnagarensis, E. parvulus E. petiolatus, E. prashanti, E. spodopterae., E. stom E. umbrocoxatus, E. utetheisae, E. viggianii, E. yaloticus E. zamoorini, E. zandanus*
41	Eurycephaloplectrus	1	*Eurycephaloplectrus natadae*
42	Farooqiella	1	*Farooqiella kashmiriensis*
43	Gautamiella	1	*Gautamiella indica*
44	Ginsiella	1	*Ginsiella indica*
45	Goetheana	2	*Goetheana shakespearei, G. tariquei*
46	Goethella	1	*Goethella asulcata*
47	Hayatiola	1	*Hayatiola macrocephala*
48	Hemiptarsenus	9	*Hemiptarsenus aditus, H. banati, H. effarkhani H. indicus, H. nuperus, H. olaticus, H. orizae H. unguicellus, H. varicornis*
49	Holcotetrastichus	1	*Holcotetrastichus manaliensis*

(Continued)

S.No	Name of Genus	Number of Species in India	Name of Species
50	Horismenus	1	*Horismenus absonus*
51	Hyssopus	3	*Hyssopus indicus, H. kamrupicus, H. licinus*
52	Idukkiella	1	*Idukkiella elattariae*
53	Kiggaella	1	*Kiggaella oryzae*
54	Kostjukovius	1	*Kostjukovius keralensis*
55	Kostjurixia	2	*Kostjurixia laharensis K. ootiyensis*
56	Lasalleola	1	*Lasalleola obsona*
57	Leptocybe	1	*Leptocybe invasa*
58	Melittobia	2	*Melittobia acasta, Melittobia assemi*
59	Mestocharella	3	*Mestocharella deltoids, M. indica, M.javensis*
60	Minotetrastichus	1	*Minotetrastichus zeasmi*
61	Mischotetrastichus	1	*Mischotetrastichus keralensis*
62	Mohaniella	1	*Mohaniella indica*
63	Narendrania	1	*Narendrania keralensis*
64	Necremnoides	1	*Necremnoides harithodaris*
65	Necremnus	1	*Necremnus leucarthros*
66	Neoaceratoneura	1	*Neoaceratoneura mangiferae*
67	Neochrysocharis	3	*Neochrysocharis horticola, N. indicus, N. liriomyzae*
68	Neogasterichus	4	*Neogasterichus achalicus, N. bhubaneswaricus N. dulciculus, N. longigastris*
69	Neomestocharella	1	*Neomestocharella keralensis*
70	Neopediobopsis	1	*Neopediobopsis setosa*
71	Neotrichoporoides	16	*Neotrichoporoides arunima, N. beonus, N. crinius N. curiosus, N. delhiensis, N. doproides, N. galia N. idukkiensis, N. mani, N. mediterraneus N. mohanae, N. nyemitawus, N. stom, N. tonimus N. viridimaculatus, N. yoopi*
72	Nesolynx	5	*Nesolynx flavipes, N. javanica, N. orientalis N. phaeosoma, N. thymus*
73	Notanisomorphella	4	*Notanisomorphella cheriani, N. flaviventris N. manjerica, N. walayarica*
74	Obesulus	2	*Obesulus indicus, Obesulus keralicus*
75	Omphale	2	*Omphale calicuti, Omphale stonia*
76	Omphalentedon	1	*Omphalentedon dasi*
77	Oomyzus	10	*Oomyzus cerococci, O. gallerucae, O. liriomyzae O. mashhoodi, O. neogallerucae, O. scaposus, O. sokolowskii, O. trounius, O. viraktamathi, O. waylaricus*
78	Ootetrastichus	1	*Ootetrastichus indicus*
79	Parachrysocharis	4	*Parachrysocharis anomalococci, P. javensis P. malabarensis, P. pantnagarensis*
80	Parahorismenus	2	*Parahorismenus infuscatipennis, P. pondicherryensis*
81	Parasecodella	3	*Parasecodella bouceki, P. nigricorpus, P. pantnagarensis*
82	Parzaommomyia	2	*Parzaommomyia malabarica, P. malabarica*
83	Pediobius	29	*Pediobius agaristae, P. anomalus, P. antennalis P. bethylicidus, P. bhimtalensis, P. bruchicida P. cydiae, P. darwini, P. deplagastrus, P. elasmi P. foveolatus, P. fraternus, P. hebbalensis, P. imbreus P. incertulus, P. indicus, P. kalpetticus, P. khani P. maduraiensis, P. metallicus, P. nigritarsis, P. parvulus P. salicifolii, P. scutilaris, P. shafeei, P. stenochoreus P. thakerei, P. viggianii*
84	Perthiola	1	*Perthiola moringae*
85	Phymastichus	1	*Phymastichus coffea*
86	Platyplectrus	9	*Platyplectrus baricus, P. bussyi, P. daricus, P. ericatus P. kuriani, P. nilamburicus, P. oricus, P. viridiceps P. yarensus*
87	Pleurotroppopsis	2	*Pleurotroppopsis dentate, Pleurotroppopsis femorata*
88	Pnigalio	6	*Pnigalio ameti, P. anoti, P. Bengali, P. cadoti, P. camicalis, P. zandani*

(*Continued*)

S.No	Name of Genus	Number of Species in India	Name of Species
89	Pomphale	3	*Pomphale atturensis, P. setosipennis, P. striptipennis*
90	Propodeochertus	1	*Propodeochertus sureshani*
91	Pseudiglyphus	1	*Pseudiglyphus orientalis*
92	Pseudosecodes	1	*Pseudosecodes calicuticus, P. malabaricus*
93	Quadrastichus	24	*Quadrastichus bararius, Q. cerolus, Q. davius, Q. delusus Q. demiticus, Q. dicleus, Q. erythrinae, Q. fagri, Q. garuda Q. getiolus, Q. lantanae, Q. lavania, Q. lekatus Q. longiclavatus, Q. longicorpus, Q. lurvi, Q. mangalae Q. narangae, Q. ovulorum, Q. plaquoi, Q. stomalis Q. sudheeri, Q. vinitia, Q. zenta*
94	Shardiella	1	*Shardiella savitri*
95	Sigmoepilachna	1	*Sigmoepilachna indica*
96	Sigmophora	3	*Sigmophora brevicornis, S. keralensis, S. polyseta*
97	Stenomesius	8	*Stenomesius aligarhicus, S. anati, S. calicuticus S. japonicas, S. modicellus, S. orientalis, S. pantnagarensis S. singularis*
98	Stenopetius	1	*Stenopetius rugosus*
99	Sureshanella	1	*Sureshanella nupera*
100	Sympiesis	14	*Sympiesis abalisis, S. acicus, S. bardisis, S. clausenis S. dolichogaster, S. elatisis, S. harithi, S. hyblaeae S. hyplosis, S. jeolikotensis, S. kazmii, S. kottiyooricus S. striatipes, S. thyrsisae*
101	Tachinobia	1	*Tachinobia repanda*
102	Tamarixia	6	*Tamarixia girishi, T. orientalis, T. pookodica T. radiate, T. sheebae, T. yoorica*
103	Tetrastichus	90	*Tetrastichus albocoxalis, T. anamikicus, T. apanteles T. arucicus, T. badius, T. bellus, T. bharathi, T. bilgiricus T. burki, T. calicuticus, T. chakrataensis, T. chindakicus T. chinicus, T. colemani, T. coorgensis, T. cotesiae, T. cupressi, T. dalotus, T. damitus, T. dasi, T. demonaxi T. dinoticus, T. dulciculus, T. epilachnae, T. flavilatus T. gadagkari, T. gardneri, T. garicus, T. girishi, T. hindicus, T. hotspoticus, T. howardi, T. hydrelliae, T. iaricus, T. inferens. T. kodaikanalensis, T. kottiyooricus T. krishnaiahi, T. krishnieri, T. kumaonensis, T. laparus T. magnicorpus, T. mangifera, T. manjumalaicus T. mindicus, T. miser, T. mohani, T. momardicae T. mukatus, T. neocupressi, T. neothoracicus, T. niger T. nilamburensis, T. obliqua, T. ootyensis, T. ophiusae T. orissaensis, T. orthagae, T. pantnagarens, T. paraponyxi, T. partellus., T. punjabensis, T. quadriseta T. ramakrishnae, T. ramamurthyi, T. santhoshi, T. schoenobii, T. shonicus, T. sinui, T. sissoo, T. spirabilis T. sringericus, T. stom, T. sumatus, T. tanjorensis T. taprobanensis, T. thalaparicus, T. thetisae, T. thonicus T. triozai, T. tunicus, T. umae, T. uniarticulata T. uttaranchalicus, T. vayalicus, T. zeyae*
104	Thripastichus	1	*Thripastichus gentilei*
105	Thripoctenus	1	*Thripoctenus javae*
106	Trichospilus	3	*Trichospilus diatraeae, T. hayati, T. pupivorus*
107	Zagrammosoma	1	*Zagrammosoma variegatum*

REFERENCES

Agrò, A.; Cusumano, A.; Pinto, M. lo 2009, Study on the parasitoid complex of Prays oleae (Bernard) and parasitism levels in an organic olive orchard in western Sicily (Italy). Protezione delle Colture 2009(1):33–39

Argov, Y. and Rössler, Y. Introduction, release and recovery of several exotic natural enemies for biological control of the citrus leafminer, Phyllocnistis citrella, in Israel. Phytoparasitica. 1996, *24*:33–38.

Askew, R.R. Considerations on speciation in Chalcidoidea (Hymenoptera). *Evolution*. 1968, *22*(3):642–645.

Askew, R.R. Sib-mating in *Nasonia vitripennis* (Walker) (Hym., Pteromalidae) and other Chalcidoidea, and its possible evolutionary significance. In Proceedings, XIII International Congress of Entomology, Moscow. 1971, 1:325.

Askew, R.R., Ruse, M. Biology and taxonomy of the genus *Enaysma* Delucchi (Hym., Eulophidae, Entedontinae) with special reference to the British fauna. *Transactions of the Royal Entomological Society of London*. 1974, *125*(3):257–294.

Askew, R.R.; Shaw, M.R. 1979, Mortality factors affecting the leaf-mining stages of Phyllonorycter (Lepidoptera: Gracillariidae) on oak and birch. I. Analysis of the mortality factors. *Zoological Journal of the Linnean Society* 67(1):31–49

Assem, J. van den. Temporal patterning of courtship behaviour in some parasitic Hymenoptera, with special reference to *Melittobia acasta* (Eulophidae). *Journal of Entomology (A)*. 1975, *50*(3):137–146.

Assem, J. van den, Gijswijt, M.J., Nübel, B.K. Characteristics of courtship and mating behaviour as classificatory criteria in Eulophidae—Tetrastichinae (Hymenoptera), with special reference to the genus *Tetrastichus* sl. *Tijdschrift voor Entomologie*. 1982, *125*:205–220.

Assem, J. van den, Maeta, Y. On a fourth species of *Melittobia* from Japan. *Kontyû*. 1980, *48*(4):477–448.

Assem, J. van den, Maeta, Y. Some observations on *Melittobia* species (Hymenoptera, Chalcidoidea - Eulophidae) collected in Japan. *Kontyû*. 1978, *46*(2):264–272.

Baitha, A., Jalali, S.K., Rabindra, R.J. and Venkatesan, T. Parasitizing efficiency of the pupal parasitoid, *Tetrastichus howardi* (Olliff) (Hymenoptera: Eulophidae) on *Chilo partellus* (Swinhoe) at different exposure periods. *Journal of Biological Control*. 2004, *18*:65–68.

Balfour-Browne, F. On the life history of *Melittobia acasta*, Walker; a chalcid parasite of bees and wasps. *Parasitology*. 1922, *14*(3/4):349–370.

Beaver, R.A. Notes on the biology of the bark beetles attacking elm in Wytham Wood, Berks. *Entomologist's Monthly Magazine*. 1966, *102*:163–170.

Belasque, Jr J., Yamamoto, P.T., Miranda, M.P.D., Bassanezi, R.B., Ayres, A.J. and Bove, J.M. Lessons from huanglongbing management in São Paulo State. Brazil. *Journal of Plant Pathology*. 2010, *92*:285–302.

Berg, E. van den, Prinsloo, G.L. and Neser, S., An unusual host association: *Aprostocetus* sp. (Eulophidae), a hymenopterous predator of the nematode *Subanguina mobilis* (Chit; Fisher, 1975) Brzeski, 1981 (Anguinidae). *Phytophylactica*. 1990,*22*:125–127.

Bosch, H.A.J. in den, Assem, J. van den. The taxonomic position of *Aceratoneuromyia granularis* Domenichini (Hymenoptera: Eulophidae) as judged by characteristics of its courtship behaviour. *Systematic Entomology*. 1986, *11*(1):19–23.

Boucek, Z. *Australian Chalcidoidea (Hymenoptera) A Biosystematic Revision of Genera of Fourteen Families, with a Reclassification of Species*. CABI, Wallingford, UK. 1988, 832 pp.

Boucek, Z. and Askew, R., Hym. Chalcidoidea. In *Palearctic Eulophidae (excl. Tetrastichinae). Index of Entomophagous Insects, 3*. 1968. 260 pp.

Boucek, Z. and Graham, M. W. R. de V. British check-list of Chalcidoidea (Hymenoptera): Taxonomic notes and additions. *Entomologist's Gazette*. 1978,*29*(4):225–235.

Bryan, G. *Achrysocharoides* species in the Netherlands (Hymenoptera: Eulophidae). *Entomologische Berichten*. 1983, *43*(11):172–176.

Cameron, E. The holly leaf-miner (*Phytomyza ilicis*, Curt.) and its parasites. *Bulletin of Entomological Research*. 1939, *30*:173–208.

Clancy, D.W. Natural enemies of some Arizona cotton insects. *Journal of Economic Entomology*. 1946, *39*(3):326–328.

Costa, D.P., Pereira, F.F., Kassab, S.O., Rossoni, C., Favero, K.R and Barbosa, H. Reprodução de *Tetrastichus howardi* em pupas de *Diatraea saccharalis* de diferentes idades. *Amazonian Journal of Agricultural and Environmental Sciences*. 2014, *57*:67–71.

Cruz, I., Redoan, A.C., Silva, R.B., Figueiredo, M.L.C. and Penteado-Dias, A.M. New record of *Tetrastichus howardi* (Olliff) as a parasitoid of *Diatraea saccharalis* (Fabr.) on maize. *Scientia Agricola*. 2011, *68*:252–254.

Dinardo-Miranda, L.L., Anjos, I.A., Costa, V.P. and Fracasso, J.V. Resistance of sugarcane cultivars to *Diatraeasaccharalis*. *Pesquisa Agropecuária Brasileira*. 2012, *47*:1–7.

Duong, C.A., Diep, D.N. and Hung, H.Q. Survey of sugarcane moth borers in southeast Vietnam. *International Sugar Journal*. 2011, *113*(13540):732–737.

Eggleton, P. and Belshaw R. 1992. Insect parasitoids an evolutionary overview. *Philosophical Transactions of the Royal Society of London*. 1992, *337*(1279):1–20.

Erwin, T.R. Biodiversity at its utmost: Tropical forest beetles. In M. L. Reaka-Kudla, D. E. Wilson, and E. O. Wilson (eds). *Biodiversity II. Understanding and Protecting our Biological Resources*. Joseph Henry Press, Washington, DC. 1993. pp. 27–68.

Erwin, T.R. *The Tropical Forest Canopy: The Heart of Biotic Diversity*. In E. O. Wilson and F. M. Peter (eds). *Biodiversity*. National Academy Press, Washington, DC. 1988. pp. 123–129.

Gauthier, N., LaSalle, J., Quicke, D.L.J. and Godfray, H.C.J., Phylogeny of Eulophidae (Hymenoptera: Chalcidoidea), with a reclassification of Eulophinae and the recognition that Elasmidae are derived eulophids. *Systematic Entomology*. 2000, *25*:521–539.

Gibson, G.A.P. Some pro- and mesothoracic structures important for phylogenetic analysis of Hymenoptera, with a review of terms used for the structures. *Canadian Entomologist*. 1985, *117*(11):1395–1443.

Gradwell, G.R. *Eulophus nigribasis* Gradwell (Hym., Chalcidoidea) the overwintering form of *E. larvarum* (L.). *Entomologist's Monthly Magazine*. 1958, *94*:234–235.

Gradwell, G.R. A new tetrastichine (Hym., Eulophidae) genus with three included species. *Entomologist's Monthly Magazine.* 1957, *93*:1–5.

Graham, M.W.R. de, V. Relationships and synonymy of Winneinana Crawford (Hymenoptera Eulophidae). *Journal of Entomology.*1975, *44*:281–282.

Greathead, D.J. Parasitoids in classical biological control. In J. K. Waage and D. J. Greathead (eds). *Insect Parasitoids.* Academic Press, London. 1986. pp. 287–318.

Grimaldi, D. and Engel, M. S., 2005. *Evolution of the Insects.* Cambridge University press, Cambrudge, New York, 755 Pp.

Hammond, P. Species inventory. In B. Groombridge (ed). *Global Biodiversity. Status of the Earth's Living Resources. A Report Compiled by the World Conservation Monitoring Centre.* Chapman and Hall, London. 1992. pp. 17–39.

Hayat, M and Shahi, M.H. Taxonomic notes on Indian Eulophidae (Hymenoptera: Chalcidoidea)—1. On the types of some Tetrastichinae. *Oriental Insects.* 2004, *38*:303–314.

Hodkinson, I. D. and Casson, D. A lesser predilection for bugs: Hemiptera (Insecta) diversity in tropical rain forests. *Biological Journal of the Linnean Society.* 1991, *43*:101–109.

Hoy, M.A., Nguyen, R., Pomerinke, M., Bullock, R., Hall, D and Knapp, J., et al. Distribution of A. citricola—A parasitoid of the citrus leaf miner. *Citrus Industry.* 1997,*78*:51–52.

Jackson, D.J. Observations on the life-history of *Mestocharis bimaculatus* (Dalman) (Hym. Eulophidae), a parasitoid of the eggs of Dytiscidae. *Opuscula Entomologica.* 1964, *29*:81–97.

Kfir, R. Parasitoids of *Plutella xylostella* (Lep.: Plutellidae) in South Africa: An annotated list. *Entomophaga.* 1997, *42*:517.

Kfir, R., Gouws, J. and Moore, S.D. Biology of *Tetrastichus howardi* (Olliff) (Hymenoptera: Eulophidae)—A facultative hyper-parasitoid of stem borers. *Biocontrol Science and Technology.* 1993, *3*:149–159.

Knap, J.L., Albrigo, L.G., Browning, H,W., Bullock, R.C, Heppner, J.B and Hall, D.G, et al. *Citrus leafminer, Phyllocnistis citrella Stainton: Current Status in Florida.* University of Florida, Florida Coop. Ext. Service, IFAS, Gainesville, FL. 1995. p. 35.

Mafi, S. and Ohbayashi, N. Biology of *Chrysocharis pentheus*, an endoparasitoid wasp of the citrus leaf miner *Phyllocnistis citrella* Stainton. *Journal of Agricultural Science and Technology.* 2010, *12*:145–154.

Majer, J. D. The conservation and study of invertebrates in remnants of native vegetation. In D. A. Saunders, G. W. Arnold, A. A. Burbridge and A. J. M. Hopkins (eds). *Nature Conservation*: The Role of Remnants of Native Vegetation. Survey Beatty and Sons, Sydney. 1987. pp. 333–335.

May, R. M. How many species? *Philosophical Transactions of the Royal Society. B. Biological Sciences.* 1990, *330*:293–304.

Moore, S.D. and Kfir, R. Host preference of the facultative hyperparasitoid *Tetrastichus howardi* (Hym: Eulophidae). *Entomophaga.* 1995,*40*:69–76.

Morris, K.R.S. and Cameron, E. The biology of *Microplectron fuscipennis* Zett. (Chalcid.), a parasite of the pine sawfly (*Diprion sertifer* Geoff.). *Bulletin of Entomological Research.* 1935, *26*(3):407–418.

Noyes, J.S. Chalcidoids and biological control. Chalcid Forum. 1985, *5*:5–10.

Noyes, J.S. 2002, Interactive catalogue of World Chalcidoidea (2001 - second edition). CDrom: Taxapad, Vancouver and The Natural History Museum, London.

Noyes, J.S. Universal Chalcidoidea Database. World Wide Web electronic publication. 2019. http://www.nhm.ac.uk/chalcidoids.

Noyes, J.S. Universal Chalcidoidea Database. World Wide Web electronic publication. 2020. http://www.nhm.ac.uk/chalcidoids.

Parra, J.R.P., Alves, G.R., Diniz, A.J.F. and Vieira, J.M. Tamarixia radiata (Hymenoptera: Eulophidae) x Diaphorina citri (Hemiptera: Liviidae): Mass rearing and potential use of the parasitoid in Brazil. *Journal of Integrated Pest Management.* 2016, *5*:1–11.

Parnell, J.R. 1963 (08 Oct 1963), Three gall midges (Dipt., Cecidomyiidae) and their parasites, found in the pods of broom (Sarothamnus scoparius (L.) Wimmner). Transactions of the Royal Entomological Society of London 115(10):261–275.

Prutz, G., Brink, A. and Dettner, K. Transgenic insect-resistant corn affects the fourth trophic level: Effects of *Bacillus thuringiensis* corn on the facultative hyperparasitoid *Tetrastichus howardi. Naturwissenschaften.* 2004, *91*:451–454.

Quicke, D. I. J. *Parasitic Wasps.* Chapman and Hall, London, UK. 1997.

Qureshi, J.A.; Rogers, M.E.; Hall, D.G.; Stansly, P.A. 2009, Incidence of invasive Diaphorina citri (Hemiptera: Psyllidae) and its introduced parasitoid Tamarixia radiata (Hymenoptera: Eulophidae) in Florida citrus. Journal of Economic Entomology 102(1):247–256.

Rodrigues, M.A.T., Pereira, F.F., Kassab, S.O., Pastori, P.L., Glaeser, D.F., Oliveira, H.N. and Zanuncio, J.C. Thermal requirements and generation estimates of *Trichospilus diatraeae* (Hymenoptera: Eulophidae) in sugarcane producing regions of Brazil. *Florida Entomologist.* 2013, *96*:154–159.

Rossato, J.A.S., Costa, G.H.G., Madaleno, L.L., Mutton, M.J.R., Higley, L.G. and Fernandes, A. Characterization and impact of the sugarcane borer on sugarcane yield and quality. *AgronomyJournal.* 2013, *105*:643–648.

Rothschild, G.H.L. Notes on two hymenopterous egg parasites of Delphacidae (Hem.). *Entomologist's Monthly Magazine.* 1966, *102*:5–9.

Shurin, J. B., D. S. Gruner, and H. Hillebrand. All wet or dried up? Real differences between aquatic and terrestrial food webs. *Proceedings of the Royal Society. B. Biological Sciences.* 2005, *273*:1–9.

Silva-Torres, C.S.A., Pontes, I.V.A.F. and Torres, J.B. New records of natural enemies of *Plutella xylostella* (L.) (Lepidoptera: Plutellidae) in Pernambuco, Brazil. *Neotropical Entomology.* 2010, *39*:835–838.

Silvestri, F. 1911 (04 Mar 1911), Di una nuova species di *Aleurodes vivente sull'olivo. Bollettino del Laboratorio di Zoologia Generale e Agraria della* R. Scuola Superiore d'Agricoltura, Portici 5:214–225

Simões, R.A., Letícia, R.G., Bento, J.M.S., Solter, L.F. and Delalibera, I. Jr. Biological and behavioral parameters of the parasitoid *Cotesia flavipes* (Hymenoptera: Braconidae) are altered by the pathogen *Nosema* sp. (Microsporidia: Nosematidae). *Biological Control.* 2012, *63*:164–171.

Stork, N. E. How many species are there? Biodiversity and Conservation. 1993, *2*:215–232.

Stork, N. E. Insect diversity: Facts, fiction and speculation. *Biological Journal of the Linnean Society.* 1988, *35*:321–337.

Svedese, V.M., Lima, E.Á.L.A. and Porto, A.L.F. Horizontal transmission and effect of the temperature in pathogenicity of *Beauveria bassiana* against *Diatraea saccharalis* (Lepidoptera: Crambidae). *Brazilian Archives of Biology and Technology.* 2013, *56*:413–419.

Tavares, W. de S.; Hansson, C.; Serrao, J.E.; Zanuncio, J.C. 2011, First report of Trichospilus pupivorus (Hymenoptera: Eulophidae) parasitizing pupae of Anticarsia gemmatalis (Lepidoptera: Noctuidae). Entomologia Generalis 33(4):281–282.

Tavares, W.S., Hansson, C., Mielke, O.H.H., Serrão, J.E. and Zanuncio, J.C. Parasitism of *Palmistichus elaeisis* Delvare & LaSalle, 1993 on pupae of *Methona themisto* (Hübner, [1818]) reared on two hosts (Lepidoptera: Nymphalidae; Hymenoptera: Eulophidae). *Shilap: Revista de Lepidopterologia*. 2013, *41*:43–48.

Vargas, E.L., Pereira, F.F., Tavares, M.T. and Pastori, P.L. Record of *Tetrastichus howardi* (Hymenoptera: Eulophidae) parasitizing *Diatraea* sp. (Lepidoptera: Crambidae) in sugarcane crop in Brazil. *Entomotropica*. 2011, *26*:135–138.

Vercher, R., Ferran, M.G., Comelles, J.C., Marzal, C. and Villalba, M. Biological control of the citrus leafminer Phyllocnistis citrella (Lepidoptera: Gracillariidae) in Spain: Native parasitoids and establishment of Citrostichus phyllocnistoides (Hymenoptera: Eulophidae). *Biological Control*. 2004, *29*: 215–226.

Viggiani, G. Centri di studi entomologici della Cenfederazione Elvetica. *Bollettino del Laboratorio di Entomologia Agraria 'Filippo Silvestri'*. 1964, *22*:123–130.

Viggiani, G. Reintroduction de *Leptomastix dactylopii* How. pour la lutte biologique contre *Planococcus citri* Risso. *Fruits*. 1975, *30*:259–260.

Walker, A and Hoy, M. A. Citrus leafminer parasitoid. University of Florida Featured Creatures. 2003. Available from: http://entnemdept.ufl. edu/creatures/beneficial/a_citricola.htm [Accessed on: Mar 20, 2020].

Westwood, J. O. On the Chalcididae. *Zoological Journal*. 1828, 3–31.

White, W.H. and Wilson, L.T. Feasibility of using an alternative larval host and host plants to establish *Cotesia flavipes* (Hymenoptera: Braconidae) in the temperate Louisiana sugarcane ecosystem. *Environmental Entomology*. 2012, *41*:275–281.

Wilson, E. O., 2003. *What is nature worth for*? Part-1 span XLIV-4: 54–59

Zanuncio, J.C., Pereira, F.F., Jaques, G.C., Tavares, M.T. and Serrão, J.E. Tenebrio molitor Linnaeus (Coleoptera: Tenebrionidae), a new alternative host to rear the pupae parasitoid *Palmistichus elaeisis* Delvare and LaSalle (Hymenoptera: Eulophidae). *Coleopterists Bulletin*. 2008. *62*:64–66.

39

Management of Diatreae *spp. through Parasitoids*

Fernando Belezini Vinha and Alexandre De Sene Pinto

CONTENTS

39.1 Introduction

Stem borers, *Diatraea* spp. (Lepidoptera: Crambidae), are considered to be the most economically important and widely distributed pests throughout the Americas, especially affecting maize and sugarcane crops (Almeida and Stingel, 2005; Mendonça et al., 1996). Sugarcane crop is attacked by *Diatraea* spp. throughout its development and the incidence is higher as the plant grows, especially at the time when the internodes are formed. The attack is quite variable, depending on the sugarcane variety, time of year, crop cycle, among other factors (Macedo & Botelho, 1988).

The borers *Diatraea saccharalis* (Fabricius, 1794) and *Diatraea flavipennella* (Box, 1931) (Lepidoptera: Crambidae) are important pests of the sugarcane crop in Brazil. Controlling *D. saccharalis* with chemicals is difficult because this insect develops in a protected location (inside the stem of the plant). Moreover, due to the phytotoxicity from some chemicals, much research was dedicated to the control of this pest by biological methods (Marconato, 1988). According to Nava et al. (2009), the success of the biological control of *Diatraea* spp. is due to the existence of great diversity of parasitoids and predators, which act mainly on the egg and larva phases of the pest.

Currently, control of this pest insect in sugarcane fields is mainly carried out by the parasitoids *Cotesia flavipes* (Cameron, 1891) (Hymenoptera: Braconidae) and *Trichogramma galloi* Zucchi (Hymenoptera: Trichogrammatidae) (Carvalho et al., 2007). While transgenic sugarcane developed by the CTC (Sugarcane Technology Center) in Brazil to control *D. saccharallis* is not commercially a reality, sugarcane mills and farmers continue their quest for the best tool to control this pest. In the Brazilian scenario this pest is now found in 100% of country sugarcane, causing losses estimated at R$ 5 billion per year, with the control cost equivalent to approximately 20% of this amount (Parra and Coelho Junior,2019).

39.2 *Diatraeae* spp.

There are about 21 species of the genus *Diatraea* occurring in sugarcane and other crops on the American continent, not all species cause economic damage, of this total, two species are important as insect pests *D. saccharalis* and *D. flavipennella*. *D. saccharalis*, commonly known as the sugarcane borer, is the most important species found in the Western hemisphere (Long & Hensley, 1972). This species, among those of the genus, is the one with the widest geographical distribution. It has been found from the southern United States to the Antilles and in all countries of Central and South America (Box, 1931; Box, 1956; Guevara, 1976).

In addition to sugarcane, *D. saccharalis* attacks other crops such as rice, oats, barley, millet, maize, sorghum, wheat, and economically important pasture species (Elias,1970; Silva et al., 1968). Many weeds or wild plants may also serve as hosts, including *Paspalum* sp., *Panicum* spp., *Holcus* sp., and *Adropogon* sp. (Jepson, 1954; Box, 1956; Randolph et al., 1967; Dinther, 1970). Quintana-Muñiz and Walker (1970) studied the preference of *D. saccharalis* oviposition in 17 host plants in Puerto Rico. The results showed maize and sugarcane being the preferred plants.

The larvae of *D. saccharalis* causes direct and indirect damage. Direct damage results from the insect feeding on plant tissues and is characterized by weight loss, opening of galleries, failure in germination, death of the apical bud, tipping of the stems, shortening of the internode, aerial rooting, and germination of the lateral buds. These damages can occur isolated or associated. Indirect damage, on the other hand, is related to the entry of opportunistic microorganisms, like *Fusarium moniliforme* and *Colletotricum falcatum* fungi, which promote the inversion of sucrose and decrease the purity of the juice, leading to a lower sugar yield and contamination of alcoholic fermentation, which microorganisms compete with yeasts resulting in lower alcohol yield (Long & Hensley, 1972; Macedo & Botelho, 1988; Gallo et al., 2002).

Esteban-Durán et al. (1998) indicate that in addition to the penetration of opportunistic microorganisms, there is also the secondary action of coleobrocas such as *Metamasius hemipterus* (Linnaeus, 1765) (Coleoptera: Curculionidae), *Xyleborus* spp., and *Sphenophorus* spp., which contribute to increasing the damage.

The reduction in the sucrose concentration is reflected in the decrease of sugar percentage recovered, this value is determined by the infestation intensity (Gomez & Lastra Borja, 1995). The intensity of infestation of *D. saccharalis* in sugarcane is a parameter that determines the percentage of internodes attacked and this is indicative of the losses occurring in a ton of cane per hectare and the sucrose content caused by *D. saccharalis*. This is an index that allows the evaluation of the borer attack situation in the cane field and allows the producer to develop control programs (Macedo & Lavorenti, 2004).

39.3 Monitoring of *D. saccharalis* in Sugarcane Crops

According to CTC (2016) *D. saccharalis* can cause losses of 0.42% in sugar, 0.21% in alcohol and 1.14% in the weight of the cane harvested for each 1% of infestation intensity index (III) [100 x (number of bored internodes / total number of internodes)]. The monitoring of *D. saccharalis* larvae from the first to the last instars, when carried out by the mills team or the farmer, is made from three to four months of planting or cutting up to eight to ten months, when it is no longer possible to enter the cane field. This sugarcane field will be at least four months waiting for the harvest. However, just if this cane is not being a "repeated-cane or bisada-cane," in other words, if it does not spend two summers in the field without harvesting (Pinto, 2016).

To explain "repeated-cane" or "bisada-cane," it is the sugarcane that should be harvested, but for lack of time it will be harvested only in the next harvest. This sugarcane goes through two developmental cycles, because of this it is a cane with more invert sugars and consequently with less potential to produce alcohol and sugar.

In a less serious situation, where sugarcane will not be monitored for four months waiting for harvest, *D. saccharalis* completes its life cycle (from egg to adult) in the field in about 60 days (Parra, 2014). This means that in a field that awaits harvesting, two generations of *D. saccharalis* may occur, damaging the internodes (buds) that were formed in these 4 months (Figure 39.1) (Pinto, 2016; Pinto et al., 2018).

When the field reaches eight to ten months, about ten internodes will be formed. When sugarcane is harvested, about 20 internodes already exist, a value that may differ between the varieties used. This means that half of the internodes formed during the harvest period will be subject to new infestations that may occur in the unmonitored sugarcane (Pinto, 2016; Pinto et al., 2018). If this additional infestation can occur in an unmonitored sugarcane field that awaits harvest, in an area of "repeated-cane", which will remain stationary for much longer, the situation is alarming.

For a well-monitored sugarcane with adequate control of *D. saccharalis*, when evaluating the sugarcane that is harvested in these areas, we will find more bored internodes in the upper than in the bottom, indicating the problem discussed. Therefore, unmonitored sugarcane awaiting harvest and cane are continuous sources of infestation of adjacent areas on a farm and are the main factors for increasing the rate of infestation intensity in well-maintained sugarcane (Pinto, 2016; Pinto et al., 2018).

There is a way to monitor and control *D. saccharalis* in awaiting harvested sugarcane fields or "repeated-cane." Monitoring can be done at "zero point," in plants around the field, provided that the area has been monitored from three or four to eight months of development and has no infestation above 1,000 larvae per hectare. Remembering that infestations most often start from the outside to inside, monitoring the "zero point" would show the exact moment of control at this location, preventing the infestation from spreading. Detecting the population increase of the pest at "zero" or in areas already infested by eight months after planting or harvesting, some control measures should be adopted (Pinto, 2016; Pinto et al., 2018).

If the area has completed the monitoring period with infestation greater than 1,000 *D. saccharalis* larvae per hectare, this "zero-point" monitoring will not be possible until harvest and the control can only be performed in full area with aircraft or drones. Chemical control in these cases is usually not the best option, in a total area would not be able to reach all the larvae inside the sugarcane and in the borders would do only partial control in this region. If zero-point infestation is initial, chemical control may have some effect on non-culm larvae, but an atomizer would be more suitable for its application as it would spread the product for at least 20 meters within the cane field (Pinto, 2016; Pinto et al., 2018).

If the population survey of *D. saccharalis* is not carried out on the farm, parasitoids must be released in areas where the infestation intensity was greater than 2% at harvest. The economic damage level varies according to the variety, time of planting, crop conditions, and other factors, it is generally adopted a value between 3 and 4% of infestation intensity index (Nava et al., 2009)

If the sugarcane waiting for the harvest or those "repeated-cane" are not monitored and *D. saccharalis* is not biologically controlled, we will continue to embark on high damage and hardly reduce the infestations (Pinto, 2016). According to Parra et al. (2014) the most appropriate management for *D. saccharalis* in sugarcane awaiting harvest or "repeated-cane" is biological control. In this case, the parasitoids *Cotesia flavipes* and *Trichogramma galloi* would be the most suitable biological products. If the cane fields awaiting the harvest or the "repeated-cane" are not monitored and *D. saccharalis* is not adequately controlled, an increase in damage may occur and will make it difficult to decrease infestations below 1% of III (CTC, 2016).

39.4 Use of Parasitoids against *D. saccharalis*

Currently, the most commonly used natural enemies in sugarcane crop in Brazil are *C. flavipes* and *T. galloi*, both agents for control of *D. saccharalis* (Parra and Coelho Junior,2019).

If *T. galloi* is the chosen parasitoid, zero-point monitoring can be done by pheromone modified delta traps with virgin *D. saccharalis* females (Figure 39.2).

Whenever more than ten males are trapped weekly in the traps for two consecutive weeks, the parasitoid should be released (Pinto et al., 2018). This release can be done by plane or drone, only on the carriers, releasing about 5,000 pupae of the parasitoid per 1,000 linear meters. If carried out manually, on foot or on motorcycles/quads, a carton containing 2,000 parasitoids every 20 meters around the field should be placed between the leaves of the plants (up to two meters into the field) (Figure 39.3).

As research progresses, trap monitoring should have virgin *D. saccharalis* females replaced with synthetic pheromone lozenges,

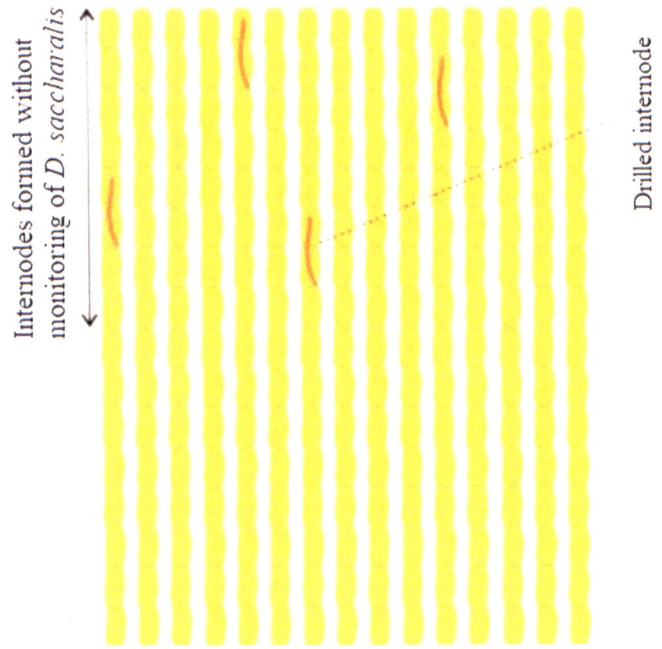

FIGURE 39.1 Illustrative scheme of drilled sugarcane internodes.

and the amount of parasitoid released, when better studied, should decrease. If the release is carried out by *C. flavipes*, monitoring should be done at least ten "zero points" where each two parallel lines, one meter each, very close to the carrier, will be evaluated by counting the number of small and big larvae (Pinto et al., 2018). Once more than 1,500 larvae are found per hectare, release should be made. In this case, a cup containing 750 parasitoids should be opened every 40 meters around the field, within two meters inside the sugarcane field (Figure 39.3). The producer can acquire the parasitoid in the pupa phase "masses" (set of pupae containing cocoons wrapped in silk threads), packed in cups containing about 1,500 parasitoids each (750, in the case of eight-point ha release).

The advantage of *C. flavipes* and *T. galloi* over insecticides is that these parasitoids within ten days will give birth to their offspring, which will continue to parasitize new hosts (*D. saccharalis* larvae or eggs) until sugarcane harvest. Therefore, these should be preferred as they are more functional in the highlighted situation. The combination of biological control agents has the advantage of speeding up the control of *D. saccharalis* and reducing errors that can be made by the field applicator. According to Botelho et al. (1999), the association of *C. flavipes* and *T. galloi* would be the best biocontrol option, with high *D. saccharalis* control rates.

39.4.1 The Larvae Parasitoid *Cotesia flavipes*

Sugarcane is now grown on about 9 million ha in Brazil, and around 3.5 million ha are treated with *C. flavipes* (Parra and Coelho Junior,2019).

The use of *C. flavipes* parasitoid to control *D. saccharalis* in sugarcane is one of the greatest biological control programs worldwide, because it is easy to rear in laboratory conditions for field release and for the extent treated area, with mass production in several different regions of the country (Vacari and De Bortoli, 2010; Parra, 2014, Parra and Coelho Junior 2019).

According to Dinardo-Miranda et al. (2014), *C. flavipes* has a smaller dispersal capacity, even when released in a higher quantity than recommended. The authors recommended that the dispersion of *C. flavipes* adults parasitoids would be released in approximately nine points per hectare, with release points to 30 m from each other because the dispersal may happen in a 15 m ray. Despite that, Pinto et al. (2009) reported that adults disperse around 18 m in cane fields, being released at least eight points per hectare in most areas. *C. flavipes* locates *D. saccharalis* by its feces. Upon reaching the feces, attracted by a water-soluble substance present in them, they begin to actively probe them with the antennae, until they find *D. saccharalis* larvae, parasitizing them (Pinto et al., 2009).

The parasitism should be monitored for 10 to 15 days after release, a new population sampling is carried out to observe parasitized *D. saccharalis* larvae or wasp "masses." The larvae

FIGURE 39.2 Plastic delta trap used to catch *D. saccharalis*.

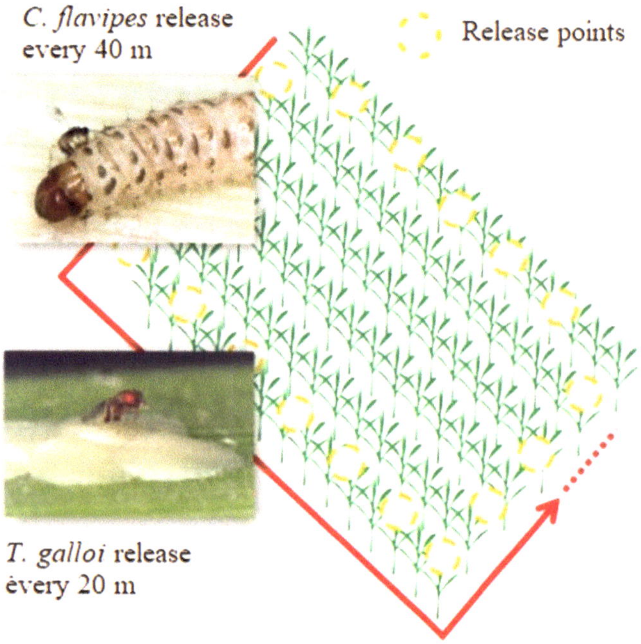

FIGURE 39.3 Illustrative scheme of parasitoids application.

collected are placed in small containers with pieces of diet and kept in an air-conditioned room for confirmation of parasitism. After calculating the parasitism [% parasitism = (total parasitized larvae and masses of the wasp / total of larvae and pupae) × 100], it is verified if the control was efficient or whether a new release should be made Nava et al. (2009).

In most farms, the dissemination of *C. flavipes* is still done manually and, depending on the size of the area, hours are spent propagating the insects. There is no doubt that manual application is more costly and less efficient than release via drones.

39.4.2 The Egg Parasitoid *Trichogramma galloi*

In Brazil, around 2 million ha are treated with *T. galloi* for *D. saccharalis* control in sugarcane crop (Parra and Coelho Junior,2019).

The small wasp *Trichogramma galloi* has a faster action than the *C. flavipes* to control *D. saccharallis*, as it acts in the egg phase of the pest, thus decreasing its proliferation, reaching up to 2% the intensity index of the *D. saccharallis* infestation. *C. flavipes* can effectively control *D. saccharallis* too, but its efficiency will take more time to appear (Botelho et. al., 1999).

In the case of *T. galloi*, most applications are made by aerial release, either via manned aircraft or drones.

39.5 Novelties in Parasitoid Release

Brazil has an enormously diverse fauna, including many ant species. These ants can prey on natural enemies when they are exposed and susceptible. predation by ants can reach 100% within in some sugarcane crops after the release. The methods of releasing natural enemies must be rapid and effective. Other means have been studied to enhance the release of natural enemies in large crops, such as the use of drones or airplanes.

In sugarcane crop, several predatory insects, some species of ants, chrysopids, ladybugs, and earwigs, among others associated with *D. saccharallis*, and other pests, exert strong pressure mainly on newly hatched eggs and larvae, causing high mortality in these pest phases. Predatory ants are the main natural enemies of *D. saccharallis* (Sousa-Silva et al., 1992), with the genera Solenopsis, Pheidole, Dorymyrmex, and Crematogaster (Hymenoptera: Formicidae) the most common in São Paulo state, Brazil (Rossi & Fowler, 2004). The biggest problem with ants is that they are not selective, so in addition to pests, ants end up attacking natural enemies like *C. flavipes* and *T. galloi* as well (Vinha et al., 2019).

New application technologies are being developed, where the masses of *C. flavipes* can be treated with repellents to avoid predators, mainly ants, so they can be spread by carriers with motorcycles/quads, airplanes or drones without losses by predation, and not disturbing the natural predation of *D. saccharallis* (Vinha et al., 2019).

39.6 Conclusion and Future Prospects

Application technology via drones is revolutionizing the use of predators, parasites, and parasitoids in agriculture, because the operational difficulty for insects release in the field is still a limiting factor for the adoption of biological control technology.

Much research still needs to be carried out to improve the parasitoids rearing in the laboratory, the rearing automation at biofactories may open more opportunities for biological control, in addition, new application technologies such as the use of drones have facilitated the release of these parasitoids in the field; however, further studies are needed for sampling methodologies and the ideal amount of unprotected parasitoids to be released with or without repellents.

REFERENCES

Almeida, L. C., Stingel, E. 2005. *Curso de monitoramento e controle de pragas da cana-de-açúcar.* Piracicaba: Centro de Tecnologia Canavieira.

Botelho, P.S.M., Parra, J.R.P., Chagas Neto, J.F. das, Oliveira, C.P.B. 1999. Associação do parasitóide de ovos *Trichogramma galloi* Zucchi (Hymenoptera: Trichogrammatidae) e do parasitóide larval *Cotesia flavipes* (Cam.) (Hymenoptera: Braconidae) no controle de *Diatraea saccharalis*, (Fabr.) (Lepidoptera: Crambidae) em cana-de-açúcar. *Anais da Sociedade Entomológica do Brasil* 28:491–496.

Box, H.E. 1931. The crambine genera Diatraea and Xantoperne (Lep., Pyral.). *Bulletin of Entomological Research* 22:1–50.

Box, H.E. 1956. The new species and records of diatraea guilding and zeadiatraea box from Mexico, Central and South America (Lepid., Pyral.). *Bulletin of Entomological Research* 47:755–776.

Carvalho, J.S., Viel, S.R., Vacari, A.M., Proença, D.C., Pereira, L.C.G., Bortoli, S.A. 2007 Parâmetro para controle de qualidade de *Cotesia flavipes* (Cameron, 1891) (Hymenoptera: Braconidae). Paper presented at VIII Congresso de Ecologia do Brasil, Sociedade de Ecologia do Brasil, Caxambu, MG, Brasil.

CTC (Centro De Tecnologia Canavieira). 2016. Causando prejuízo bilionário, broca da cana é um dos maiores problemas das usinas. em: https://ctc.com.br/causando-prejuizo-bilionario-broca-da-cana-e-um-dos-maiores-problemas-das-usinas (Accessed December 12, 2019).

Dinardo-miranda, L.L., Fracasso, J.V., Costa, V.P., Lopes, D.O.T. 2014. Dispersal of *Cotesia flavipes* in sugarcane field and implications for parasitoid releases. *Bragantia* 73:163–170.

Dinther, J.B.M. 1970. A method of assessing rice yield losses caused by the stem borer Rupela albinilla and *Diatraea saccharalis* in Surinam and the aspect of economic thresholds. *Entomophaga* 16:320–326.

Elias, L.A. 1970. Maize resistance to stalk borer in Zeadiatraea Box and *Diatraea Guilding* (Lepidoptera: Pyralidae) at five localities in Mexico. PhD thesis, Kansas State University.

Esteban-Durán, J., Yela, J.L., Beitia-Crespo, F., Jiménez-Álvarez, A. 1998. Curculiónidos exóticos susceptibles de ser introducidos en España y otros países de la Unión Europea a través de vegetales importados (Coleóptera: Curculionidae: Rhynchophorinae). *Boletín de Sanidad Vegetal Plagas* 24:23–40.

Gallo, D., Nakano, O., Silveira Neto, S., Carvalho, R.P.L., Baptista, G.C., Berti-Filho, E., Parra, J.R.P., Zucchi, R.A., Alves, S.B., Vendramim, J.D., Marchini, L.C., Lopes, J.R.S., Omoto, C. 2002. *Agricultural Entomology (Entomologia Agrícola)* Piracicaba: FEALQ.

Gómez, L.L.A., Lastra Borja, L.A. 1995. *Insectos asociados con la caña de azúcar en Colombia*. Cali, Valle del Cauca: Cenicaña.

Guevara, L.A.C. 1976. Aspectos da biologia em condições naturais e freqüência de acasalamento da *Diatraea saccharalis* (Fabr., 1974) (Lepidoptera: Crambidae) a broca da cana de açúcar. Ph.D diss, University of São Paulo.

Jepson, W.F. 1954. A critical review of the world literature on the lepidopterous stalk borers of tropical graminaceous crops. London: Commonwealth Institute of Entomology, 127 p.

Long, W.H., Hensley, S.D. 1972. Insect pests of sugar cane. *Annual Review of Entomology* 17:149–176.

Macedo, N., Botelho, P.S.M. 1988. Controle integrado da broca da cana-de-açúcar *Diatraea saccharalis* (Fabr. 1794) (Lepidoptera: Pyralidae). *Brasil Açucareiro* 106:2–12.

Macedo, N., Lavorenti, N. 2004. Novo método de amostragem de intensidade de infestação da broca da cana-de-açúcar (*Diatraea saccharalis*). *STAB* 22:32–41.

Mendonça, A.F.1996. *Guia das principais pragas da cana-de-açúcar na América Latina e Caribe*. Maceió: Insetos & Cia.

Nava, D.E.N., Pinto, A.D.S., Silva, D.A.S. 2009. *Controle biológico da broca da cana-de açúcar*. Embrapa Clima Temperado.

Parra, J.R.P. 2014. Biological control in Brazil an overview. Scientia Agricola 71:420–429.

Parra, J.R.P., Coelho Junior,A. 2019 Applied Biological Control in Brazil: From Laboratory Assays to Field Application. Journal of Insect Science 19:1–6.

Pinto, A.S. 2016. Broca-da-cana: você pode estar dormindo com o inimigo. G.bio, Grupo de Pesquisa e Extensão em Controle Biológico. https://gebio.com.br/campo/broca-da-cana-voce-pode-estar-dormindo-com-o-inimigo/ (Accessed December 12, 2019).

Pinto, A.S., Botelho, P.S.M.B., Oliveira, H.N.O. 2009. *Guia ilustrado de pragas da cana-de-açúcar*. Piracicaba: CP2.

Pinto, A.S., Rodrigues, L.R., Oliva, M.B. 2018. Uso de armadilhas de feromônio para a broca-da-cana, Diatraea SPP., em canaviais. Piracicaba: Editora Occasio.

Quintana-muñiz, V. Walker, D.W. 1970. Oviposition preferente by gravid sugarcane borer moths in Puerto Rico. *Journal of Economic Entomology* 63:987–988.

Randolph, N.M., Teetes, G.L., Jeter, B.E. Jr. 1967. Insecticide sprays and granules for control of the sugarcane borer on grain sorghum. *Journal of Economic Entomology* 60:762–765.

Rossi, M.N., Fowler, H. G. Predaceous ant fauna in new sugarcane fields in the state of São Paulo, Brazil. 2004. *Brazilian Archives of Biology and Technology* 47:805–811.

Silva, A.G.A., Gonçalves, C.R., Galvão, D.M., Gonsalves, J.L., Gomes, J., Silva, M.N., Simoni, L. 1968. Quarto catálogo dos insetos que vivem nas plantas do Brasil. Brasília: Embrapa Algodão CNPA.

Sousa-silva, C.R., Sgrillo, R.B., Oliveira, A.R., Pacheco, J.M. 1992. Uso do P-32 no estudo de predadores de *Diatraea saccharalis* (Fabricius,1794) (Lepidoptera: Pyralidae). *Anais da Sociedade Entomológica do Brasil* 21:133–138.

Vacari, A.M., De Bortoli, S.A. 2010. Tópicos em Entomologia Agrícola III. Jaboticabal: Multipress.

Vinha, F.B., Rodrigues, L.R., Pinto, A.S. 2019. Use of repellents in *Cotesia flavipes* (cam.) (Hymenoptera: Braconidae) pupae for aerial release. *Nucleus* 15:347–352.

40

Challenges in the Applied Use of Parasitoids to Control Agricultural Pests

Simone Mundstock Jahnke, Kássia Cristina Freire Zilch, Gisele De Souza Da Silva, Cleder Pezzini, and Roberta Agostini Rohr

CONTENTS

40.1 Introduction

Parasitoids are one of the most important groups of biological control agents in the control of pests in agricultural systems (Hanson & Heydon, 2006). They are common in wasps (Hymenoptera), and less common in flies (Diptera) and species of beetles (Coleoptera), moths, and butterflies (Lepidoptera), and lacewing order (Neuroptera) (Hajek, 2004; Rafael et al., 2012).

Among flies, Tachinidae are the most diverse family, with more than 8,000 species with this behavior. Most flies do not have an ovipositor to inject their eggs into the host (Rafael et al., 2012) and so attack exposed hosts and not those that live in hidden locations in plants or galls (Hajek, 2004). Hymenopteran families who have more species used as biological control agents are Aphelinidae, Braconidae, Encyrtidae, Eulophidae, Ichneumonidae, Pteromalidae, Platygastridae, and Trichogrammatidae (Costa & Perioto, 2006). These females have an elongated tubular egg-laying structure, using to inject eggs into the hosts or lay eggs on top of the hosts' bodies (Gullan & Cranston, 2012). There are many examples of species that have been used successfully.

Trichogramma spp. are exclusively parasitoids of eggs and have a wide geographical distribution and a large number of hosts (Haji et al., 2002). The most used species include *Trichogramma evanescens* Westwood, *Trichogramm adendrolimi* Matsumura, and *Trichogramma pretiosum* Riley, due to the plasticity in the selection of habitats and hosts. Another important egg parasitoid is *Telenomus podisi* Ashmead (Hymenoptera, Platygastridae), a natural controller of the Pentatomidae species in soybeans (Peres & Corrêa-Ferreira, 2004) and in rice culture (Riffel et al., 2010).

The larval parasitoid *Cotesia flavipes* Cameron (Hymenoptera: Braconidae) originated from Southeast Asia is used in biological control programs in Brazil in sugarcane crop to control *Diatraea saccharalis* (Fabricius) (Lepidoptera: Crambidae) (Nava et al., 2009). In the successful Biological Control Program of Wheat Aphids (BCPWA) in Brazil in the 1970s to 1980s, many parasitoid wasps were released, including *Aphidius colemani* Viereck (Hymenoptera: Braconidae) capable of growing in several aphid species (Tomanovic et al., 2014).

40.2 Challenges of the applicability of parasitoids in pest management

The misuse of species concepts and the lack of continuous checking of the identification of the agents is still one of the failure factors of the biological control programs (Gokhman, 2018). The novelties in systematics and taxonomy involve new data sources for analysis (molecular data, on partial nucLSU rDNA sequences, isoenzyme analyses, and morphological data with non-destructive anatomy, etc.) (Baracchi et al., 2010; Gokhman, 2018) for processing and disseminating information that allows more accurate and efficient interpretation and dissemination of results.

For the establishment of parasitoid rearing, often living organisms are used as hosts and, although in vitro breeding techniques are being made possible, these are still controversial results (Cônsoli & Parra, 2000). Size, sex ratio, rate of development, egg production, and mobility are typically monitored to define quality control on each situation (Fernandes & Nentwig, 1997). IOBC (International Organization for Biological Control) designed quality control guidelines to keep organisms produced with high quality and to avoid failures in applied biological control (Morales-Ramos et al., 2014).

The search for parasitoids by hosts is an innate behavior; however, this can be modified through successive contact with a host, which can result in learning (Tognon et al., 2017, Masry et al., 2018). Complex mixtures of volatile organic compounds, known as herbivore-induced plant volatiles (HIPVs) or egg-laying oviposition-induced plant volatiles (OIPVs), also serve as chemical clues (Dicke, 2009; Kessler &Heil, 2011). The elicitors of these compounds are often derived from oral insect secretions (Alborn et al., 1997) or compounds present in membranes or chorium (Tan et al., 2019). These compounds can have their production stimulated in plants through exogenous application of compounds such as salicylic acid or methyl jasmonate, which are elicitors of plant resistance responses (Patt et al., 2018).

Another relevant aspect is competition. Interspecific competition occurs between adult parasitoids (extrinsic competition), interactions between females looking for the same hosts, as well as between immature ones (intrinsic competition), between larvae developing in the same host (Wang et al., 2008). Some of the studied examples refer to competition between exotic fruit fly (Tephritidae) parasitoids, such as *Diachasmimorpha longicaudata* (Ashmead), (Hymenoptera: Braconidae) and parasitoids native to the neotropical region like *Doryctobracon* spp. (Wang et al., 2008; Miranda et al., 2015).

Mathematical modeling is used to explore optimal timing for releasing parasitoids in biological control programs, as in the work carried out in Italy to control *Drosophila suzukii* Matsumura (Diptera, Drosophilidae) (Pfab et al., 2018).

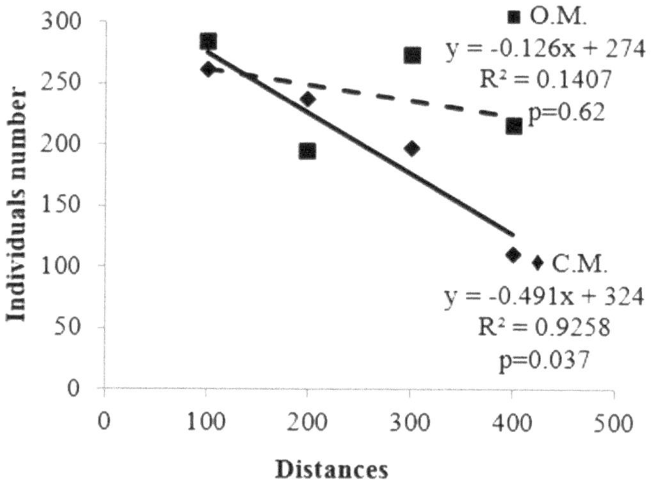

FIGURE 40.1 Correlation between distances in rice crop season (A: 0, 100 m, 200 m, 300 m, and 400 m from native vegetation (O.M. $p > 0.05$ and C.M. $p < 0.05$. Pearson Test). Adapted from Silva et al. (2019).

40.3 Parasitoids Associated with Conservative Biological Control

Conservation biological control works with other environmental factors and involves habitat management that may occur within crops, within farms, or across landscapes to increase the effectiveness of natural enemies (Landis et al., 2000). The intensification of agriculture often undermines ecological services that could benefit crop production (Jonsson et al., 2015; Tscharntke et al., 2005). Furthermore, the increase in agricultural fields with continuous extension and few crop species leads to homogenization landscape that enhances the resources available to crop pests (Tscharntke et al., 2012) and decreases non-crop habitats that support natural enemies (Chaplin-Kramer et al., 2011). A study of similarity in the composition of parasitoid families between an area of native vegetation and organic rice crop showed that 39% of morphospecies were shared between areas (Da Silva et al. 2016).

In an agroecosystem, bottom-up control refers to how much a parasitoid (or group of predators or parasitoids) influences the structure or population dynamics of the ecosystem (Pywell et al., 2011). According to Zaho et al. (2015), Letourneau et al. (2012) also found a positive association between the richness and abundance of Tachinidae with permanent areas of vegetation and the negative relationship with cultivated areas.

There is a group of researchers who defend the remodeling of agriculture on a large scale (Gliessman, 2001; Merck & Pereira, 2014). In large areas of rice crops next to natural vegetation, an abundance of parasitoids is observed (Da Silva et al., 2019) (Figure 40.1). In an organic management area, the correlation was not demonstrated (Da Silva et al., 2019). In this case, levees can serve as corridors extending the distances traveled by parasitoids (Simões-Pires et al., 2016).

40.4 Parasitoids in the Control of Fruit Flies Tephritidae

Fruit flies (Diptera: Tephritidae) are one of the largest groups of phytophagous insects with worldwide economic importance, with a significant impact on the production and export of fresh fruits (dos Santos &Guimarães, 2018). Most of the pest species belong to the genera *Anastrepha* Schiner, *Bactrocera* Macquart, *Ceratitis* Macleay, *Dacus* Fabricius, and *Rhagoletis* Loew (Norrbom, 2006; Uramoto, 2007). The damage caused by fruit flies is mainly generated by the larvae and females, through oviposition (Montes, 2006).

Parasitoids of Braconidae and Figitidae's family are koinobiont endoparasites of larvae mainly and some of pupae. Species of Pteromalidae and Diapriidaeare pupae parasitoids (Canal & Zucchi, 2000; Ovruski et al., 2000). In different areas of fruit production, species of parasitoids have been introduced to control flies, such as the larval-pupal koinobiont endoparasitoid, *Tetrastichus giffardianus* Silvestri (Hymenoptera: Eulophidae), the koinobiont endoparasite of larva-pupa, *D. longicaudata* (Hymenoptera: Braconidae), and the koinobiont egg parasitoid, *Fopiusarisanus* (Sonan) (Hymenoptera: Braconidae).

Diachasmimorpha longicaudata is native to the Indo-Australian region, where it parasitizes at least 14 species of *Bactrocera* (Wharton & Gilstrap, 1983). It is widely used as a biological control agent worldwide (Devescovi et al., 2017), being able to parasitize *Ceratitiscapitata* (Wiedemann) and several species of *Anastrepha* (Carvalho & Nascimento, 2002). Usually parasites larvae of the second and third instar, but it can also parasite the first instar and pupae (Sime et al., 2006; Rohr et al., 2019).

Doryctobracona reolatus is a koinobiont endoparasitoid native to the neotropical region, parasitizing Tephritidae larvae (Nunes et al., 2011; Uchôa, 2012). This parasitoid exhibits diapause, which allows it to expand its reach to regions with little plant diversity (Eitam et al., 2004; Ovruski et al., 2004), has a preference for parasitizing end-of-second larvae and the beginning of a third instar (Carvalho et al., 2000) and has a spatial distribution closely related to that of its hosts (Jahnke et al., 2014).

Aganaspis pelleranoi Brèthes (Hymenoptera: Figitidae), attacking a wide variety of hosts belonging to the Tephritidae and Lonchaeidae families in native and exotic fruits (Ovruski et al., 2000; Sivinski & Aluja, 2003). Females are able to forage and attack their hosts inside fallen fruits, penetrating through holes or cracks in the surface and infiltrating the pulp to find the host larva (Guimarães et al., 2000; Sivisnki et al., 2001). Preferably parasitize larvae of the third instar and this stage presents the best emergency rate (Gonçalves et al., 2013).

In order to control this group of pests, it is necessary to use biological control measures integrated with other like the sterile insect techniques (SIT), for fruit flies in Costa Rica (Messing, 1996), southern Mexico (Sivinski et al., 1996), Guatemala (Rendon et al., 2006) and Hawaii (Kaplan, 2008).

FIGURE 40.3 Female of *Habrobracon hebetor.*

40.5 Parasitoids in Biological Control in Stored Products

Stored products as grains, seeds, flour, coffee, tobacco, among others, can be infested by a variety of insects such as weevils, beetles, and moths (Ramos & Nanya, 2009). The most common pests are coleopterans (Ptinidae, Curculionidae, and Bostrichidae) and lepidopterans (Pyralidae and Gelechiidae) (Pacheco & Paula, 1995).

Many natural enemies of stored product pests are already known (Lorini et al., 2015). Parasitoid wasps of Pteromalidae, Bethylidae, Ichneumonidae, Braconidae, and Trichogrammatidae are frequently found (Athié & Paula, 2002). The most studied parasitoids for the control of beetles in storage environments are: *Anisopteromalus calandrae* (Howard) (Hymenoptera: Pteromalidae), *Theocolax elegans* (Westwood) (Hymenoptera: Pteromalidae), and *Lariophagus distinguendus* (Forster) (Hymenoptera: Pteromalidae). Parasitoids for the control of moths are: *Habrobracon hebetor* (Say) (Hymenoptera: Braconidae), *Venturia canescens* (Gravenhorst) (Hymenoptera: Ichneumonidae), and *Trichogramma evanescens* Westwood (Hymenoptera: Trichogrammatidae) (Schöller et al., 2018).

Anisopteromalus calandrae (Figure 40.2) attacks last-instar larvae and pupae of coleopteran pests that occur in high densities (Sasakawa et al., 2013), including: *Sitophilus oryzae* (L.), *S. granarius* (L.), *S. zeamais* (Motschulsky) (Coleoptera: Curculionidae), *Rhyzoperthadominica* (Fabricius) (Coleoptera: Bostrichidae), and *Lasiodermaserricorne* (Fabricius) (Coleoptera: Ptinidae) (Menon et al., 2002; Belda & Riudavets, 2010; Benkhellat et al., 2015).

The female uses some clues to locate the host's larvae and can also detect the movement or sound of the host's feeding in grains infested through the antennas (Menon et al., 2002). As soon as the host's larvae are located, the female uses her ovipositor to perforate the surface of the grain, paralyzes the larva, and lays an egg on it (Ozelame et al., 2011). Only one parasitoid develops from each host larva (Belda & Riudavets, 2010) and the total development is completed in about 13 days (Zilch et al., 2017).

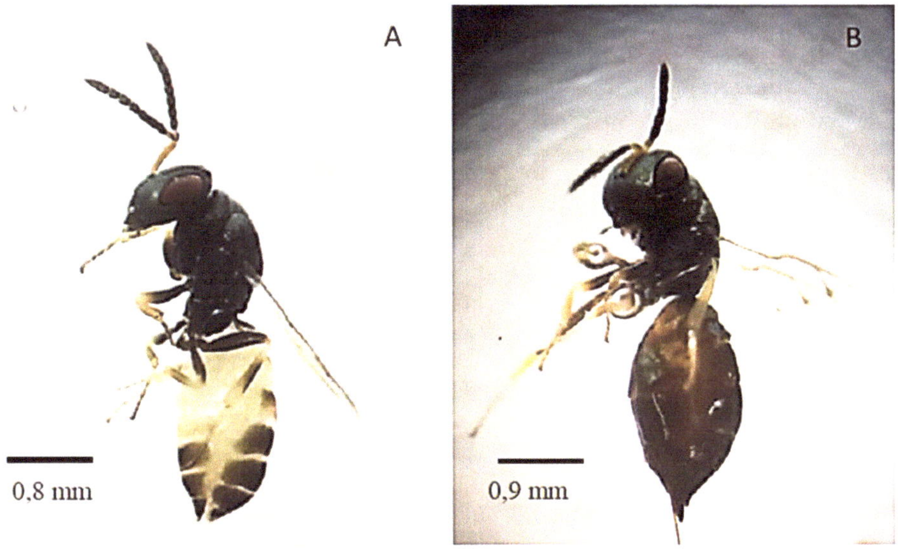

FIGURE 40.2 Male (A) and female (B) of *Anisopteromalus calandrae.*

Habrobracon hebetor (Figure 40.3) is a cosmopolitan larval ectoparasitoid and parasitizes *Ephestia kuehniella*, *Ephestiaelutella* (Hübner), *Ephestia cautella* (Walker), and *Plodia interpunctella* (Hübner) (Lepidoptera: Pyralidae), among others (Athié & Paula, 2002; Eliopoulos & Stathas, 2008). *Habrobracon hebetor* has already been commercialized by some biofactories, especially in the European continent, for the control of moths that affect various stored products (Solà et al., 2018).

The cycle of this parasitoid begins with the moving away from the already dried larva to become pupa (Athié & Paula, 2002; Pezzini et al., 2017). There is a positive correlation between the increase in the average total number of eggs throughout the female's life in relation to the increase in the light period in which the females were created (Pezzini et al., 2019).

REFERENCES

Alborn, T. C. J.; Turlings, T. H.; Jones, G.; Stenhagen, J. H.; Loughrin, J. H.; Tumlinson, J. H. An elicitor of plant volatiles from beet armyworm oral secretion. *Science*. 1997, 276(5314), 945–949.

Athié, I.; Paula, D. C. de. *Insetos de grãos armazenados: aspectos biológicos e identificação*. 2nd ed. São Paulo: Varela, 2002; p. 244.

Baracchi, D.; Dapporto, L.; Teseo, S.; Hashim, R.; Turillazzi, S. Medium molecular weight polar substances of the cuticle as tools in the study of the taxonomy, systematics and chemical ecology of tropical hover wasps (Hymenoptera: Stenogastrinae). *J. Zool. Syst. Evol. Res.* 2010, 48(2), 109–114.

Belda, C.; Riudavets, J. Attraction of the parasitoid *Anisopteromal uscalandrae*(Howard) (Hymenoptera: Pteromalidae) to odors from grain and stored product pests in a Y-tube olfactometer. *Biol. Control*. 2010, 54(1), 29–34.

Benkhellat, O.; Jaloux, B.; Moali, A.; Chevrier, C.; Monge, J. P. Host discrimination and egg laying in *Anisopteromalus calandrae* (Hymenoptera: Pteromalidae) ectoparasitoid of *Callosobruchus maculatus* (Coleoptera: Chrysomelidae). *J. Stored Prod. Res.* 2015, 61, 48–53.

Canal D. N. A.; Zucchi, R. A. 15. Parasitoides—Braconidae. In: Malavasi, A.; Zucchi, R. A., eds. *Moscas-das-frutas de importância econômica no Brasil: conhecimento básico aplicado*. Ribeirão Preto: Holos, 2000; p. 327.

Carvalho, R. S.; Nascimento, A. S. Criação de *Diachasmimorpha longicaudata* para controle biológico de moscas-das-frutas (Tephritidae). In: Parra, J. R. P.; Botelho, P. S. M.; Corrêa-Ferreira, B. S.; Bento, J. M. S., eds. *Controle biológico no Brasil: Parasitoides e Predadores*. Ribeirão Preto: Holos, 2002; p. 635.

Carvalho, R. S.; Nascimento, A. S.; Matrangolo, W. J. R. 14. Controle Biológico. In: Malavasi, A.; Zucchi, R. A., eds. *Moscas-das-frutas de importância econômica no Brasil: conhecimento básico aplicado*. Ribeirão Preto: Holos, 2000; p. 327.

Chaplin-Kramer, R.; O'Rourke, M. E.; Blitzer, E. J.; Kremen, C. A meta-analysis of crop pest and natural enemy response to landscape complexity. *Ecol. Lett.* 2011, 14(9), 922–932.

Cônsoli, F. L.; Parra, J. R. P. Effect of the age of the pupal holo tissue on the nutritional quality of artificial diets for *Trichogramma* spp. (Hymenoptera: Trichogrammatidae). *An. Soc.* Entomol. Bras. 2000, 29(3), 555–563.

Costa, V. A.; Perioto, N. W. Insetos parasitoides. In: Batista Filho A. (Org.). *Controle biológico de insetos e ácaros*. 1st ed. São Paulo: Instituto Biológico, 2006; pp. 52–68.

Da Silva, G. de S.; Jahnke, S. M.; Ferreira, M. L. G. Hymenoptera parasitoids in protected area of Atlantic Forest biomes and organic rice field: 2 compared assemblages. *Rev. Colomb. Entomol.* 2016, 42(2), 110–117.

Da Silva, G. S.; Jahnke, S. M.; Johnson, N. Riparian forest fragments in rice fields under different management: Differences on hymenopteran parasitoids diversity. *Braz. J. Biol.* 2019, 79, 1–11.

Devescovi, F.; Bachmann, G. E.; Nussenbaum, A. L.; Viscarret, M. M.; Cladera, J. L.; Segura, D. F. Effects of superparasitism on immature and adult stages of *Diachasmimorphalongicaudata* Ashmead (Hymenoptera: Braconidae) reared on *Ceratitis capitata* Wiedemann (Diptera: Tephritidae). *Bull. Entomol. Res.* 2017, 107(06), 756–767.

Dicke, M. Behavioural and community ecology of plants that cry for help. *Plant Cell Environ.* 2009, 32(6), 654–665.

Dos Santos, J. P.; Guimarães, J. A. Parasitoids associated with *Anastrepha fraterculus* (Diptera: Tephritidae) in native fruits: first record of *Aganaspisnordlanderi* (Hymenoptera: Figitidae) in the State of Santa Catarina. *Rev. Bras. Frutic.* 2018, 40(3), e-414.

Eitam, A.; Sivinski, J.; Holler, T.; Aluja, M. Biogeography of braconid parasitoids of the Caribbean fruit fly (Diptera: Tephritidae) in Florida. *Ann. Entomol. Soc. Am.* 2004, 97(5), 928–939.

Eliopoulos, P. A.; Stathas, G. J. Life tables of *Habrobracon hebetor* (Hymenoptera: Braconidae) parasitizing *Anagasta kuehniella* and *Plodia interpunctella* (Lepidoptera: Pyralidae): Effect of host density. *J. Econ. Entomol.* 2008, 101(3), 982–988.

Fernández, C.; Nentwig, W. Quality control of the parasitoid *Aphidius colemani* (Hymenoptera: Aphidiidae) used for biological control in greenhouses. J. Appl. Entomol.1997,121 (1–5),447–456.

Gliessman, S. R. *Agroecologia: Processos ecológicos em agricultura sustentável*. 2nd ed. Porto Alegre: Editora da UFRGS, 2001; p.653.

Gokhman, V. E. Integrative taxonomy and its implications for species-level systematics of parasitoid Hymenoptera. *Entomol. Rev.* 2018, 98(7), 834–864.

Gonçalves, R. S.; Nava, D. E.; Pereira, H. C.; Lisbôa, H.; Grützmacher, A. D.; Valgas, R. A. Biology and fertility life table of *Aganaspispelleranoi* (Hymenoptera: Figitidae) in larvae of *Anastrepha fraterculus* and *Ceratitis capitata* (Diptera: Tephritidae). *Ann. Entomol. Soc. Am.* 2013, 106(6), 791–798.

Guimarães, J. A.; Dias, N. B.; Zucchi, R. A. Parasitóides—Figitidae (Eucoilinae). In: Malavasi, A.; Zucchi, R. A., eds. *Moscas-das-frutas de importância econômica no Brasil: conhecimento básico e aplicado*. Ribeirão Preto: Holos, 2000; pp. 127–134.

Gullan, P. J.; Cranston, P. S. *Os Insetos: um resumo de entomologia*. 4th ed. São Paulo: Roca, 2012; p. 496.

Hajek, A. E. *Natural Enemies: An Introduction to Biological Control*. Cambridge: Cambridge University Press, 2004; p. 378.

Haji, F. N. P.; Prezotti, L., Carneiro, J. D. S.; Alencar, J. D. *Trichogramma pretiosum* para o controle de pragas no tomateiro industrial. In: Parra, J. R. P.; Botelho, P. S. M.;

Corrêa-Ferreira, B.S.; Bento, J. M. S., eds. *Controle Biológico no Brasil: parasitoides e predadores*. São Paulo: Manole, 2002; pp. 477–494.

Hanson, P.E.; Heydon, S.L. Família Pteromalidae. In: Fernández, F.; Sharkey, M. J., eds. *Introdución a los Hymenoptera de la Región Neotropical*. Bogotá: Sociedad Colombiana de Entomologia y Universidad Nacional de Bogotá, 2006; pp. 673–681.

Jahnke, S. M.; Ponte, E. M. D.; Redaelli, L. R.; Rego, D. R. G. P. Spatial patterns and associations of *Anastrepha fraterculus* (Diptera: Tephritidae) and its parasitoid *Doryctobracon areolatus* (Hymenoptera: Braconidae) in organic orchards of Psidium guajava and Acca sellowiana. *Fla. Entomol.* 2014, 97(2), 744–752.

Jonsson, M.; Straub, C. S.; Didham, R. K.; Buckley, H. L.; Case, B. S.; Hale, R. J.; Gratton, C.; Wratten, S. D. Experimental evidence that the effectiveness of conservation biological control depends on landscape complexity. *J. Appl. Ecol.* 2015, 52(5), 1274–1282.

Kaplan, J.K. Fruitful returns from fruit fly management project. *Agric. Res.* 2008, 56, 10–13.

Kessler, A.; Heil, M. The multiple faces of indirect defences and their agents of natural selection. *Funct. Ecol.* 2011, 25(2), 348–357.

Landis, D. A.; Wratten, S. D.; Gurr, G. M. Habitat management to conserve natural enemies of arthropod pests in agriculture. *Annu. Rev. Entomol.* 2000, 45(1), 175–201.

Letourneau, D. K.; Allen, S. G. B.; Stireman III, J. O. Perennial habitat fragments, parasitoid diversity and parasitism in ephemeral crops. *J. Appl. Ecol.* 2012, 49(6), 1405–1416.

Lorini, I.; Krzyzanowski, F. C.; França-Neto, J. B.; Henning, A. A.; Henning, F. A. *Manejo Integrado de Pragas de Grãos* e Sementes Armazenadas. Brasília: Embrapa, 2015; p. 84.

Masry, A.; Clarke, A. R.; Cunningham J. P. Learning influences host versus nonhost discrimination and post alighting searching behavior in the Tephritid fruit fly parasitoid *Diachasmimorpha kraussii* (Hymenoptera: Braconidae). *J. Econ. Entomol.* 2018, 111(2), 787–794.

Menon, A.; Flinn, P. W.; Barry, A.; Dover, B. A. Influence of temperature on the functional response of *Anisopteromalus calandrae* (Hymenoptera: Pteromalidae), a parasitoid of *Rhyzopertha dominica* (Coleoptera: Bostrichidae). *J. Stored Prod. Res.* 2002, 38(5), 463–469.

Merck, T.; Pereira, H. M. Reshaping agri-environmental subsidies: from marginal farming to large-scale rewilding. *Basic Appl. Ecol.* 2014, 16(2), 95–103.

Messing, R. H. Status and needs of biological control research for Tephritid flies. In: McPheron, B. A.; Steck, G. J., eds. *Fruit Fly Pests: A World Assessment of Their Biology and Management*. Florida: St. Lucie Press, 1996; pp. 365–367.

Miranda, M.; Sivinski, J.; Rull, J.; Cicero, L.; Aluja, M. Niche breadth and interspecific competition between *Doryctobracon crawfordi* and *Diachasmimorpha longicaudata* (Hymenoptera: Braconidae), native and introduced parasitoids of *Anastrepha* spp. fruit flies (Diptera: Tephritidae). *Biol. Control.* 2015, 82, 86–95.

Montes, S. M. N. M. Moscas-das-frutas—Importante praga da fruticultura. *Pesqui. Tecnol.* 2006, 3(2), 1–5.

Morales-Ramos, J. A.; Rojas, M. G.; Shapiro-Ilan, D. I. *Mass Production of Beneficial Organisms, Invertebrates and Entomopathogens*. Cambridge, MA: Academic Press, 2014; 764p.

Nava, D. E.; Pinto, A. S.; Silva, S. D. dos A. *Controle biológico da broca-da-cana-de-açúcar*. Pelotas: Embrapa Clima Temperado, 2009.

Norrbom, A. L. Fruit fly (Diptera: Tephritidae) classification and diversity. 2006. http://www.sel.barc.usda.gov:591/diptera/Tephritidae/TephHosts/search.html. (accessed Dec. 2019).

Nunes, A. M., Nava D. E.; Müller F. A.; Gonçalves R. S.; Garcia M. S. Biology and parasitic potential of *Doryctobracon areolatus* on *Anastrepha fraterculus* larvae. *Pesqui. Agropecu. Bras.* 2011, 46(6), 669–671.

Ovruski, S.; Aluja, M; Sivinski, J.; Wharton, R. Hymenopteran parasitoids on fruit-infesting Tephritidae (Diptera) in Latin America and the southern United States: Diversity, distribution, taxonomic status and their use in fruit fly biological control. *Integr. Pest Manage. Rev.* 2000, 5, 81–107.

Ovruski, S. M. Schliserman, P.; Aluja, M. Indigenous parasitoids (Hymenoptera) attacking *Anastrepha fraterculus* and *Ceratitis capitata* (Diptera: Tephritidae) in native and exotic host plants in Northwestern Argentina. *Biol. Control.* 2004, 29(1), 43–57.

Ozelame, A. L.; Nörnberg, S. D.; Nava, D. E. Exigências térmicas e número de gerações de *Anisopteromaluscalandrae*, em *Sitophiluszeamais. 12º SICONBIOL, Simpósio de Controle Biológico*- 18 a 21 de julho de 2011.

Pacheco, I. A.; Paula, D. C. de. *Insetos de grãos armazenados: identificação e biologia*. Campinas: Fundação Cargill, 1995; p. 237.

Patt, J. M.; Robbins, P. S.; Niedz, R.; Mccollum, G.; Alessandro, R. Exogenous application of the plant signalers methyl jasmonate and salicylic acid induces changes in volatile emissions from citrus foliage and influences the aggregation behavior of Asian citrus psyllid (*Diaphorinacitri*), vector of Huanglongbing. *PLoS ONE*. 2018, 13(3), e0193724.

Peres, W. A. A.; Corrêa-Ferreira, B. S. Methodology of mass multiplication of Telenomus podisi Ash. and Trissolcus basalis (Woll.) (Hymenoptera: Scelionidae) on eggs of Euschistus heros (Fab.) (Hemiptera: Pentatomidae). Neotrop. Entomol. 2004, 33(4),457–462.

Pezzini, C.; Jahnke, S. M.; Köhler, A. Capacidade de paralisação, parasitismo e viabilidade de Habrobracon hebetor sob o efeito do fotoperíodo. *Rev. Ciênc. Agrár.Amaz. J. Agric. Environ. Sci.* 2019, 62, 1–8.

Pezzini, C.; Jahnke, S. M.; Köhler, A. Morphological characterization of immature stages of *Habrobracon hebetor* (Hymenoptera: Braconidae) ectoparasitoid of *Ephestia kuehniella* (Lepidoptera, Pyralidae). *J. Hymenopt. Res.* 2017, 60, 157–171.

Pfab, F.; Stacconi, M. V. R.; Anfora, G. Grassi, A.; Walton, V.; Pugliese, A. Optimized timing of parasitoid release: A mathematical model for biological control of *Drosophila suzukii*. *Theor. Ecol.* 2018, 11, 489–501.

Pywell, R. F. Meek, W. R.; Loxton, R. G.; Nowakowski, M.; Carvell, C.; Woodcock, B.A. Ecological restoration on farmland can drive beneficial functional responses in plant and invertebrate communities. *Agric., Ecosyst. Environ.* 2011, 140(1–2), 62–67.

Rafael, J. A. Melo, G. A.; Carvalho, C. J. B. de.; Casari, S. A.; Costantino, R. *Insetos do Brasil: diversidade e taxonomia*. Ribeirão Preto: Holos, 2012; p. 796.

Ramos, A. L.; Nanya, S. Caracterização anatomo-morfológica do sistema digestório de *Lasiodermaserricorne* (Fb., 1792), (Coleoptera; Anobidae), praga de grãos e produtos armazenados. In: VI EPCC Encontro Internacional de Produção Científica Cesumar, October 27–30, 2009, pp. 27–30, Maringá.

Rendon, P.; Sivinski, J.; Holler, T.; Bloem, K.; Lopez, M.; Martinez, A.; Aluja, M. The effects of sterile males and two braconid parasitoids, *Fopiusarisanus* (Sonan) and *Diachasmimorphakrausii* (Fullaway) (Hymenoptera), on caged populations of Mediterranean fruit flies, *Ceratitiscapitata* (Wied.) (Diptera: Tephritidae) at various sites in Guatemala. *Biol. Control.* 2006, 36, 224–231.

Riffel, C. T.; Prando, H. F.; Boff, M. I. C. Primeiro relato de ocorrência de Telenomus podisi (Ashmead) e Trissolcus urichi (Crawford) (Hymenoptera: Scelionidae) como parasitóides de ovos do percevejo-do-colmo-do-arroz, Tibraca limbativentris (Stål) (Hemiptera: Pentatomidae), em Santa Catarina. Neotrop.Entomol. 2010, 39(3), 447–448.

Rohr, R. A.; Jahnke, S. M.; Redaelli, L. R. Does Diachasmimorpha longicaudata (Hymenoptera: Braconidae) have a preferential instar to parasitize Tephritidae (Diptera)? *Iheringia, Sér.* Zool. 2019, 109, e2019014.

Sasakawa, K.; Uchijima, K.; Shibao, H.; Shimada, M. Different patterns of oviposition learning in two closely related ectoparasitoid wasps with contrasting reproductive strategies. *Naturwissenschaften.* 2013, 100, 117–124.

Schöller, M., Prozell, S., Suma, P., Russo, A., Biological control of stored-product insects. In: Athanassiou, C. G., Arthur, F. H., eds. *Recent Advances in Stored Product Protection.* Springer: Berlin, 2018; pp. 183–210.

Sime, K. R.; Daane, K. M.; Nadel, H.; Funk, C. S.; Messing, H.; Andrews Jr, J. W.; Johnson, M. W.; Pickett, C. H. *Diachasmimorpha longicaudata* and *D. kraussii* (Hymenoptera: Braconidae), potential parasitoids of the olive fruit fly. *Biocontrol Sci Technol.* 2006, 16(2),169–179.

Simões-pires, P. R.; Jahnke.; S. M.; Redaelli, L. R. Influence of the vegetation management of the leeves in irrigated rice organic in diversity of Hymenoptera parasitoids. *Braz. J. Biol.* 2016, 76(3), 774–781.

Sivinski, J.; Aluja, M. The evolution of ovipositor length in the parasitic hymenoptera and the search for predictability in Biological Control. *Fla. Entomol.* 2003, 86(2),143–150.

Sivinski, J. M.; Calkins, C. O.; Baranowski, R. M.; Harris, D.; Brambila, J.; Diaz, J.; Burns, R. E.; Holler, T.; Dodson, G. Suppression of caribbean fruit fly (*Anastrepha suspensa* (Loew) Diptera: Tephritidae) population through releases of the parasitoid *Diachasmimorpha longicaudata* (Ashmead) (Hymenoptera: Braconidae). *Biol. Control.* 1996, 6,177–185.

Sivinski, J.; Vulinec, K.; Aluja, M. Ovipositor length in a guild of parasitoids (Hymenoptera: Braconidae) attacking *Anastrepha* spp. fruit flies (Diptera: Tephritidae) in Southern Mexico.Ann. Entomol. Soc. Am. 2001, 94 (6),886–895.

Solà, M., Castañé, C.; Lucas, E.; Riudavets, J. Optimization of a banker box system to rear and release the parasitoid *Habrobracon hebetor* (Hymenoptera: Braconidae) for the control of stored-product moths. *J. Econ. Entomol.* 2018, 20(20), 1–6.

Tan, C. W.; Peiffer, M.; Hoover, K.; Rosa, C.; Felton, G.W. Parasitic wasp mediates plant perception of insect herbivores. *J. Chem. Ecol.* 2019, 45, 972–981.

Tognon, R.; Aldrich, J. R.; Buffington, M. L.; Talamas, E. J.; Sant'ana, J.; Zalom, Frank G. *Halyomorpha halys* (Heteroptera: Pentatomidae) egg surface chemicals inhibit North American *Telenomus* and *Trissolcus* (Hymenoptera: Scelionidae) parasitism. *Biol. Control.* 2017, 114, 39–44.

Tomanović Ž.; Petrović, A.; Mitrović, M.; Kavallieratos, N. G.; Starý, P.; Rakhshani, E.; Ivanović, A.Molecular and morphological variability within the Aphidius colemani group with a redescription of *Aphidius platensis* Brethes (Hymenoptera: Braconidae: Aphidiinae). B. Entomol. Res. 2014,104(5), 552–565.

Tscharntke, T.; Klein, A. M.; Kruess, A.; Steffan-Dewenter, I.; Thies, C. Landscape perspectives on agricultural intensification and biodiversity—ecosystem service management. *Ecol. Lett.* 2005, 8(8), 857–874.

Tscharntke, T.; Tylianakis, J. M.; Rand, T. A.; Didham, R. K.; Fahrig, L.; Batáry, P.; Bengtsson, J.; Clough, Y.; Crist, T. O.; Dormann, C. F.; Ewers, R. M. Fründ, J.; Holt, R. D.; Holzschuh, A.; Klein, A. M.; Kleijn, D.; Kremen, C.; Landis, D. A.; Laurance, W.; Lindenmayer, D.; Scherber, C.; Sodhi, N.; Steffan-Dewenter, I.; Thies, C.; van der Putten, W. H; Westphal, C. Landscape moderation of biodiversity patterns and processes-eight hypotheses. *Biol. Rev.* 2012, 87(3), 661–685.

Uchôa, M. A. 2012. Fruit flies (Diptera: Tephritoidea): 12. Biology, host plants, natural enemies, and the implications to their natural control. In: Soloneski, S.; Larramendy, M. L., eds. *Integrated Pest Management and Pest Control.* Rijeka, Croatia: InTech. http://cdn.intechopen.com/pdfs-wm/29609.pdf (accessed Jan. 2018).

Uramoto, K. Diversidade de moscas-das-frutas (Diptera, Tephritidae) em pomares comerciais de papaia e em áreas remanescentes da Mata Atlântica e suas plantas hospedeiras nativas, no município de Linhares, Espírito Santo. Tese de Doutorado. Piracicaba – SP, 2007, 106p.

Wang, X.-G.; Bokonon-Ganta, A. H.; Messing, R. H. Intrinsic interspecific competition in a guild of tephritid fruit fly parasitoids: Effect of co-evolutionary history on competitive superiority. *Biol. Control.* 2008, 44(3), 312–320.

Wharton, R. A.; Gilstrap, F. E. Key to and status of opiine braconid (Hymenoptera) parasitoids used in biological control of *Ceratitis* and *Dacuss.l.* (Diptera: Tephritidae). *Ann. Entomol. Soc. Am.* 1983, 76, 721–742.

Zhao, Z.H.; Hui, C.; Li, Z.H.; Li, B. L. Habitat heterogeneity stabilizes the spatial and temporal interactions between cereal aphids and parasitic wasps. Basic Appl. Ecol.2015, 16, 510–518.

Zilch, K. C. F.; Jahnke, S. M.; Köhler, A.; Bender, E. Effect of diet, photoperiod and host density on parasitism of *Anisopteromalus calandrae* on the tobacco beetle and biological parameters of the parasitoid. *Am. J. Plant Sci.* 2017, 8(12), 3218–3232.

Section XVII

Management of Crop Diseases through Microbial Biopesticides

41

Microbial Inoculants for Insect Pest Management in Organic Agriculture

Shekharappa and C. R. Patil

CONTENTS

41.1 Introduction

Biopesticides are biological materials and organisms formulated and/or used as pesticide to control obnoxious pests. They include living organisms or plant extracts used strategically for crop protection and are safe to both humans and the environment. The types of biopesticides include microorganisms such as bacteria, fungi, viruses, protozoa, entomopathogenic nematodes, biorationals, minerals, and plant-incorporated protectants.

In all 281 biopesticides involving insects (38.17%), bacteria (37%), nematodes (15.7%), fungi (4.7%), viruses (2.85%), and protozoa (2.14%) are available in the market. The most important aspect of biopesticides is that they could be region or ecosystem specific and some like botanicals are mostly based on indigenous knowledge. While some biopesticides such as predators, parasitoids, etc., are specific to ecosystems, the use of botanicals for pest control is the wisdom of farmers and not all botanicals are available under all situations. The microbial group of biopesticides includes viruses, bacteria, fungi, protozoa, and nematodes capable of reducing insect pest load below economic threshold level (ETL) by their action.

41.2 Microbial Biopesticides

The microbial insecticides consist of living cells or spores of microorganisms in various preparations intended for use as spray, dust, or paste applicable under wide environmental conditions. So far over 3000 microorganisms capable of causing disease in insects are known. Some of them can be easily mass-produced and are reported to be utilized as microbial insecticides in the management of insect pests. Being very diverse this group of biopesticides show specificity to insects and vary in their mode of action. Predation, parasitization, production of toxins, etc., are some action modes of microbial insecticides.

41.2.1 Bacterial Biopesticides

Bacteria are one of the important groups of microorganisms playing a pivotal role in agriculture. Broadly two groups of bacteria the spore formers and non-spore formers have been identified for their insecticidal activities. The most widely used bacterial biocontrol agent is *Bacillus thuringiensis* (Patil et al., 2018a and 2018b) with fast larvicidal activity and minimal to no effects on beneficial insects and other non-target organisms. The crystal (*cry*) protein from *Bacillus thuringiensis* (Bt) normally acts as a stomach toxin and hence is often referred to as a biorational insecticide or biopesticide (Lacey & Goettel, 1995). The possibilities for integrating *B. thuringiensis* with other biological control agents, cultural practices, and conventional pesticides make it a versatile bioagent. Some bacteria with biocontrol potentiality such as *Bacillus popillae* and *Bacillus lentimorbus* cannot be grown in fermentation tanks. They are "cultivated" in laboratory-reared insect larvae. Products containing *Bacillus popillae* and *Bacillus lentimorbus* can be applied to control the larval (grub) stage of the Japanese beetle and less effectively some other beetle grubs. *Bacillus sphaericus* and *Serratia entomophila* are other bacteria that control respectively, Culicidae (Davidson, 1985) and *Costelytra* spp., a scarabaeid (Jackson et al., 1992) insect pest.

41.2.2 Viral Biopesticides

There are at least seven families of viruses containing insect pathogenic representatives. One family, however, the *Baculoviridae* is unique in that it is exclusive to invertebrates and these viruses bear no resemblance either structurally or biochemically, to invertebrate pathogens. The family is divided into two main groups: the Eubaculovirinae (nuclear polyhedrosis viruses, NPV) and granulosis viruses (GV) and the Nudibaculovirinae (non-occluded viruses). The NPV against Heliothis (*Helicoverpa*) has been widely used in pulse cultivation (Lindsey & Belnavis, 2009). The Baculoviruses can be applied to control insect pests such as the larvae of lepidoptera and coleoptera. A baculovirus that controls the rhinocerous beetle (*Oryctes rhinoceros* L.) in coconut is found to be a very effective management option.

Nuclear Polyhedrosis Virus (NPV) is found effective against *H. armigera* and *S. litura*, Red Headed Hairy Caterpillar (RHHC), and armyworm. Recent improvements in commercial production of infective juveniles and formulations that have shown extended shelf life and enabled application using a broad range of equipment would boost application of viruses in biological control (Lacey & Goette, 1995).

41.2.3 Protozoal Biopesticides

Entomogenous protozoa exhibit host specificity, act slowly, most often producing chronic infections leading to sluggishness, irregular or slowed growth, and reduced feeding, vigor, fecundity, and longevity. Spores must be eaten in sufficient quantity by the host to initiate disease. Although there have been several attempts to develop protozoans as microbial control agents of insects, only one, *Nosema locustae* has been registered and is commercially available for use against grasshoppers (Henry & Onsager, 1982). The organism has been registered in the United States for use as a microbial insecticide against grasshoppers and is currently produced and sold commercially.

41.2.4 Nematodal Biopesticides

Nematodes such as *Steinernema carpocapsae* and *Heterorhabditis bacteriophora* are available as talc- and gel-based formulations (Malvika, 2016). The most promising nematodes belong to the family Steinernematidae. These nematodes are characterized by their association with a bacterium of the *Xenorhabdus* genus. When they enter their insect host, they release the bacterial cells which propagate and kill the insect within 48 hours. The most commonly and commercially used nematode for biocontrol is *Steinernema feltiae*.

41.2.5 Fungal Biopesticides

More than 950 species of entomopathogenic fungi belonging to 100 genera have been recorded, and only ten species have been commercially exploited, of which *Beuveria bassiana* (Balsomo), *Metarhizium anisopliae* (Met.), *Metarhizium (Nomuraea) rileyi* (Farlow), *Lecanicillium lecanii* (Zimmerman), *Hirsutella thomposonii*, *Paecilomycis* sp., etc., are widely used in insect management. Fungi are the only insect pathogens that are capable of invading insects by penetrating the cuticle. This is the most useful means of infection. Most successes with fungi have been with deuteromycetes which cause epizootic on foliage feeding insects in tropical environments. The important fungi useful in control of Lepidopteran, Coleopteran, and Dipteran insect pests are *Beauveria bassiana*, *Beauveria brongniartii*, *Entomophaga maimaiga*, *Hirsutella thompsonii* (on mites), *Metarhizium anisopliae*, *Metarhizium flavoviride*, *Nomuraea rileyi*, *Paecilomyces fumosoroseus*, etc., and *Verticillium lecanii*. Throughout the world, four prominently used mycoinsecticides belong to efficient strains of *Metarhizium rileyi*, *Lecanicillium lecanii*, *Metarhizium anisopliae*, and *Beauveria bassiana*.

41.2.5.1 *Metarhizium anisopliae*

Metarhizium anisopliae is a potential pathogen on insect pests and is explored for mycobiocontrol of notorious insect pests. *Metarhizium anisopliae* (Metschnikoff) can infect more than 200 insect hosts. The potential is proved beyond doubt in the management of different species of grasshoppers, termites, root grub, pyrilla of sugarcane BPH in paddy, coconut rhinocerous beetle, coffee berry borer, etc., and is known to produce cuticle degrading enzymes (Leger et al., 1987). A complete bioactivity of *M. anisopliae* has been tested on teak skeletonizer *Eutectona machaeralis* and was found to be a potential mycobiocontrol agent of teak pest (Sandhu et al., 2000).

Inoculation of 7 g dry conidia of this fungus into mature termite mounds caused 80 percent mortality of termites. Similarly, 10 g of mycelial fragments in sawdust bait caused 60 percent mortality and in 20 g rice bait caused 58 percent mortality at 7 weeks after treatment (Nasr & Moein, 1997). Application of 100 g per m^2 of *M. anisopliae* granules (2 mm) prepared by mixing vegetable oil to bran pellet cultures was at par with recommended chemical pesticide (fenthion) in lowering the root damage in sweet potato (Yokoyama et al., 1993). Application of *M. anisopliae* isolate F1-147 at a higher rate was effective in reducing the root damage caused by root grub, *Lepidiota negatoria* Wied, and produced sugarcane yield comparable to chemical insecticides (Samson et al., 1999). Aerial spray containing 6×10^{11} to 12×10^{11} conidia caused 40, 55, and 65 percent in mortality at 30, 60, and 90 days, respectively. This clearly indicated that commercial products of *M. anisopliae* should have a minimum load of spores and or hyphal bits to cause the desired effect.

41.2.5.2 *Lecanicillium lecanii*

Lecanicillium lecanii, also known as *Verticillium lecanii*, is another promising mycoinsecticide and offers protection against sucking pests viz., thrips, mites, aphids, mealy bugs, scales, and leafhoppers. This fungus attacks nymphs and adults and stocks to the leaf underside by means of a filamentous mycelium. *Verticillium lecanii* was considered to be a major parasite which caused a massive decline of cereal-cyst nematode populations in monocultures of susceptible crops (Kerry, et al., 1982). *Verticillium chlamydosporium* has a wide host range amongst cyst and root-knot nematodes, but it is very variable and only some isolates may have potential as commercial biocontrol agent.

41.2.5.3 Beauveria bassiana

Beauveria bassiana is a fungus that grows naturally in soils throughout the world and acts as a pathogen on various insect species, causing white muscardine disease. It is being used as a biological insecticide to control a number of pests such as termites, whitefly, and malaria-transmitting mosquitoes (Elsworth & Grove, 1977). *B. bassiana* is the anamorph (asexually reproducing form) of *Cordyceps bassiana*. Like many species of entomogenous fungi, *B. bassiana* is composed of many genetically distinct variants associated with geographical location and hosts which differ substantially in their ability to produce pathogenesis. As an insecticide, the spores are sprayed on affected crops as an emulsified suspension or wettable powder. *B. bassiana* parasitizes a wide range of arthropod hosts and therefore is considered as a non-selective biological insecticide.

41.2.5.4 Nomuraea

Nomuraea sp., currently renamed as *Metarhizium (Nomuraea) rileyi*, is another dimorphic hyphomycete with potential entomopathogenic activity. It can cause epizootic death in various insects. It has been demonstrated that many insect species belonging to Lepidoptera including *Spodoptera litura* and some belonging to Coleoptera are susceptible to *Metarhizium (Nomuraea) rileyi*. The host specificity of this fungus and its ecofriendly nature encourage its use in insect pest management. Its mode of infection and development have been reported for several insect hosts such as *Trichoplusia*, *Heliothis zea*, *Plathypena scabra*, *Bombyx mori*, *Pseudoplusia*, and *Anticarsia gemmatalis*. Another insect, *Spilosoma*, was found to be severely attacked by *Nomuraea rileyi*. Similarly, an epizootic of *Nomuraea rileyi* was observed on *Junonia orithya*, which was proved to be the best alternative to manage the hedge plant eater *Junonia orithya*.

A number of studies aimed at mass production of these entomopathogenic fungi, assessing their field efficacy, have been conducted at the Department of Agril Entomology, College of Agriculture, University of Agricultural Sciences, Dharwad during 1999 to 2003 (Lingappa & Patil, 2002).

41.3 Culturing and Application

Based on the reports of Lingappa et al., (2003) the following points for culturing and applications can be followed:

- Based on fixed plot survey, the period between August and October was found to be the most favorable for its epizootics.
- Groundnut and soybean crop ecosystems are most amenable for establishment of early epizootics.
- *Nomuraea rileyi* isolates (eight) collected from different places exhibited similar insecticidal activity but differed in molecular characteristics as revealed from RAPD analysis.
- Pathogenicity of *Nomuraea rileyi* (Dharwad isolate) was confirmed against lepidopterous pests viz., *Helicoverpa armigera*, *Spodoptera litura*, *Thysanoplusia (= Plusia) orichalcea*, *Mythimna separata*, *Chilo partellus*, *Agrotis* sp., *Spilarctia obliqua*, *Euproctis* sp., and *Achroentia styx*.
- Lucern and groundnut as intercrops in cotton provide a congenial microclimate for the development of entomopoathogenic fungus on American bollworm, and lower the pest load on the principal crop.
- Early sowing with closer spacing in soybean and groundnut resulted in higher mycopathogen incidence on *S. litura*.
- Twofold increase in mycosis of *S. litura* due to *N. rileyi* was evident with one irrigation.
- Low-cost mass production technology for the entomopathogenic fungus has been developed. Rice proved to be the best and cost-effective substrate for the mass multiplication of the fungus.
- Fungicides were highly detrimental to the mycopathogen *in vitro* than insecticides and botanicals.
- *Nomuraea rileyi* was compatible with *Vitex negundo* and NSKE and combined application proved economical for management of *S. litura* in groundnut.
- Performance of *N. rileyi* against *Spodoptera litura* and *Helicoverpa armigera* on groundnut in the coastal agroecosystem of Karnataka was comparable to that in the transitional zone.
- Conidia of *N. rileyi* persisted for up to eight months in soil and longer if soil moisture was maintained.
- Conidia of *N. rileyi* persisted for up to ten days on groundnut foliage during kharif and was reduced to three days during summer.
- In potato, application of mycopathogen at 2×10^8 conidia /L was as good as spraying SlNPV and Bt to suppress leaf-eating caterpillar, *S. litura*.
- In sorghum, *N. rileyi* at 2×10^8 conidia/L proved as effective as recommended insecticide consistently and could be substituted totally for insecticide.
- Performance of *N. rileyi* against *H. armigera* in chickpea during rabi (November–January) was next to NPV and B.t.
- *Nomuraea rileyi* based *H. armigera* management module was next best to adoptable IPM.
- Large-scale evaluation of the mycopathogen in soybean (4.4 ha) and groundnut (0.4 ha) revealed that *N. rileyi* was as effective as insecticide and yielded higher benefits per rupee invested in insect control.
- Hasan et al. (2002) have tested spore production of *M. anisopliae* by solid state fermentation.

REFERENCES

Davidson, E. W. *Bacillus sphaericus* as a microbial control agent for mosquito larvae. In: *Integrated Mosquito Control Methodologies* (M. Laird & J. Miles, eds). Academic Press, London, 1985, vol. 2, pp. 213–226.

Elsworth J. F. & Grove, J. F. "Cyclodepsipeptides from *Beauveria bassiana* bals. Part 1. Beauverol-ides H and I," *Journal of the Chemical Society, Perkin Transactions*. 1977, 1(3): 270–273.

Hasan, S. Bhamra, A. K. Sil, K. Rajak R. C, & S. S. Sandhu, "Spore production of Metarhizium anisopliae (ENT-12) By Solid State fermentation," Journal of Indian Botanical Society, 2002, vol. 8, pp. 85–88.

Henry, J. E. & Onsager, J. A. Large-scale test of control of grasshoppers on rangeland with *Nosema locustae. Journal of Economic Entomology.* 1982, **75**: 31–35.

Jackson, T. A., Pearson, J. F., O'Callaghan, M. & Willocks, M. J. Pathogen to product M Development of *Serratia entomophila* (Enterobacteriaceae) as a commercial biological control agent for the New Zealand grass grub (*Costelytra zealandica*). In: *Use of Pathogens in Scarab Pest Management* (T. A. Jackson & T. R. Glare, eds). Intercept, Andover, 1992, pp. 191–198.

Kerry, B. R Crumph, D. & Mullen, A. "Studies of the cereal-cyst nematode, Heterodera avenue under continuous cereals, 1975–1978. II. Fungal parasitism of nematode females and eggs," *Annals of Applied Biology.* 1982, 100, 489–499.

Lacey L. A. & Goettel M. S. "Current developments in microbial control of insect pests and prospects for the early 21st century," *Entomophaga.* 1995, **40**(I): 3–27.

Leger, R. J. S. Charnley, A. K. & Cooper, R. M. "Characterization of cuticle-degrading proteases produced by the entomopathogen *Metarhizium anisopliae*," *Archives of Biochemistry and Biophysics.* 1987, 253(1), 221–232.

Lindsey Flexner J. & Belnavis D. L. Microbial insecticides. In: *Biological and Biotechnological Control of Insect Pests* (J. E. Rechcigl & A. R. Rechcigl, eds). CRC Press, Boca Raton, FL, 2009, pp. 35–62.

Lingappa, S. & Patil, R. K. *Nomuraea rileyi—A Potential Mycoinsecticides.* University of Agricultural Sciences, Dharwad and Department of Bio Technology, Govt. of India, New Delhi, 2002, pp. 30–10.

Lingappa, S. Patil, R. K., Kulkarni J. H. & Krisnaraj, P. U. National Agricultural Technology Project Report on "Mass multiplication and utilization of entomopathogenic fungi in integrated pest management," Department of Agril Entomology, MARS, UAS, Dharwad submitted to Indian Council of Agricultural Research, New Delhi, 2003.

Malvika, C. *Indian Biopesticide Market—Present Trends and Development Trends.* Asia Plant Wise, Bangkok, Thailand, 2016.

Nasr, F. N. & Moein, S. I. M. "New trend of the use of *Metarhizium anisopliae* (Metschinkoff) Sorokin and *Verticillium indicum* (Petch) genus as entomopathogens to the termite, *Cryptotermes brevis* (Walker) (Isoptera: Kalotermitidae)," *Anziger fur Schadlingskunde Pflanzenschutz,* Umweschutz. 1997, 70: 13–16.

Samson, P. R., Milner, R. J. & Bullard, G. K. "Development of *Metarhizium* based biopesticides for sugarcane pest management current progress and future prospects," *Proceedings of the Australian Society of Sugarcane Technology.* 1999, 21: 156–163.

Sandhu, S. S., Rajak, R. C & Hasija, S. K. "Potential of entomopathogens for the biological management of medically important pest: Progress and prospect," *Glimpses in Plant Sciences.* 2000, 2000: 110–117.

Yokoyama, T. Fujikata, M. & Fujiie, A. "Improvement of infectivity of *Metarhizium anisopliae* (Metschnikoff) to *Anomala cuprea* Hope," *Applied Entomology and Zoology.* 1993, 28: 451–461.

42

Management of the Diseases of Crops through Microbial Biopesticides

Rafaela Cristina dos Santos

CONTENTS

42.1 Introduction

Biopesticides demonstrate more specificity with the target organism than synthetic pesticides, are ecofriendly and biodegradable, have low-residue and high performance, less chance to pathogens and pests develop resistance (Kumar et al., 2012). Therefore, applications of microbe-based pesticides in agricultural crops have been expanded and encouraged as an alternative control (Chandler et al., 2011; Prabha et al., 2016).

42.2 What Are Microbial Biopesticides?

Microbial biopesticides have been used for insect and disease control in several crops. Bacteria, fungi, viruses, and protozoans have been used as active ingredients and have shown similar efficiency in pest control compared to chemical pesticides (MacGregor, 2006). The term "microbial biopesticide" has been applied to the use of microorganisms as the active ingredient in biopesticides used for pests control (Morán-Diez & Glare, 2016). A lot of species and strains of fungi and bacteria have been shown a huge potential to use as microbial biopesticides (Dominguesa et al., 2000; Glick et al., 1998; Kloepper et al., 2004). Because a lot of microorganisms can be used as components in microbial biopesticide, pathogen suppression can result from different types of interactions with the target pathogen (Duijff et al., 1997; Maurhofer et al. 1994; Ryu et al 2004; Zhao et al., 2017).

42.3 Microbial Biopesticides Mode of Action

Beneficial rhizosphere microorganisms that could be used as biocontrol agent present diverse ways to pathogen suppression (Table 42.1). The control can be a result of direct antagonism with physical contact, mainly involving the mechanism known as hyperparasitism with a high degree of selectivity for the target pathogen (Pal & Gardener, 2006). On the other hand, some bacteria and fungi can produce siderophores or exudates that compete with plant pathogens (Sharma & Johri, 2003). Furthermore, another indirect mechanism involved in plant defense is known as induced systemic resistance (ISR), which stimulate plant host defense pathways (Choudhary & Johri, 2009; Ryu et al., 2005). There are some mechanisms described as mixed-path antagonism, such as antibiotic production and secretion, toxin, hydrogen cyanide (HCN), and lytic enzymes (Compant et al., 2005; Pal & Gardener, 2006; Whipps, 2001).

42.3.1 Hyperparasitism

The term hyperparasitism, have been applied when the parasite is also damaged by other organisms through parasitism interaction (Kohl et al., 2019). In general, hyperparasites colonize their hosts and kill biological structures as mycelium, spores, and cells (Ghorbanpour et al., 2018).

Pasteuria penetrans is an endoparasite bacteria (Sayre & Wergin, 1977), which has been reported to present potential suppression of juveniles of *Meloidogyne arenaria* race 1, *M. incognita*, and *M. javanica* (Chen et al., 1996; 1997; Dube & Smart Jr., 1987; Weibelzahl-Fulton et al., 1996). *P. penetrans* spores connect to the cuticle of second-stage juveniles when they are in the soil, and consequently the nematode reproduction will be blocked (Bird, 1986).

Jeffries (1995) described 30 antagonistic species belonging to 16 different genera that are able to control *Rhizoctonia solani*, which cause damping-off in too many crops. In addition, around 30 fungal species have shown hyperparasitism against rust pathogen (Traquair et al., 1984; Zheng et al., 2017).

TABLE 42.1

Biocontrol Agents and Mechanisms of Action to Target Pathogen Suppression

Biocontrol Agent	Mechanism of Action	Target Pathogen	Reference
Pasteuriapenetrans	Hyperparasitism	*Meloidogyne arenaria* race 1, *M. incognita*, and *M. javanica*	Chen et al., 1996; 1997; Dube & Smart Jr, 1987; Weibelzahl-Fulton et al., 1996
Pichia guilliermondii	Competition	*Penicillium digitatum*	Droby et al., 1989
Pichia guilliermondii	Competition	*Colletotrichum* spp.	Chanchaichaovivat et al., 2008
Pseudomonas fluorescens strains A1, BK1, TL3B1, and B10	Competition by siderophore production	*Erwinia carotovora*	Kloepper et al., 1980
Trichoderma asperellum	Competition by siderophore production	*Fusarium* wilt	Segarra et al., 2010
Metschnikowia pulcherrima strain MACH1	Competition by siderophore production	*Botrytis cinerea*, *Alternaria alternata*, and *Penicillium expansum*	Saravanakumar et al., 2007
Bacillus subtilis	Iturin production	*Aspergillus flavus*	Moyne et al., 2001
Bacillus subtilis	Iturin production	*Rhizoctonia solani*	Yu et al., 2002
Bacillus subtilis	Iturin production	*Colletotrichum dematium*	Hiradate et al., 2002
Pseudomonas sp. strain F113	antibiotic 2,4-Diacetylphloroglucinol (DAPG)	*Pythium ultimum*, *Phoma beta*, *Rhizopus stolonifera*, and *Fusarium oxysporum*	Shanahan et al., 1992
Serratia marcescens	Chitinase production	*Sclerotium rolfsii*	Ordentlich et al., 1988
Lysobacter enzymogenes strain C3	β-1,3-glucanases synthesis	*Bipolaris* leaf spot of tall fescue and *Pythium* damping-off	Palumbo et al., 2005
Trichoderma virens	Antibiotic Gliotoxin	*Rhizoctonia solani*	Wilhite et al., 2001
Pseudomonas fluorescens CHA0	Hydrogen cyanide (HCN)	*Thielaviopsis basicola*	Voisard et al., 1989
*T. harzianum*Tr6 and *Pseudomonas* sp. Ps14	Induced systemic resistance (ISR)	*Fusarium oxysporum* f. sp. *radiciscucumerinum*	Alizadeh et al., 2013
Pseudomonas aeruginosa 7NSK2	ISR	*Botrytis cinerea*	De Meyer & Hofte, 1997
Bacillus sp.	ISR	Cucumber mosaic virus (CMV)	Zehnder et al., 2000
Pseudomonas putida strain BTP1	ISR by Z-3-hexenal production	*Botrytis cinerea*	Ongena et al., 2004
Bacillus amyloliquefaciens strains GB03 and IN937a	ISR by volatile organic compounds production	*Erwinia carotovora* ssp. carotovora	Ryu et al., 2004
Pseudomonas fluorescens strain CHA0	ISR by siderophore production	tobacco necrosis virus (TNV)	Maurhofer et al., 1994

42.3.2 Competition

Microbial control by competition for micronutrients, such as iron has also been studied, due to iron availability in the soil is extremely limited (Lindsay, 1982). To survive in this environment, some microorganisms developed the capacity to produce siderophores that are compounds with high-affinity iron-chelating (Neilands, 1981). The plant growth-promoting rhizobacteria *Pseudomonas fluorescens* strains A1, BK1, TL3B1, and B10 were the first microorganisms to demonstrate to use siderophore production as a mechanism to control the bacteria *Erwinia carotovora* (Kloepper et al., 1980). Later, Raaijmakers et al., (1995) proved that some isolates of *Pseudomonas* spp. reduced pathogen populations in the soil through the siderophore production. Several fungi have been demonstrated the potential to use siderophores to control pathogens. For example, *Trichoderma asperellum* has been reported to controls *Fusarium* wilt via siderophore production (Segarra et al., 2009). Similarly, *Metschnikowia pulcherrima* converts pulcherriminic acid and iron ions in the red pigment pulcherrim in inhibiting development of *B. cinerea*, *Alternaria alternata*, and *Penicillium expansum* (Saravanakumar et al., 2008).

42.3.3 Microbial Metabolites Production

A wide range of antibiotics have been studied and identified and have been shown to be effective in suppressing the growth of a target pathogen *in vitro* or *in situ*, consequently with potential to decrease the disease severity in the field (Islam et al., 2005; Koumoutsi et al., 2004; Moyne et al., 2001; Sandra et al., 2001; Shanahan et al., 1992). According to Bérdy et al. (2005), around 8,700 different antibiotics are produced by actinomycetes, 2,900 by bacteria and 4,900 by fungi.

Microbial metabolites production has been detected mainly in bacteria species belonging to the genera *Agrobacterium*, *Bacillus*, *Pseudomonas*, and *Streptomyces* (Fravel, 2005). Some families of lipopeptides as iturin (Desjardins et al., 1992; Hiraoka et al., 1992), surfactin (Kakinuma et al., 1969; Kowall et al., 1998), and fengycin (Vanittanakom et al., 1986) occupy a prominent place within the group of antibiotics to which they belong. These families are produced especially by *Bacillus* species (Bonmatin et al., 2003; Huszcza & Burczyk, 2006; Kim et al., 2004; Koumoutsi et al., 2004). Iturins are produced by *B. subtilis* (Bonmatin et al., 2003) and *B. amyloliquefaciens* (Koumoutsi et al., 2004); this family has exhibited an efficient in-vitro action against *Aspergillus flavus* (Moyne et al., 2001), *Rhizoctonia solani* (Yu et al., 2002), *Colletotrichum dematium* (Hiradate et al., 2002), and a wide range of phytopathogenic microorganisms (Phae et al., 1990). Surfactin is a potent family produced by *B. coagulans* (Huszcza & Burczyk, 2006), *B. pumilus*, and *B. licheniformis* (Peypoux et al., 1999), and this lipopeptide can quickly associate and strictly penetrate into lipid layers of phytopathogens damaging membrane integrity (Carrillo et al., 2003; Heerklotz & Seelig, 2004). Fengycin production was detected in *B. cereus* (Tsuge et al., 1999), *B. thuringiensis* (Kim et al., 2004), *B. subtilis* (Jacques et al., 1999), and *B. amyloliquefaciens* (Koumoutsi et al., 2004); however, there is not much information is yet available on the regulation mechanism of fengycin biosynthesis (Ongena & Jacques, 2008).

Genomic information shows the potential of Pseudomonas genus to produce a huge number of unknown secondary metabolites with possible use in biocontrol (Kohl et al., 2019). For example, the antibiotic 2,4-diacetylphloroglucinol (DAPG) isolated from from Pseudomonas sp. strain F113 contributes to faster colonization of roots, which and improve the bacteria ability to compete and suppress pathogen activity in the rhizosphere (Raaijmakers et al., 2002; Shanahan et al., 1992).

Lytic enzymes damage virulence factors or components present in pathogen cell-wall. Such enzymes can hydrolyze polymeric compounds, as is the case with chitin, proteins, cellulose, hemicellulose, and DNA; they have also been used by biocontrol agents to suppress pathogen activities (Whipps, 2001). For example, some studies have been reported the importance of chitinase as mediator of interactions between *Serratia marcescens* and pathogenic microorganisms (Jones et al., 1986; Oppenheim & Chet, 1992; Ordentlich et al., 1988; Shapira et al., 1989; Sundheim, 1990; 1992). The control of *Sclerotium rolfsii* by *S. marcescens* seems to be mediated by chitinase expression (Ordentlich et al., 1988). In addition, mutants of *S. marcescens* with silenced gene expression of *chiA*, the chitinase gene, demonstrated decrease in biocontrol capacity (Jones et al., 1986). In other studies, was observed incidence decreased of diseases caused by *Sclerotium rolfsii* and *Rhizoctonia solaniw* when a recombinant *Escherichia coli* expressed the *chiA* gene from *S. marcescens* (Oppenheim & Chet, 1992; Shapira et al., 1989). At the same time, *Pseudomonas* sp. have also shown great results to control *F. oxysporum f.* sp. *redolens* and *Gauemannomyces graminis* var. *tritici* (Sundheim, 1990; 1992).

The production β-1,3-glucanase by bacteria (Fleuri & Sato, 2008), yeast (Nebreda et al., 1987; Daenen et al., 2008), and fungi (Vásquez-Garcidueñas et al., 1998) has distinct applications. These enzymes may lyse or pressure cell walls from other living organisms and decompose plant residues and non-living organic matter. The bacteria Lysobacterenzymogenes strain C3 synthesize β-1,3-glucanasesencoded by the gluA, gluB, and gluC genes showed antimicrobial activity against Bipolaris leaf spot of tall fescue and Pythium damping-off of sugar beet (Palumbo et al., 2005). Thus, beneficial microorganisms that tend to colonize plant pathogens can be considered to have good potential for use in biocontrol.

42.3.4 Induction of Plant Resistance

Over time, plants have improved their ability to defend themselves against adverse stimuli, after the primary pathogen infection they produce an immune response which is known as systemic acquired resistance (SAR) (Ross, 1961; van Loon et al. 1998). Plants can associate with beneficial microorganisms, which enhance the plant defense capacity against pathogen infection through biochemical changes, this process is called induced systemic resistance (ISR) (Kuc, 1982; van Loon et al., 1998). Salicylic acid (SA), jasmonic acid (JA), and ethylene (ET) are important plant hormones that regulate both referred plant defense system (Thomma et al., 2001). The PR proteins consist of a range of enzymes with different roles in the plant; some of them may act straightly to lyse parasitizing cells, improve cell-wall borders against infections, or induce localized cell death (Vallad & Goodman, 2004; Vlot et al., 2009). The production of the hormones JA and/or ET under pathogen attack is related to association with beneficial microorganisms, triggering the ISR (DeVleesschauwer & Hoft, 2009) Figure 42.1.

Since the beginning of researches linking beneficial microorganisms with the ISR triggering (Alstrom, 1991; Wei et al., 1991; Van Peer et al., 1991) studies in this area have grown widely. A wide number of soilborne microorganisms, including *Pseudomonas*, *Bacillus*, *Trichoderma*, and mycorrhiza species, have been shown potential to enhance ISR (Choudhary & Johri, 2009; Harman, 2004; Haas & Defago, 2005; Ryu et al., 2004; Van Loon et al. 1998). In cucumber *T. harzianum* Tr6 and *Pseudomonas* sp. Ps14, both isolated from the rhizosphere of cucumber, when used in combination triggered ISR against

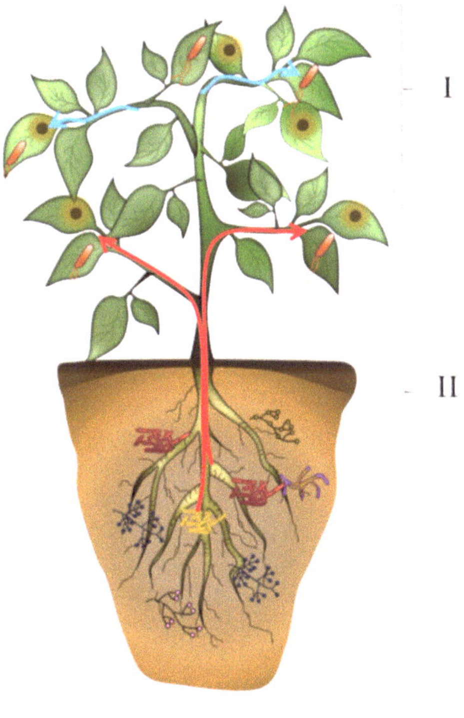

FIGURE 42.1 Schematic representation of (I) systemic acquired resistance (SAR) triggered by pathogen infection (blue arrow) and (II) induced systemic resistance (ISR) by soilborne beneficial microorganisms (red arrows). ISR occur when, under pathogen attack, soilborne beneficial microorganisms enhance plant defense, involving long-distance signals that are carried in vascular system or as airborne signals, spreading systemically throughout the plant.

TABLE 42.2

Some Microbial Biopesticides Available in Indian Market-to-Plant Disease Control

Biocontrol Agent	Commercial Name	Company	Target Pathogen
Paecilomyces lilacinus	Bioniconema®	Nico Orgo Manures	*Nematode*
Trichoderma harzianum	Bioderma H®	Biotech International Ltd.	*Phytophthora*, *Fusarium*, *Cercospora*, *Colletotrichum*, *Alternaria*, *Ascochyta*, *Macrophomina*, *Myrothecium*, and *Ralstonia*
Trichoderma harzianum	ECO-T®	Plant Health Products	*Rhizoctonia*, *Pythium*, *Fusarium*, and *Phytophthora*
Trichoderma viride	Nicoderma®	Nico Orgo Manures	*Soil pathogens*
Bacillus subtilis	Biosubtilin®	Biotech International Ltd	*Fusarium*, *Verticillium Pythium*, *Cercospora*, *Colletotrichum*, *Alternaria*, *Ascochyta*, *Macrophomina*, *Myrothecium*, *Ramularia*, *Xanthomonas*, and *Erysiphe polygoni*

Fusarium oxysporum f. sp. *radices cucumerinum* (Alizadeh et al., 2013). In bean, *P. aeruginosa* 7NSK2 induced systemic resistance against *B. cinerea*, via production of SA (De Meyer & Hofte, 1997). Similar results have been observed in bean, whereby *P. putida* strain BTP1 induced against *B. cinerea* by production of the fungitoxic final product Z-3-hexenal, which was just detected in leaves from BTP1-inoculated beans (Ongena et al., 2004). Zehnder et al., (2000) verified in a greenhouse experiment. Volatile blends from *Bacillus amyloliquefaciens* strains GB03 and IN937a activated ISR in *Arabidopsis* seedlings and consequently significantly reduced disease severity caused by *Erwinia carotovora* subsp. *carotovora* compared with seedlings not exposed to bacterial volatiles (Ryu et al., 2004).

Siderophores produced by rhizosphere bacteria can also trigger ISR, as in the case of *P. fluorescens* strain CHA0, whereby production of the siderophore pyoverdine by CHA0 is involved in the induction of resistance in tobacco leaves against infection of tobacco necrosis virus (TNV) (Maurhofer et al., 1994, Press et al., 1997, Bezemer & van Dam, 2005).

42.4 Microbial Biopesticides Market

Biopesticide annual growth rate is estimated at more than 15% (Marrone, 2014; Olson, 2015). Commercial microbial biopesticides available in the Indian market are shown in Table 42.2.

Discovering new molecules for synthesizing pesticides has becoming an increasingly difficult challenge for companies, because they normally test about 140,000 chemicals to find just one new, efficient, commercially appropriate one (Bakry et al., 2016). Thereby, it is estimated that the cost to develop a new synthetic pesticide is more than $250 million and that it takes ten years (Baesso et al., 2014; Khot et al., 2012). Thus, the advantages to biopesticide products involve the cost of development, around $3 to $7 million, in addition to the short time, three years in United States, to make them commercially available (Agrawal & Rathore, 2014).

Some multinational companies are developing microbial biopesticide programs, such as Monsanto and Novozymes (Durán et al., 2013) and FMC and Chr. Hansen (Gogos et al., 2012). In the global microbial biopesticide market, most available products are bacteria-based products. In addition, about 50% of biopesticides is employed in horticulture (Mishra et al., 2017).

Some new technologies, such as microencapsulation and nanoformulation can increase the capacity of biopesticide products to be stable and consequently improve the capacity to use in field crops and facilitate commercialization (Damalas & Koutroubas, 2018). The acceptance of these products in the market, such as analyze different modes of action that each microorganism present and regulations that are increasing in complexity and cost in some regions (Marrone, 2014).

42.5 Challenges and Future Prospects

The use of microbial biopesticides have been increased, due to a safer alternative to chemicals and potentially less risk to humans, animals, and the environment, Moreover, must be understood which modes of action each microorganism has as a controlling agent and after spread the knowledge about how the products work and how it can be integrated into pest and disease management programs.

REFERENCES

Agrawal, S.; Rathore, P. Nanotechnology pros and cons to agriculture: A review. *Int. J. Curr. Microbiol. Appl. Sci.* 2014, *3*, 43–55.

Alizadeh, H.; Behboudi, K.; Ahmadzadeh, M.; Javan-Nikkhah, M.; Zamioudis, C.; Pieterse, C. M.; Bakker, P. A. Induced systemic resistance in cucumber and *Arabidopsis thaliana* by the combination of *Trichoderma harzianum* Tr6 and Pseudomonas sp. Ps14. *Biol. Control.* 2013, *65*(1), 14–23.

Alstrom, S. Induction of disease resistance in common bean susceptible to halo blight bacterial pathogen after seed bacterization with rhizosphere pseudomonads. *J. Gen. Appl. Microbiol.* 1991, *37*, 495–501.

Baesso, M. M.; Teixeira, M. M.; Ruas, R. A. A.; Baesso, R. C. E. Pesticide application technologies. *Rev. Ceres.* 2014, *61*, 780–785.

Bakry, A. M.; Abbas, S.; Ali, B.; Majeed, H.; Abouelwafa, M. Y.; Mousa, A.; Liang, L. Microencapsulation of oils: A comprehensive review of benefits, techniques, and applications. *Compr. Rev. Food Sci. Food Saf.* 2016, *15*, 143–182.

Bérdy, J. Bioactive microbial metabolites. *J. Antibiot.* 2005, *58*, 1–26.

Bezemer, T. M.; van Dam, N. M. Linking aboveground and belowground interactions via induced plant defenses. *Trends Ecol. Evol.* 2005, *20*(11), 617–624.

Bird, A. F. The influence of the actinomycete, *Pasteuriapenetrans*, on the host-parasite relationship of the plant-parasitic nematode, *Meloidogyne javanica*. *Parasitology*. 1986, *93*, 571–580.

Bonmatin, J. M.; Laprévote, O.; Peypoux, F. Diversity among microbial cyclic lipopeptides: Iturins and surfactins. Activity-structure relationships to design new bioactive agents. *Comb. Chem. High Throughput Screening*. 2003, *6*(6), 541–556.

Carrillo, C.; Teruel, J. A.; Aranda, F. J.; Ortiz, A. Molecular mechanism of membrane permeabilization by the peptide antibiotic surfactin. *Biochim. Biophys. Acta Biomembr.* 2003, *1611*(1–2), 91–97.

Chanchaichaovivat, A., Panijpan, B., & Ruenwongsa, P. (2008). Putative modes of action of Pichia guilliermondii strain R13 in controlling chilli anthracnose after harvest. Biological control, 47(2), 207–215.

Chandler, D.; Bailey, A. S.; Tatchell, G. M.; Davidson, G.; Greaves, J.; Grant, W. P. The development, regulation and use of biopesticides for integrated pest management. *Philos. Trans. R. Soc. B*. 2011, *366*(1573), 1987–1998.

Chen, Z. X.; Dickson, D. W.; McSorley, R.; Mitchell, D. J.; Hewlett, T. Suppression of *Meloidogyne arenaria* race 1 by soil application of endospores of *Pasteuria penetrans*. *J. Nematol.* 1996, *28*(2), 159.

Chen, Z. X.; Dickson, D. W.; Mitchell, D. J.; McSorley, R.; Hewlett, T. Suppression mechanisms of *Meloidogyne arenaria* race 1 by *Pasteuria penetrans*. *J. Nematol.* 1997, *29*(1), 1.

Choudhary, D. K.; Johri, B. N. Interactions of *Bacillus* spp. and plants–with special reference to induced systemic resistance (ISR). *Microbiol. Res.* 2009, *164*(5), 493–513.

Compant, S.; Duffy, B.; Nowak, J.; Clément, C.; Barka, E. A. Use of plant growth-promoting bacteria for biocontrol of plant diseases: Principles, mechanisms of action, and future prospects. *Appl. Environ. Microbiol.* 2005, *71*(9), 4951–4959.

Daenen, L.; Saison, D.; Sterckx, F.; Delvaux, F. R.; Verachtert, H.; Derdelinckx, G. Screening and evaluation of the glucoside hydrolase activity in *Saccharomyces* and *Brettanomyces* brewing yeasts. *J. Appl. Microbiol.* 2008, *104*, 478–488.

Damalas, C.; Koutroubas, S. Current status and recent developments in biopesticide use. *Agriculture*. 2018, *8*, 1–6.

De Meyer, G.; Höfte, M. Salicylic acid produced by the rhizobacterium *Pseudomonas aeruginosa* 7NSK2 induces resistance to leaf infection by *Botrytis cinerea* on bean. *Phytopathol.* 1997, *87*(6), 588–593.

De Vleesschauwer, D.; Höfte, M. Rhizobacteria-induced systemic resistance. *Adv. Bot. Res.* 2009, *51*, 223–281.

Desjardins, A. E.; Plattner, R. D.; Shackelford, D. D.; Leslie, J. F.; Nelson, P. E. Heritability of fumonisin B1 production in *Gibberella fujikuroi* mating population A. *Appl. Environ. Microbiol.* 1992, *58*(9), 2799–2805.

Dominguesa, F. C.; Queiroza, J. A.; Cabralb, J. M. S.; Fonsecab, L. P. The influence of culture conditions on mycelial structure and cellulose production by *Trichoderma reesei* rut C-30. *Enzyme Microb. Technol.* 2000, *26*, 394–401.

Droby, S., Chalutz, E., Wilson, C. L., Wisniewski, M. (1989). Characterization of the biocontrol activity of Debaryomyces hansenii in the control of Penicillium digitatum on grapefruit. Canadian Journal of Microbiology, 35(8), 794–800.

Dube, B.; Smart Jr, G. C. Biological control of *Meloidogyne incognita* by *Paecilomyces lilacinus* and *Pasteuria penetrans*. *J. Nematol.* 1987, *19*(2), 222.

Duijff, B. J.; Gianinazzi-Pearson, V.; Lemanceau, P. Involvement of the outer membrane lipopolysaccharides in the endophytic colonization of tomato roots by biocontrol *Pseudomonas fluorescens* WCS417r. *New Phytol.* 1997, *135*, 325–334.

Durán, N.; Marcato, P. D. Nanobiotechnology perspectives. Role of nanotechnology in the food industry: A review. *Int. J. Food Sci. Technol.* 2013, *48*, 1127–1134.

Fleuri, L. F.; Sato, H. H. β-1,3 Glucanases e quitinases: aplicaçãonalise de leveduras e inibição de fungos. *Ciênc. Agrotecnol.* 2008, *32*, 1224–1231.

Fravel, D. R. Commercialization and implementation of biocontrol. *Annu. Rev. Phytopathol.* 2005, *43*, 337–359.

Ghorbanpour, M.; Omidvari, M.; Abbaszadeh-Dahaji, P.; Omidvar, R.; Kariman, K. Mechanisms underlying the protective effects of beneficial fungi against plant diseases. *Biol. Control.* 2018, *117*, 147–157.

Glick, B. R.; Penrose, D. M.; Li, J. A model for the lowering of plant ethylene concentrations by plant growth-promoting bacteria. *J. Theor. Biol.* 1998, *190*, 63–68.

Gogos, A.; Knauer, K.; Bucheli, T.D. Nanomaterials in plant protection and fertilization: Current state, foreseen applications, and research priorities. *J. Agric. Food Chem.* 2012, *60*, 9781–9792.

Haas, D.; Defago, G. Biological control of soil-borne pathogens by fluorescent pseudomonads. *Nat. Rev. Microbiol.* 2005, *3*, 307–319.

Harman, G. E.; Howell, C. R.; Vitarbo, A.; Chet, I.; Lorito, M. *Trichoderma* species—Opportunistic, avirulent plant symbionts. *Nat. Rev. Microbiol.* 2004, *2*, 43–56.

Heerklotz, H.; Wieprecht, T.; Seelig, J. Membrane perturbation by the lipopeptide surfactin and detergents as studied by deuterium NMR. *J. Phys. Chem. B.* 2004, *108*(15), 4909–4915.

Hiradate, S.; Yoshida, S.; Sugie, H.; Yada, H.; Fujii, Y. Mulberry anthracnose antagonists (iturins) produced by *Bacillus amyloliquefaciens* RC-2. *Phytochemistry*. 2002, *61*(6), 693–698.

Hiraoka, H.; Asaka, O.; Ano, T.; Shoda, M. Characterization of *Bacillus subtilis* RB14, coproducer of peptide antibiotics iturin A and surfactin. *J. Gen. Appl. Microbiol.* 1992, *38*(6), 635–640.

Huszcza, E.; Burczyk, B. Surfactin isoforms from *Bacillus coagulans*. *Z. Naturforsch. C.* 2006, *61*(9–10), 727–733.

Islam, T. M.; Hashidoko, Y.; Deora, A.; Ito, T.; Tahara, S. Suppression of damping-off disease in host plants by the rhizoplane bacterium *Lysobacter* sp. strain SB-K88 is linked to plant colonization and antibiosis against soilborne peronosporomycetes. *Appl. Environ. Microbiol.* 2005, *71*, 3786–3796.

Jacques, P.; Hbid, C.; Destain, J.; Razafindralambo, H.; Paquot, M.; De Pauw, E.; Thonart, P. Optimization of biosurfactant lipopeptide production from *Bacillus subtilis* S499 by Plackett-Burman design. In *Twentieth Symposium on Biotechnology for Fuels and Chemicals* (pp. 223–233). Humana Press, Totowa, NJ, 1999.

Jeffries, P. Biology and ecology of mycoparasitism. *Can. J. Bot.* 1995, *73*, 1284–1290.

Jones, J. D. G.; Grady, K. L.; Suslow, T. V.; Bedbrook, J. R. Isolation and characterization of genes encoding two chitinase enzymes from *Serratia marcescens*. *EMBO J.* 1986, *5*(3), 467–473.

Kakinuma, A.; Sugino, H.; Isono, M.; Tamura, G.; Arima, K. Determination of fatty acid in surfactin and elucidation of the total structure of surfactin. *Agric. Biol. Chem.* 1969, *33*(6), 973–976.

Khot, L. R.; Sankaran, S.; Maja, J. M.; Ehsani, R.; Schuster, E. W. Applications of nanomaterials in agricultural production and crop protection: A review. *Crop Prot.* 2012, *35*, 64–70.

Kim, P. I.; Bai, H.; Bai, D.; Chae, H.; Chung, S.; Kim, Y., Park, R.; Chi, Y. T. Purification and characterization of a lipopeptide produced by *Bacillus thuringiensis* CMB26. *J. Appl. Microbiol.* 2004, *97*(5), 942–949.

Kloepper, J. W.; Leong, J.; Teintze, M.; Schroth, M. N. *Pseudomonas* siderophores: A mechanism explaining disease suppression in soils. *Curr. Microbiol.* 1980, *4*, 317–320.

Kloepper, J. W.; Ryu, C. M.; Zhang, S. Induced systemic resistance and promotion of plant by *Bacillus* spp. *Phytopathology.* 2004, *94*, 1259–1266.

Köhl, J.; Kolnaar, R.; Ravensberg, W. J. Mode of action of microbial biological control agents against plant diseases: Relevance beyond efficacy. *Front. Plant Sci.* 2019, *10*(845), 1–19.

Koumoutsi, A.; Chen, X. H.; Henne, A.; Liesegang, H.; Hitzeroth, G.; Franke, P.; Vater, J.; Borriss, R. Structural and functional characterization of gene clusters directing nonribosomal synthesis of bioactive cyclic lipopeptides in *Bacillus amyloliquefaciens* strain FZB42. *J. Bacteriol.* 2004, *186*(4), 1084–1096.

Kowall, M.; Vater, J.; Kluge, B.; Stein, T.; Franke, P.; Ziessow, D. Separation and characterization of Surfactin isoforms produced by *Bacillus subtilis* OKB 105. *J. Colloid Interface Sci.* 1998, *204*(1), 1–8.

Kuc, J. Induced immunity to plant disease. *Bioscience.* 1982, *32*, 854–860.

Kumar, S. Biopesticides: A need for food and environmental safety. *J. Biofert. Biopestic.* 2012, *3*(4), 1–3.

Lindsay, W. L.; Schwab, A. P. The chemistry of iron in soils and its availability to plants. *J. Plant Nutr.* 1982, *5*(4–7), 821–840.

MacGregor, J. T. Genetic toxicity assessment of microbial pesticides: Needs and recommended approaches. *Int. Assoc. Environ. Mutagen Soc.* 2006, *43* 1–17.

Marrone, P. G. The market and potential for biopesticides. In *Biopesticides: State of the Art and Future Opportunities*; Gross, A. D.; Coats, J. R.; Duke, S. O.; Seiber, J. N. Eds. American Chemical Society, Washington, DC, 2014, 245–258.

Maurhofer, M.; Hase, C.; Meuwly, P.; Metraux, J. P.; Defago, G. Induction of systemic resistance to tobacco necrosis virus by the root-colonizing *Pseudomonas fluorescens* strain CHA0: Influence of the *gacA* gene and of pyoverdine production. *Phytopathol.* 1994, *84*, 139–146.

Mishra, S.; Keswani, C.; Abhilash, P. C.; Fraceto, L. F.; Singh, H. B. Integrated approach of agri-nanotechnology: Challenges and future trends. *Front. Plant Sci.* 2017, *8*, 471.

Morán-Diez, M. E.; Glare, T. R. What are microbial-based biopesticides? In *Microbial-Based Biopesticides.* Humana Press, New York, NY, 2016, 1–10.

Moyne, A. L.; Shelby, R.; Cleveland, T. E.; Tuzun, S. Bacillomycin D: An iturin with antifungal activity against *Aspergillus flavus. J. Appl. Microbiol.* 2001, *90*, 622–629.

Nebreda, A. R.; Vazquez, C. R.; Villa, T. G.; Villanueva, J. R.; Del Rey, F. Heterogeneous glycosylation of the EXG1 gene product accounts for the two extracellular exo-β-glucanases of Saccharomyces cerevisiae. *FEBS Lett.* 1987, *220*(1), 27–30.

Neilands, J. B. Microbial iron compounds. *Annu. Rev. Biochem.* 1981, *50*, 715–731.

Olson, S. An analysis of the biopesticide market now and where is going. *Outlooks Pest Manage.* 2015, *26*, 203–206.

Ongena, M.; Jacques, P. *Bacillus* lipopeptides: Versatile weapons for plant disease biocontrol. *Trends Microbiol.* 2008, *16*(3), 115–125.

Oppenheim, A. B.; Chet, I. Cloned chitinases in fungal plant-pathogen control strategies. *Trends Biotechnol.* 1992, *10*, 392–394.

Ordentlich, A.; Elad, Y.; Chet, I. The role of chitinase of *Serratia marcescens* in biocontrol of *Sclerotium rolfsii. Phytopathology.* 1988, *78*(1), 84–88.

Pal, K. K., Gardener, B. M. (2006). Biological control of plant pathogens.

Palumbo, J. D.; Yuen, G. Y.; Jochum, C. C.; Tatum, K.; Kobayashi, D. Y. Mutagenesis of β-1, 3-glucanase genes in *Lysobacter enzymogenes* strain C3 results in reduced biological control activity toward *Bipolaris* leaf spot of tall fescue and *Pythium* damping-off of sugar beet. *Phytopathology.* 2005, *95*(6), 701–707.

Peypoux, F.; Bonmatin, J. M.; Wallach, J. Recent trends in the biochemistry of surfactin. *Appl. Microbiol. Biotechnol.* 1999, *51*(5), 553–563.

Phae, C. G.; Shoda, M.; Kubota, H. Suppressive effect of *Bacillus subtilis* and its products on phytopathogenic microorganisms. *J. Ferment. Bioeng.* 1990, *69*(1), 1–7.

Prabha, S.; Yadav, A.; Kumar, A.; Yadav, A.; Yadav H. K.; Kumar, S.; Yadav, R.S.; Kumar, R. Biopesticides-an alternative and eco-friendly source for the control of pests in agricultural crops. *Plant Arch.* 2016, *16*(2), 902–906.

Press, C. M.; Wilson, M.; Tuzun, S.; Kloepper, J. W. Salicylic acid produced by *Serratia marcescens* 90–166 is not the primary determinant of induced systemic resistance in cucumber or tobacco. *Mol. Plant-Microbe Interact.* 1997, *10*, 761–768.

Raaijmakers, J. M., Sluis, L. V. D., Bakker, P. A., Schippers, B., Koster, M., Weisbeek, P. J. (1995). Utilization of heterologous siderophores and rhizosphere competence of fluorescent Pseudomonas spp. Canadian Journal of Microbiology, 41(2), 126–135.

Raaijmakers, J. M.; Vlami, M.; De Souza, J. T. Antibiotic production by bacterial biocontrol agents. *Anton. van Leeuw.* 2002, 81, 537–547.

Ross, A. F. Systemic acquired resistance induced by localized virus infections in plants. *Virology.* 1961, *14*, 340–358.

Ryu, C. M.; Farag, M. A.; Hu, C. H.; Reddy, M. S.; Kloepper, J.W.; Paré, P. W. Bacterial volatiles induce systemic resistance in *Arabidopsis. Plant Physiol.* 2004, *134*,1017–1026.

Ryu, C. M.; Farag, M. A.; Paré, P. W.; Kloepper, J. W. Invisible signals from the underground: Bacterial volatiles elicit plant growth promotion and induce systemic resistance. *Plant Pathol. J.* 2005, *21*, 7–12.

Sandra, A. I.; Wright, C. H.; Zumoff, L. S.; Steven, V. B. *Pantoea agglomerans* strain EH318 produces two antibiotics that inhibit *Erwinia amylovora* in vitro. *Appl. Environ. Microbiol.* 2001, *67*, 282–292.

Saravanakumar, D.; Ciavorella, A.; Spadaro, D.; Garibaldi, A.; Gullino, M. L. *Metschnikowia pulcherrima* strain MACH1 outcompetes *Botrytis cinerea*, *Alternaria alternata* and *Penicillium expansum* in apples through iron depletion. *Postharvest Biol. Technol.* 2008, *49*(1), 121–128.

Saravanakumar, D., Vijayakumar, C., Kumar, N., & Samiyappan, R. (2007). PGPR-induced defense responses in the tea plant against blister blight disease. Crop Protection, 26(4), 556–565.

Sayre, R. M.; Wergin, W. P. Bacterial parasite of a plant nematode: Morphology and ultrastructure. *J. Bacteriol.* 1977, *129*, 1091–1101.

Segarra, G., Casanova, E., Avilés, M., & Trillas, I. (2010). Trichoderma asperellum strain T34 controls Fusarium wilt disease in tomato plants in soilless culture through competition for iron. Microbial ecology, 59(1), 141–149.

Segarra, G.; Van der Ent, S.; Trillas, I.; Pieterse, C. M. J. 2009. MYB72, a node of convergence in induced systemic resistance triggered by a fungal and a bacterial beneficial microbe. *Plant Biol.* 2009, *11*, 90–96.

Shanahan, P.; O'Sullivan, D. J.; Simpson, P.; Glennon, J. D.; O'Gara, F. Isolation of 2,4-Diacetylphloroglucinol from a fluorescent pseudomonad and investigation of physiological parameters influencing its production. *Appl. Environ. Microbiol.* 1992, *58*, 353–358.

Shapira, R.; Ordentlich, A.; Chet, I.; Oppenheim, A. B. Control of plants diseases by chitinases expressed from cloned DNA in *Escherichia coli*. *Phytopathology*, 1989, *79*, 1246–1249.

Sharma, A.; Johri, B. N. Growth promoting influence of siderophore-producing *Pseudomonas* strains GRP3A and PRS9 in maize (*Zea mays* L.) under iron limiting conditions. *Microbiol. Res.* 2003, *158*(3), 243–248.

Sundheim, L. Biocontrol of *Fusarium oxysporum* with a chitinase encoding gene from *Serratia marcescens* on a stable plasmid in *Pseudomonas*. *J. Cell Biochem.* 1990, *13A*, 171.

Sundheim, L. Effect of chitinase encoding genes in biocontrol *Pseudomonas* spp. In *Biological Control of Plant Diseases*. Springer, Boston, MA, 1992, 331–333.

Thomma, B. P.; Penninckx, I. A.; Cammue, B. P.; Broekaert, W. F. The complexity of disease signaling in *Arabidopsis*. *Curr. Opin. Immunol.* 2001, *13*(1), 63–68.

Traquair, J. A.; Meloche, R. B.; Jarvis, W. R.; Baker, K. W. Hyperparasitism of *Puccinia violae* by *Cladosporium uredinicola*. *Can. J. Bot.* 1984, *62*, 181–184.

Tsuge, K.; Ano, T.; Hirai, M.; Nakamura, Y.; Shoda, M. The genes *degQ*, *pps*, and *lpa-8* (sfp) are responsible for conversion of *Bacillus subtilis* 168 to plipastatin production. *Antimicrob. Agents Chemother.* 1999, *43*(9), 2183–2192.

Vallad, G. E.; Goodman, R. M. Systemic acquired resistance and induced systemic resistance in conventional agriculture: Review and interpretation. *Crop Sci.* 2004, *44*, 1920–1934.

Van Loon, L. C.; Bakker, P. A. H. M.; Pieterse, C. M. J. Systemic resistance induced by rhizosphere bacteria. *Annu. Rev. Phytopathol.* 1998, *36*(1), 453–483.

Van Peer, R.; Niemann, G. J.; Schippers, B. Induced resistance and phytoalexin accumulation in biological control of *Fusarium* wilt of carnation by *Pseudomonas* sp. strain WCS 417 r. *Phytopathology.* 1991, *81*(7), 728–734.

Vanittanakom, N.; Loeffler, W.; Koch, U.; Jung, G. Fengycin—a novel antifungal lipopeptide antibiotic produced by *Bacillus subtilis* F-29-3. *J. Antibiot.* 1986, *39*(7), 888–901.

Vázquez-Garcidueñas, S.; Leal-Morales, C. A.; Herrera-Estrella, A. Analysis of the β-1, 3-glucanolytic system of the biocontrol agent *Trichoderma harzianum*. *Appl. Environ. Microbiol.* 1998, *64*(4), 1442–1446.

Vlot, A. C.; Dempsey, D. A.; Klessig, D. F. Salicylic acid, a multifaceted hormone to combat disease. *Annu. Rev. Phytopathol.* 2009, *47*,177–206.

Wei, G.; Kloepper, J. W.; Tuzun, S. Induction of systemic resistance of cucumber to *Colletotrichum orbiculare* by select strains of plant-growth promoting rhizobacteria. *Phytopathology.* 1991, *81*, 1508–1512.

Weibelzahl-Fulton, E.; Dickson, D. W.; Whitty, E. Suppression of *Meloidogyne incognita* and *M. javanica* by *Pasteuria penetrans* in field soil. *J. Nematol.* 1996, *28*(1), 43.

Whipps, J. M. Microbial interactions and biocontrol in the rhizosphere. *J. Exp. Bota.* 2001, *52*(suppl_1), 487–511.

Wilhite, S. E.; Lumsden, R. D.; Straney, D. C. Peptide synthetase gene in *Trichoderma virens*. *Appl. Environ. Microbiol.* 2001, *67*(11), 5055–5062.

Yu, G. Y.; Sinclair, J. B.; Hartman, G. L.; Bertagnolli, B. L. Production of iturin A by *Bacillus amyloliquefaciens* suppressing *Rhizoctonia solani*. *Soil Biol. Biochem.* 2002, *34*(7), 955–963.

Zehnder, G. W.; Yao, C.; Murphy, J. F.; Sikora, E. J.; Kloepper, J. W. Induction of resistance in tomato against cucumber mosaic cucumovirus by plant growth-promoting rhizobacteria. *Biocontrol.* 2000, *45*, 127–137.

Zhao, Y.; Qian, G.; Chen, Y.; Du, L.; Liu, F. Transcriptional and antagonistic responses of biocontrol strain *Lysobacter enzymogenes* OH11 to the plant pathogenic oomycete *Pythium aphanidermatum*. *Front. Microbiol.* 2017, *8*, 1025.

Zheng, L.; Zhao, J.; Liang, X.; Zhan, G.; Jiang, S.; Kang, Z. Identification of a novel *Alternaria alternata* strain able to hyper parasitize *Puccinia striiformis* f. sp. tritici, the causal agent of wheat stripe rust. *Front. Microbiol.* 2017, *8*, 71.

43

Quality Control of Fungal Biocontrol Agents with Emphasis on Trichoderma

María Alejandra Correa, Massiel Pinto, Marcelo Molinatti, Oscar Valbuena, and Domenico Pavone

CONTENTS

43.1 Introduction

Biological control of pests and diseases is a reliable option in modern agriculture, due to the efficacy and contribution to decreasing the use of toxic agrochemicals, thereby minimizing environmental risks (Syed Ab Rahman et al., 2018; Adnan et al., 2019; Wyckhuys et al., 2019). Actually, the challenge in biocontrol is to overcome the research barrier and positioning not only these products on the market, but also the philosophy of organic agriculture. Although a small industry, organic agriculture is growing in importance. Though only a small percentage of farmers are expected to become organic producers, consumer demand for organically produced food provides new market opportunities (Francis, 2013, Markets & Markets, 2020).

Biological control agents (BCA) production based on microorganisms require several operational steps, including: (a) isolation and identification; (b) biocontrol evaluation; (c) mass production and formulation; and (d) field evaluation. In order to obtain a good effective commercial product, it is necessary to guarantee quality in each step of these processes.

Quality control does not only refer to the final end-use product, but also to the production processes in which internal procedures are checked to determine whether inputs comply with their specifications (Ravensberg, 2010). The non-fulfillment of the quality requirements is the first step to the failure of BCA as an alternative in agriculture (Kumar et al., 2014).

This chapter describes some methodologies used in quality control of end products based on microbial biocontrol agents with emphasis on *Trichoderma*. This information will allow verifying the efficiency of products before their commercialization, ensuring to the farmer a secure and high-quality product.

43.2 Standards in Quality Control of Biocontrol Agents

43.2.1 Spore Concentration

To determine spore concentration, a quantitative method is used to evaluate the spore number per weight or volume employing a Neubauer chamber. For solid formulations it is necessary to suspend a representative quantity of the product using a known volume of water. The adequate use of Neubauer chamber is of crucial importance to obtain accurate results. Tensoactive substances like Tween 80® are used to improve spore separation. In products containing very hydrophobic spores, the use of Tween 80® is a must.

43.2.2 Viability and Purity

When spores in a product lose viability, the capacity to exert its biological effect will decrease, resulting in failed applications (Faria & Wraight, 2007; Oliveira et al., 2015). The most common way to determine viability is to estimate the number of colony-forming units (CFU). This method requires the use of suitable dilution and culture medium for the microorganism (Stentelaire et al., 2001; Laengle et al., 2005).

Although the CFU methodology seems to be simple, several factors affecting results have been described (Jarvis et al., 1983). Fungal colonies are cellular aggregates from mycelium fragments and spores (individual or grouped), growing on solid medium. A colony count could not be exactly equal to the spore number in the formulation, because mycelium fragmentation and spore separation or aggrupation will result in different numbers of colonies (Stentelaire et al., 2001). It is also possible to find a competitive inhibition when a high number of spores are inoculated on the plate (Jarvis et al., 1983). When determining the number of CFU it is important to take into account that error will increase with dilutions and when using few replicates (Sutton, 2011).

Purity tests are used to determine the proportion of the BCA in relation to other microorganisms present in the formulation. As contaminants could affect the product quality as well as to be a risk to applicators and consumers, low numbers of contaminants and absence of human pathogens only are accepted. In this way, the CFU determination is used to establish a counting of contaminant colonies.

Another method used to determine viability in a formulation is a direct counting of germinated spores inoculated on a solid medium by microscopic observation (Odell & Smith, 1982; Goettel & Inglis, 1997). For this purpose, a petri dish disposed as a humid chamber with a slide covered with agar can be used. An adequate dilution of the BCA formulation is spread on the agar layer. After few hours, the number of germinated/non-germinated spores is determined.

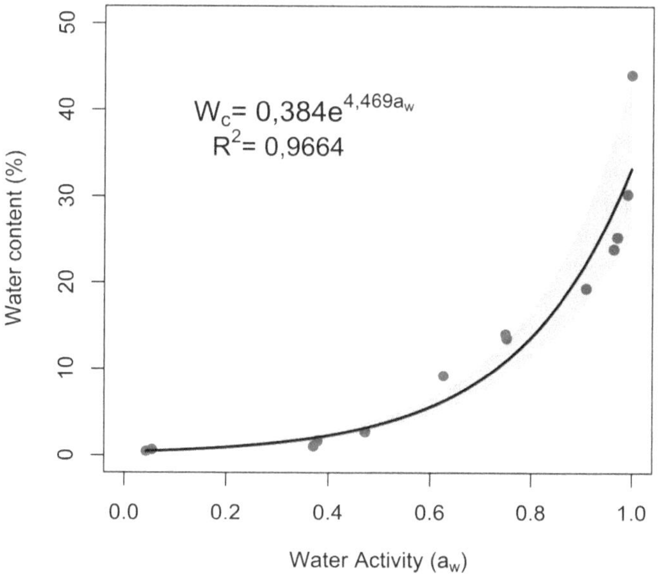

FIGURE 43.1 Water adsorption isotherm for a wettable granular starch formulation of *Trichoderma asperellum*.

43.2.3 Water Content (Wc) and Water Activity (a_w)

In mass production of BCA it is neccesary to maintain optimal storage conditions in order to obtain a product with long shelf life, with temperature and humidity the key parameters in determining spore viability (Jenkins & Grzywacz, 2010). High humidity can promote growth of contaminants competing with BCA, reducing viability and effectiveness. A fast and precise method to determine water content is the thermogravimetric comparing initial and dried samples weights.

Water activity (a_w) is related to availability (Schiraldi et al., 2012) or the ratio of the fugacity of water in a sample and fugacity in pure water at the same temperature (Reid, 2007). High water activity in storage could promote BCA growth and contaminant proliferation, affecting viability and quality (Honeycutt & Benson, 2001). Figure 43.1 shows a water adsorption isotherm for a wettable granular starch formulation of *T. asperellum*, which requires an optimal a_w between 0.980 and 0.995, but at 0.93 spore germination could occur (Begoud et al., 2007). The minimal a_w value for *T. asperellum* growth (0.93 according to Begoud et al., 2007), is reached at a water content of 20 %. From data presented in Figure 43.1, and using a log transformation over Wc, it was possible to derive an equation that relates the Wc to a_w. In this way, determining water content, it is possible to know the corresponding a_w value.

43.2.4 Biological Efficacy

The components and processes used to produce a BCA must not affect the efficacy of the final product..The same bioassays employed for strain selection can be used in quality control tests. Despite the field studies are important they cannot be conducted for each batch produced because of time and costs issues. In this sense, several methodologies are available to determine efficacy of a BCA based on antagonistic fungi being the most used the *in vitro* dual culture (Soares & Faull, 2000; Szekeres et al., 2006; Rahman et al., 2009; Pakdaman et al., 2013; Bunbury-Blanchette & Walker, 2019). In this method, opposite borders of an agar plate are inoculated with the pathogen and the antagonistic fungi in order to observe how the colonies grow. Although this method does not simulate real field conditions and no final conclusions can be done about biocontrol efficacy based only on this methodology, it can give information about the biocontrol mechanisms used by the antagonistic fungus.

TABLE 43.1

Growth of Phytopathogenic Fungus *Rhizoctonia solani* (R) and *Trichoderma* spp. (TV) in Dual Cultures on PDA Plates

	Colony Area (cm^2) 24 h	
Strain	**TV**	**R**
Rs	–	6.01 ± 1.14a
TV21	6.32 ± 0.31	2.53 ± 0.29b
TV30	3.75 ± 0.3	4.53 ± 0.6ab
TV35	5.08 ± 0.3	2.72 ± 0.63b
TV72	5.4 ± 0.19	6.24 ± 1.54ab
TV113	5.02 ± 0.14	5.37 ± 0.56ab
TV118	4.73 ± 0.19	3.78 ± 0.52b
TV147	5.4 ± 0.34	5.92 ± 1.43ab
TV165	6.05 ± 0.26	4.29 ± 1.1ab
TV190	3.68 ± 0.18	4.8 ± 0.58ab
TV219	13.24 ± 0.52	6.27 ± 1.47a

Values represent mean and standard error of colony areas. Five replicates were used. Letters indicate statistical differences ($p < 0.05$).

Several methods can be used to analyze a dual culture assay (Magan & Lacey, 1984; Szekeres et al., 2006). Using the software Image J (Rueden et al., 2017) or ImageTool® 3.0 (University of Texas Health Science Center, San Antonio, USA; free download at: http://compdent.uthscsa.edu/ITDownload.asp) colonies areas can be measured. In our laboratory this analysis is focused on pathogen colony area. Table 43.1 shows data collected from dual cultures of *Rhizoctonia solani* (R) and *Trichoderma* spp (TV) at 24 h, at which point the colonies had not reached each other, giving early evidence of reduced *R. solani* growth compared to a control where *R. solani* was alone on the plate. These data allowed knowledge of whether *Trichoderma* produced any substance spreading on the agar or volatile capable to reduce *R. solani* growth. When colonies touch each other, different mechanisms like parasitism could act.

The colony area for *R. solani* alone (Rs) was 6.01 cm^2 (Table 43.1), whereas the *R. solani* area confronted with TV21 (*Trichoderma harzianum*) was 2.53 cm^2, which suggests strong evidence of reduced growth. On the other hand, the *R. solani* area was 6.27 cm^2 in presence of TV219 (*Trichoderma reesei*). Colony areas measured each 24 h can be used to study mycelia interaction, inhibition, and superposition between colonies.

The fast growth ability on soil provides the antagonistic fungi an important advantage against phytopathogens for space and nutrients (Benítez et al., 2004). Growth in plates with sterile soil from fields in which *Trichoderma* will be used can reveal incompatibility with some physicochemical parameters or agrochemical residues present in these soils. For *Trichoderma*-based products other desired effects like stimulation on plant growth (Lopez-Bucio et al., 2015; Naik et al., 2019) can be determined in greenhouse assays.

43.2.5 Storage Stability

Stability is defined as the ability of a formulation to maintain its physical, chemical, and biological characteristics in time. Many factors affect stability, including interactions between active ingredients and excipients, fabrication process, storage and transportation conditions, and time from fabrication until final application (Abadías et al., 2001; Sriram et al., 2011; Fernandez-Sandoval et al., 2012). Viability is probably the parameter most affected during storage in which time and temperature are determinant factors (Melin et al., 2006; Kumar et al., 2014). A tolerance range for stability loss is usually accepted for biopesticides. The permitted lost activity should not be higher than 25% of the original value (Ravensberg, 2010).

43.3 Conclusions and Recommendations

Quality control is a pivotal step in BCA production because it guarantees high and consistent efficient products avoiding risks to human health and the environment. A quality control plan should be applied in all production steps including processes and end-product. Handling, shipping, and storage recommendations should be informed to distributors and users. Controlling the production process allows verifying raw material quality, equipment maintenance, and early contamination, which together improve the quality of the product. Quality control protocols are generally described in local and international organizations, being the main parameters determined: spore concentration, water content, viability, purity, and biological efficacy. To apply these techniques, it is necessary a well-trained technician in order to execute experiments adequately, choosing sample size and volume to complete suspension, appropriate dilutions, media selection, adequate use of Neubauer chamber and calculations. For *in vitro*, greenhouse, and field biological efficacy, several tests are available, each one giving important information. Obviously, experiments simulating as much as possible real field conditions are preferred. Quality control must be a natural process in the whole company in which continuous improvement should become automatic, ensuring satisfaction and benefit to the customer.

Acknowledgments

The authors would like to express their gratitude to Dr. Blas Dorta (IBE-UCV) for his support in Wc and a_w determinations.

REFERENCES

Abadias, M., Teixidó, N., Usall, J., Benabarre, A., & Viñas, I. 2001. Viability, efficacy and storage stability of freeze-dried biocontrol agent Candida sake using different protective and rehydration media. *Journal of Food Protection* 64(6): 856–861. DOI: 10.4315/0362-028x-64.6.856.

Adnan, M., Islam, W., Shabbir, A., Khan, K., Ghramh, H., Huang, Z., Chen, H. & Lu, G. 2019. Plant defense against fungal pathogens by antagonistic fungi with *Trichoderma* in focus. *Microbial Pathogenesis* 129: 7–18. DOI: 10.1016/j.micpath.2019.01.042.

Begoude, B., Lahlali, R., Friel, D., Tondje, P. & Jijakli, M. 2007. Response surface methodology study of the combined effects of temperature, pH, and aw on the growth rate of *Trichoderma asperellum*. *Journal of Applied Microbiology* 103: 845–854. DOI: 10.1111/j.1365-2672.2007.03305.x.

Benítez, T., Rincón, A., Limón, M. & Codón, A. 2004. Biocontrol mechanisms of *Trichoderma* strains. *International Microbiology* 7: 249–260. DOI: 10.2436/im.v7i4.9480.

Bunbury-Blanchette, A. & Walker, A. 2019. *Trichoderma* species show biocontrol potential in dual culture and greenhouse bioassays against *Fusarium* basal rot of onion. *Biological Control* 130: 127–135. DOI: doi.org/10.1016/j.biocontrol.2018.11.007.

Faria, M. & Wraight, S. 2007. Mycoinsecticides and mycoacaricides: a comprehensive list with worldwide coverage and international classification of formulation types. *Biological Control* 43: 237–256. DOI: 10.1016/j.biocontrol.2007.08.001.

Fernández-Sandoval, M., Ortiz-García, M., Galindo, E. & Serrano-Carreón, L. 2012. Cellular damage during drying and storage of *Trichoderma harzianum* spores. *Process Biochemistry* 47: 186–194. DOI:10.1016/j.procbio.2011.10.006.

Francis, C. 2013. Organic farming. In: *Reference Module in Earth Systems and Environmental Sciences*. DOI: doi.org/10.1016/B978-0-12-409548-9.05237-4.

Goettel, M. & Inglis, G. 1997. Fungi: Hyphomycetes. In: *Manual of Techniques in Insect Pathology*, ed. L.A. Lacey, 213–250. Academic Press, Cambridge, MA.

Honeycutt, E. & Benson, D. 2001. Formulation of binucleate *Rhizoctonia* spp. and biocontrol of *Rhizoctonia solani* on impatiens. *Plant Disease* 85(12): 1241–1248. DOI: 10.1094/PDIS.2001.85.12.1241.

Jarvis, B., Seiler, D., Ould, A. & Williams, A. 1983. Observations on the enumeration of moulds in food and feeding stuffs. Journal of Applied Bacteriology 55: 325–336. DOI: doi.org/10.1111/j.1365-2672.1983.tb01329.x.

Jenkins, N. & Grzywacz, D. 2010. Quality control of fungal and viral biocontrol agents—Assurance of product performance. *Biocontrol Science and Technology* 10(6): 753–777. DOI: doi.org/10.1080/09583150020011717.

Kumar, S., Manibhushan T. & Rani, A. 2014. *Trichoderma*: Mass production, formulation, quality control, delivery and its scope in commercialization in India for the management of plant diseases. *African Journal of Agricultural Research* 9(53): 3838–3852. DOI: 10.5897/AJAR2014. 9061.

Laengle, T., Pernfuss, B., Strasser, H. & Raffalt, J. 2005. Appendices: Risk assessment of fungal biocontrol agents standards. A: Quality control of fungal biological control agents. *Sydowia* 57: 67–77.

López-Bucio, J., Pelagio-Flores, R. & Herrera-Estrella, A. 2015. *Trichoderma* as biostimulant: Exploiting the multilevel properties of a plant beneficial fungus. *Scientia Horticulturae* 196, 30: 109–123. DOI: doi.org/10.1016/j.scienta.2015.08.043.

Magan, N. & Lacey, J. 1984. Effect of water activity, temperature and substrate on interactions between field and storage fungi. *Transactions of the British Mycological Society* 82: 83–93. DOI: doi.org/10.1016/S0007-1536(84)80214-4.

Markets and Markets. 2020. Biopesticides market by type (bioinsecticides, biofungicides, and bionematicides), source (microbials, plant extracts, and beneficial insects), mode of application, formulation, crop application, and region - global forecast to 2023. In: https://www.marketsandmarkets.com/Market-Reports/biopesticides-267.html (Accessed: January 6, 2020).

Melin, P., Hakansson, S., Eberhard, T. & Schnurer, J. 2006. Survival of the biocontrol yeast *Pichia anomala* after long-term storage in liquid formulations at different temperatures, assessed by flow cytometry. *Journal of Applied Microbiology* 100: 264–271. DOI: 10.1111/j.1365-2672.2005.02778.x.

Naik, K., Mishra, S., Srichandan, H., Kumar, P. & Kumar, P. 2019. Plant growth promoting microbes: Potential link to sustainable agriculture and environment. *Biocatalysis and Agricultural Biotechnology* 21: 101326. DOI: doi.org/10.1016/j.bcab.2019.101326.

Odell, D. & Smith, J. 1982. A simple method for spore germination studies. *Bulletin of the British Mycological Society* 16(2): 159–161. DOI: doi.org/10.1016/S0007-1528(82)80012-6.

Oliveira, D., Pauli, G., Moura, G. & Delalibera, I. 2015. A protocol for determination of conidial viability of the fungal entomopathogens *Beauveria bassiana* and *Metarhizium anisopliae* from commercial products. *Journal of Microbiological Methods* 119: 44–52. DOI: dx.doi.org/10.1016/j.mimet.2015.09.021.

Pakdaman, B., Goltapeh, E., Soltani, B., Talebi, A., Nadepoor, M., Kruszewska, J., Piłsyk, S., Sarrocco, S. & Vannacci, G. 2013. Toward the quantification of confrontation (dual dulture) test: A case study on the biological control of *Pythium aphanidermatum* with *Trichoderma asperelloides*. *Journal of Biofertilizers and Biopesticides* 4: 137. DOI: 10.4172/2155-6202.1000137.

Rahman, M., Begum, M. & Alam, M. 2009. Screening of *Trichoderma* isolates as a biological control agent against *Ceratocystis paradoxa* causing pineapple disease of sugarcane. *Mycobiology* 37(4): 277–285. DOI: 10.4489/MYCO.2009.37.4.277.

Ravensberg, W. 2010. The development of microbial pest control products for control of arthropods: A critical evaluation and a roadmap to success. PhD thesis. Wageningen University, Wageningen, NL.

Reid, D. 2007. Water activity: Fundamentals and relationships. In: *Water Activity in Foods. Fundamentals and Applications*, eds. G. Barbosa-Cánovas, A. Fontana, S. Schmidt, & T. Labuza, 15–28. Blackwell Publishing and the Institute of Food Technologists. Available in: http://www.amac.md/Biblioteca/data/30/14/10/60.2.pdf (Accessed January 15, 2020).

Rueden, C., Schindelin, J. & Hiner, M. 2017. ImageJ2: ImageJ for the next generation of scientific image data. *BMC Bioinformatics* 18: 529. DOI: 10.1186/s12859-017-1934-z.

Schiraldi, A., Fessas, D. & Signorelli, M. 2012. Water activity in biological systems—A review. *Polish Journal of Food and Nutrition Sciences* 62(1): 5–13. DOI: doi.org/10.2478/v10222-011-0033-5.

Soares, I. & Faull, J. 2000. Parasitism of *Rhizoctonia solani* by strains of *Trichoderma* spp. *Scientia Agricola* 57(1): 55–59. DOI: http://dx.doi.org/10.1590/S0103-90162000000100010.

Sriram, S., Panchakshari, K. & Jeyamma, M. 2011. Extended shelf-life of liquid fermentation derived talc formulations of *Trichoderma harzianum* with the addition of glycerol in the production medium. *Crop Protection* 30: 1334–1339. DOI: 10.1016/j.cropro.2011.06.003.

Stentelaire, C., Antoine, N., Cabrol, C., Feron, G. & Durand, A. 2001. Development of a rapid and highly sensitive biochemical method for the measurement of fungal spore viability. An alternative to the CFU method. *Enzyme and Microbial Technology* 29: 560–566. DOI: doi.org/10.1016/S0141-0229(01)00432-X.

Sutton, S. 2011. Accuracy of plate counts. *Journal of Validation Technology* 17(3): 42–46.

Syed Ab Rahman, S., Singha, E., Pieterseb, C. & Schenk, P. 2018. Emerging microbial biocontrol strategies for plant pathogens. *Plant Science* 267: 102–111. DOI: doi.org/10.1016/j.plantsci.2017.11.012.

Szekeres, A., Leitgeb, B., Kredics, L., Manczinger, L. & Vágvölgyi, C. 2006. A novel, image analysis-based method for the evaluation of in vitro antagonism. *Journal of Microbiological Methods* 65: 619–622. DOI: 10.1016/j.mimet.2005.09.014.

Wyckhuys, K., Hughes, A., Buamas, C., Johnson, A., Vasseur, L., Reymondin, L., Deguine, J. & Sheil, D. 2019. Biological control of an agricultural pest protects tropical forests. *Communications Biology* 2: 10. DOI: doi.org/10.1038/s42003-018-0257-6.

Section XVIII

Integrated Pest Management (IPM), a Noble Biopesticidal Approach

44

Integrated Pest Management in Major Legume Crops

Jagdish Jaba, Jatin,K, Reddy Ramya Sree, Sathish Kota, and Suraj Prasad Mishra

CONTENTS

44.1 Introduction

Integrated pest management (IPM) is the basic concept in the containment of a pest below economically damaging levels using the combination of feasible control measures with environmental and operational safety. The term was earlier used as "integrated control" by Bartlett (Bartlett, 1956) and was further elucidated by Stern and coworkers (Stern et al., 1959). Various national, international organizations and NGOs support and promote IPM because of its safety for the environment and public health. However, there is no universally agreed definition of IPM. (Alstair, 2003). In principle, IPM may be defined as a principle and holistic approach but in a recent IPM pyramid presented by (Stenberg, 2017) identified the lack of a holistic approach that uses traditional and modern tools. Previous models of IPM mainly focused on ecological and to some extent on the evolutionary aspects of pest management (Peterson et al., 2018).

44.2 Extent of Losses Due to Insect Pests in Grain Legumes

In India, insect pests lead to an approximate economic loss in yield of 15.00% worth $2285.29 million (Dhaliwal et al., 2015). Pod borer, *H. armigera*—the single largest yield shrinking factor in food legumes—causes an estimated loss of $317 million in pigeon pea and $328 million in chickpea (ICRISAT 1992). Worldwide, it causes an estimated loss of over $2 billion annually, despite over $1 billion value of insecticides used to control *H. armigera* (Sharma 2005). In general, the estimates of yield losses vary from 50 to 100% in the tropics and 5 to 10% in the temperate regions (van Emden et al., 1988). Another pod borer, *M. vitrata*, causes loss estimated to be $30 million annually (Saxena et al., 2002). In pigeon pea, yield losses due to pod borer 25–70%, pod fly 10–50%, *Maruca* 5–25%, and pod bug 10–30% have been reported (Sharma et al., 2010). Legume flower thrips (LFT), *M. sjostedti* Trybom, and cowpea *V. unguiculata* in tropical Africa causes yield losses ranging from 20 to 100% (Karungi et al., 2000). The avertable losses in grain/food legumes at current production levels of 60.45 mt would be nearly 18.14 mt (at an average loss of 30%), worth nearly $10 billion (Sharma et al., 2008). The loss in legumes due to various pests and pathogens is estimated to be up to 20% on average annually (Dhaliwal et al., 2010). Up to 100% of losses have been observed in various legume crops in Asia and Africa in the circumstances that favor diseases and pests (Vijay et al., 2015). In a survey conducted by ICRISAT, *Helicoverpa armigera* has been reported to cause yield loss varying from 4.2 to 39.7% in 12 major chickpea growing states and in case of pigeon pea *Melanagromyza obtusa* was reported to cause 2.5% yield loss in north India (Krishnamurthy Rao and Murthy 1983) (Table 44.1).

TABLE 44.1
Economically Important Insect Pests in Pulses in Asia and Their Threshold Levels

Common Name	Scientific Name	ETLs	Existing Control Methods
Pigeon Pea			
Pod borer	*Helicoverpa armigera* (Hubner)	5 eggs or 3 small larvae per plant	IPM
Podfly	*Melanagromyza obtusa* (Malloach)	In all endemic locations	Chemical
Leaf webber	*Maruca vitrata* (Geyer)	5 webs per plant	Chemical
Pod sucking bugs	*Clavigralla gibbosa Spinola*	One egg mass per plant	Chemical
Chick Pea			
Pod borer	*Helicoverpa armigera* (Hubner)	3 eggs or 2 small larvae per plant	IPM
Cutworm	*Agrotis ipsilon* (Hufnagel)	5% plant mortality	Chemical
Aphids	*Aphis craccivora* Koch	25.0% of plants infested at seedling stage in dry situations only in rainy season	IPM
Green Gram, Black Gram and Cowpea			
Hairy caterpillar	*Spilosoma obliqua* (Walker)	5 larvae per meter row	Chemical
Leaf miner	*Liriomyza trifoli*	5 infested leaves per plant at 30 days of crop age	IPM
Tobacco caterpillar	*Spodoptera litura* (Fab)	20–25% defoliation at 40 days	IPM
Thrips	*Scirtothrips dorsalis* Hood	5 thrips/terminal at seedling stage	Chemical
Aphids	*Aphis craccivora* Koch	5–10 aphids per terminal at seedling stage in dry spells only in rainy season	IPM
Bruchids	*Callosobruchus chinensis*	Immediately after first observation	Sanitation, Triplelayer bags

44.3 Role of IPM in Organic Farming

Organic farming and sustainable farming systems are the best alternatives to conventional farming (Shetty et al., 2014). The main aim of organic farming is to promote natural enemies (Kristiansen, 2006). The use of botanical pesticides to protect crops from insect pests has assumed greater importance throughout the world in recent years (Hire math, 1994).

44.4 Pest Management Tactics in IPM

The following are common pest management techniques, which can be used at different stages of crop production to prevent, reduce, minimize, or treat pest infestations. Each may provide a certain level of control, but their additive effect may be significant in preventing loss of yield.

44.4.1 Host Plant Resistance

It is described as the best strategy that uses insect resistant and insect tolerant cultivars produced by conventional breeding or genetic engineering (Douglass, 2018). Among various alternatives of IPM, growing genotypes with tolerance to insects would be the most economical way of reducing the yield losses. Painter (1958) classified plant resistance to insects into three categories based on the mechanisms involved: non-preference (for oviposition, food, or shelter), antibiosis (adverse effect of plants on the biology of insects), and tolerance (repair, recover, or ability to withstand infestation). The HPR is the first line of defense in IPM program.

44.4.2 Cultural Control

These maneuvers are location-specific and must be designed to suit local practices and customs (Ranga Rao and Shanower, 1999). A reduced infestation of pod bugs and aphids was recorded in cowpea due to early planting (Karungi et al., 2000) and the pod borer (*Maruca vitrata*), and the pod bug (*Clarigralla tomentosicollis*) in West Africa. A high-density planting system reduced root maggot (Delia spp.) invasions in canola crop of Canada region (Dosdall et al., 1996), and aphid invasions in cowpea in Uganda. Crop rotation with non-host crops or tolerant varieties disrupts the pest life cycles therefore, crop rotation tactic has been widely used in controlling many insect pests all over the world (Mohler and Johnson, 2009). Seed priming in chickpea (soaking overnight) improves the chickpea establishment and reduces losses due to pod borer damage (Musa et al., 1999). Intercropping coriander with chickpea may afford nectar sources for adult parasitoid improving natural control of *Helicoverpa* in chickpea. It was reported that in a sole crop of chickpea, mean parasitism by *Campoletis chlorideae* on the larvae of *H. armigera* was 60.61% and in the chickpea-coriander intercropping ecosystem, it was 76.96% (Jaba et al., 2016).

44.4.3 Biological Control

Insect pests of legume crops have many natural enemies, viz. predatory arthropods (Hajek and Ellenberge 2018) (Figure 44.1).

Predators: Many predators feed on insect pests that attack pigeon pea. For example, >60 species of arthropods have been reported to prey on *H. armigera*. In addition

FIGURE 44.1 Generalist predators that are common in pest management.

to predatory insects, several species of spiders, lizards, and birds feed on *H. armigera*.

Parasitoids: Many parasitoids have been reported to feed on pigeon pea pests. For example, more than 75 insect parasitoids (orders: Hymenoptera and Diptera) attack various life stages of *H. armigera*. Some parasitoids can be mass-reared and released into an infested field. In addition, a number of commercial companies are marketing parasitoids, the most common being the egg parasitoid, *Trichogramma* spp., which attack eggs of *H. armigera* and other lepidopterons (Kenis et al., 2017) (Figure 44.2).

44.4.4 Behavioral Control

Pest behavior can be utilized to monitor and regulate the pests through bait, trap, and mating disruption methods. Baits containing poisonous material will attract and kill the pests when scattered in the field or placed in traps. Certain colors, lights, odors of attractants or pheromones draw the attention of the pest by manipulating their behavior. Pheromone lures confuse adult insects and disrupt their mating and thus there will not be any progeny. The techniques used for behavior control were described by Fooster et al., 1997. Various baits mixed with chemicals are used for killing stored grain pests (Figures 44.3 and 44.4).

44.4.5 Physical/Mechanical Methods

It refers to the use of a variety of physical or mechanical techniques for pest exclusion, trapping, removal, and destruction of the pests (Web and Linda, 1992).

44.4.6 Microbial Control

Microbial control refers to the use of disease-causing microorganisms to reduce the population of insect pests below the damaging levels. Using entomopathogenic bacteria, fungi, spore farming unicellular parasites, nematodes, or viruses, and fermentation byproducts of some microbes against arthropod pests, plant-parasitic nematodes, and plant pathogens usually come under microbial control (Mankau, 1981). When controlling lepidopteran pests, the bacteria like *Bacillus thuringiensis*, baculoviruses are successful in IPM. Nucleopolyhedrosis virus (Ha-NPV) is used exclusively for controlling *H. armigera* affecting pigeon pea and chickpea (Figure 44.5).

44.4.7 Chemical Control

Chemical method of pest management usually refers to the use of synthetic chemical pesticides (Pimental, 2009). Although botanical extracts such as azadirachtin and pyrethrins, and microbe-derived toxic metabolites such as avermectin and spinosad are considered as biorationals (Lasota and Dybas, 1991). Chemical pesticides are characterized into different groups based on their mode of action (IRAC, 2018) and switching chemicals pesticides with different modes of action will reduce the risk of resistance development by the targeted pests. (Sparks and Nauen, 2015).

44.5 IPM Modules for Managing Legume Pests

- Seed treatment with fungicide before sowing.
- Deep summer plowing is helpful to eliminate quiescent pupa from the field.
- The use of tolerant/resistant varieties for *Helicoverpa* management.
- Installation of *Helicoverpa* pheromone traps for intensive monitoring and tuning of control strategies.
- Mixing of sorghum/maize seed at 250 g/ha to function as live bird perches. These plants also help in conserving natural enemies.

FIGURE 44.2 Generalist parasitoids that are common in pest management.

FIGURE 44.3 Monitoring by pheromone traps.

FIGURE 44.4 Monitoring adult lepidopterans by light traps.

- Intensive weed management in the early stages of the crop.
- Increased monitoring for pod borers at the flower initiation stage.
- Fixing of bird perches after crop establishment.
- Application of 5% neem seed powder extract (NSKE) at flower initiation.
- Early sowing of pigeon pea short duration varieties is helpful against pest attack.
- Polyethylene mulch in peanut reduced thrips damage to 4.6% compared to 24.3% leaflets damaged in an unmulched area. A reduction in injury caused by jassids was recorded as 11% in unmulched ones when compared to mulched ones.
- Pigeon pea varieties ICPL 332 WR, ICPL 84060, ICPL 187-1, and ICP 7203 were promisingly found to be resistant to pod borer *Helicoverpa*, and ICPL 98003 was found to be resistant to *Maruca vitrata* (Sharma et al., 2017).
- Application of HNPV at 500 LE/ha when the egg/larval population reaches ETL.
- The pod borer complex may be managed effectively by applying *Trichogramma chlionis* in the field at weekly intervals at 1.5 lakh/ha/week three to four times.

FIGURE 44.5 *Helicoverpa* larva infected with HaNPV under field conditions.

- Manual shaking of pigeon pea plants to dislodge larvae, and destruction of larvae, in "outbreak" situations.
- Adopt soil application of sulfur at 20 kg/ha (e.g., through SSP, gypsum, or elemental) and zinc as $ZnSO_4$, which will lead to healthy crops and impart general resistance.
- Increased monitoring for pod borers from flower initiation stage.
- Intercrop with linseed/coriander/mustard and "sprinkle" with sunflower to promote bioagent activity. Legumes must be planted before mid-October to save plants from pod borer damage.
- Application of 5% neem seed kernel extract at flower initiation.
- If the infestation by larvae (pod borer or webber or blue butterfly) is causing higher damage spray green label pesticides, e.g., Emamectin benzoate or Chlorantraniliprole (Rynaxypyr) to save the crop. Repeat spray if necessary.
- Cautious application of appropriate chemical pesticides, if the other controls measures recommended above do not contain pest population below ETL's. Appropriate drying, application of inert dusts during storage, conversion of seed into dhal, and fumigation with celphos at 3 g per 40 kg seed against storage pests.
- Spraying of effective molecules like Indoxacarb 14.5 SC% at 300 ml/ha or Emamectin benzoate 5% SG at 11 a.i. gm/ha or spinosad 45 SC at 56–73 a.i. gm/ha is effective against pod borer complex (Sujithra, M, 2013).

44.6 IPM Success Stories

44.6.1 An ICRISAT Success Story

44.6.1.1 Reviving Chickpea through Integrated Pest Management in Nepal

IPM is a new concept in Nepal. Nearly 75% of the population in Nepal has no knowledge about the system while the rest of them have a piece of basic knowledge in pest control through chemical techniques. Chickpea has great potential in Nepal with chickpea occupying a large area of about 390,000 ha. Therefore, an opportunity to arrest and reverse the declining trend in the production of chickpeas by extending the potential benefits of the IPM package to farmers and, indeed with the farmer. The introduction of IPM has revived high demand for the crop. Promotion of IPM was carried out in collaboration with Nepal Agricultural Research Complex and various NGO and farmers self-help groups and ICRISAT. The recommendations suggested by IPM team ICRISAT like, need- and weather-based applications of fungicide for the management of various diseases and insecticide for the management of pod borer, wide spacing, seed treatment with insecticide, installation of pheromone traps, etc were followed. Through IPM, the pod borer incidence was reduced to a six fold increase in grain yield (up to 4 tons per ha was obtained). In the 1998/1999 season, 110 farmers tried IPM with chickpea. The following season there was a five fold increase in IPM adoption. The good news spread, and by the end of 2000/2001 and 2001/2002, 1,100 and 7,000 farmers, respectively, adapted IPM techniques. At the end of 2003, 11,000 farmers have adapted the IPM technology (Icrisat success stories—SS2-Jul 2003).

44.6.2 ICAR Success Story on Implementation of IPM in India

44.6.2.1 Intensive Application of IPM for Increasing Productivity of Pulse Crops

Pigeon pea and chickpea are widely cultivated in the states of Karnataka, Maharashtra, Andhra Pradesh, Uttar Pradesh, Madhya Pradesh and Jharkhand. Farmers in these regions are facing bigger problems with pod borers, plume moths, and cutworms. Farmers were taught new techniques like field scouting, installation of pheromone traps, monitoring population dynamics and pest forecasting through "National Pest Reporting and Alert system e-surveillance" through http://www.ncipm.org.in/A3P/UI/HOME/ Login.aspx. Crop yields have increased and infestation by pod borer decreased after following all the IPM practices recommended by ICAR (Sharma et al., 2016).

44.7 Conclusion and Future Plan

IPM strategies recognize that many environmental factors interact and work together to affect the abundance, distribution, movement, and phenotype of insect pest populations. The IPM system approach draws on an in-depth knowledge of how management practices affect the biology of insects and other trophic interactions. Advances in genomics make it possible to better understand the dynamics of pest populations (e.g., patterns of movement or biotypes that respond differently to host plant resistance traits) that may provide information needed to develop new approaches for insect management. There must be a flow of information between the scientists and users who apply the strategies. Agricultural researchers, instructors, sociologists, economists, professional analysts, managers, growers, pest management professionals, agricultural input manufacturers, retailers, and consumers play a key role in food production. The new IPM model serves as a template for focusing on the different areas of the Paradigm and to encourage collaboration between different disciplines. This new model is expected to develop IPM strategies around the world for implementing sustainable agricultural practices to ensure profitability for farmers, affordability for consumers, and food security for the growing world population.

REFERENCES

Alastair, O. Integrated pest management for resource-poor African farmers: Is the emperor naked? *World Development* 2003, 31(5): 831–845.

Barlett, B. R. Natural predators. Can selective insecticides help to preserve biotic control. *Agric. Chem.* 1956, 11: 42–44.

Dhaliwal, G. S., Jindal, V., Dhawan, A. K. Insect pest problems and crop losses: Changing trends. *Indian J. Entomol.* 2010, 37: 1–7.

Dhaliwal G. S., Jindal, V., Mohindru, B. Crop losses due to insect pests: Global and Indian scenario. *Indian J. Entomol.* 2015, 77(2): 165–168.

Dosdall, L. M., Herbut, M. J., Cowle, N. T., Micklich, T. M. The effect of seeding date and plant density on infestations of root maggots, Delia spp. (Diptera: Anthomyiidae), in canola. Can. J. Plant Sci. 1996,76: 169–177.

Douglas, A. E. Strategies for enhanced crop resistance to insect pests. *Annu. Rev. Plant Biol.* 2018, 69: 637–660.

Foster, S., Harris, Ma. Behavioural manipulation methods for pest-management. *Annu. Rev. Entomol.* 1997, 42: 123–146.

Hajek, A. E., Eilenberg, J. *Natural Enemies: An Introduction to Biological Control.* Cambridge University Press, Cambridge, UK, 2018.

Hiremath, I. G. Isolation and identification of pesticides from selected Indian and African plants. Post-Doctoral Res. Doc., Seoul National University, South Korea, 1994, pp. 109.

International Crops Research Institute for the Semi-Arid Tropics. *The Medium-term Plan.* ICRISAT, Patancheru, 1992.

IRAC (Insecticide Resistance Action Committee). *IRAC mode of action classification scheme.* 2018, https://www.irac-online.org/documents/ moa-classification/?ext=pdf.

Jaba, J., Meena, A., Snehel, C. Biology, relative parasitization of larval endoparasitoid campoletis chlorideae uchida on *Heliocoverpa Armigera* Hübner under sole and chickpea-coriander intercropping system. *J. Biol. Control.* 2016, 30(2):60–67. DOI: 10.18641/jbc/0/0/94441.

Karungi, J., Adipala, E., Ogenga-Latigo, M. W., Kyamanywa, S., Oyobo, N. Pest management in cowpea. Part 1. Influence of planting time and plant density on cowpea field pests infestation in eastern Uganda. *Crop Prot.* 2000, 19: 231–236.

Kenis, M., Hurley, B. P., Hajek, A. E., Cock, M. J. W. Classical biological control of insect pests of trees: Facts and figures. *Biol. Invasions.* 2017, 19: 3401–3417.

Krishnamurthy Rao, B. H., Murthy, K. S. R. K. Entomological society of India. In Proceedings National Seminar on Crop Losses Due to Insect Pests (Vols. I–II). Entomological Society of India, Hyderabad, India, 1983.

Kristiansen, P. *Organic Agriculture: A Global Perspective.* CSIRO, Clayton, Victoria, 2006, pp. 484.

Lasota, J. A., R. A. Dybas. Avermectins, a novel class of compounds: Implications for use in arthropod pest control. *Annu. Rev. Entomol.* 1991, 36: 91–117.

Mankau, R. Microbial control of nematodes. In Zuckerman, B., Rohde, R. A. (eds.), *Plant Parasitic Nematodes.* Academic Press, New York, NY, 1981, 3, 475–478.

Mohler, C. L., Johnson, S. E. *Crop Rotation on Organic Farms a Planning Manual.* Natural Resource, Agriculture, and Engineering Service, Ithaca, NY. 2009, 3, 27–32.

Musa, A. M.; Johansen, C.; Kumar, J.; Harris, D. Response of chickpea to seed priming in the High Barind Tract of Bangladesh. *Int. Chickpea Pigeonpea Newslett.* 1999, 6: 20–22.

Peterson, R. K. D., Higley, L. G., Pedigo, L. P. Whatever happened to IPM? Am. Entomol. 2018, 64: 146–150.

Pimentel, D. Pesticides and pest control. *Integrated Pest Management: Innovation-Development Process.* Springer, Dordrecht, 2009. 83–87.

Ranga Rao, G.V., Shanower, T. G. Cultural practices in insect management with reference to groundnut. *Indian J. Plant Prot.* 1999, 27(1&2): 87–108.

Saxena, K. B., Lateef, S. S., Ariyaratne, H. P., Fenseka, H. H. D., Dharmasena, C. M. D. *Maruca testulalis* damage in determinate and indeterminate lines of pigeonpea. *Int. Pigeonpea Newslett.* 2002, 3: 91–93.

Sharma, H. C, Dhillon, M. K., Arora, R. Effects of *Bacillus thuringiensis* delta-endotoxin-fed *Helicoverpa armigera* on the survival and development of the parasitoid, *Campoletis chlorideae. Entomol. Exp. Appl.* 2008, 126: 1–8.

Sharma, H. C., Gaur, P. M., Hoisington, D. A. Physico-chemical and molecular markers for host plant resistance to *Helicoverpa armigera.* In: Saxena, H., Rai, A. B., Ahmad, R. et al. (eds.), *Recent Advances in Helicoverpa Management Indian Society of Pulses Research and Development.* Indian Institute for Pulses Research, Kanpur, 2005, 108–121.

Sharma, O. P, Gopali, J. B, Yelshetty S., Bambawale, O. M., Garg, D. K., Bhosle, B.B. *Pests of Pigeonpea and Their Management.* National Centre for Integrated Pest Management, IARI Campus, New Delhi, 2010.

Sharma, O. P., Yelshetty, S, Rajappa, V. *Intensive application of IPM for production of pulse crops: Success stories of IPM.* NCIPM, New Delhi, 2016, 1: 23–38.

Shetty, P. K., Alvares, C., Yadav, A. K. (eds). *Organic Farming and Sustainability*. National Institute of Advanced Studies, Bangalore, 2014, 3–7.

Sparks, T. C., R. Nauen. IRAC: Mode of action classification and insecticide resistance management. *Pestic. Biochem. Physiol.* 2015, 121: 122–128.

Stenberg, J. A. A conceptual framework for integrated pest management. *Trends Plant Sci.* 2017, 22: 749–769.

Stern, V. M., Smith, R. F., van den Bosch, R., Hagen, K. S. The integration of chemical and biological control of the spotted alfalfa aphid (the integrated control concept). *Hilgardia*. 1959, 29: 81–101.

Sujithra, M. Major insect pests of pigeonpea and their managment 1988. https://www.krishisewa.com/disease-management/268-pigeon-pea-insect-pests.html

Van Emden, H. F., Ball, S. L., Rao, M.R. Pest and disease problems in pea, lentil, faba bean, and chickpea. In: Summerfield, R. J. (ed) *World Crops: Cool Season Food Legumes*. Kluwer, Dordrecht, 1988, 519–534.

Vijay, P., Anil, K. C., Anchal, D., Bana, R.S., Rana K.S., Rana, D.S., Tyagi, V.K., Puniya, M. M. Improved crop management practices for sustainable pulse production: an Indian perspective. *Indian J. Agric. Sci.* 2015, 85(6): 747–758.

Webb, S. E., Linda, S. B. Evaluation of spunbonded polyethylene row covers as a method of excluding insects and viruses affecting fall grown squash in Florida. *J. Econ. Entomol.* 1992, 85: 2344–2352.

45

Integrated Insect Pest Management of Economically Important Crops

Surendra K. Dara

CONTENTS

45.1 Introduction

With the gradual increase of organic food production around the world, the need for improved agronomic and pest management practices has also grown. Integration of various practices that reduce, delay, or suppress pest infestations and their damage has been the key principle of integrated pest management (IPM), which promoted sustainable farming long before the popularity of organic agriculture. IPM practices have been encouraged to reduce environmental and human health risks, development of pesticide resistance, and other associated problems from excessive use of chemical pesticides or other agricultural inputs. Now, organic farming very much resembles conventional farming except that only organically approved inputs, sometimes very similar to their conventional counterparts, are to be used. As a result, organic agriculture also has the same risks of negative environmental and human effects as well as pesticide resistance from indiscriminate use of organically approved inputs. IPM is no longer limited to the use of various pest management strategies, but also involves the business, social, and environmental aspects of food production (Dara 2019a). The new model has redefined IPM as an economically viable, socially acceptable, and environmentally safe pest management, which includes, in addition to various pest control options, the use of modern technology, organization of information and decision making, and communication and education at various levels in the management aspect. Production of agricultural goods, their marketing, and consumer preference, as well as the social and economic aspects of farm profitability, food security, and affordability are also discussed in this new model. This chapter will focus on presenting some integrated insect pest management strategies, with specific examples from economically important crops, based on this new model that apply to a variety of crops.

45.2 Various IPM Options

Several options, from the selection of a resistant cultivar to rotating pesticides with different modes of action, can help reduce pest infestations and their damage. Some of the most common options will be discussed here.

45.2.1 Host Plant Resistance

While genetically modified cultivars of corn, cotton, soybean, and other crops with toxins from the bacterium *Bacillus thuringiensis* resist damage from various lepidopteran pests, organic farming uses cultivars that resist or tolerate insect damage through selected natural traits rather than genetic modification (Lefebvre et al. 2020). Using resistant cultivar reduces the need for pesticide applications and their potential negative impact on natural enemies. Silva et al. (2013) evaluated several soybean entries for their antibiosis towards the redbanded stink bug, *Piezodorus guildinii*, and thought the physical characteristics, presence of the anti-nutritional and/or secondary metabolites, thickness of the epidermis, and other traits of the entries reduced the nymphal survival and affected their longevity. Trichomes on soybean leaves have been known to impart resistance against the soybean aphid, *Aphis glycines* (Todd et al. 2016), the kudzu bug, *Megacopta cribraria* (Larhiri et al. 2020), and the armyworm,

Spodoptera litura (Krisnawati et al. 2017) among other pests. Several soybean varieties that have *Resistance to Aphis glycines* (*Rag*) genes are used against *A. glycines* (McCarville et al. 2012). These varieties suppress the growth and reproduction of aphids, delay the population growth, and prevent them from reaching economically damaging levels. However, aphids can overcome this resistance mechanism (Crossley and Hogg 2015) warranting judicious use of any option and the need for integrating multiple options.

Lahiri et al. (2019) reported that two of the sorghum hybrids were highly resistant to the sugarcane aphid, *Melanaphis sacchari*, and one of them did not need any insecticide sprays to maintain the yield potential. Breeding rice cultivars resistant to the brown planthopper, *Nilaparvata lugens*, has been a major part of controlling this pest, which causes significant yield losses through direct feeding and transmitting viral diseases. Liu et al. (2014) found that *Bph3*, a cluster of three genes responsible for leptin receptor kinase, in rice gives durable, broad-spectrum resistance to *N. lugens* and the whitebacked planthopper, *Sogatella furcifera*. Mechanisms that suppress the production of serotonin biosynthesis appear to confer resistance in rice against *N. lugens* and the striped stem borer, *Chilo suppressalis* (Lu et al. 2018). Using cultivars possessing one or more of these traits will be a critical part of IPM for many crops.

45.2.2 Induced Resistance

Various non-pathogenic microbes, microbial metabolites, botanical extracts, and some minerals induce resistance in plants against pests, diseases, and/or abiotic stressors (Sequeira 1983; Oostendorp et al. 2001; Conrath 2009). For example, methyl jasmonate increased the activities of proteinase inhibitors and/or polyphenol oxidase in tomato leaves, reducing the larval growth rate of the beet armyworm, *Spodoptera exigua* (Stout et al. 1998). Jasmonic acid-induced resistance against chewing insects and pathogens was also found in rice (Karban and Chen 2001). Pest suppression through silica-based materials has been reported in corn, rice, sorghum, sugarcane, and several other crops against aphids, borers, maggots, planthoppers, thrips, worms, and other pests (Keeping et al. 2010; Vasanthi et al. 2014; Sithanantham et al. 2020). Silicon improves the structural strength and production of defensive chemicals or induces resistance through other interactions.

Several beneficial microbes improve nutrient and water absorption, impart salt and drought resistance, and promote plant growth. These beneficial microbes are increasingly used in agriculture for building soil structure and as biostimulants or biofungicides. Many of these plant-associated beneficial microbes can also have an impact on herbivores by influencing their selection of host plants (Grunseich et al. 2020). Arbuscular mycorrhizae such as *Rhizophagus irregularis*; foliar endophytes such as *Acremonium loliae* and *Neotyphodium* spp.; and plant growth-promoting rhizobacteria (PGPR) such as *Bacillus* spp., *Brevibacillus* spp., and *Paenibacillus* spp. result in plant repellence against aphids, beetles, and worm pests (Grunseich et al. 2020). Beneficial microbes activate induced systemic resistance by triggering the jasmonic acid and ethylene pathways of the plant immune system against a wide range of pests and diseases (Pieterse et al. 2014). Species of *Agrobacterium*, *Bacillus*, *Burkholderia*, *Photorhabdus*, *Psudomonas*, *Streptomyces*, *Xenorhabdus*, and others are some of those microbes that are associated with insect control, either by inducing systemic resistance and production of secondary metabolites, or as insect pathogens (Disi et al. 2019). Root colonization of the fungal endophyte *Trichoderma atroviridae* triggered plant defenses in corn and negatively impacted the feeding of the fall armyworm, *Spodoptera frugiperda* (Contreras-Cornejo et al. 2018). In addition to their pathogenicity to arthropods through direct contact, entomopathogenic fungi also appear to have a negative impact on foliar pests through soil application or seed treatment, either through induced resistance or as endophytes (Dara 2019b).

Mycorrhizae play an important role in crop growth and health as well as that of herbivores (Vannette and Hunter 2009). While mycorrhizae directly improve plant nutrition and resulting growth, improved nutritional quality can also support herbivore populations. However, certain plant defenses are also activated in the presence of mycorrhizae, making these interactions very complex. While the arbuscular mycorrhizal fungus *Glomus mosseae* reduced the survival of the cotton bollworm, *Helicoverpa armigera*, and triggered the plant defenses in response to insect damage in tomato (Song et al. 2013), parasitism of the leafminer *Chromatomyia syngenesiae* by the parasitoid *Diglyphus isaea* in ox-eye daisy was reduced in the presence of mycorrhizae (Gange et al. 2003). These examples show that the interactions among plants, mycorrhizae, and herbivores are very complex and good knowledge of these interactions is necessary for successfully using mycorrhizae for pest management.

45.2.3 Cultural Practices

Good cultural practices are the first line of defense against pests and diseases. Adjusting planting dates to avoid the occurrence of pests during vulnerable stages of the crop growth, using crop rotation with non-host plants to reduce the pest pressure during the following season, growing trap crops to divert pests from crop plants, adjusting the plant density to promote plant growth and withstand pest damage, modifying irrigation practices and optimizing nutrient inputs to reduce pest buildup, sanitation practices to remove and reduce infestations, and utilizing other such practices have a significant impact on pest management and reduce the reliance on other options, such as pesticide applications.

Significant reduction in pest populations, improved system robustness, and increased crop yields and soil structure from crop rotation have been documented by several studies around the world and crop rotation is especially important in organic agriculture (Zohry and Ouda 2018; He et al. 2019; Li et al. 2019).

While nutrient inputs are essential for maintaining optimal plant growth and yields, they also have an impact on herbivores and their natural enemies. Low rates of nitrogen and phosphorous affected the survival and fitness of *N. lugens* in rice (Yang et al. 2019). Excessive nitrogen, on the other hand, increased the egg size of *N. lugens* (Zhu et al. 2020a). The predatory mirid bug *Cyrtorhinus lividipennis* also benefited from nitrogen that resulted in increased survival and developmental rate of the nymphs, and the weight, longevity, and fecundity of adult females. However, it negatively affected the predator's efficiency by prolonging the handling time and reducing the number of prey eggs attacked. Similarly, nitrogen application to rice prolonged the developmental time and reduced the adult female fecundity, host-searching efficiency, and overall parasitism of *Anagrus flaveolus*, an egg parasitoid of *N. lugens* (Zhu et al. 2020b). While excessive nitrogen increases a plant's attractiveness to pests, potassium increases a plant's fitness to herbivore damage (Vinay and Sood 2017; SongHe et al 2019). Increased levels of potassium reduce carbohydrate content and increase the production of defensive secondary metabolites in plants (Bala et al. 2018). Similarly, phosphorus can have an impact on herbivores by decreasing the populations of some insects while supporting others (Facknath and Lalljee 2005; Bala et al. 2018). Micronutrients such as copper, iron, manganese, nickel, and zinc are also important for plant health and their ability to withstand biotic and abiotic stressors (Poschenrieder et al. 2006; Cabot et al. 2019). As discussed in the previous section, minerals such as silicon also play a critical role in a plant's fitness to withstand pest pressure.

Appropriate plant spacing is another important strategy for managing pests. Increased plant spacing reduced insect numbers and their damage in okra in Sierra Leone (Norman et al. 2019), increased yields in cucumber in Nigeria (Oso and Awe 2019), and reduced insect numbers in cotton in Bangladesh (Momtaz et al. 2018), while optimal spacing reduced insect numbers and increased yields in tomato in Nigeria (Degri et al. 2019). Increasing plant density reduced the incidence of the leafminer *Leucoptera coffeella* in coffee in Brazil (Assis et al. 2012).

Irrigation management is also an important tactic for maintaining crop health and controlling pests (Rousselin et al. 2016; Murrell 2017). Improved control of the silverleaf whitefly, *Bemisia tabaci*, in cucumber (Oso and Awe 2019) and the tomato leafminer, *Tuta absoluta*, in tomato (Han et al. 2018) were seen with improved irrigation practices. Rendon and Walton (2019) found that more adult spotted-wing drosophila (*Drosophila suzukii*) emerged from pupa in blueberries irrigated with overhead sprinklers compared to drip irrigated blueberries. Compared to unirrigated coffee, drip irrigated coffee had 2.2 times fewer leaf mines from *L. coffeella* (Assis et al. 2012).

45.2.4 Mechanical or Physical Control

Several mechanical or physical pest control tactics can be used in greenhouse or field production systems (Sorensen et al. 2016). Row covers with or without the support of metal hoops are used to exclude pests in multiple vegetable crops (Cannon et al. 2016). Organic mulches, such as straws or compost, support natural enemies and suppress pests, while colored plastic mulches alter insect behavior (Sorensen et al. 2016; Jabran 2019). Mulches also conserve moisture, suppress weeds, promote plant growth, and offer other benefits. Using metalized reflective mulch significantly reduced the number of melon thrips, *Thrips palmi*, in cucumber, eggplant, Jalapeño pepper, snap bean, squash, and tomato (Razzak et al. 2019). Ultraviolet-reflective mulch was very effective, especially early in the crop development, to deter the western flower thrips, *Frankliniella occidentalis*, in vegetables (Demirozer et al. 2012). Single- and double-charged dipolar electric field screens effectively trapped *B. tabaci* and prevented their infestation in greenhouse tomato (Takikawa et al. 2019). In addition to the electric field screens to the greenhouse windows, using an electrostatic insect sweeper during routine monitoring was helpful in controlling *B. tabaci* and the viral diseases it transmits in tomato (Takikawa et al. 2015). Modifying greenhouse temperature, relative humidity, and airflow are also some of the physical control strategies for managing pests in greenhouses (Sorensen et al. 2016).

45.2.5 Biological Control

Biological control with predators and parasitoids is a popular practice in many cropping systems (El-Wakeil et al. 2019; Souza et al. 2019). While importing natural enemies from the native region of the invasive pests and releasing them in pests' new habitats is successfully used in several classical biological control efforts, inundative release of native natural enemies is also a part of IPM, especially in greenhouse production.

Parasitoids play a critical role in controlling the cabbage moth, *Mamestra brassicae*, in Switzerland (Barloggio et al 2015), the diamondback moth, *Plutella xylostella* worldwide (Sarfraz et al. 2007), *T. absoluta* in South America (Gervassio et al. 2019), the coffee berry borer, *Hypothenemus hampei*, in Brazil and Colombia (Rodríguez et al. 2017) and other important pests around the world. Similarly, numerous predatory arthropods contribute to the biological control of several insect pests all over the world (Follett et al. 2016; El-Wakeil and Gaafar 2019).

Where the release of natural enemies is not possible, conserving natural enemies helps promote biological control. Natural enemies are influenced by agronomic practices and pesticides including those approved for organic farming. Avoiding practices that harm natural enemies, such as using reduced-risk pesticides, timing pesticide applications to reduce the negative impact to natural enemies, or managing the habitat to conserve and promote natural enemies, are some of the strategies that enhance biological control (Landis et al. 2000; Messelink et al. 2014; Arora et al. 2019; Koch et al. 2019; Quesada and Sadof 2019).

45.2.6 Microbial Control

Microbial control refers to the use of entomopathogens such as bacteria, fungi, nematodes, and viruses for controlling arthropod pests (Lacey 2017). Each of these has a specific mode of infection and can be used in IPM programs against a wide range of chewing and sucking insects that attack both underground and aboveground plant parts (Dara 2017a). Entomopathogenic fungi, such as *Beauveria bassiana*, *Cordyceps* (=*Isaria*) *fumosorosea*, and *Metarhizium* spp., have been successfully used against pests of citrus (Chow et al. 2018), potato (Wraight and Ramos 2017), rice (Rizwan et al. 2019), strawberry (Dara et al. 2018), vegetables (Dara 2017b; Liu et al. 2020), and other crops either alone, in combination with other entomopathogens, or in combination or rotation with botanicals and other insecticides. Entomopathogenic fungi cause infections when conidia come in contact with the host insects and are ideal for both sucking and chewing pests. Bacteria (e.g., *Bacillus thuringiensis*—Bt) and viruses (granuloviruses and nucleopolyhedroviruses) are ideal for chewing pests, as they need to be ingested by their hosts. Nematodes (e.g., *Heterorhabditis* spp. and *Steinernema* spp.), which actively seek their hosts, are typically used for soil-dwelling insects. Several commercial formulations of bacteria (Glare et al. 2017), nematodes (Shapiro-Ilan et al. 2017), and viruses

(Grzywacz 2017) are used worldwide for controlling a variety of insect pests. While transgenic Bt crops play a significant part in addressing lepidopteran pests in conventional agriculture, approved Bt formulations can be used for the same purpose in organic agriculture.

45.2.7 Control with Botanical and Microbial Metabolites

Insecticides containing essential oils and other botanical extracts and microbial metabolites as active ingredients are successfully used for pest and disease management. Similar to the synthetic chemical pesticides in conventional agriculture, botanical and microbial metabolite-based pesticides are useful in organic agriculture for quick knockdown effect and complement microbial biopesticides for suppressing pest populations. Essential oils are extracted from aromatic plants (mainly from the families Asteraceae, Lamiaceae, Lauraceae, and Myrtaceae) which contain terpenes and phenylpropanoids, among other volatile compounds, that act as insecticides, antifeedants, growth regulators, repellents, or fumigants (Regnault-Roger 1997; Regnault-Roger et al. 2012). The fumigant property of essential oils can be used to control stored product pests and greenhouse pests. Choi et al. (2003) evaluated 53 essential oils and found ovicidal, nymphicidal, and adulticidal properties against the greenhouse whitefly, *Trialeurodes vaporariorum*, through fumigation activity. Essential oils from *Mentha pulegium* and *Thymus mastichina* were highly efficacious as fumigants to control *F. occidentalis* (Stepanycheva et al. 2019). Toledo et al. (2020) found that clove (*Syzygium aromaticum*) essential oil was highly toxic to the corn leaf aphid, *Rhopalosiphum maidis* while exhibiting low toxicity to the spotted lady beetle, *Coleomegilla maculata*. Essential oils from *Pimpinella anisum* and *Trachyspermum ammi* were effective in controlling the olive fruit fly, *Bactrocera oleae*, when used in protein baits (Rizzo et al. 2020).

Various other botanical extracts also possess insecticidal properties and have been used for pest management for a long time. Pyrethrins from the flowers of *Chrysanthemum cinerariaefolium* and azadirachtin from the seeds of *Azadirachta indica* are successfully used for pest management in many crops (Mordue and Blackwell 1993; Cox 2002). Pyrethrins contain six insecticidal esters that are neurotoxic and have an excellent knockdown effect against several pests (Schleier III and Peterson 2011). Despite their negative impact on non-target organisms such as natural enemies, pyrethrin-based products are very popular, especially in organic agriculture. In recent years, several azadirachtin-based products have entered the pesticide market and have been widely used against several pests. Azadirachtin has been effectively used as an insecticide and antifeedant against the European corn borer, *Ostrinia nubilalis* (Arnason et al. 1985), an insecticide and growth regulator against the cotton leafworm, *Spodoptera litura* (Shue et al. 2018), and as a growth regulator against the vegetable leafminer, *Liromyza sativae* (Khorshidi et al. 2017). Azadirachtin is also used in combination with entomopathogenic fungi for controlling *L. hesperus* and other insect pests in strawberry (Dara et al. 2018; Dara 2020a), root aphids in celery (Dara 2015), and as a part of IPM with other control options against *F. occidentalis* (Gholam and Sadeghi 2016). There are several examples of other botanical extracts in insect pest management. Extracts of *Artemisia argyi*, *Cannabis indica*, and *Citrullus colocynthis* demonstrated high efficacy against the cabbage aphid, *Brevicoryne brassicae*, an important pest of several brassica crops (Ahmed et al. 2020).

There are multiple biopesticides based on microbial metabolites, which provide effective pest management in many cropping systems. Spinosad derived from the bacterial fermentation of *Saccharopolyspora spinosa* (Sparks 1999) is one of the examples. Spinosad is used all over the world for controlling a number of pests including the Mediterranean fruit fly (*Ceratitis capitate*) and the Caribbean fruit fly (*Anastrepha suspensa*) in citrus in Florida (Burns et al. 2001), and *D. suzuskii* in sweet cherries in the Pacific Northwest (Beers et al. 2011) in the United States to a variety of vegetable pests in China (Wang et al. 2009), Thailand (Premachandra et al. 2005), Europe (Schoonejans and Van der Staaij 2001), and humid tropics (Kumar and Poehling 2007). Products based on the fermentation solids and heat-killed bacterial cells of *Chromobacterium subtsugae* (Martin et al. 2007; Koivunen et al. 2009) and *Burkholderia rinojensis* (Cordova-Kreylos et al. 2013) are also available for pest management in the United States and other countries.

45.2.8 Mating Disruption

The use of synthetic insect sex pheromones to confuse adult insects by interfering with their chemical communication and reducing their mating and reproduction is referred to as mating disruption. It is a highly effective IPM strategy used for managing several species of pests, especially lepidopteran pests (Welter et al. 2005). There are several pheromones that are approved for use in organic production systems. The codling moth (*Cydia pomonella*) and the oriental fruit moth (*Grapholita molesta*) in fruit orchards, the pink bollworm (*Pectinophora gossypiella*) in cotton, the tomato pinworm (*Keiferia lycopersicella*) in vegetables, the omnivorous leafroller (*Platynota stultana*) in grapes, the oriental beetle (*Anomala orientalis*) in blueberries and cranberries, and the navel orangeworm (*Amyelois transitella*) in almonds are some examples in which mating disruption is effectively used for their management (Rice and Kirsch 1990; Cardé and Minks 1995; Shorey et al. 1995; Trimble et al. 2001; Brunner et al. 2002; Welter et al. 2005; Wenninger and Averill 2006; Rodriguez-Saona et al. 2009; Burks et al. 2019).

45.3 Social Aspects

While the traditional IPM models were limited to maintaining ecological balance, exploring the potential of various control tactics, and making treatment decisions based on pest thresholds, the new model also covers how the preference of consumers influences sustainable production practices (Dara 2019a). The preference of consumers for clean, perfectly shaped, and uniformly sized fruits and vegetables directly or indirectly contributes to increased pest management efforts. For example, *L. hesperus* feeding causes fruit deformity in strawberry. While the deformity does not decrease the nutritional value, deformed strawberries are considered unmarketable, so a significant amount of pesticides is applied, or other efforts are made to control *L. hesperus* in strawberry. Consumer acceptance for deformed strawberry will have a major impact on strawberry IPM (Dara 2020b). Deformed (popularly known as "ugly") fruits and vegetables are a major source of food waste in the world. In the United States alone, the value of fruits and vegetables wasted at retail and consumer levels was $43 billion in 2008 (Buzby et al. 2011). While food loss on the farm is different from the loss off-farm, educating consumers about the negative impact of their choice on the environment and building trust between consumers and producers are important in marketing deformed produce and contributing to sustainable food production (Tu et al. 2018). Since farming operations are focused both on sustainability and profitability, an increased market for deformed produce can lessen the financial burden on farmers. There are several efforts for secondary markets for deformed produce (Barba and Díaz-Ruiz 2015) or to reduce food wastage through multiple means (Mourad 2016), and all of them help contribute to food security and sustainability.

45.4 Conclusions and Future Opportunities

Integrating various agronomic and pest control tactics improves pest management efficacy. IPM is not only about various ways of suppressing pests, but also involves various social, economic, and environmental aspects. Building the knowledge base through continuous research and outreach and educating consumers about the impact of their food choices on the environment in addition to educating the farmers on sustainable practices will also have a major impact on global food production. Integration, implementation, and education are key elements of this effort.

REFERENCES

Ahmed, M., Q. Peiwen, Z. Gu, Y. Liu, A. Sikandar, D. Hussain, A. Javeed, J. Shafi, M. F. Iqbal, R. An, H. Guo, Y. Du, W. Wang, Y. Zhang, and M. Ji. 2020. Insecticidal activity and biochemical composition of *Citrullus colocynthis*, *Cannabis indica* and *Artemisia argyi* extracts against cabbage aphid (*Brevicoryne brassicae* L.). *Scientific Reports* 10: 522. https://doi.org/10.1038/s41598-019-57092-5.

Arnason, J. T., B. J. R. Philogène, N. Donskov, M. Hudon, C. McDougall, G. Fortier, P. Morand, D. Gardner, J. Lambert, C. Morris, and C. Nozzolillo. 1985. Antifeedant and insecticidal properties of azadirachtin to the European corn borer, *Ostrinia nubilalis*. *Entomologia Experimentalis et Applicata* 38: 29–34. https://doi.org/10.1111/j.1570-7458.1985.tb03494.x.

Arora, S., M. Sehgal, D. S. Srivastava, S. Arora, and S. K. Sarkar. 2019. Rice pest management with reduced risk pesticides in India. *Environmental Monitoring and Assessment* 191: 241. https://doi.org/10.1007/s10661-019-7384-5.

Assis, G. A., F. A. Assis, M. S. Scalco, F. J. T. Parolin, I. Fidelis, J. C. Moraes, and R. J. Guimarães. 2012. *Pesquisa Agropecuária Brasileira* 47: 157–162.

Bala, K., A. K. Sood, V. P. Pathania, and S. Thakur. 2018. Effect of plant nutrition in insect pest management: A review. *Journal of Pharmacognosy and Phytochemistry* 7: 2737–2742.

Barba, M. and R. Díaz-Ruiz. 2015. Switching imperfect and ugly products to beautiful opportunities. In *Envisioning a Future without Food Waste and Food Poverty: Societal Challenges*, eds. L. E. San-Epifanio and M. D. R. Scheifler, 279–284. Wageningen: Wageningen Academic Publishers. https://doi.org/10.3920/978-90-8686-820-9.

Barloggio, G., L. Tamm, T. Oberhänsli, P. Nagel, and H. Luka. 2015. The egg parasitoid *Telenomus* sp. as a novel biocontrol agent to prevent the cabbage moth. *Acta Fytotechnica et Zootechnica* 18: 47–49. http://dx.doi.org/10.15414/afz.2015.18.si.47-49.

Beers, E. H., R. A. V. Steenwyk, P. W. Shearer, W. W. Coates, and J. A. Grant. 2011. Developing *Drosophila suzukii* management programs for sweet cherry in the western United States. *Pest Management Science* 67: 1386–1395. https://doi.org/10.1002/ps.2279.

Brunner, J., S. Welter, C. Calkins, R. Hilton, E. Beers, J. Dunley, T. Unruh, A. Knight, R. V. Steenwyk, and P. V. Buskirk. 2002. Mating disruption of codling moth: A perspective from the Western United States. *IOBC/WPRS Bulletin* 25: 11–20.

Burks, C. S. and D. R. Thomson. 2019. Optimizing efficiency of aerosol mating disruption for navel orangeworm (Lepidoptera: Pyralidae). *Journal of Economic Entomology* 112: 763–771. https://doi.org/10.1093/jee/toy417.

Burns, R. E., D. L. Harris, D. S. Moreno, and J. E. Eger. 2001. Efficacy of spinosad bait sprays to control Mediterranean and Caribbean fruit flies (Diptera: Tephritidae) in commercial citrus in Florida. *Florida Entomologist* 84: 672–678.

Buzby, J. C., J. Hyman, H. Stewart, H. F. Wells. 2011. The value of retail- and consumer-level fruit and vegetable losses in the United States. *Journal of Consumer Affairs* 45: 492–515. https://doi.org/10.1111/j.1745-6606.2011.01214.x.

Cabot, C., S. Martos, M. Llugany, B. Gallego, R. Tolrà, and C. Poschenrieder. 2019. A role for zinc in plant defense against pathogens and herbivores. *Frontiers in Plant Science* 10: 1171. https://doi.org/10.3389/fpls.2019.01171.

Cannon, C., B. Bunn, D. Alston, and M. Murray. 2016. Caterpillar pests of brassica vegetables. *All Current Publications* Paper 1595. https://digitalcommons.usu.edu/extension_curall/1595 (Accessed 25 March 2020).

Cardré, R. T. and A. K. Minks. 1995. Control of moth pests by mating disruption: Successes and constraints. *Annual Review of Entomology* 40: 559–585. https://doi.org/10.1146/annurev.en.40.010195.003015.

Choi, W.-I., E.-H. Lee, B.-R. Choi, H.-M. Park, and Y.-J. Ahn. 2003. Toxicity of plant essential oils to *Trialeurodes vaporariorum* (Homoptera: Aleyrodidae). *Journal of Economic Entomology* 96: 1479–1484. https://doi.org/10.1093/jee/96.5.1479.

Chow, A., C. A. Dunlap, M. A. Jackson, P. B. Avery, J. M. Patt, and M. Sétamou. 2018. Field efficacy of autodissemination and foliar sprays of an entomopathogenic fungus, *Isaria fumosorosea* (Hypocreales: Cordycipitaceae), for control of Asian citrus psyllid, *Diaphorina citri* (Hemiptera: Liviidae), on residential citrus. *Journal of Economic Entomology* 111: 2089–2100. https://doi.org/10.1093/jee/toy216.

Conrath, U. 2009. Priming of induced plant defense responses. *Advances in Botanical Research* 51: 361–395. https://doi.org/10.1016/S0065-2296(09)51009-9.

Contreras-Cornejo, H. A., L. Macías-Rodríguez, E. del-Val, and J. Larsen. 2018. The root endophytic fungus *Trichoderma atroviride* induces foliar herbivory resistance in maize plants. *Applied Soil Ecology* 124: 45–53. https://doi.org/10.1016/j.apsoil.2017.10.004.

Cordova-Kreylos, A. L., L. E. Fernandez, M. Koivunen, A. Yang, L. Flor-Weiler, and P. G. Marrone. 2013. Isolation and characterization of *Burkholderia rinojensis* sp. nov., a non-*Burkholderia cepacia* complex soil bacterium with insecticidal and miticidal activities. *Applied and Environmental Microbiology* 79: 7669–7678. https://doi.org/10.1128/AEM.02365-13.

Cox, C. 2002. Pyrethrins/pyrethrum. *Journal of Pesticide Reform* 22: 14–20.

Crossley, M. S. and D. B. Hogg. 2015. Rag virulence among soybean aphids (Hemiptera: Aphididae) in Wisconsin. *Journal of Economic Entomology* 108: 326–338. https://doi.org/10.1093/jee/tou022.

Dara, S. K. 2015. Reporting the occurrence of rice root aphid and honeysuckle aphid and their management in organic celery. *UCANR E-Journal of Entomology and Biologicals* August 21, 2015. https://ucanr.edu/blogs/blogcore/postdetail.cfm?postnum=18740 (Accessed on March 27, 2020).

Dara, S. K. 2017a. Entomopathogenic microorganisms: Modes of action and role in IPM. *UCANR E-Journal of Entomology and Biologicals* May 20, 2017. https://ucanr.edu/blogs/blogcore/postdetail.cfm?postnum=24119 (Accessed March 26, 2020).

Dara, S. K. 2020a. Implementation of IPDM in strawberries and other berries. In *Integrated Pest and Disease Management in Greenhouse Crops. Plant Pathology in the 21st Century*, eds. M. Gullino, R. Albajes, and P. Nicot, 597–624. Cham: Springer. https://doi.org/10.1007/978-3-030-22304-5_21.

Dara, S. K. 2017b. Microbial control of arthropod pests in small fruits and vegetables in temperate regions. In *Microbial Control of Insect and Mite Pests: From Theory to Practice*, ed. L. A. Lacey, 209–221. New York: Academic Press. https://doi.org/10.1016/B978-0-12-803527-6.00014-7.

Dara, S. K. 2019a. The new integrated pest management paradigm for the modern age. *Journal of Integrated Pest Management* 10: 1–9. https://doi.org/10.1093/jipm/pmz010.

Dara, S. K. 2019b. Non-entomopathogenic roles of entomopathogenic fungi in promoting plant health and growth. *Insects* 10: 277. https://doi.org/10.3390/insects10090277.

Dara, S. K. 2020b. Learn the new IPM model for producing strawberries. *Western Fruit Grower* 140: 6–8.

Dara, S. K., D. Peck, and D. Murray. 2018. Chemical and non-chemical options for managing twospotted spider mite, western tarnished plant bug and other arthropod pests in strawberries. *Insects* 9: 156. https://doi.org/10.3390/insects9040156.

Degri, M. M., A. E. Samaila, and C. S. Oaya. 2019. Effects of weeding frequency and plant spacing on the infestation and damage of the major insect pests of tomato (*Solanum lycopersicum* l.). *African Journal of Agriculture Technology and Environment* 8: 124–132.

Demirozer, O., K. Tyler-Julian, J. Funderburk, N. Leppla, and S. Reitz. 2012. *Frankliniella occidentalis* integrated pest management programs for fruiting vegetables in Florida. *Pest Management Science* 68: 1537–1545.

Disi, J., J. Simmons, and S. Zebelo. 2019. Plant growth-promoting rhizobacteria-induced defense against insect herbivores. In *Sustainable Management by PGPR*, eds. D. Maheshwari and S. Dheeman, 385–410. Cham: Springer.

El-Wakeil, N. and N. Gaafar. 2019. Predacious insects and their efficiency in suppressing insect pests. In *Cottage industry of biocontrol agents and their applications. Practical Aspects to Deal Biologically with Pests and Stresses Facing Strategic Crops*, eds. N. El-Wakeil, M. Saleh, and M. Abu-hashim, 133–155. Cham: Springer.

El-Wakeil, N., M. Saleh, and M. Abu-hashim. 2019. *Cottage Industry of Biocontrol Agents and Their Applications. Practical Aspects to Deal Biologically with Pests and Stresses Facing Strategic Crops*. Cham: Springer.

Facknath, S. and B. Lalljee. 2005. Effect of soil-applied complex fertilizer on an insect-host plant relationship: *Liriomyza trifolii* on *Solanum tuberosum*. *Entomologia Experimentalis et Applicata* 15: 67–77.

Follett, P. A., A. Kawabata, R. Nelson, G. Asmus, J. BUrt, K. Goschke, C. Ewing, J. Gaertner, E. Brill, and S. Geib. 2016. Predation by flat bark beetles (Coleoptera: Silvanidae and Laemophloeidae) on coffee berry borer (Coleoptera: Curculionidae) in Hawaii coffee. *Biological Control* 1010: 152–158. https://doi.org/10.1016/j.biocontrol.2016.07.002.

Gange, A. C., V. K. Brown, and D. M. Aplin. 2003. Multitrophic links between arbuscular mycorrhizal fungi and insect parasitoids. *Ecology Letters* 6: 1051–1055. https://doi.org/10.1046/j.1461-0248.2003.00540.x.

Gervassio, N. G. S., D. Aquino, C. Vallina, A. Biondi, and M. G. Luna. 2019. A re-examination of *Tuta absoluta* parasitoids in South America for optimized biological control. *Journal of Pest Science* 92: 1343–1357. https://doi.org/10.1007/s10340-018-01078-1.

Gholami, Z. and A. Sadeghi. 2016. Management strategies for western flower thrips in vegetable greenhouses in Iran: A review. *Plant Protection Science* 52: 87–98.

Glare, T. R., J.-L. Jurat-Fuentes, and M. O'Callaghan. 2017. Basic and applied research: Entomopathogenic bacteria. In *Microbial Control of Insect and Mite Pests: From Theory to Practice*, ed. L. A. Lacey, 47–67. New York: Academic Press. https://doi.org/10.1016/B978-0-12-803527-6.00004-4.

Grunseich, J. M., M. N. Thompson, N. M. Aguirre, and A. M. Helms. 2020. The role of plant-associated microbes in mediating host-plant selection by insect herbivores. *Plants* 9: 6. https://doi.org/10.3390/plants9010006.

Grzywacz, D. 2017. Basic and applied research: Baculovirus. In *Microbial Control of Insect and Mite Pests: From Theory to Practice*, ed. L. A. Lacey, 27–46. New York: Academic Press. https://doi.org/10.1016/B978-0-12-803527-6.00003-2.

Han, P., N. Desneux, C. Becker, R. Larbat, J. Le Bot, S. Adamowicz, J. Zhang, and A.-V. Lavoir. 2018. Bottom-up effects of irrigation, fertilization and plant resistance on *Tuta absoluta*: Implications for integrated pest management. *Journal of Pest Science* 92: 1359–1370. https://doi.org/10.1007/s10340-018-1066-x.

He, H.-M., L.-N. Liu, S. Munir, N. H. Bashir, Y. Wang, J. Yang, and C.-Y. Li. 2019. Crop diversity and pest management in sustainable agriculture. *Journal of Integrative Agriculture* 18: 1945–1952. https://doi.org/10.1016/S2095-3119(19)62689-4.

Jabran, K. 2019. Mulches for insect pest and disease management. In *Role of Mulching in Pest Management and Agricultural Sustainability. SpringerBriefs in Plant Science*, 27–32. Cham: Springer.

Karban, R. and Y. Chen. 2001. Induced resistance in rice against insects. *Bulletin of Entomological Research* 97: 327–335. https://doi.org/10.1017/S0007485307005056

Keeping, M. G., S. A. Mcfarlane, N. Sewpersad, and R. S. Rutherford. 2010. Effects of silicon and plant defence inducers on sugarcane yield parameters, *Eldana saccharina* Walker (Lepidoptera: Pyralidae) and thrips, *Fulmekiola serrata* Kobus (Thysanoptera: Thripidae). In Proceedings of the Annual Congress-South African Sugar Technologists' Association 83: 271–275.

Khorshidi, M., M. J. Hejazi, and S. Iranipour. 2017. Effect of azadirachtin, chlorantraniliprole and some insect growth regulators on vegetable leafminer, *Liriomyza sativae* (Blanchard) (Diptera: Agromyzidae). *Journal of Crop Protection* 6: 115–123.

Koch, R. L., O. S. Queiroz, R. C. Aita, E. W. Hodgson, B. D. Potter, T. Nyoike, and C. D. Ellers-Kirk. 2019. Efficacy of afidopyropen against soybean aphid (Hemiptera: APhididae) and toxicity to natural enemies. *Pest Management Science* 76: 375–383. https://doi.org/10.1002/ps.5525

Koivunen, M., L. Chanbusaraku, L. Fernández, R. Asolkar, E. Tan, D. Wallner, and P. Marrone. 2009. Development of a new microbial insecticide based on *Chromobacterium subtsugae*. *Insect Pathogens and Insect Parasitic Nematodes*, IOBC/WPRS Bulletin 45: 183–186.

Krisnawati, A., M. S. Y. I. Bayu, and M. M. Adie. 2017. Identification of soybean genotypes based on antixenosis and antibiosis to the armyworm (*Spodoptera litrua*). *Nusantara Bioscience* 9: 164–169. https://doi.org/10.13057/nusbiosci/n090210

Kumar, P. and H. M. Poehling. 2007. Effects of azadirachtin, abamectin, and spinosad on Sweetpotato whitefly (Homopter: Aleyrodidae) on tomato plants under laboratory and greenhouse conditions in the humid tropics. *Journal of Economic Entomology* 100: 411–420. https://doi.org/10.1093/jee/100.2.411

Lacey, L. A. 2017. *Microbial Control of Insect and Mite Pests: From Theory to Practice*. New York: Academic Press.

Lahiri, S., D. D. Reisig, F. P. F. Reay-Jones, J. K. Greene, T. E. Carter, Jr., R. Mian, and B. D. Fallen. 2020. Soybean host plant resistance to *Megacopta cribraria* (Hemiptera: Plataspidae) and the potential role of leaf trichome density. *Environmental Entomology* 49: 88–97. https://doi.org/10.1093/ee/nvz158

Lahiri, S., X. Ni, G. D. Buntin, S. Punnuri, A. Jacobson, F. P. F. Reay-Jones, and M. D. Toews. 2019. Combining host plant resistance and foliar insecticide application to manage *Melanaphis sacchari* (Hemiptera: Aphididae) in grain sorghum. *International Journal of Pest Management* 1–10. https://doi.org/10.1080/09670874.2019.1660830.

Landis, D., S. D. Wratten, and G. Gurr. 2000. Habitat management to conserve natural enemies of arthropod pests in agriculture. *Annual Review of Entomology* 45: 175–201. https://doi.org/10.1146/annurev.ento.45.1.175.

Lefebvre, V., N. Boissot, and J. L. Gallois. 2020. Host plant resistance to pests and pathogens, the genetic leverage in integrated pest and disease management. In *Integrated Pest and Disease Management in Greenhouse Crops. Plant Pathology in the 21st Century*, eds. M. Gullino, R. Albajes, and P. Nicot, 259–283. Cham: Springer.

Li, J., L. Huang, J. Zhang, J. A. Coulter, L. Li, and Y. Gan. 2019. Diversifying crop rotation improves system robustness. *Agronomy for Sustainable Development* 39: 38. https://doi.org/10.1007/s13593-019-0584-0.

Liu, J.-F., Z.-Q. Zhang, J. R. Beggs, E. Paderes, X. Zou, and X.-Y. Wei. 2020. Lethal and sublethal effects of entomopathogenic fungi on tomato/potato psyllid, *Bactericera cockerelli* (Šulc) (Hemiptera: Triozidae) in capsicum. *Crop Protection* 129: 10523. https://doi.org/10.1016/j.cropro.2019.105023.

Liu, Y., H. Wu, H. Chen, Y. Liu, J. He, H. Kang, Z. Sun, G. Pan, Q. Wang, J. Hu, F. Zhou, K. Zhou, X. Zheng, Y. Ren, L. Chen, Y. Wang, Z. Zhao, Q. Lin, F. Wu, X. Zhang, X. Guo, X. Cheng, L. Jiang, C. Wu, H. Wang, and J. Wan. 2014. A gene cluster encoding lectin receptor kinases confers broad-spectrum and durable insect resistance in rice. *Nature Biotechnology* 33: 301–305.

Lu, H.-P., T. Luo, H.-W, Fu, L. Wang, Y.-Y. Tan, J.-Z. Huang, Q. Wang, G.-Y. Ye, A. M. R. Gatehouse, Y.-G. Lou, and Q.-Y. Shu. 2018. Resistance of rice to insect pests mediated by suppression of serotonin biosynthesis. *Nature Plants* 4: 338–344.

Martin, P. A. W., A. D. S. Shropshire, D. E. Gundersen-Rindal, and M. B. Blakburn. 2007. *Chromobacterium subtsugae* sp. nov. for control of insect pests. U.S. Patent 7,244,607.

McCarville, M., M. O'Neal, and E. Hodgson. 2012. Sybean aphid-resistant soybean varieties for Iowa. Iowa State University of Science and Technology. https://lib.dr.iastate.edu/cgi/viewcontent.cgi?article=1090&context=extension_ag_pubs (Accessed March 23, 2020).

Messelink, G. J., J. Bennison, O. Alomar, B. L. Ingegno, L. Tavella, L. Shipp, E. Palevsky, F. L. Wäckers. 2014. Approaches to conserving natural enemy populations in greenhouse crops: current methods and future prospects. *BioControl* 59: 377–393. https://doi.org/10.1007/s10526-014-9579-6.

Momtax, M., K. Yeasmin, M. R. Khatun, and M. Ahmad. 2018. Impact of plant spacing on population dynamics of sucking pest of cotton. *Journal of Environmental Science and Natural Resources* 11: 241–243. https://doi.org/10.3329/jesnr.v11i1-2.43391.

Mordue, A. J. and A. Blackwell. 1993. Azadirachtin: An update. *Journal of Insect Physiology* 39: 903–924. https://doi.org/10.1016/0022-1910(93)90001-8.

Mourad, M. 2016. Recycling, recovering and preventing "food waste": competing solutions for food systems sustainability in the United States and France. *Journal of Cleaner Production* 126: 461–477. https://doi.org/10.1016/j.jclepro.2016.03.084.

Murrell, E. G. 2017. Can agricultural practices that mitigate or improve crop resilience to climate change also manage crop pests? *Current Opinion in Insect Science* 23: 81–88.

Norman, J. E., D. D. Quee, P. J. Kamanda, and A. E. Samura. 2019. Influence of plant spacing on insect population, growth and yield of okra in Sierra Leone. *American Journal of Entomology* 3: 49–55. https://doi.org/10.11648/j.aje.20190302.14

Oostendorp, M., W. Kunz, B. Dietrich, and T. Staub. 2001. Induced disease resistance in plants by chemicals. *European Journal of Plant Pathology* 107: 19–28.

Oso, A. A. and G. O. Awe. 2019. Effects of different water regimes and spacing on insect pest infestation and efficacy of control measures in cucumber (*Cucumis sativus*). *Journal of Experimental Agriculture International* 41: 1–8. https://doi.org/10.9734/jeai/2019/v41i530415.

Pieterse, C. M. J., C. Zamioudis, R. L. Berendsen, D. M. Weller, S.C.M. Van Wees, and P.A.H.M. Bakker. 2014. Induced systemic resistance by beneficial microbes. *Annual Review of Phytopathology* 52: 347–375.

Poschenrieder, C., R. Tolrà, and J. Barceló. 2006. Can metals defend plants against biotic stress? *Trends in Plant Science* 11: 288–295. https://doi.org/10.1016/j.tplants.2006.04.007.

Premachandra, D. W. T. S., C. Borgemeister, and H.-M. Poehling. 2005. Effects of neem and spinosad on *Ceratothripoides claratris* (Thysanoptera: Thripidae), an important vegetable pest in Thailand under laboratory and greenhouse conditions. *Journal of Economic Entomology* 98: 438–448. https://doi.org/10.1093/jee/98.2.438.

Quesada, C. R. and C. S. Sadof. 2019. Field evaluation of insecticides and application timing on natural enemies of selected armored and soft scales. *Biological Control* 133: 81–90. https://doi.org/10.1016/j.biocontrol.2019.03.013.

Razzak, M. A., D. R. Seal, P. A. Stansly, O. E. Liburd, and B. Schaffer. 2019. Host preference and plastic mulches for managing melon thrips (Thysanoptera: Thripidae) on field-grown vegetable crops. *Environmental Entomology*, 48: 434–443. https://doi.org/10.1093/ee/nvz010.

Regnault-Roger, C. 1997. The potential of botanical essential oils for insect pest control. *Integrated Pest Management Reviews* 2: 25–34. https://doi.org/10.1023/A:1018472227889.

Regnault-Roger, C., C. Vincent, and J. T. Arnason. 2012. Essential oils in insect control: low-risk products in a high-stakes world. *Annual Review of Entomology* 54: 405–424. https://doi.org/10.1146/annurev-ento-120710-100554.

Rendon, D. and V. M. Walton. 2019. Drip and overhead sprinkler irrigation in blueberry as acultural control for *Drosophila suzukii* (Diptera: Drosophilidae) in Northwestern United States. *Journal of Economic Entomology* 112: 745–752. https://doi.org/10.1093/jee/toy395.

Rice, R. E. and P. Kirsch. 1990. Mating disruption of oriental fruit moth in the United States. In *Behavior-Modifying Chemicals for Insect Management. Applications of Pheromones and Other Attractants*, eds. R. L. Ridgway, R. M. Silverstein, and M. N. Inscoe, 193–211. New York: Marcel Dekker, Inc.

Rizwan, M., B. Atta, A. M. Sabir, M. Yaqub, and A. Qadir. 2019. Evaluation of the entomopathogenic fungi as a non-traditional control of the rice leaf roller, *Cnaphalocrocis medinalis* (Guenee) (Lepidoptera: Pyralidae) under controlled conditions. *Egyptian Journal of Biological Pest Control* 29: 10. https://doi.org/10.1186/s41938-019-0111-2.

Rizzo, R., G. L. Verde, M. Sinacori, F. Maggi, L. Cappellacci, R. Petrelli, S. Vittori, M. R. Morshedloo, N. B. Y. Fofie, and G. Benelli. 2020. Developing green insecticides to manage olive fruit flies? Ingestion toxicity of four essential oils in protein baits on *Bactrocera oleae*. *Industrial Crops and Products* 143: 111884. https://doi.org/10.1016/j.indcrop.2019.111884.

Rodríguez, D., J. R. Cure, A. P. Gutierrez, and J. M. Cotes. 2017. A coffee agroecosystem model: III. Parasitoids of the coffee berry borer (*Hypothenemus hampei*). *Ecological Modelling* 363: 96–110. https://doi.org/10.1016/j.ecolmodel.2017.08.008.

Rodriguez-Saona, C. R., D. F. Polk, and J. D. Barry. 2009. Optimization of pheromone deployment for effective mating disruption of oriental beetle (Coleoptera: Scarabaeidae) in commercial blueberries. *Journal of Economic Entomology* 102: 659–669. https://doi.org/10.1603/029.102.0226.

Rousselin, A., M. H. Sauge, M. O. Jordan, G. Vercambre, F. Lescourret, and D. Bevacqua. 2016. Nitrogen and water supplies affect peach tree-green peach aphid interactions: the key role played by vegetative growth. *Agricultural and Forest Entomology* 18: 367–375.

Sarfraz, M., A. B. Keddie, and L. M. Dosdall. 2007. Biological control of the diamondback moth, *Plutella xylostella*: A review. *Biocontrol Science and Technology* 15: 763–789. https://doi.org/10.1080/09583150500136956.

Schleier III, J. J. and R. K. D Peterson. 2011. Pyrethrins and pyrethroid insecticides. In *Green Trends in Insect Control*, eds. O. López and J. G. Fernández-Bolaños, 94–131. Cambridge: Royal Society of Chemistry.

Schoonejans, T. and M. Van der Staaij. 2001. Spinosad, a new tool for insect control in vegetables cultivated in greenhouses. *Mededelingen (Rijksuniversiteit te Gent. Fakulteit van de Landbouwkundige en Toegepaste Biologische Wetenschappen)* 66: 375–386.

Sequeira, L. 1983. Mechanisms of induced resistance in plants. *Annual Review of Microbiology* 37: 51–79.

Shapiro-Ilan, D., S. Hazir, and I. Glazer. 2017. Basic and applied research: entomopathogenic nematodes. In *Microbial Control of Insect and Mite Pests: From Theory to Practice*, ed. L. A. Lacey, 91–105. New York: Academic Press. https://doi.org/10.1016/B978-0-12-803527-6.00006-8.

Shorey, H. H., C. B. Sisk, and R. G. Gerber. 1995. Disruption of pheromone communication in *Platynota stultana* (Lepidoptera: Tortricidae) in grape vineyards. *Environmental Entomology* 24: 1270–1274. https://doi.org/10.1093/ee/24.5.1270.

Shu, B. J. Zhang, G. Cui, R. Sun, X. Yi, and G. Zhong. 2018. Azadirachtin affects the growth of *Spodoptera litura* Fabricius by inducing apoptosis in larval midgut. *Frontiers in Physiology* 9: 137. https://doi.org/10.3389/fphys.2018.00137.

Silva, J. P. G. F., E. L. L. Baldin, E. S. Souza, V. F. Canassa, and A. L. Lourenção. 2013. Characterization of antibiosis to the redbanded stink bug *Piezodorus guildinii* (Hemiptera: Pentatomidae) in soybean entries. *Journal of Pest Science* 86: 649–657.

Sithanantham, S., M. Prabakaran, and P. Narayanasamy. 2020. Potential of silica sources including fly ash as green technology inputs to induce resistance to biotic and abiotic stresses in crop plants: Overview. In *Circular Economy and Fly Ash Management*, eds. S. K. Ghosh and V. Kumar, 135–144. Singapore: Springer.

Song, Y. Y., M. Ye, C. Y. Li, R. L. Wang, X. C. Wei, S. M. Luo, and R. S. Zeng. 2013. Defense in tomato by arbuscular mycorrhizal fungus and involvement of the jasmonate pathway. *Journal of Chemical Ecology* 39: 1036–1044.

SongHe, X., L. LiNa, B. Xue, G. MeiLan, and R. Qin. 2019. Effect of potassium application and aphid feeding-induced salicylic acid on insect resistance in potato. *Acta Entomologica Sinica* 62: 343–350.

Sorensen, K. A., S. Mohankumar, and S. R. Thangaraj. 2016. Physical, mechanical and cultural control of vegetable insects. In *Integrated pest management of tropical vegetable crops*, eds. R. Muniappan and E. Heinrichs, 131–148. Dordrecht: Springer.

Souza, B., L. Vázquez, and R. Marucci. 2019. *Natural Enemies of Insect Pests in Neotropical Agroecosystems. Biological Control and Functional Biodiversity*. Cham: Springer.

Sparks, T. C., G. D. Thompson, H. A. Kirst, M. B. Herlein, J. S. Mynderse, J. R. Turner, and T. V. Worden. 1999. Fermentation-derived insect control agents. In *Biopesticides: Use and Delivery. Methods in Biotechnology, vol. 5*, ed. F. R. Hall and J. J. Menn, 171–188. Totowa: Humana Press.

Stepanycheva, E., M. Petrova, T. Chermenskaya, and R. Pavela. 2019. Fumigant effect of essential oils on mortality and fertility of thrips *Frankliniella occidentalis* Perg. *Environmental Science and Pollution Research* 26: 30885–30892. https://doi.org/10.1007/s11356-019-06239-y.

Stout, M. J., K. V. Workman, R. M. Bostock, and S. S. Duffey. 1998. Stimulation and attenuation of induced resistance by elicitors and inhibitors of chemical induction in tomato (*Lycopersicon esculentum*) foliage. *Entomologia Experimentalis et Applicata* 86: 267–279.

Takikawa, Y., K. Kakutani, Y. Matsuda, T. Nonomura, S. Kusakari, and H. Toyoda. 2019. A promising physical pest-control system demonstrated in a greenhouse equipped with simple electrostatic devices that excluded all insect pests. *Preprints* https://doi.org/10.20944/preprints201905.0256.v1.

Takikawa, Y., Y. Matsuda, K. Kakutani, T. Nonomura, S.-I. Kusakari, K. Okada, J. Kimbara, K. Osamura, and H. Toyoda. 2015. Electrostatic insect sweeper for eliminating whitelies colonizing host plants: a complementary pest control device in an electric field screen-guarded greenhouse. *Insects* 6: 442–454. https://doi.org/10.3390/insects6020442.

Todd, J. C., M. A. R. Mian, E. A. Backus, J. J. Finer, and M. G. Redinbaugh. 2016. Feeding behavior of soybean aphid (Hemiptera: Aphididae) biotype 2 on resistant and susceptible soybean. *Journal of Economic Entomology* 109: 426–433. http://dx.doi.org/10.1093/jee/tov315.

Toledo, P. F. S., L. O. V. Jumbo, S. M. Rezende, K. Haddi, B. A. Silva, T. S. Mello, T. M. C. D. Lucia, R. W. S. Aguiar, G. Smagghe, and E. E. Oliveira. 2020. Disentangling the ecotoxicological selectivity of clove essential oil against aphids and non-target ladybeetles. *Science of The Total Environment* 718: 137328. https://doi.org/10.1016/j.scitotenv.2020.137328.

Trimble, R. M., D. J. Pree, and N. J. Carter. 2001. Integrated control of oriental fruit moth (Lepidoptera: Tortricidae) in peach orchards using insecticide and mating disruption. *Journal of Economic Entomology* 94: 476–485. https://doi.org/10.1603/0022-0493-94.2.476.

Tu, J.-C., Y.-L. Lee, and M.-Y. Wei. 2018. Analysis and research on the key success factors of marketing ugly fruits and vegetables. *Sustainability* 10: 2783. https://doi.org/10.3390/su10082783.

Vannette, R. L. and M. D. Hunter. 2009. Mycorrhizal fungi as mediators of defence against insect pests in agricultural systems. *Agricultural and Forest Entomology* 11: 351–358. https://doi.org/10.1111/j.1461-9563.2009.00445.x.

Vasanthi, N., L. M. Saleena, and S. A. Raj. 2014. Silicon in crop production and crop protection: A review. *Agricultural Reviews* 35: 14–23.

Vinay, S. and A. K. Sood. 2017. Plant nutrition: A tool for the management of hemipteran insect pests: A review. *Agricultural Reviews* 38: 260–270. https://doi.org/10.18805/ag.R-1637.

Wang, D., W.-C. Zhang, and X.-X. Ren. 2009 Determination of toxicity of spinosad against vegetable pests. *Agrochemicals Research and Application* 1.

Welter, S. C., C. Pickel, J. G. Millar, F. Cave, R. V. Steenwyk, and J. Dunley. 2005. Pheromone mating disruption offers selective management options for key pests. *California Agriculture* 59: 16–22. https://doi.org/10.3733/ca.v059n01p16.

Wenninger, E. J. and A. L. Averill. 2006. Mating disruption of oriental beetle (Coleoptera: Scarabaeidae) in cranberry using retrievable, point-source dispensers of sex pheromone. *Environmental Entomology* 35: 458–464. https://doi.org/10.1603/0046-225X-35.2.458.

Wraight, S. P. and M. E. Ramos. 2017. Characterization of the synergistic interaction between *Beauveria bassiana* strain GHA and *Bacillus thuringiensis morrisoni* strain tenebrionis applied against Colorado potato beetle larvae. *Journal of Invertebrate Pathology* 144: 47–57. https://doi.org/10.1016/j.jip.2017.01.007.

Yang, M., W. He, Y. Li, J. Qiu, F. Liu, R. Li, and Y. Qiu. 2019. Effects of low levels of nitrogen or phosphorus provided in hydroponic culture on brown planthopper feeding and survival. *International Journal of Pest Management* 1–10. https://doi.org/10.1080/09670874.2019.1682210.

Zhu, P., X. Zheng, H. Xu, A. C. Johnson, K. L. Heong, G. M. Gurr, and Z. Lu. 2020a. Nitrogen fertilization of rice plants improves ecological fitness of an entomophagous predator but dampens its impact on prey, the rice brown planthopper, *Nilaparvata lugens*. *Journal of Pest Science* 93: 747–755. https://doi.org/10.1007/s10340-019-01174-w.

Zhu, P., X. Zheng, H. Xu, A. C. Johnson, K. L. Heong, G. M. Gurr, and Z. Lu. 2020b. Nitrogen fertilizer promotes the rice pest *Nilaparvata lugens* via impaired natural enemy, *Anagrus flaveolus*, performance. *Journal of Pest Science* 93: 757–766. https://doi.org/10.1007/s10340-019-01177-7.

Zohry, A. and S. Ouda. 2018. Crop rotation defeats pests and weeds. In *Crop Rotation: An Approach to Secure Future Food*, eds. S. Ouda, A. E. Zohry, and T. Noreldin, 77–88. Springer: Cham.

46

Integrated Pest Management (IPM): Concepts, Practices, and Advances Focused on Organic Crops

Katherine Girón Pérez

CONTENTS

46.1 Introduction

Agriculture is a human invention to feed the world; consequently, biotic and abiotic challenges need to be overcome to feed the increasing world population and preserve the environment concomitantly. Crop management strategies need to be grouped to attend market demands, so innovation and science components play a determinant role. After the Second World War, agricultural systems in Europe and Asia were severely disrupted, so diverse strategies to increase production should be developed (Kogan and Bajwa, 1999). Intensive agriculture began to develop as one of the stronger alternatives to feed the world. Characterized by getting high yields per area unit, managing biotical stresses by using pesticides is highly efficient, and these pesticides were adopted without considering environment care or prior pest-level monitoring. Negative impacts on nature resulted from DDT, a chemical substance used in the 1950s, until today classified as harmful and ecologically adverse to mammals, humans, and the environment (Metcalf, 1986). Since then, pest control strategies have been developed based on sustainability and environmentally friendly tools in agriculture and urban places. Over the years, a more conservative production system has been spread and adopted, the organic agriculture, characterized for be sustainable, not polluting and environmentally respectful, but in many cases distant from supply the growing food demand when compared with conventional agriculture, dividing opinions between scientific community and society about the effectiveness and its potential to migrate from conventional farmland to an organic system. In this chapter, I compile scientific facts about integrated pest management (IPM) advances and the evidence about the challenges to be overcome by inserting IPM principles into the organic scenario.

46.2 Principles and Evolution of Integrated Pest Management

Integrated pest management evolved from an integrated control concept created in the 1950s by a group of entomologists from the University of California–Riverside, emphasizing the selective use of chemicals and conserving natural enemies in the ecosystem. Then, in the 1960s, the green revolution was rising, and integrated control concept migrated to pest management and finally to integrated pest management (IPM). As its name suggests, the fundamental objective of IPS is to integrate different management tactics, using low residual pesticides, narrow organism spectrum, diverse mode of action delaying resistance evolution and also the risk of secondary pest outbreak and highly effective against target control, considering economic injury threshold, crop and pest knowledge (habitat, life cycle, seasonality, alternative hosts, etc.). The main goal of IPM is to maintain harmony and complementation among different strategies, keeping pest populations (insects, weeds, disease pathogens, nematodes) below economic damage levels since a unique technique is not enough to manage them. According to the FAO, IPM is the preferred approach to crop protection of both sustainable intensification of crop production and pesticide risk reduction.

46.3 Conventional versus Organic Integrated Pest Management Scenarios

Conventional or intensive agriculture still occupies a predominant place in food production and has a strong preponderance level in farmlands worldwide, focused on increasing crop yield warranting food and fuel demands. Since in most of cases markets and consumers do not tolerate any visual injury caused by pests, not only preventive but curative actions (most of them synthetic-based) should be taken frequently to reduce the pest population drastically, sometimes before reaching economic damage. In this way, conventional agriculture may induce biodiversity loss, massive soil erosion and degradation, eutrophication and oceanic dead zones, pesticide effects on humans and wildlife, greenhouse gas emissions, and regime shifts in hydrological cycling (Gordon et al., 2008; Gil et al., 2012; Ponisio et al., 2015).

Recently, governments have strengthened politics about organic farming, resulting in organics areas increases. At the end of 2017, 69.8 million hectares were organically managed, representing 20%.

Organic principles suggest that not all IPM components could be inserted into an organic production system (i.e., synthetic fertilizers and chemical pesticides, genetically modified organisms) and efforts have been made to include compatible tools in many studies and examples about IPM tools to be applied in organic systems. In the case of pathogens, Bruggen et al. (2015) listed the general practices to manage pathogens in organic farming making a comparison against conventional farming. These approaches suggest that general guidance is available, but some strategies should be done according to crop, season, farmland location, pest complexes, policies, and regulations.

Biodiversity conservation is probably one of the most important components in organic agriculture, indirectly if biodiversity is conserved, its habitat should be preserved and the environment surrounding, so ecofriendly integrated pest management strategies are welcome. Biopesticides are an alternative that offers an ecologically effective solution to pest management in organic farming (Gupta and Dikshit, 2010). Chandler et al. (2011) defined them as a mass-produced agent manufactured from a living microorganism or a natural product to control plant pests. Three major biopesticides groups are classified: Microbial, biochemicals (secondary metabolites extracts), and semiochemicals (sex pheromones) (Gupta and Dikshit, 2010); all of them are less threatening to the environment and human health than previous pesticides, degrade quickly, and are compatible with biodiversity conservation (Chandler et al., 2011). The US Environmental Protection Agency (EPA) also classifies some transgenes as biopesticides, but in organic farming, genetically modified organisms (GMOs) are banished (FAO).

Another aspect to be considered is how pest populations react to management practices. In insects, natural populations can overcome any control tools, not only synthetic tools but those considered harmless to the environment: crop rotation (Weisberger et al., 2019), biological control (Carton et al., 2008), and cultural practices (Onstad, 2008) thus add the pest resistance monitoring into IPM. In corn, the coleopteran *Diabrotica barbieri* evolved resistance to crop rotation that involves planting the same primary crop every second year in the same location by extending egg diapause for more than one winter (Spencer et al., 2014).

46.4 Does Integrated Pest Management Fit in Organic Farming?

IPM was defined by Stenberg (2017) as a holistic "approach" or "strategy" to combat plant pests and diseases using all available methods while minimizing applications of chemical pesticides, a valuable effort to reduce ecosystems disbalance in humans and wildlife security. From this point of view, a "holistic approach or strategy" makes it necessary to consider as many detectable factors as possible to preserve the ecosystem.

Wyss et al. (2005) proposed a four-level IPM model for organic crops. The first level includes cultural practices like crop rotations, resistant varieties, and soil enhancement. In the second level, habitat management takes an important place, reflected in diversity increase and pest antagonists through the incorporation of natural alternative hosts into or nearby the crop (Pfiffnerand Wyss, 2003). Third- and fourth-level programs include the incorporation of direct measures such as biocontrol agents and also insecticides approved for organic crops. According to the authors, if the strategies for pest prevention implemented in the first two levels, the need for direct control measures will be obviated.

However, as Dara (2019) mentioned, IPM is not a strict and uniform principle to be applied but a philosophy that can guide the practitioner to use it as appropriate for their situation. In some places, host plant resistance is effectively used in some crops with pest and disease resistant or tolerant varieties, but not in other crops because pest strains or populations susceptibility can differ between locations. In other cases, practitioners do not have access to resistant cultivars. Pheromones are widely used for mating disruption, mass trapping, or monitoring of certain lepidopteran and coleopteran pests, but not for several others pests.

From a holistic view, some practices seem to be the best solution for crop management. Indeed they are part of a big puzzle where pieces have to fit into the production system. For example, crop rotation is recommended for weeds management, but what about the other biotic factors?

Weisberger et al. (2019) based their conclusions on 298 paired observations from 54 studies across six continents. They realized that more important than simple rotation or multiple rotation system is the management given: mechanical intervention, altered crop sowing density or row spacing configurations, or the selection of more competitive crop genotypes, sowing dates, and reduced soil tillage or zero-tillage condition are characteristics with a great capacity to reduce weeds populations (Gaba et al., 2014; Cordeaou et al., 2017; Westwood et al., 2018). Multicropping and crop rotations substantially reduce the yield gap between organic and conventional farming (Ponisio et al., 2015). But for nematodes management, crop rotation is considered a limited strategy against those species with cosmopolitan host ranges, such as *Meloidogyne incognita*, which may parasitize up to 3,000 plant species (Abad et al., 2003). That example shows that previous knowledge of the environment, crops, and pest complex interactions is a key factor, is part of the holistic strategy.

The most remarkable challenge in IPM and its new approaches is aware of the balance that must exist between conservation and food production: production needs to be sustainable, while conservation cannot be totally at the expense of crop yield; in less productive agricultural landscapes, biodiversity benefit can be gained by concentrating organic farms into hotspots without a commensurate reduction in yield (Gabriel et al., 2013). According to recent meta-analysis, yield gap between both production system range between 19–25%, depending on the crop (Seufert et al., 2012; Ponisio et al., 2015), this scenario may reduce the interest in adopting or make a transition to organic farming. In the United States, a list of approved products for organic crops is available in the USDA's national organic program.*

46.5 IPM Advances and Tools in Organic Farming

46.5.1 Plant Breeding

Breeding goals converge at aiming for higher productivity, incorporation of resistance or tolerance to biotic and abiotic factors, and higher resource-use efficiency (water, nutrients, light, etc.), but in organic farming plant breeding is restricted to crossing methods that do not break the reproductive barriers between species. Since GMOs are banished from organic systems, Ryffel (2012) proposed a new gene manipulation approach to be used in organic farms, organic plants contain genes that are naturally exchanged between species or are sterile to avoid outcrossing if they received a transgene from a different species, but until today is not implemented or accepted as a pest management tool in organic farming.

Local adaptation may be more important for organic agriculture as the resource recycling and quality of the inputs that are used can vary from region to region, although more than 95% of organic production is still based on crop varieties that were bred for the conventional sector and consequently lack important traits required under organic and low-input production conditions (Lammerts van Bueren et al., 2011). Crespo-Herrera and Ortiz (2015) identified some inputs to be investigated in order to develop resistant cultivars within organic farming regulations: (i) broad multilocation testing to better exploit Genotype × Environment; (ii) shuttle breeding implementation between organic farming conventional farming to develop adapted cultivars; (iii) screening of plant materials deposited in gene-banks to identify useful genetic resources; (iv) evaluate the possibility to implement prediction of germplasm performance using genotyping platforms and phenotypic information derived from multilocation testing; and (v) evaluate the use of cultivar mixtures to deploy host plant resistance or increase resilience to abiotic stress.

Murphy et al. (2007) demonstrated that highest yielding wheat genotypes in conventional systems were not the highest yielding genotypes in organic systems, suggesting that increasing yield in organic systems through breeding will require direct selection within organic systems rather than indirect selection in conventional systems. The conventional farming trend is to use nearly genetically identical plants allowing insect pests and pathogens to move easily from plant to plant and decimate crop fields. Still, studies confirm that increasing plant diversity in agricultural fields will reduce pest abundance and damage, as demonstrated for rusts and powdery mildews of small grain crops (Garret and Mundt, 1999; Tooker and Frank, 2007).

* www.ams.usda.gov/about-ams/programs-offices/national-organic-program

46.5.2 Semiochemicals

Pheromones and other semiochemicals are extensively used to manage agricultural stored products pest, are important in pests monitoring, mass trapping, mating disruption, attract-and-kill, and push-pull strategies, harmless and pest-specific. Recently, semiochemicals have been discovered to be involved in the interspecific chemical communications between plants, insect pests, and parasitoids can also aid in the successful dispersal of entomopathogenic microbials (Sharma et al., 2019).

46.5.3 Entromopathogenics

Most commercial products employ Ascomycetes in the genera *Metarhizium* and *Beauveria*. However, their utilization has been limited by inconsistent field results due to sensitivity to abiotic stresses and naturally low virulence. Other naturally occurring biocontrol agents also face these hurdles to a successful application. Still, the availability of complete genomes and recombinant DNA technologies have facilitated the design of multiple fungal pathogens with enhanced virulence and stress resistance. These biological insecticides are environmentally friendly and cost-effective insect pest control options (Lovett and Leger, 2018).

There is now substantial evidence that some endophytic fungal entomopathogens, particularly *Beauveria bassiana* and *Lecanicillium* spp. (formerly *Verticillium lecanii*), may also demonstrate antagonistic activity against plant pathogens in addition to their well-known biocontrol activity against insect pests (Jaber and Ownley, 2018). This suggests that these entomopathogens have a promising potential to be developed as biopesticides for multiple purposes in IPM strategies (Goettel et al., 2008).

46.6 Conclusion

As an expanding food production alternative, organic farming must face challenges in several external and internal areas: political laws that strengthen and contribute to environmental protection and social acceptance, land use and agricultural practices transition pursuing sustainability, yield increases, and competitive prices to feed the world. At the same time, a macro scenario is interconnected. Organic farming internal practices are also. Within a changing world, agricultural systems become dynamics and integrated pest management refers to the importance of building strong but flexible systems, as plunging ecosystems knowledge consequence in order to adopt most organic permitted practices, converging in resource protection, local practices, and food production where scientific studies are essential.

REFERENCES

Abad P, Favery B, Rosso MN, Castagnone-Sereno P. 2003. Root-knot nematode parasitism and host response: Molecular basis of a sophisticated interaction. *Molecular Plant Pathology*, *4*: 217–224.

Buainain AM, Garcia JR. 2019. Agriculture and the environment. In: *Agricultural Development in Brazil: The Rise of a Global Agro-food Power.* Routledge: Taylor & Francis Group, pp. 139–146.

Carson R. 1962. *Silent Spring.* Boston: Houghton Mifflin.

Carton Y, Poirié M, Nappi A. 2008. Insect immune resistance to parasitoids. *Insect Science, 15*: 67–87.

Chandler D, Bailey AS, Tatchell GM, Davidson G, Greaves J, Grant, WP. 2011. The development, regulation and use of biopesticides for integrated pest management. *Philosophical Transactions of the Royal Society B, 366*: 1987–1998. doi:10.1098/rstb.2010.0390.

Clark MM, Gwinn KD, Ownley BH, 2006. Biological control of *Pythium myriotylum. Phytopathology*, 96, S25.

Cordeau S, Smith RG, Gallandt ER, Brown B, Salon P, DiTommaso A. 2017. Timing of tillage as a driver of weed communities. *Weed Science, 65*(4): 504–514.

CrespoHerrera L, Ortiz R. 2015. Plant breeding for organic agriculture: Something new? *Agriculture and Food Security, 4*(25): 1–7. doi: 10.1186/s40066-015-0045-1.

Gaba S, Fried G, Kazakou E, Chauvel B, Navas ML. 2014. Agroecological weed control using a functional approach: A review of cropping systems diversity. *Agronomy for Sustainable Development, 34*(1): 103–119.

Gill RJ, Ramos-Rodriguez O, Raine NE. 2012. Combined pesticide exposure severely affects individual- and colony-level traits in bees. *Nature, 491*: 105–108.

Goettel MS, Koike M, Kim JJ, Aiuchi D, Shinya R, Brodeur J. 2008. Potential of *Lecanicillium* spp. for management of insects, nematodes and plant diseases. *Journal of Invertebrate Pathology, 98*: 256–261.

Gordon LJ, Peterson GD, Bennett EM. 2008. Agricultural modifications of hydrological flows create ecological surprises. *Trends in Ecology Evolution, 23*: 211–219.

Gupta S, Dikshit AK. 2010. Biopesticides: An ecofriendly approach for pest control. *Journal of Biopesticides, 3*(Special Issue): 186–188.

Jaber LR, Ownley BH. 2018. Can we use entomopathogenic fungi as endophytes for dual biological control of insect pests and plant pathogens? *Biological Control, 116*: 36–45.

Kogan M, Bajwa WI. 1999. Integrated pest management: A global reality? *Anais da Sociedade Entomológica do Brasil, 1*(28): 1–25.

Le Campion A, Oury FX, Heumez E, Rolland B.2019. Conventional versus organic farming systems: dissecting comparisons to improve cereal organic breeding strategies. *Organic Agriculture, 10*: 1–12.

Lovett B, St. Leger RJ. 2018. Genetically engineering better fungal biopesticides. *Pest Management Science, 74*: 781–789.

Metcalf RL. 1986. The ecology of insecticides and the chemical control of insects. In: *Ecological theory and integrated pest management practice.* New York: Wiley, pp. 251–297 (362 p).

Murphy KM, Campbell KG, Lyon SR, Jones SS. 2007. Evidence of varietal adaptation to organic farming systems. *Field Crops Research, 3*(102): 172–177.

Onstad DW. 2008. Major issues in insects resistance management. In: *Insect Resistance Management.* A.P Elsevier, pp. 1–16.

Pfiffner L, Wyss E. 2003. Use of sown wildflower strips to enhance natural enemies of agricultural pests. In: Gurr GM, Wratten SD, Altieri MA (eds.) *Ecological Engineering for Pest Management: Advances in Habitat Manipulation for Arthropods.* Collingwood, Australia: CSIRO, pp. 165–186.

Ponisio LC, M'Gonigle LK, Mace KC, Palomino J, de Valpine P, Kremen C. 2015. Diversification practices reduce organic to conventional yield gap. *Proceedings of the Royal Society B: Biological Sciences*, 282: 20141396.

Ryffel, GU. 2012. Orgenic plants: Gene-manipulated plants compatible with organic farming. *Biotechnology Journal, 11*: 1328–1331.

Sharma A, Sandhi RK, Reddy GVP. 2019. A review of interactions between insect biological control agents and semiochemicals. *Insects, 10*: 1–16.

Spencer J, Hughson SA, Levine E. 2014. Insect resistance to crop rotation. In: *Insect Resistance Management* (Second Edition), Cambridge, MA: Academic Press, pp. 233–278.

Weisberger D, Nichols V, Liebman M. 2019. Does diversifying crop rotations suppress weeds? A meta-analysis. *PLoS ONE, 7*(14): e0219847. https://doi.org/10.1371/journal.pone.0219847.

Westwood JH, Charudattan R, Duke SO, Fennimore SA, Marrone P, Slaughter DC, Swanton C, Zollinger R. 2018.Weed management in 2050: Perspectives on the future of weed science. *Weed Science, 66*: 275–285.

Wyss E, Luka H, Pfiffner L, Schlatter C, Uehlinger G, Daniel C. 2005. Approaches to pest management in organic agriculture: a case study in European apple orchards. Paper presented at a Symposium Entitled "IPM in Organic Systems", XXII International Congress of Entomology, Brisbane, Australia, 16 August 2005, 33–36.

47

Integrated Pest Management against Arthropod Pests of Cannabis sativa *in the United States of America*

Matthew Russell Gates

CONTENTS

47.1 Introduction

In order to supplement the ebullient and burgeoning interest of *Cannabis* cultivation, the following brief chapter on various pests and pathogens known to infest *Cannabis* crops has been composed. Many of these organisms are common pests in other agricultural contexts, and there are many more that exist which are not displayed in this collection.

47.2 Arthropoda

Arthropods, and particularly the recently proposed Cenocondyla (Aria 2019) are organisms with an exoskeletonized body, encompassing mites, spiders, centipedes, crustaceans, insects, and related groups. Many micro-arthropods are very common pests both among foliage (phyllosphere), and soil (pedosphere) and may be hard to observe in small populations early without knowing the symptoms of their presence and where they tend to exist.

47.3 Tetranychidae (Spider Mites)

The spider mite family Tetranychidae contains many common pest species, and are so named because the vast majority are able to spin silklike spiders, both hailing from the group arachnomorpha, and use it to create nests that protect themselves from predation as well as to catch wind and travel passively by air currents as aeroplankton. The most common spider mite encountered in *Cannabis* is also thought to be the most common spider mite affecting many other crops: *Tetranych usurticae*, the two-spotted spider mite.

Tetranychus urticae is the most commonly encountered spider mite, adapted to feeding on over 1,000 species of plants, owed to its highly adaptive genome which can process many different potentially hazardous defense compounds (Grbić et al., 2011, Bensoussan et al., 2018). Adults are ~400 micometers and often called the two-spotted spider mite because of the two significantly sized caeca, or intestinal pouches, visible from the dorsal side. Caeca may also disappear if they are empty as when feeding has stopped. Additionally, *Tetranychus urticae* comes in two main color forms: red, and green, and can alternatively appear orange and even pink. This variability in coloration can be due to differences in carotenoid biosynthesis, feeding behavior, and reactions to the environment such as during diapause.

Other spider mites are known to have similarly visible caeca or coloration such as *Tetranychus mcdanieli* and *Tetranychus marianae*, which makes true identification technically very difficult or virtually impossible from color and markings alone in the field (Ros and Breeuwer, 2007). However, certain contextual information can increase the likelihood that an encountered tetranychid is the two-spotted spider mite. For example, it is the most prolifically polyphagous tetranychid both in terms of hosts species documented and overall detection rate, and much research has been devoted to it specifically (Ros and Breeuwer, 2007). Knowing the host species allows the identifier to cross-reference with documented hosts of other species.

Development is difficult to generalize since it is mediated through complex mite-plant interactions, but on three different cultivars of for example Hyacinth Bean, *Lablab purpureus*, life table parameters such as eggs/female, gross reproductive rate, and longevity were highest on Kartica and lowest on IPSA Seam 1 (Ros and Breeuwer, 2007). Individuals were documented to develop over a span of ~9 days at 25 degrees Celsius, but over time, familiarization with a population of hosts may induce preference and increased adaptation with prolonged interaction (Magowski et al., 2003). *Tetranychus urticae* disperses by active walking or by passive transport in the wind, on plants, tools, or people. Diapause is induced by short day length, lack of food supply, and low temperature, and is normally terminated by a fixed period of lower temperatures. Gravid females seek a protected niche at the end of summer. Diapausing adults are often orange/red in color as a consequence of carotenoid biosynthesis (Altincicek et al., 2012).

Phytoseiulus persimilis and *Neoseiulus californicus* predatory mites are often used both separately and in tandem to control spider mite populations, with the former being a Type Ia specialist of *Tetranychus urticae* and the latter being a Type II specialist that consumes several other pest species (McMurtry et al., 2013). Two-spotted spider mites are known to be resistant to several commercial agricultural chemistries like Clofentazine and Hexythiazox due to their adaptive genome which processes xenobiotics exceedingly well (Grbić et al., 2011). There is some efficacy of *Beauveria bassiana* documented against spider mites (Wu et al., 2016). *Bryobia praetiosa*, one of a few species known as a clover mite, is an agricultural pest of many grasses, including agriculturally important crop species (Zhang et al., 2016). They have also been recorded on both coniferous and deciduous trees like cedar and apple, low-growing herbaceous plants like iris and strawberry, as well as gooseberry, onion, and walnut (Station, 1959). Despite being a spider mite in the family Tetranychidae, it does not produce silk, and its larger size of ~700 micometers in length distinguishes it from other spider mites. Similar to many predatory mites, the first two legs are much longer than the others, used to probe the clover mite's surroundings.

47.4 Eriophyidae (Russet Mites)

The russet mite family Eriophyidae are generally innocuous, and the vast majority of the more than 3000 known species exist on their specialized hosts without causing much noticeable damage or symptomatic response (de Lillo et al., 2018). The pestiferous species however can be very deleterious to the health of their host, and are known as russet, rust, gall, and blister mites, due to the various sorts of damage and structures their infestation can produce. In *Cannabis*, however, only one species is known: *Aculops cannabicola*, the hemp russet mite (Petanovic et al., 2007).

The hemp russet mite, *Aculops cannabicola*, is one of the most deleterious pests in *Cannabis* cultivation, due to its microscopic size, fast reproductive rate, and ability to travel passively on air currents as aeroplankton (Petanovic et al., 2007). Unlike some other more generalist mite pests, the hemp russet mite is not documented on other hosts besides *Cannabis*, in keeping with the typical specialist trend in Eriophyid mites in general, although it is considered closely related to *Aculops lycopersici*, the tomato russet mite in the same genus.

Management of eriophyid mites is challenging even under the less-restrictive treatment protocols allowed for other crops, like *Aloe vera* (Villavicencio et al., 2014). For this reason, *Cannabis* cultivation spaces often put additional effort into preventative measures, especially where physical barriers can be employed, as well as a comprehensive crop scouting procedure in order to minimize response time and maximize treatment efficacy. Preventative measures often include the establishment of commercially available predatory mites and/or disruptive spray applications at regular intervals. Curative measures must be aggressive and quick when a population is detected and may include securely bagging and removing whole individual plants if they are heavily infested, in order to quickly reduce the pest population. Due to recent prohibition, academic literature on biocontrol agents effective against *Aculops cannabicola* specifically is lacking, so literature related to similar species and anecdotal information from commercial entities is often relied upon, but this information will be supplemented over time as new research becomes available.

47.5 Aleyrodidae (Whiteflies)

The silverleaf whitefly, *Bemisia tabaci*, is not a single species, but a complex of many physically indistinguishable whitefly species. It has a multiple-decades-long international history of agricultural documentation starting around the mid-1970s and is perhaps most famous for these myriad species that have been ascribed to its designation, particularly the former B-biotype, which was once redescribed as *Bemisia argentifolii*, and the former insecticide-resistant Q-biotype which is now referred to as the Mediterranean group (Vyskočilová et al., 2019). More recently, it has been proposed that what encompasses "*Bemisia tabaci*" is actually at least 44 morphologically indistinguishable species. In such case, in-crop identification is impossible visually and muddles historical bionomic data that does not take this into account—which would be the vast majority (Vyskočilová et al., 2019).

Bemisia tabaci has several very important details associated with its pest status. For example, it is a known supervector of more than 150 plant viruses, with a large amount existing in three separate genera, of three separate viral families: Begomoviruses of the family Geminiviridae, Criniviruses of family Closeroviridae, and Ipomoviruses of family Potyviridae. Some of the most destructive examples include Tomato Yellow Leaf Curl Virus, which can be devastating to tomato crops while symptomless in other hosts, and the African Cassava Mosaic Virus, both of which are in *Begomoviruses.* The most relevant virus vectored for *Cannabis* is the typically incurable *Lettuce Chlorosis Virus* a *Crinivirus* which has been documented to be transmitted in *Cannabis* by the silverleaf whitefly (Duffus et al., 1996). Like most whiteflies in the group Sternorrhyncha, they produce copious amounts of honeydew over their life span, which is deposited onto leaves and can be substrate for sooty molds, especially in dense populations. Apart from causing cosmetic damage, the mold growth blocks photosynthesis if it develops on lower leaves, which is common. They are also phoretic hosts for broad mites, *Polyphagotarsonemus latus* (Soroker et al., 2003), which sometimes seek and attach themselves to whitefly bodies, which transport them greater distances than they would otherwise travel ambulatorily. Additionally, the common name "silverleaf" whitefly comes from the special silvering damage that used to be associated with certain populations like what is now known as *Bemisia argentifolii*, when they feed on potato and nightshade among others. "Argentifolii" literally means, "silver leaves" in Latin.

Bearing in mind the fact that many of the biotypes of *Bemisia tabaci* are really separate species, but also conceding that their physiological similarity gives them broadly similar strengths and weaknesses, recent research data about various whiteflies can still be generally useful. Whiteflies are polyphagous and feed on over 900 documented species (Soroker et al., 2003), many of which being important human crops. Since populations of *Bemisia tabaci* tend to feed almost exclusively on the underside of leaves and do well in dense, humid foliage, an unnoticed population is able to grow quickly if unchecked. *Bemisia tabaci* and others in its complex have four nymphal instars, a pupal form, and then a final adult form. After hatching from eggs, what are referred to as "crawlers" will move across the plant and find a suitable place to feed, wherein they will stay in the same place until they eclose as an adult. Adults often lay eggs in a ring-like pattern, and the process repeats. Different species seem to have variable success at interbreeding, but such assessments are inconclusive in their entirety (Soroker et al., 2003).

Biocontrol agents like *Beauveria bassiana* are used to great effect against the larvae primarily, colonizing and killing the bodies and multiplying itself in the cultivation space, provided it is humid enough for successful sporulation, in turn killing other nearby larvae, and sometimes, adults as well. The spores are released in the cultivation space and will start the colonization process when they come into contact with the next suitable host. This entomopathogenic fungus can be very effective as well for stubborn and dense populations. The biocontrol agent known as the whitefly predatory beetle *Delphastus catalinae* is used to control large populations but is very sensitive to insecticides and fungicides. Parasitic wasps like *Eretmocerus eremicus* are available commercially to manage several whitefly species, including *Bemisia tabaci. Amblyseius swirskii* and *Neoseiulus cucumeris* mites can have a good suppressant effect but might not eradicate populations completely. Removing habitually highly infested plants can be a viable strategy.

47.6 Thysanoptera (Thrips)

Most of the thousands of documented thrips species from the family Thysanoptera are not very pestiferous, but there is a small ratio that intensely are. Perhaps the most common is *Frankliniella occidentalis*, the western flower thrips, and it is unsurprisingly also found on *Cannabis* crops as well as its closest relative, hops. Other thrips species have been documented to feed on *Cannabis* at minimum, like the onion thrips, *Thrips tabaci* and tobacco thrips, *Frankliniella fusca* in eastern Colorado, United States of America (Cranshaw et al., 2019).

Life cycle development is influenced greatly by host plant and temperature. For instance, a 1995 evaluation of western flower thrips development under different temperatures clearly showed how much a change of just a few degrees can have (van Rijn et al., 2009). The difference between 12 degrees Celsius and 15 degrees Celsius can be the difference between first instar larval development taking over a month and taking about half a month on cucumber. An increase of five more degrees cuts developmental time to larvae in half, with the developmental sweet spot residing between 20 and 30 degrees Celsius, wherein larvae can be expected in as little as five days or so post-oviposition. Temperatures higher than 30 degrees Celsius start to have a severe deleterious effect on population growth and survival, with complete instantaneous mortality occurring around 35 degrees Celsius.

In a 2007 study that aimed to understand different host plant development times at a constant 25 degrees Celsius, documented western flower thrips can reach adulthood in as little as 9 days on cucumber, or approximately 13 days on tomato (Dong et al., 2019). An individual female thrips could produce about 21 females in its lifespan on cucumber and cabbage, about 11 on bean, 7 on tomato, but only 1 or so on *Capsicum* peppers. This matches up with longevity rates, which were about 16 days on cabbage, but only 8 on capsicum, implying that the latter is a lower-quality food source than the former. There was another 2003 study that documented the developmental changes of cucumber-fed western flower thrips with various supplemental pollen at 25 degrees Celsius. Cucumber leaf with no pollen had thrips producing about 32 females per female (Cao et al., 2019). The pollen that affected net reproduction rate the least was the pollen of the common cattail *Typha latifolia*. Pollen from *Pinus sylvestris*, the Scotch pine, increased the production of females the most, with females producing a staggering 141 females per female—interestingly, this pollen is toxic to the predatory mite *Neoseiulus cucumeris*, which is used for control of western flower thrips. The takeaway from this study is that access to pollen can have a substantially positive supplementary effect for western flower thrips.

At 20 degrees Celsius, western flower thrips can complete their lifecycle in ~22 days: ~6 days as eggs, ~8 days as larvae, ~3 days as pupae, and ~3 days to become adults. Western flower thrips can be found feeding on the flowers and the leaves of plants, causing cosmetic damage. Western flower thrips adults can be found between the petals of flowers, where their presence causes petals to become scarred, wilted, or wrinkled. They can be dislodged by turning a flower downwards and knocking the back of the flower with force. This is one way to check for the presence of western flower thrips.

Biopesticides such as *Beauveria bassiana* can have a suppressing effect on several thrip populations. Biocontrol agents like the predatory mites *Amblyseius swirskii* and *Amblyseius cucumeris* are popular controls for western flower thrips. In addition to this, nematodes such as *Steinernema feltiae* can be applied as an additional controlling force (Mouden et al., 2017).

REFERENCES

Altincicek, Boran, Jennifer L. Kovacs, and Nicole M. Gerardo. 2012. "Horizontally transferred fungal carotenoid genes in the two-spotted spider mite *Tetranychus urticae*." *Biology Letters* 8 (2):253–257. doi: 10.1098/rsbl.2011.0704.

Aria, Cédric. 2019. "Reviewing the bases for a nomenclatural uniformization of the highest taxonomic levels in arthropods." *Geological Magazine* 156 (8):1463–1468. doi: 10.1017/S0016756819000475.

Bensoussan, Nicolas, Vladimir Zhurov, Sota Yamakawa, Caroline H. O'Neil, Takeshi Suzuki, Miodrag Grbić, and Vojislava Grbić. 2018. "The digestive system of the two-spotted spider mite, *tetranychus urticae* koch, in the context of the mite-plant interaction." *Frontiers in Plant Science* 9:1206.

Cao, Yu, Hong Yang, Jun Li, Guozhou Zhang, Yawei Wang, Can Li, and Yulin Gao. 2019. "Population development of *Frankliniella occidentalis* and *Thrips hawaiiensis* in constant and fluctuating temperatures." *Journal of Applied Entomology* 143 (1–2):49–57. doi: 10.1111/jen.12571.

Cranshaw, Whitney, Melissa Schreiner, Kadie Britt, Thomas P. Kuhar, John McPartland, and Jerome Grant. 2019. "Developing insect pest management systems for hemp in the United States: A work in progress." *Journal of Integrated Pest Management*, (2019) 10 (1): 26; 1–10. doi: 10.1093/jipm/pmz023.

de Lillo, Enrico, Alberto Pozzebon, Domenico Valenzano, and Carlo Duso. 2018. "An intimate relationship between eriophyoid mites and their host plants: A review." *Frontiers in Plant Science* 9:1786–1786. doi: 10.3389/fpls.2018.01786.

Dong, Fan, Nan Zhang, Zhijuan Xie, Xiangkun Meng, Kun Qian, Caihong Ji, Mingxing Lu, Yuzhou Du, and Jianjun Wang. 2019. "Characterization and in vitro expression of arginine kinase gene in the invasive western flower thrips, *Frankliniella occidentalis*." *Comparative Biochemistry and Physiology Part B: Biochemistry and Molecular Biology*. 229:51–57. doi: 10.1016/j.cbpb.2019.01.003.

Duffus, James E., Hsing Yeh Liu, Gail C. Wisler, and Ruhui Li. 1996. "Lettuce chlorosis virus —A new whitefly-transmitted closterovirus." *European Journal of Plant Pathology* 102 (6):591–596. doi: 10.1007/BF01877027.

Magowski, Wojciech, Martijn Egas, Jan Bruin, and Maurice W. Sabelis. 2003. "Intraspecific variation in induction of feeding preference and performance in a herbivorous mite." *Experimental & Applied Acarology* 29 (1):13–25. doi: 10.1023/A:1024268306409.

McMurtry, James A., Gilberto J. De Moraes, and Nazer Famah Sourassou. 2013. "Revision of the lifestyles of phytoseiid mites (Acari: Phytoseiidae) and implications for biological control strategies." *Systematic and Applied Acarology 18* (4): 24.– doi: 10.11158/saa.18.4.1.

Miodrag Grbić, Thomas Van Leeuwen, Richard M. Clark, Stephane Rombauts, Pierre Rouzé, Vojislava Grbić, Edward J. Osborne, Wannes Dermauw, Phuong Cao Thi Ngoc, Félix Ortego, Pedro Hernández-Crespo, Isabel Diaz, Manuel Martinez, Maria Navajas, Élio Sucena, Sara Magalhães, Lisa Nagy, Ryan M. Pace, Sergej Djuranović, Guy Smagghe, Masatoshi Iga, Olivier Christiaens, Jan A. Veenstra, John Ewer, Rodrigo Mancilla Villalobos, Jeffrey L. Hutter, Stephen D. Hudson, Marisela Velez, Soojin V. Yi, Jia Zeng, Andre Pires-daSilva, Fernando Roch, Marc Cazaux, Marie Navarro, Vladimir Zhurov, Gustavo Acevedo, Anica Bjelica, Jeffrey A. Fawcett, Eric Bonnet, Cindy Martens, Guy Baele, Lothar Wissler, Aminael Sanchez-Rodriguez, Luc Tirry, Catherine Blais, Kristof Demeestere, Stefan R. Henz, T. Ryan Gregory, Johannes Mathieu, Lou Verdon, Laurent Farinelli, Jeremy Schmutz, Erika Lindquist, René Feyereisen & Yves Van de Peer. 2011. "The genome of *Tetranychus urticae* reveals herbivorous pest adaptations." *Nature* 479 (7374):487–92. doi: 10.1038/nature10640.

Mouden, Sanae, Kryss Facun Sarmiento, Peter Gl Klinkhamer, and Kirsten A. Leiss. 2017. "Integrated pest management in western flower thrips: Past, present and future." *Pest Management Science* 73 (5):813–822. doi: 10.1002/ps.4531.

Petanović, Radmila, Biljana Magud, and Dragica Smiljanić. 2007. "The hemp russet mite *Aculops cannabicola* (Farkas, 1960) (Acari: Eriophyoidea) found on Cannabis sativa L. in Serbia: Supplement to the description." *Archives of Biological Sciences* 59 (1):81–85.

Ros, Vera ID, and Johannes AJ Breeuwer. 2007. "Spider mite (Acari: Tetranychidae) mitochondrial COI phylogeny reviewed: Host plant relationships, phylogeography, reproductive parasites and barcoding." *Experimental and Applied Acarology* 42 (4):239–262. doi: 10.1007/s10493-007-9092-z.

Skirvin, David J., and Michael De Courcy Williams. 1999. "Differential effects of plant species on a mite pest (Tetranychus utricae) and its predator (Phytoseiulus persimilis): implications for biological control." *Experimental and Applied Acarology* 23 (6):497–512. doi: 10.1023/a:1006150521031.

Soroker, Victoria, Dennis R. Nelson, Ornit Bahar, Saadia Reneh, Sarah Yablonski, and Eric Palevsky. 2003. "Whitefly wax as a cue for phoresy in the broad mite, Polyphagotarsonemus latus(Acari: Tarsonemidae)." *Chemoecology* 13 (4):163–168. doi: 10.1007/s00049-003-0243-3.

Station, Connecticut Agricultural Experiment. 1959. *Plant Pest Handbook: A Compendium Based on Findings of Research in Plant Sciences for Three-Quarters of a Century*: Connecticut Agricultural Experiment Station: New Haven, CT.

van Rijn, Paul CJ, Chris Mollema, and Greet M. Steenhuis-Broers. 2009. "Comparative life history studies of *Frankliniella occidentalis* and *Thrips tabaci* (Thysanoptera: Thripidae) on cucumber." *Bulletin of Entomological Research* 85 (2):285–297. doi: 10.1017/S0007485300034386.

Villavicencio, LE, JA Bethke, B Dahlke, B Vander Mey, and L Corkidi. 2014. "Curative and preventive control of *Aceria aloinis* (Acari: Eriophyidae) in Southern California." *Journal of Economic Entomology* 107 (6):2088–2094. doi: 10.1603/EC14234.

Vyskočilová, Soňa, Susan Seal, and John Colvin. 2019. "Relative polyphagy of 'Mediterranean' cryptic Bemisia tabaci whitefly species and global pest status implications." *Journal of Pest Science* 92 (3):1071–1088. doi: 10.1007/s10340-019-01113-9.

Weeks, A. R., and J. A. Breeuwer. 2001. "Wolbachia-induced parthenogenesis in a genus of phytophagous mites." *Proceedings of the Royal Society of London. Series B: Biological Sciences* 268 (1482):2245–2251. doi: 10.1098/rspb.2001.1797.

Wu, Shengyong, Haicui Xie, Maoye Li, Xuenong Xu, and Zhongren Lei. 2016. "Highly virulent *Beauveria bassiana* strains against the two-spotted spider mite, *Tetranychus urticae*, show no pathogenicity against five phytoseiid mite species." *Experimental and Applied Acarology* 70 (4):421–435. doi: 10.1007/s10493-016-0090-x.

Zhang, Yan-Kai, Ya-Ting Chen, Kun Yang, and Xiao-Yue Hong. 2016. "A review of prevalence and phylogeny of the bacterial symbiont Cardinium in mites (subclass: Acari)." *Systematic and Applied Acarology* 21 (7):978–990. –doi: 10.11158/saa.21.7.11.

48

Recent Advances in Integrated Insect Pest Management Strategies in Morocco

Elaini Rachid

CONTENTS

48.1 Overview of Moroccan Policy for Production and Exportation of Vegetables

The total production of tomato and vegetables in Morocco was 2.3 million tons in the 2018 crop season, showing an evolution of 36% compared to 2008. More than 83% (1,179 million tons) of Moroccan exportations was originated from the Souss region to the EU and Russia. The value generated by these crops was more than US$1.04 billion in the same season (Figure 48.1). Agriculture is considered to be a key sector for socioeconomic development. Moroccan agriculture is split into a modern industrial sector, which primarily produces food for export, and smallholdings, which produce food for local markets. Farm size, technical level, and market destination (domestic or export) are by far the factors influencing the adoption of integrated pest management (IPM).

48.2 Current Status of Vegetable Key Pests

48.2.1 Whiteflies

The most important whiteflies (Homoptera: Aleyrodidae) in Morocco are *Bemisia tabaci* and *Trialeurodes vaporariorum*. The direct symptom caused by sucking sap from the phloem of host plants is the excretion of honeydew, which contains sugars sucked as well as their metabolites, which are qualified as ideal substrates for the development of sooty mold *Capnodium* spp. (Hendrix et al., 1992). Indeed, the most serious damage caused by *B. tabaci* is its role as a vector of more than 200 viruses (Polston et al., 2014). Currently, *B. tabaci* in Morocco is a key pest to be managed either in protected crops and open field since it is a vector of TYLCV and New Delhi TYLCV. Its control is based mainly on biocontrol (*Nesidiocoris tenuis*), chemical control, virus-resistant cultivars, and physical exclusion by improving tightness of greenhouses (Bouharroud, 2007). However, resistance to some commonly used insecticides was already reported (Bouharroud et al., 2006)

48.2.2 Tomato Leafminer

T. absoluta (Lepidoptera: Gelechiidae) represents a global threat to commercial tomato production, both in open field and greenhouse. Native to South America, it spread over the Mediterranean Basin, Europe, Africa, and part of Asia in 12 years (Ferracini et al., 2019).

In Morocco, *T. absoluta* was first reported in 2008, and 100% damage was observed in the following season in some farms. The

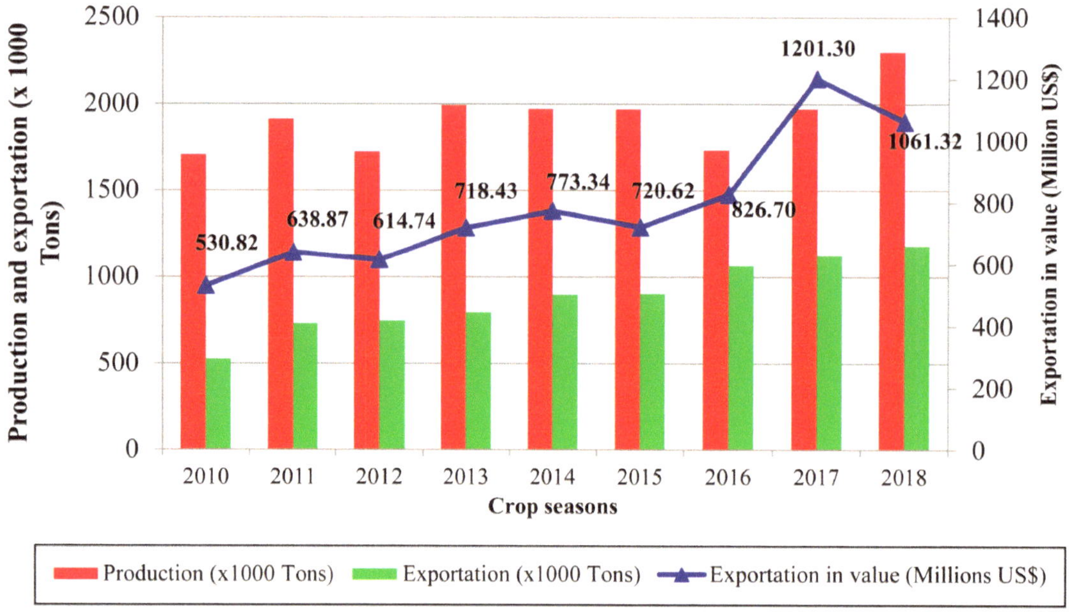

FIGURE 48.1 Evolution of tomato/vegetable production and exportations from the Souss valley of Morocco.

feeding habits of *T. absoluta* make its presence difficult to detect in the early infestation stage, resulting in severe damage to young plants. The cost related to control solely this pest was ranged from 19% to 29% of total costs in Morocco (Bouharroud, 2015). Nevertheless, after three years, the incidence of *T. absoluta* has become less severe thanks to several factors contributing to this decline, such as the increase in growers' knowledge of pest behavior/biology and the correct application of integrated pest control strategies (Urbaneja et al., 2012).

The use of "soft" pesticides was the predominant method deployed by farmers to control the pest during the first two years. However, *T. absoluta* is not easily manageable by using pesticides for two reasons. First, the larval stage completes its development inside the leaves (mesophyll). Therefore, larvae are not easily exposed to insecticides (Ayalew, 2015). The second challenge is the resistance ability of insects toward different pesticides used.

Nowadays, *T. absoluta* is managed basically in Morocco by inundative releases of Biological Control Agents (BCA) combined with the use of synthetic and biological pesticides and mating disruption. In addition, implementing and integrating other approaches, such as prophylactic and cultural practices, use of pheromone-based traps for early detection are commonly adopted. Monitoring and/or mass trapping have improved concretely the management of this pest. Thus, the number of sprays to control *T. absoluta* in the Souss valley of Morocco was dropped from 42, including 10 *Bacillus thuringiensis* (Bt)-based products to 14, including 12 Bt per tomato crop cycle under IPM strategy compared to a conventional approach. Currently, under a good installation of the predatory mirid bug *N. tenuis* and suitable conditions for its development, the number of chemical sprays is less than four, including two Bt.

To reach such results, the control strategy begins before transplanting by seedling procurement from certified nurseries that supply pest-free material. Delta-type traps with pheromone provide valuable information about the presence of the pest and its potential risk. Traps need to be counted and maintained regularly. Pheromone lures have a shelf life of four to eight weeks depending on climate conditions. The male mass trapping (water traps) using pheromone results in an imbalance of the sex ratio, which impacts the mating pattern of the pest.

For economic reasons, growers usually start with a low number of traps (10–20 per hectare) and when the pest is abundant they increase the number to reach 40/Ha. Recently, sticky rolls with the pheromone gradually released from the adhesive layer are also available in the market with three film colors: clear that are recommended in greenhouses containing BCA, yellow to be used to catch *T. absoluta* and other pests and black sticky roll that are more specific to *T. absoluta*.

The mating disruption technique adopted in 2019, aims to create sexual confusion of males by saturating the atmosphere with a female pheromone that prevents the pest mating and, consequently, reduces the population. The dispensers are applied at an application rate of 800-1000/ha at the beginning or few days before plantation.

In tomato greenhouses, inoculative releases of *N. tenuis* (Hemiptera: Miridae) are usually conducted several weeks after transplanting. *N. tenuis* was reported to be an effective predator of pests of tomato crops including *T. absoluta* in the Mediterranean area. The growers adopted the integration of *N. tenuis* on their *T. absoluta* control strategy readily as they had already enough proficiency using this predator to control whiteflies. Over time, polyphagy has been considered an advantage because it allows the control of more than one pest and different pest stages. It also allows the early establishment of the predator in the crop when the target pest is still at low densities, and helps to maintain a resident population of the natural enemy.

N. tenuis is introduced three to four weeks after transplanting at an average rate of 0.75 to 2 individuals/m². In order to increase chances of mating to speed up its establishment, it is recommended to release *N. tenuis* in spots of approximately 50 individuals on a leaf instead of spreading it over all the plants.

As development of *N. tenuis* is influenced by temperature and photoperiod it is recommended to increase the rate of *N. tenuis* to 2 individuals/m² for late transplantations.

As opposed to the Almeria region in Spain, where the release of *N. tenuis* in the nursery is common, a minority of Moroccan growers transplant tomato on which *N. tenuis* individuals have already laid eggs in the nursery. In fact, five to seven days prior to transplanting, a dose of 0.5 *N. tenuis* individuals per plant are released in the nursery, with *Ephestia kuehniella* eggs as an alternative prey. Thanks to this technique, the population of *N. tenuis* can rapidly reach the density required to control *T. absoluta* in a good distribution in the greenhouse. Despite this advantage, there are some practical issues to manage, e.g., the multitude of contributors (growers, nursery, BCA suppliers), which implies tight coordination. Also, the early release of *N. tenuis* limits the number of active ingredients that may be used to control other pests/diseases (side effects). In addition, if the number of *N. tenuis* is too high, it is possible to cause damage by feeding directly on plants, causing necrotic rings on the stem, shoots, leaf petioles, and flower stalks, which then affects plant growth and crop yield. The integration of biopesticides with inoculative releases of *N. tenuis* has resulted in very satisfactory control of *T. absoluta*. Azadarchtina, *Bt*, and Spinosad showed acceptable compatibility to this predator.

48.2.3 Two-Spotted Mite

The two-spotted spider mite, *Tetranychus urticae* (Acari: Tetranychidae), is a polyphagous pest that is distributed worldwide. This pest can cause significant yield losses in many agricultural crops, including fruits, cotton, vegetables, and ornamentals (>1000 species). Direct plant damage is due to feeding punctures. This mite pierces the epidermis and feeds on the contents of mesophyll cells that results in chlorosis, owing to a decrease in total chlorophyll content and loss of photosynthetic capacity. Adaptive strategies of spider mites are based on high fecundity of females and a female-biased off-spring sex ratio, which can lead to a rapid population increase. This mite is considered a key pest in Morocco for tomato, raspberries, and other vegetables. The strategy adopted to control *T. urticae* was based mainly on accurate surveillance since it usually starts by fireplaces. Then, biological or chemical control is adopted depending on the farmers' convictions and market requirements, but the majority use sulfur as a dusting to control this mite.

48.2.4 Tomato Russet Mite

Tomato russet mite *Aculops lycopersici* (Acari: Eriophyidae) symptoms can be observed as coppery yellow then brown coloring on various parts of the plant, slowing down of the vegetation, wilting and drying of the leaves, and finally the premature death of the plants. The attack comes usually from the bottom and extends toward the apex. The mite lives directly on the epidermal cells that depletes. The affected fruits remain small, with hard and cracked skin. The most important host plants are tomato, potato, and beans. In Morocco, this mite is considered to be a serious pest that can damage the crops in both protected and open fields. The control of this mite is usually based on the early detection of the spot infestation then the chemical control based mainly on sulfur.

48.3 Control Management of Key Pests in Morocco

Based on the damages and control strategies adopted to control key pests in Morocco, Table 48.1 summarizes the importance of these key pests on each crop.

48.3.1 Chemical Control

The National Plant Protection Organization in Morocco named "Office National de Sécurité Sanitaires des Produits Alimentaires" abbreviated ONSSA. By law (Article 2, loi 25-08), the main mission among others of this organism is inspection and proceed with the registration of all plant protection products including non-toxic and natural products (Table 48.2).

The examination of data collected in the field and the analyses in laboratories making it possible to draw up agricultural alerts bulletins thus providing farmers with the technical information necessary to better manage the protection of their crops and to be able to act duly. The Moroccan NPPO provides also technical support and expertise in the implementation of new approaches to control harmful new pests/diseases, namely integrated pest management, biocontrol, and good agricultural practices. Also, demonstration trials of pest control are conducted annually and workshops/FFS are organized in collaboration with national extension services (ONCA) for farmers to help them better reason their interventions to control pests.

TABLE 48.1

Importance of Key Pests on Different Crops in Morocco

Pests	Tomato	Pepper	Strawberry	Blueberry	Raspberry
B. tabaci	+++++	+++++			
T. vaporariorum	++++	++++			
T. absoluta	+++++				
F. occidentalis		+++++	++++	+++	+++
A. gossypii		++++	+++	+++++	
M. persicae		++++	+++	+++++	
A. lycopersici	+++++				
T. urticae	+++	++++	+++		++++

TABLE 48.2

Biopesticide Active Ingredients Registered in Morocco (January 2020)

Active Ingredient	Concentration	Dose
Azadirachtine	31.95 g/l	1 l/ha
Bacillus thuringiensis subsp, aizawai	15000 UI	500 g/ha
Bacillus thuringiensis var. Kurstaki EG2348	32000 UI/mg	0.5 kg/ha
*Beauveria bassiana*strain GHA	(11.3%) (2 × 10^{10}cfu/ml)	125 cc/hl
Beauveria bassiana	7.16%	1 l/ha
Clove oil/Cottonseed oil	25%/45%	1 l/ha
Capsicum oleoresin extract/Garlic extract/Soya oil	7.60% w/w/23.40%/59.30%	100 cc/hl
Helicoverpa armigera nucleopolyhedro virus	7.5 × 10^{12}/l	200 cc/ha
Petroleum oil	80%	2 l/hl
Vaseline oil	817 g/l	1 l/hl
Paraffinic mineral oil	78%	1 l/hl
Maltodextrine	27%	1.6 l/hl
Paecilomyces fumosoroseus strain FE 9901	(18%) (2 × 10^9 cfu/g)	200 g/hl
Potassium fatty acid salts (C7-C18)	1%	300 cc/hl
Spinetoram	120 g/l	25 cc/hl
Spinosade	480 g/l	75 cc/hl
Spodoptera littoralis nucleopolyhedro virus	5 × 10^{11}/l	200 cc/ha
D-limonène	6% [95%]	0.4 l/hl

48.3.2 Cultural Practices

Several cultural practices have been adopted in Morocco to control pests of protected crops. Generally, and in the case of mites in particular, cultural practices are very useful since they contribute to reducing the population of these formidable pests. By order of importance, the following measures are adopted:

1. Avoid water stress.
2. Avoid excess nitrogen.
3. Plant windbreaks to reduce wind speed and dust deposition in open field.
4. Mulch to control weeds and thrips (break life cycle development in the soil).
5. Weed part of the plots while keeping a section for the activity of natural enemies.

The plantation dates need to be chosen to avoid periods of high heat and overlapping of host crops. In the Souss valley region, this measure proved to be difficult since the exporters aim to export from October, and are led to plant in summer at the end of neighboring crop cycles. The most adopted practice is to create a crawl space in order to break the life cycle of pests. Suppressing sources of inoculum is an important component to the success of an integrated pest management program. In fact, the most less cost cultural practice adopted is the use of resistant and grafted varieties. These varieties limit the application of insecticides especially TYLCV resistance. Soil solarization is not common in Morocco since this tool is time cost, at least four weeks, which is very important for the precocity of production.

48.3.3 Physical/Mechanical Control

It has been found that a physical barrier is the easiest way to exclude pests from crops. The physical control using insect nets became necessary only after the introduction of the TYLC virus in 1998 (Bouharroud, 2007). Anti-UV insect nets with special mesh sizes are frequently used in Morocco in protected crops. The insect nets have considerably reduced pesticide applications since producers did not have to spray to control pests. For organic crops, the use of these nets prevents the migration of natural enemies, usually released for pest control (Hanafi et al., 2007). In the Souss region, known for its windy and dusty climate, the maintenance of insect nets is a common practice that farmers plan to keep their nets in good condition. Wind and dust work by blocking the mesh, reducing greenhouse ventilation, and increasing the temperature inside the greenhouse. Repairing net tears is a daily concern of farmers in the Souss region.

As a practice to control *T. absoluta*, the farmers remove the infested tomato leaves at the beginning of infestation, then, burn it or bury it in the ground.

48.4 Biocontrol Agent Regulation in Morocco

The importation and release of BCA in Morocco are subject to regulatory measures reflected by the instruction of the Minister of Agriculture, Public Works and the Environment No. 2730-97 of 31 October 1997. Annex 1 in this instruction includes a list of BCA exempted from prior importation authorization issued by the phytosanitary authorities. This list includes 78 agents that some are commonly used in Morocco, such as *A. colemani*, *Amblyseius cucumeris*, *Eretmocerus eremicus*, *Orius laevigatus*, and *Phytoseiulus persimilis* (Table 48.3).

Other species not included in annex 1 are subject to technical import authorization. In this case the importer has to prepare a file for presentation to the authority, which mainly includes a clear identification of the importer, details related to the BCA (identification, packaging, substrate, exporter references, point of entry, quantity, etc.), advice on efficient use, environment

TABLE 48.3

Primary Natural Enemies Currently Used to Control Key Pests in Morocco

Pests	*B. tabaci* *T. vaporariorum*	*T. absoluta*	*T. tabaci* *F. occidentalis*	*A. gossypii* *M. persicae*	*T. urticae*
BCA	*N. tenuis* *E. eremicus* *A. swirskii*	*N. tenuis*	*Orius laevigatus* *A. swirskii* *A. cucumeris*	*A. colemani* *A. matricariae*	*P. persimilis*

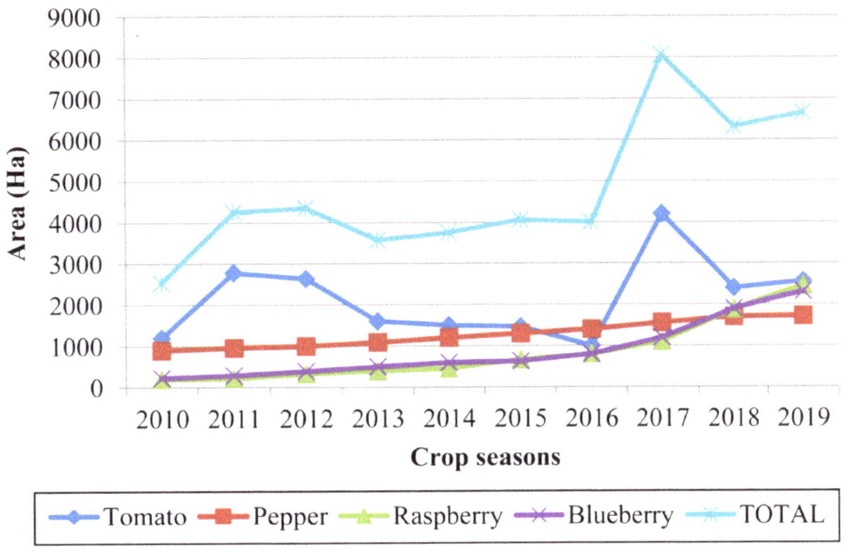

FIGURE 48.2 Evolution of crop area under biocontrol in Morocco.

compatibility, and guidelines for use. The consequent impact of this strategy was observed on the increase of the areas conducted under biocontrol (Figure 48.2).

48.5 Side Effects of Plant Protection Products on Natural Enemies

The success of released or naturally occurring BCA in preventing pest outbreaks in protected crops has led the greenhouse industry to be particularly conscious of the necessity of applying selective pesticides (Blümel et al., 1999). Thus, it is crucial to take into consideration the understanding of the effects of pesticides on natural enemies and how to mitigate those effects through the choice of the mode of action, timing, or location of pesticide application.

The effect of pesticides on natural enemies was evaluated by measuring mortality rates (using LC_{50} statistics) 24–48 h following topical application or exposure to residues. However, for other pesticides, natural enemies may experience more sublethal effects, such as reduced fecundity or male-biased sex ratios, which are more difficult to predict. A demographic approach makes it possible to combine both acute and sublethal effects (survivorship, development rate, fecundity, sex ratio, etc.) associated with exposure to pesticides into a single index, which in turn makes it possible to compare the consequences of pesticide application on population growth in both pest and natural enemy populations (Stavrinides and Mills, 2009).

In Morocco, the use of bumblebees has gone through rapid development, and now 100% of tomato greenhouses are pollinated using bumblebees, which are imported or produced locally by five private companies. By using bumblebees, farmers were constrained to use less pesticide with minimum side effects. Hence bumblebees have played a major role in the development of IPP and opened the way to biocontrol in tomato greenhouses in Morocco (Hanafi and El-Fadl, 2004). The use of BCA has also contributed to improving farmers' awareness regarding the choice of compatible pesticides. Recently, more biopesticides and selective pesticides have been registered in Morocco, contributing significantly to the success of IPM.

48.6 Challenges and Constraints

48.6.1 Complexity versus Simplicity

In Moroccan circumstances, although substantial efforts to implement IPM have been made, pesticides continue to be predominantly used because of their ease of use and their prompt efficacy. The method is easily understood; it is effective (at least in the short term), reduces labor costs, and is considered as a practice that the farmer could control and decide upon independently of suppliers, advisors, or extension officers. This finding is more perceptible after the introduction of new pests resulting in panic reactions resulting in extensive chemical control programs, which upsets ongoing IPM programs. The introduction of TYLC virus in 1998 and *T. absoluta* in 2008 and its impact on the increased need for pesticide applications are two considerable examples. Furthermore, IPM philosophy recognizes that the optimal choice of management options in a given situation depends on a thorough knowledge of the biology and ecology of the pest involved, including its life history, population dynamics, and the form as well consequences of its interactions with other components of its environment. Thus, the effective implementation of IPM is heavily knowledge-dependent that is arduous for growers.

48.6.2 Direct Cost Approach

From an economic point of view, the adoption of alternative tactics, such as pesticide rotation/diversification or using selective insecticide, meet with grower reluctance if the grower is not convinced it will improve. Implementation of IPM requires potential input costs, as well as the acceptance of economic risk and uncertainty that comes with decisions made while juggling a multitude of complexities. The motivation to simplify and/or reduce the risk of economic loss by sticking with conventional, fast-acting, "proven" tactics (chemical control) is often very strong. In addition, items needed to implement IPM such as traps, pheromones, BCA, etc., financial incentives are not given to growers especially when we know that several IPM-adapted inputs are more expensive than the basic ones.

48.6.3 One Pest Management Approach

While chemical control gives the possibility to manage each pest separately, IPM implies the simultaneous control of multiple pests, their monitoring, and the use of thresholds or economic injury levels to apply pesticides. In fact, various pests must be monitored, different control methods must be integrated and situation-specific prescriptions must be applied. Integration of BCA, require a degree of knowledge and sophistication much greater than pesticide technology. During decision making, the grower is thinking holistically, because he/she is managing the entire farm at once and must make timely decisions on all aspects of crop production that each has ripple effects on other aspects of the operation.

Therefore, IPM is considered highly time-consuming, complex, and needs close supervision of crops which urges growers to opt for one by one pest and disease management.

48.6.4 Extension Lack

One of the challenges is the difficulty in getting the IPM program accepted by growers probably because of their lack of familiarity with the idea of biocontrol, but also from cultural inertia and the associated difficulty of introducing new concepts to a relatively conservative people. The demand has increased for well-trained human resources who are able to integrate multifaceted controls in a rigorous IPM program. Hence, extension services need to be strengthened. However, the efforts made by the relevant authorities are not enough to make up the deficit. Thus, the private sector, including suppliers, provides assistance with farmers' technical requirements, and is more involved. Independent advisors support growers to monitor pests and BCA through trained objective observers, who are committed to providing recommendations based on the suitable grower benefits. For an efficient IPM program, greenhouse workers have to be trained to recognize nutrient deficiencies, disease, and insect problems, and to take appropriate decisions. Workers have the opportunity to visit every plant of the greenhouse on several occasions (pruning, harvesting, etc.). Their observations could be of great help in early detection of spot infestations. Personal protective gear, disinfectants, disposal bins, markers, etc., have to be made available to workers so that they can play their role in an IPP program. It is simply the cheapest way of monitoring the health of each plant.

Acknowledgments

We greatly appreciate the support of Said Bahouch for his critical and scientific review.

REFERENCES

Ayalew, G. 2015. Efficacy of selected insecticides against the South American tomato moth, *Tuta absoluta* Meyrick (Lepidoptera: Gelechiidae) on tomato in the Central Rift Valley of Ethiopia. *African Entomology* 23: 410–417.

Blümel, S., Matthews G.A., Grinstein A. and Elad Y. 1999. Pesticides in IPM: Selectivity, side-effects, application and resistance problems. In *Integrated Pest and Disease Management in Greenhouse Crops*, Springer: Berlin, 150–167.

Bouharroud R. 2015. Control of tomato leafminer as a new invasive species: Costs and lessons. *Watch Letter* 33: 63–67.

Bouharroud R. 2007. Resistance of whitefly *Bemisia tabaci* to commonly used insecticide on tomato crop in Morocco. PhD thesis dissertation, 160 p. University Ibn Zohr, Agadir, Morocco.

Bouharroud R., Hanafi A., Brown J. K. and Serghini M. A. 2006. Resistance and cross-resistance of *Bemisia tabaci* to commonly used insecticides in the tomato greenhouses of the Souss valley of Morocco. *European Journal of Scientific Research* 14(4): 587–594.

Ferracini, C., Bueno, V. H. P., Dindo, M. L., Ingegno, B. L., Luna, M. G., Salas Gervassio, N. G, Sânchez, N. E., Siscaro, G, van Lenteren, J. C., Zappalà, L. and Tavella, L. 2019. Natural enemies of *Tutaabsoluta* in the Mediterranean basin, Europe and South America. *Biocontrol Science and Technology* 29(6): 578–609.

Hanafi, A. and El-Fadl, A. 2004. Integrated production and protection of greenhouse tomato in Morocco. In International Symposium on Mediterranean Horticulture: Issues and Prospects, Cairo, Egypt, 582.

Hanafi, A., Bouharroud, R., Amouat, S. and Miftah, S. 2007. Efficiency of insect nets in excluding whiteflies and their impact on some natural biological control agents. *Acta Horticulturae* 747: 383–387.

Hendrix, D. L., Wei, Y. and Leggett, J. E. 1992. Homopteran honeydew sugar composition is determined by both insect and plant species. *Compendium of Biochemistry and Physiology* 101: 23–27.

Polston J. E., De Barro, P. J., Boykin, L. M. 2014. Transmission specificities of plant viruses with the newly identified species of the *Bemisia tabaci* species complex. *Pest Management Science* 70(10): 1547–1552.

Stavrinides, M. C. and Mills, N. J. 2009. Demographic effects of pesticides on biological control of Pacific spider mite (*Tetranychus pacificus*) by the western predatory mite (*Galendromus occidentalis*). *Biological Control* 48(3): 267–273.

Urbaneja, A., Gonzalez, J., Arno, J. and Gabarra, R. 2012. Prospects for the biological control of *Tuta absoluta* in tomatoes of the Mediterranean basin. *Pest Management Science* 68(9): 1215–1222.

Section XIX

Biopesticides Weapons against Agricultural Mite Pests

49

Biopesticide Weapons Against Agricultural Mite Pests

Muhammad Haroon Sarwar

CONTENTS

49.1 Introduction

Most mite species causing damage to important crops worldwide are included in the families Eriophyidae, Tetranychidae, Tenuipalpidae, and Tarsonemidae. Red spider mites (Tetranychus species, Figure 49.1) are polyphagous, parenchyma cell feeding pests and have a serious economic impact on many crops. They generally live on the undersides of plant leaves, where they may spin protective silk webs and can cause damage by puncturing the plant cells to feed. Spider mites are known to feed on several hundred species of plants (Bolland et al., 1997). Also, the fact that they weave small webs around the leaves as well helps to notice them more easily. That is why they are found indoors all year round and in the garden during summer (Sarwar, 2014),

Two common arthropod pests of cosmopolitan importance are the tetranychid mites: carmine spider mite *Tetranychus cinnabarinus* Boisd and two-spotted spider mite and *Tetranychus urticae* Koch (Sarwar, 2013b, c). These mites feed on numerous cultivated and wild plants (Tomczyk, 1989). The economic threat posed by these mites is constantly increasing because of the development of pesticide resistance, and the resurgence of mite populations due to the non-selectivity of synthetic pesticides, which damage and may even destroy natural enemies, such as predaceous mites and spiders (Sarwar et al., 2011a, b). Many of the plant botanicals used as insecticides in commercial agriculture are not very selective because they target a broad range of insect pests (Ahmed et al., 2011; Sarwar, 2012a). Vegetables that repel red spider mite are lettuce, chili pepper, and wild tomato (Sarwar and Salman, 2015; Sarwar, 2019a).

Biopesticides are secondary metabolites, which cause insects on the plants to reduce their food intake and hence starve them to death; as oviposition deterrents, by preventing pests from laying eggs; or as inhibitors by interfering with the life cycle of the pests (Sarwar, 2013a). Botanicals are cheap, easy to prepare, and in most cases readily available and have more than one active ingredient which work synergistically, making it difficult for pests to develop resistance (Khan et al., 2010; Sarwar, 2015c). Azadirachta and Melia are found to be highly effective against insects and mites (Sarwar, 2015b).

Neem oil, obtained by cold-pressing seeds, can be effective against soft-bodied insects and mites specifically spider mite *T. urticae* (Bernardi et al., 2013; Dimetry et al., 1993; Sarwar, 2016c). Natural enemies are susceptible via direct contact, predators and parasitoids reinvading a treated crop one or more days after treatment are unlikely to be poisoned by residue contact with pesticides (Ismail et al., 2011).

49.2 Alternative Control Methods against Phytophagous Mites

The potential of crude extracts prepared from four plant species *Artemisia herba-alba*, *A. leucodes*, *Eruca sativa*, and *Sinapisalba*

FIGURE 49.1 Red spider mite.

FIGURE 49.2 Rosemary.

is measured in vitro using the vital mitochondrial dye 3-[4,5-dim ethyliazol-2-yl]-2,5-diphenyltetrazolium bromide assay. The assay revealed that the extracts from A. herba-alba, A. leucodesor and S. alba are not toxic to mites cells up to concentrations of 100 μg/ ml. Extracts from E. sativa at concentrations > 100 μg/ ml are found to significantly reduce cell viability and therefore are toxic to pest cells. The mortality and repellency potential of the different extracts showed that A. leucodes or E. sativa at a concentration of 50 μg/ ml resulted in more than 80% mortality and repellency. font italic A. herba-alba, A. leucodesor, S. alba, E. sativa. These results indicate that further investigations are required to characterize and identify the bioactive ingredients from *A. leucodes* and *E. sativa* that revealed greater potential as acaricidal sources for mite control (Azaizeh et al., 2007). The compounds showed ovipositional deterrence, suppressed the emergence of nymphs, and caused reduction in leaf damage (Mansour et al., 2004; Makundi and Kashenge, 2002).

The plant will absorb some of the components of these products, which will form a barrier to any additional mites that may try to feed on it. Rosemary (Figure 49.2) can be used on

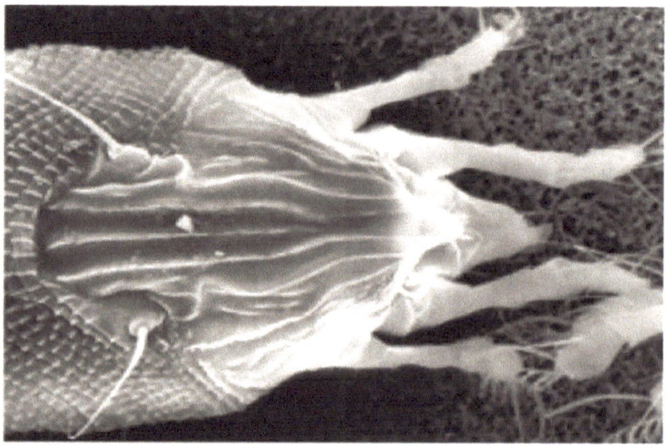

FIGURE 49.3 Coconut mite.

delicate new growth, as well as clones, tomatoes, and roses (Miresmailli and Isman, 2007).

Kheradmand et al. (2015) reported ecologically safe acaricide, effects of essential oils derived from *Cuminum cyminum* (cumin), *Syzygium aromaticum* (clove), and *Mentha spicata* (spearmint) determined on *T. urticae*. According to repellency tests, by increasing concentration of oils, the repellency effects are increased. On the basis of different parameters, the results regarding efficacy ranking specified that *M. azedarach* appeared to be the most effective plant material to cause the highest mortality of adults in comparison to the other treatments (Sarwar et al., 2013). Tea (*Camellia sinensis*) is subjected to the attack of insects and mites, and its production is greatly hindered due to these maladies. These botanical materials can be used as an alternative to chemical pesticides (Mamun and Ahmed, 2013). Such botanical extracts will help in controlling major pests including red spider mite as a component of integrated pest management (IPM) program (Sarwar, 2012b; 2017a, b; 2020a).

Plant oils have harmless effects on the predacious mites as reported by Amer and Momen (2002, 2005), who studied effects of some essential oils on *Amblyseius swirskii*. Similarly, the effect of French lavender essential oil has been found almost harmless on some predacious mites of the family Phytoseiidae, which contains one of the most efficient predatory mites (Sarwar, 2016a; Sarwar et al., 2009; 2010; 2012). The glandular trichomes of wild tobacco contain complexes of either glucose or sucrose esters (sometimes both). This product is contact pesticide that kills small insects and mites through suffocation (by blocking the spiracles) or disruption of cuticular waxes and membranes in the integument as recorded by Buta et al. (1993), leading to desiccation (Isman, 2006). Coconut eriophyid mite *Aceria guerreronis* (Keifer) (Figure 49.3) is a serious threat to the coconut industry in terms of yield losses. whereas the least percent reduction of mite population is noticed in calotropis leaf extract (17.46%) (Balaji and Hariprasad, 2015).

Jute yellow mite *Polyphagotarsonemus latus* (Bank) (Figure 49.4) feeds on plants and if left unchecked, its infestations build up quickly, and can destroy plants. Three botanical oils and microbial derivatives of an acaricide (Abamectin) are evaluated for their efficacy in controlling jute yellow mite *P. latus*. Among the botanicals, neem oil and mahogany oil showed

FIGURE 49.4 Jute yellow mite.

FIGURE 49.5 Bee propolis.

better efficacy by showing reduction of 80.25% mite infested plant (Rahman et al., 2016).

The acaricidal activity of plant extracts offers a promising means to control pests in both open and closed environments (Vasquez et al., 2016).

The pesticidal activity from extracts of two Eucalyptus plant species (red and spotted gum) in comparison with some biocides and conventional pesticides evaluated against spider mite and Bemisia tabaci infesting cotton fields showed very promising toxic activity on both pests (Radwan et al., 2000). Abdel-Khalek et al. (2010) studied repellency and toxicity of extract from Francoeria crispa to Eutetranychus orientalis. italic font Bemisia tabaci Francoeria crispa Eutetranychus orientalis

49.3 Acaricidal Effect of Bee Propolis Extracts

Propolis (bee glue) (Figure 49.5) is a strongly adhesive resinous bee-hive product collected by honeybees (*Apis mellifera* Linnaeus) from leaf buds and cracks in the bark of various plants and is used in the hives to exclude draughts, to protect against external invaders and to mummify their carcasses. It contains phenolic acids, esters, flavanones, flavones, and cinnamic acids (Bankova, 2005).

Bee propolis extracts could be used as a safe insecticide in the control of red spider mites. The insecticidal activity is thought to be due to bioactive phytochemicals of plant origin ingested by the bees during pollination (Kareru et al., 2013).

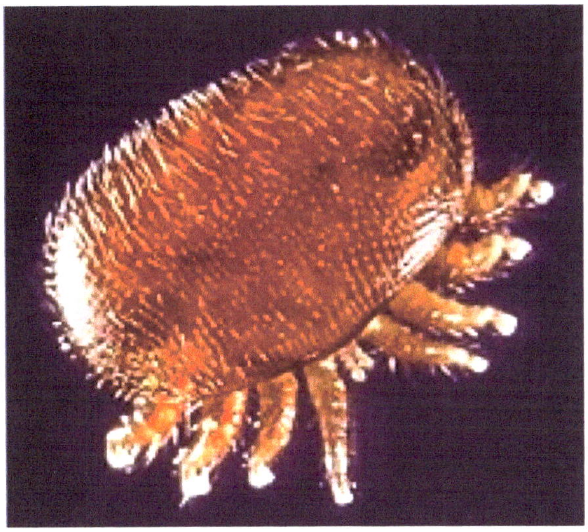

FIGURE 49.6 Varroa mite.

FIGURE 49.7 Two-spotted spider mite.

The mite *Varroa destructor* (Anderson and Trueman) is an external parasitic mite that attacks and feeds on the honeybees *Apis cerana* and *Apis mellifera*. The disease caused by the mites is called varroosis. The Varroa mite (Figure 49.6) can reproduce only in a honeybee colony. It attaches to the body of the bee and weakens the bee by sucking of bodies. Another important use of plant essential oil constituents is in fumigation of beehives to manage economically important honey bee parasites, the Varroa mite (*Varroa jacobsoni*) (Sarwar, 2016b). Fumigation of beehives menthol is widely used for this purpose (Delaplane, 1992) and thymol is most used as well (Floris et al., 2004).

49.4 Use of Botanical Pesticides for Mite Populations

49.4.1 Tetranychid Mites

Family Tetranychidae Donnadieu includes economically important pests species and Tetranychus Dufour is one of the five

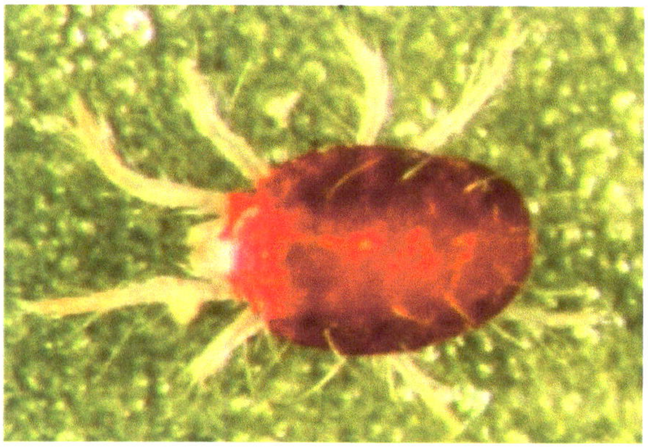

FIGURE 49.8 Tomato red spider mite.

FIGURE 49.9 Carmine spider mite.

largest genera of Tetranychidae. Damage by tetranychid mites is inflicted at the feeding point by disruption to the mesophyll cells of plants, affecting transpiration and photosynthesis and, consequently, plant growth and fruit production (Seeman and Beard, 2011; Sarwar, 2019b).

49.4.1.1 *Tetranychus urticae* Koch

The two-spotted spider mite is considered one of the most important agricultural pests, infesting crops including cotton, tomatoes, and grapes. The four stages of a two-spotted spider mite life cycle (egg, larva, nymph, and adult) are shown (Figure 49.7). Among the more sustainable strategies to manage tetranychid mite populations that have been tested are plant extracts (Debrowski and Seredynska, 2007). Although azadirachtin presented lower biological persistence (7 days) than abamectin (21 days), it is as efficient against *T. urticae* as abamectin, but did not cause significant mortality of adult predatory mites [*Neoseiulus californicus* (McGregor) and *Phytoseiulus macropilis* (Banks)] when used in strawberry crop. Similarly, the toxicity on *T. urticae* and two other predatory mites (*Euseiusa latus* De Leon and *P. macropilis*) is studied (Brito et al., 2006). The persistence of azadirachtin declines 5 days after its application due to the effects of temperature, ultraviolet light, and rainfall. Thus, a seven-day interval between applications is advisable in order to ensure efficient mite control in strawberry crop (Liu et al., 2006).

Leaf ethanolic extract from *Croton sellowii* Baill induced a higher mortality rate (69%) and a greater decrease in fecundity (≈ 85%) than extracts obtained from other Croton species used as acaricides against *T. urticae* (Soto et al., 2011).

FIGURE 49.10 Citrus red mite.

49.4.1.2 *Tetranychus evansi* Baker and Pritchard

The tomato red spider mite *Tetranychus evansi* (Figure 49.8) is an important pest of solanaceous plants, especially tomatoes. The effects of various neem-based products are evaluated on the mortality, fertility, and fecundity of the tomato red mite. The efficiency in reducing the *T. evansi* population is above 95% five days after spraying. Direct mortality one day after application is responsible for a higher reduction of *T. evansi* at a concentration corresponding to LC_{95} (Soto et al., 2010).

49.4.1.3 *Tetranychus cinnabarinus* (Boisduval)

The carmine spider mite *Tetranychus cinnabarinus* (Figure 49.9) is reported on plant species of economic importance, including vegetables, fruit-trees, and ornamentals. Wild oregano (*Lippia origanoides* Kunth.) and gliricidia (*Glirici diasepium* (Jacq.) Kunth ex Walp.), are commonly found in arid environmental conditions, presenting a valuable source of extracts that exert pesticide activity. At 5% concentrations, oregano and gliricidia extracts produced diminutions of 43.7% and 57% in *T. cinnabarinus* oviposition, respectively, while at 10%, the extracts caused mortality of 42.2% and 72.5%, respectively (Sivira et al., 2011). The uses of gliricidia and wild oregano seem to constitute a promising strategy for the management of *T. cinnabarinus*. Since both plant species are abundant in tropical countries they could be made readily available for small-scale growers to be included among other carmine mite control practices (Soto et al., 2011).

49.4.1.4 *Panonychus citri* (McGregor)

The citrus red mite *Panonychus citri* (Figure 49.10) is considered a secondary pest of citrus crops in commercial plantations. Acetogenins, a series of natural products isolated exclusively from the Annonaceae species widely distributed in tropical and subtropical regions are being employed for pest control. Acetogenins constitute a potent inhibitor of the mitochondrial respiratory chain complex I and exhibit a broad range of biological activities such as cytotoxic, antitumoural, immunosuppress,

FIGURE 49.11 False spider mite.

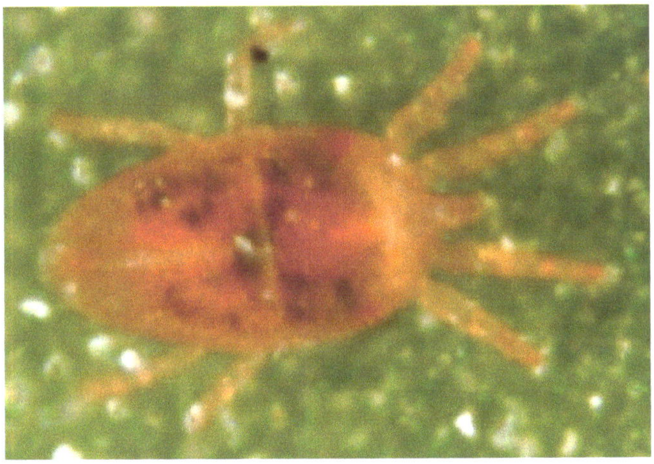

FIGURE 49.12 Chilean false red mite.

or antimicrobial and antiparasitic, and also pesticidal (Bermejo et al., 2005).

49.4.2 Tenuipalpid Mites

The family Tenuipalpidae Berlese presents a worldwide distribution, wherein Brevipalpus Donnadieu is a new world group with the most diversity. The most frequently reported species in this genus are *Brevipalpus californicus* (Banks), *Brevipalpus obovatus* Donnadieu, and *Brevipalpus phoenicis* (Geijskes) (Sarwar, 2020b). These mites feed on host plant species including fruits, ornamental plants, and forest plants, and the main damage associated with Brevipalpus mites is indirect, as vectors of plant viruses (Sarwar, 2015a, 2015d, 2015e; Sarwar et al., 2020).

49.4.2.1 *Brevipalpus phoenicis* (Geijskes)

The false spider mite *Brevipalpus phoenicis* (Figure 49.11) is a cosmopolitan tenuipalpid mite that feeds on a variety of host plants. In citrus species it plays a crucial role in the transmission of the citrus leprosis virus (CiLV), an RNA virus found mainly on sweet orange and mandarin trees, which has been

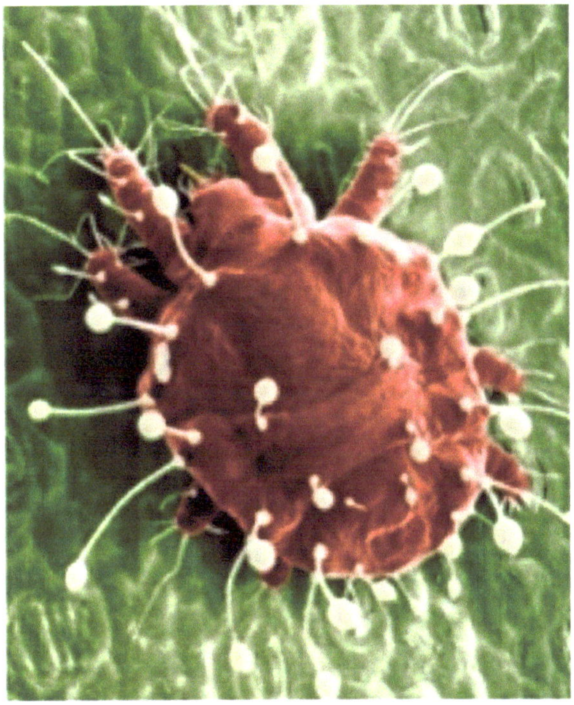

FIGURE 49.13 Red palm mite.

causing economic losses. A high toxicity of *Eremanthus goyazensis* (Gardner) leaf essential oils on *B. phoenicis* is observed even within a short exposure timeframe of 72 h. Additionally, the essential oils are sufficiently toxic to kill the entire spider mite population at the lower dose of 0.5 μL/L of air (Leon, 2012).

The toxic effect of *Eremanthus goyazensis* (Gardner) essential oil is mainly due to the presence of several monoterpenes that have been reported as toxic to other mite species. Thus, field dilutions of *E. goyazensis* oils could serve as a possible 'resistance-free' approach to stop the northward movement of *B. phoenicis* and thereby avoid endangering of orange tree fields (Baldin et al., 2012).

49.4.2.2 *Brevipalpus chilensis* Baker

The Chilean false red mite *Brevipalpus chilensis* (Figure 49.12) constitutes a quarantine pest that causes serious economic problems to grapes (*Vitis vinifera* L.), lemons (*Citrus limon* (L.) Osbeck), kiwifruits (*Actinidia deliciosa* (A. Chev.) C. F. Liang and A. R. Ferguson), persimmons (*Diospyros kaki* Thunb.), and various flowers and ornamental crops (Gonzalez, 1989).

High mortality rates (82%) are reported in *B. chilensis* adults exposed to different concentrations of *Acantholippia deserticola* (Phil.) Moldenke (Verbenaceae) essential oils containing a rich fraction of α and β-thujones (88.4%) (Benites et al., 2014). The toxic effect of essential oils may be related to impairment of vital functions in target organisms since most of monoterpenes are cytotoxic to plant and animal tissue, by causing a drastic reduction in the number of intact mitochondria and golgi bodies, impairing respiration and photosynthesis, and decreasing cell membrane permeability (Tripathi et al., 2000).

49.4.2.3 *Raoiella indica* Hirst

The red palm mite *Raoiella indica* (Figure 49.13) has been considered a serious pest in coconut (*Cocos nucifera* L.), areca palm (*Areca catechu* L.), and date palm (*Phoenix dactylifera* L.) (Daniel, 1981). In the new world, *R. indica* inflicts serious damage to Arecaceae, primarily coconut, but also to Musaceae and to plants of a few other botanical families. Coconut plantlets may die from the attack, whereas older plants show discoloration and consequent yield reduction (Rodrigues and Irish, 2012).

The essential oil of *Melaleuca quinquenervia* (Cav.) S. T. Blake (Myrtaceae) proved to be highly toxic to females of several mite species including *R. indica*, causing 100% mortality (Pino et al., 2011). The application of different doses of ethanolic extract of *Cymbopogon citratus* (DC.) Stapf caused a higher mortality rate (92.5%) and reduction of egg production (100%) in *R. indica* under laboratory conditions, when compared to the control. In this context, it is suggested that this extract may have the potential to serve as a biopesticide to control *R. indica* populations, although an evaluation of its effectiveness is required to better understand its performance under field conditions (Fernandez et al., 2016).

49.5 Bioassay Test for Acaricidal Activity of Botanics

Briefly, to investigate the effect of plant extracts on spider mite, a standard bioassay is used (Mansour et al., 2004). Bean leaf discs 25 mm in diameter taken from about 2-week-old plants are dipped for 5 sec in different concentrations of the tested extract solution. Each bean leaf disc is placed bottom side up on a filter paper in a tray and air-dried at room temperature before being tested. In parallel, comparable control discs are prepared and dipped in the extraction solution only. Each filter paper with the leaf disk is placed on a sheet of plastic foam, floating on water in a Petri dish (90 mm diameter). After drying, 15 adult female mites (about 5 days old) are placed on each leaf disc in triplicates per treatment. The treated mites are incubated under controlled conditions at 25±1°C temperature, 65±5% RH humidity and a photoperiod of 16:8 (L:D) hour. Number of live, dead or repelled (mites that have left the leaf disks are considered as repelled) are determined after 48 h of incubation.

49.6 Conclusions and Future Prospects

The current chapter evaluates the potential of plant materials as sources of bioactive ingredients with significant acaricidal activity that could lead to the development of new bio-safe products to control the mites. Although an alternative to pesticides could be achieved through organic farming and integrated pest management practices, the future lies in investing in technology, which will afford safe application of insecticides. There is an increasing expectation about the research on enzyme inhibition by compounds of botanical origin that could serve as leading compounds for the development of important substances with agrochemical properties. Therefore, by studying the plant complexes that protect them against pest attack, the researchers can learn to control this incidence in an ecological way and in addition can get pesticide active substances. A consolidated and continuous search of natural products may yield safer alternative control measures comparable to azadirachtin and pyrethryoids, which are being used in different parts of the world.

REFERENCES

Abdel-Khalek, A.A.; Amer, S.A.; Momen, F.M. Repellency and toxicity of extract from *Francoeria crispa* (Forssk) to *Eutetranychus orientalis* (Klein) (Acari: Tetranychidae). *Archiv. Phytopathol. Plant Prot.* 2010, 44: 441–445.

Ahmad, N.; Sarwar, M.; Khan, G.Z.; Tofique, M.; Salam, A. Efficacy of some plant products and synthetic chemicals to manage the outbreak of mealy bug (*Maconellicoccus hirsutus*) in Cotton. *J. Agric. Biol. Sci.* 2011, 3 (1): 16–21.

Amer, S.A.; Momen, F.M. Effect of French lavender essential oil on some predacious mites of the family Phytoseiidae. *Acta Phytopathol. Entomol. Hung.* 2005, 40: 409–415.

Amer, S.A.; Momen, F.M. Effect of some essential oil on predacious mite *Amblyseius swirskii* (Phytoseiidae). *Acta Phytopathol. Entomol. Hung.* 2002, 37: 281–286.

Azaizeh, H.; Kobaisy, M.; Dakwar, S.; Saadl, B.; Shaqir, I.; Said, O. Botanical pesticides as a source of safe bioacaricides for the control of *Tetranychus cinnabarinus. Acta Phytopathol. Entomol. Hung.* 2007, 42(1): 143–152.

Balaji, K.; Hariprasad, Y. Efficacy of botanicals on the management of coconut mite *Aceria guerreronis* (Keifer) (Acaridae: Eriophyidae). *J. Biopestic.* 2015, 8(1): 13–18.

Baldin, E. L. L.; Pavarini, D. P.; Souza, E. S.; Silva, J. P. G. F.; Souza, G. H. B.; Lopes, N. P.; Lopes, J. L. C. Towards new botanical pesticides: the toxic effect of Eremanthus goyazensis (Asteraceae) leaves essential oil against Brevipalpus phoenicis (Acari: Tenuipalpidae). Química Nova 2012, 35(11): 2254–2257. italic font Eremanthus goyazensi Brevipalpus phoenicis.

Bankova, V. Perspective paper chemical diversity of propolis and the problem of standardization *J. Ethnopharmacol.* 2005, 100: 114–117.

Benites, J.; Tello, V.; Flores, N.; Tapia, D.; Lopez, J.; Rios, D.; Buc-Calderon, P. *Acantholippia deserticola* essential oil as a natural pesticide against agricultural plagues woolly whitefly [*Aleurothrixus floccosus* (Maskell)], Chilean false red mite (*Brevipalpus chilensis* Baker) and two-spotted mite (*Tetranychus urticae* Koch). *Bol. Latinoam. Caribe Plantas Med. Aromat.* 2014, 13(3): 297–304.

Bermejo, A.; Figadère, B.; Zafra-Polo, M.C.; Barrachina, I.; Estornellc, E.; Cortes, D. Acetogenins from Annonaceae: Recent progress in isolation, synthesis and mechanisms of action. *Nat. Prod. Rep.* 2005, 22(2): 269–303.

Bernardi, D.; Botton, M.; Da Cunha, U.; Bernardi, O.; Malausa, T.; Garcia, T.; Navae, D. Effects of azadirachtin on *Tetranychus urticae* (Acari: Tetranychidae) and its compatibility with predatory mites (Acari: Phytoseiidae) on strawberry. *Pest Manage. Sci.* 2013, 69(1): 75–80.

Bolland, H.R.; Gutierrez, J.; Flechtmann, C.H.W. Introduction. In *World Catalogue of the Spider Mite Family (Acari: Tetranychidae).* Brill Publishers, Leiden, The Netherlands. 1997. pp. 1–3.

Brito, H.M.; Gondim, M.G.J.; De Oliveira, J.V.; Da Câmara, C.A. Toxicity of neem (*Azadirachta indica* A. Juss) formulations for two spotted spider mite and *Euseius alatus* de leon and *Phytoseiulus macropilis* (Banks) (Acari: Phytoseiidae). *Neotrop. Entomol.* 2006, 35(4): 500–505.

Buta, J.G.; Lusby, W.R.; Neal, J.W.J.; Waters, R.M.; Piuarelli, G.W. Sucrose esters from *Nicotian agossei* active against the greenhouse whitefly, *Trialeurodes vaporatiorwll. Phytochemistry.* 1993, 22: 859–864.

Daniel, M. Bionomics of the predaceous mite *Amblyseius channabasavanni* (Acari: Phytoseiidae), predaceous on the palm mite *Raoiella indica.* In: *Contributions to Acarology in India*, G.P. Channabasavanna (Ed.), Anubhava, Printers, India.1981. pp. 167–172.

Debrowski, Z.T.; Seredynska, U. Characterization of the two-spotted spider mite (*Tetranychus urticae* Koch, Acari: Tetranychidae) response to aqueous extracts from selected plant species. *J. Plant Prot. Res.* 2007, 47, 113–124.

Delaplane, K.S. Controlling tracheal mites (Acari: Tarsonemidae) in colonies of honey bees (Apidae) with vegetable oil and menthol. *J. Econ. Entomol.* 1992, 85: 2118–2124.

Dimetry, N.Z.; Amer, S.A.A. Reda, A.S. Biological activity of 2 Neem seed kernel extracts against the 2-spotted spider mite *Tetranychus urticae. J. Appl. Entomol.* 1993, 116: 308–312.

Fernandez, O.; sandoval, M.F.; Sanabria, M.E.; Vasquez, C. *In vitro* effectiveness ethanol extract of *Cymbopogon citratus* (D.L.) Stapf and hexythiazox *on Raoiella indica* Hirst. *Idesia.* 2016, 34(2): 77–84.

Floris, I.; Satta, A.; Cabras, P.; Garau, V.L.; Angioni, A. Comparison between two thymol formulations in the control of *Varroa destructor*: Effectiveness, persistence and residues. *J. Econ. Entomol.* 2004, 97: 187–191.

Gonzalez, R. *Insectos y ácaros de importanciaagrícola y cuarentenaria en Chile.* Universidad de Chile, Santiago de Chile. 1989. P. 210.

Ismail, M.; Ghallab, M.; Soliman, M.; Abo-Ghalia, A. Acaricidal activities of some essential and fixed oils on the two-spotted spider mite, *Tetranychus urticae. Egypt. Acad. J. Biolog. Sci.* 2011, 3(1): 41–48.

Isman, M.B. Botanical insecticides, deterrents, and repellents in modern agriculture and an increasingly regulated world. *Annu. Rev. Entomol.* 2006, 51: 45–66.

Kareru, P.; Rotich, Z.K.; Maina. E.W. Use of botanicals and safer insecticides designed in controlling insects: The African case. In: *Insecticides - Development of Safer and More Effective Technologies.* InTech, London. 2013. pp. 297–309.

Khan, M.H., M. Sarwar, A. Farid and F. Syed. Compatibility of pyrethroid and different concentrations of neem seed extract on parasitoid *Trichogramma chilonis* (Ishii) (Hymenoptera: Trichogrammatidae) under laboratory conditions. *Nucleus.* 2010, 47(4): 327–331.

Kheradmand, K.; Beynaghi, S.; Asgari, S.; Garjan, A. Toxicity and repellency effects of three plant essential oils against two-spotted spider mite, *Tetranychus urticae* (Acari: Tetranychidae). *J. Agric. Sci. Technol.* 2015, 17: 1223–1232.

Leon, M.G.A. Current status of the Citrus leprosis virus (CiLV-C) and its vector *Brevipalpus phoenicis* (Geijskes). *Rev. Fac. Agron.* 2012, 30(2): 242–250.

Liu, T.X.; Xu, H.H.; Luo, W.C., *Opportunities and Potentials of Botanical Extracts and Products for Management of Insect Pest in Cruciferous Vegetables*, Rai, M. and Carpinella, M. C. (Eds.), Elsevier, Amsterdam, The Netherlands. 2006, p. 502.

Makundi, R. H.; Kashenge, S. Comparative efficacy of neem, *Azadirachta indica*, extract formulations and the synthetic acaricide, Amitrac (Mitac), against the two-spotted spider mites, *Tetranychus urticae* (Acari: Tetranychidae), on tomatoes, *Lycopersicum esculentum. Z. Pflanzenkrankh. Pflanzenschutz.* 2002, 109, 57–63.

Mamun, M.; Ahmed, M. Prospect of indigenous plant extracts in tea pest management. *Int. J. Agric. Res. Innovation Technol.* 2013, 1(1–2): 16–23.

Mansour, F.; Azaizeh, H.; Saad, B.; Tadmor, Y.; Abo-Moch, F.; Said, O. The potential of plant extracts of Middle Eastern flora as a source of new safe bio-acaricides. *Phytoparasitica.* 2004, 32, 66–72.

Miresmailli, S.; Isman, M. Efficacy and persistence of rosemary oil as an acaricide against twospotted spider mite (Acari: Tetranychidae) on greenhouse tomato. *J. Econ. Entomol.* 2007, 99: 2015–2023.

Pino, B.O.; Sanchez, Y.; Rojas, M.M.; Rodríguez, H.; Abreu, Y.; Duarte, Y.; Martínez, B.; Peteira, B.; Correa, T.M.; Martínez, D. Composicion Quimica Y Actividad Plaguicida Del Aceite Esencial De *Melaleuca quinquenervia* (Cav) S.T. blake. *Rev.* Proteccion Veg. 2011, 26(3): 177–186.

Radwan, S.M.; Zidan, Z.H.; El-Hammady, A.; Aly, M.M. Field performance of tested Eucalyptus plant extracts, biocides and conventional pesticides, against key pests infesting cotton in Egypt. *Annal. Agri. Science, Cairo.* 2000, 45, 777–791.

Rahman, K.S. Islam, M.J.; Islam, N. Efficacy of three botanicals and a microbial derivatives acaricide (Abamectin) on the control of jute yellow mite, *Polyphagotarsonemus latus* (Bank). *J. Bangladesh Agric. Univ.* 2016, 14(1): 1–6.

Rodrigues, J.V.; Irish, B.M. Effect of coconut palm proximities and Musa spp. germplasm resistance to colonization by *Raoiella indica* (Acari: Tenuipalpidae). *Exp. Appl. Acarol.* 2012, 57(3–4): 309–316.

Sarwar, M. biology and ecology of some predaceous and herbivorous mites important from the agricultural perception. In: *Pests Control and Acarology*, Haouas, D., Hufnagel, L. (Eds.), IntechOpen Ltd., London, UK. 2019a, p. 29.

Sarwar, M. Comparative life history characteristics of the mite predator *Neoseiulus cucumeris* (Oudemans) (Acari: Phytoseiidae) on mite and pollen diets. *Int. J. Pest Manage.* 2016a, 62(2): 140–148.

Sarwar, M. Comparing abundance of predacious and phytophagous mites (Acarina) in conjunction with resistance identification between Bt and non-Bt cotton cultivars. *Afr. Entomol.* 2013a, 21(1): 108–118.

Sarwar, M. Competency of natural and synthetic chemicals in controlling gram pod borer, *Helicoverpa armigera* (Hubner) on Chickpea Crop. *Int. J. Agric. Sci.* 2012a, 2(4): 132–135.

Sarwar, M. Feasibility for development of comparative life histories and predation of predatory mites in Phytoseiidae complex and their experimental manipulations for pests control. *Int. J. Anim. Biol.* 2015a, 1(5): 150–157.

Sarwar, M. Frequency of insect and mite fauna in chilies *Capsicum annum* L., onion *Allium cepa* L. and garlic *Allium sativum* L. cultivated areas, and their integrated management. *Int. J. Agron. Plant Prod.* 2012b, 3(5): 173–178.

Sarwar, M. House dust mites: Ecology, biology, prevalence, epidemiology and elimination. In: *Parasitology and Microbiology Research.* InTech, London, UK. 2020a, p. 26.

Sarwar, M. Influence of host plant species on the development, fecundity and population density of pest *Tetranychus urticae* Koch (Acari: Tetranychidae) and predator *Neoseiulus pseudolongispinosus* (Xin, Liang and Ke) (Acari: Phytoseiidae). *N. Z. J. Crop Hortic. Sci.* 2014, 42(1): 10–20.

Sarwar, M. Information on activities regarding biochemical pesticides: An ecological friendly plant protection against insects. *Int. J. Eng. Adv. Res. Technol.* 2015b, 1(2): 27–31.

Sarwar, M. The inhibitory properties of organic pest control agents against aphid (Aphididae: Homoptera) on canola *Brassica napus* L. (Brassicaceae) under field environment. *Int. J. Sci. Res. Environ. Sci.* 2013b, 1(8): 195–201.

Sarwar, M. Integrated control of insect pests on canola and other brassica oilseed crops in Pakistan. In: *Integrated Management of Insect Pests on Canola and Other Brassica Oilseed Crops*, Gadi, V.P. Reddy,(Ed.), CABI, UK. 2017a, pp. 193–221.

Sarwar, M. The killer chemicals for control of agriculture insect pests: The botanical insecticides. *Int. J. Chem. Biomol. Sci.* 2015c, 1(3): 123–128.

Sarwar M. Management of spider mite *Tetranychus cinnabarinus* (Boisduval) (Tetranychidae) infestation in cotton by releasing the predatory mite *Neoseiulus pseudolongispinosus* (Xin, Liang and Ke) (Phytoseiidae). *Biol. Control.* 2013c, 65(1): 37–42.

Sarwar, M. Mite (Acari, Acarina) vectors involved in transmission of plant viruses. In: *Applied Plant Virology: Advances, Detection, and Antiviral Strategies*, Awasthi, L.P. (Ed.), Elsevier Inc., London. 2020b, pp. 257–273.

Sarwar, M. Mite pests (Acari) in mango (*Mangifera indica* L.) plantations and implementation of control strategy. *Biosci. Bioeng.* 2015d, 1(3): 41–47.

Sarwar, M. Mites (Acarina) as vectors of plant pathogens and relation of these pests to plant diseases. *Agric. Biol. Sci. J.* 2015e, 1(4): 150–156.

Sarwar, M. Mites—The tiny killers to push honeybee colonies into collapse and integrated pest management. *Int. J. Res. Appl. Phys.* 2016b, 1(7): 12–21.

Sarwar, M. Multiple integrated pest management (IPM) and area-wide management (AWM) approaches for insect pests of chickpea *Cicer arietinum* (L.) Walp crop. In: *Handbook of Chickpeas: Nutritional Value, Health Benefits and Management*, Lund, A.T., Schultz, N.D. (Eds.), Nova Science Publishers, Inc., New York. 2019b, pp. 237–263.

Sarwar, M. A potent folklore of botanical plant materials against insect pests together with their preparations and applications. *Sky J. Biochem. Res.* 2016c, 5(4): 58–62.

Sarwar, M. *Predatory Mites (Acari: Phytoseiidae) Culturing for their Releasing.* Lap Lambert Academic Publishing, Germany. 2017b, 140 p.

Sarwar, M.; Salman, M. Success stories of eco-friendly organically acceptable insecticides as natural products discovery. *Int. J. Mater. Chem. Phys.* 2015, 1(3): 392–398.

Sarwar, M.; Ashfaq, M.; Ahmad, N.; Randhawa, M.A.M. Assessing the potential of assorted plant powders on survival of *Caloglyphus* grain mite (Acari: Acaridae) in wheat grain. *Int. J. Agric. Sci. Bioresour. Eng. Res.* 2013, 2(1): 1–6.

Sarwar, M.; Kongming, W.; Xuenong, X. Evaluation of biological aspects of the predacious mite, *Neoseiulus cucumeris* (Oudemans) (Acari: Phytoseiidae) due to prey changes using some selected Arthropods. *Int. J. Acarol.* 2009, 35(6): 503–509.

Sarwar, M.; Kongming, W.; Xuenong, X.; Wang, E. Evaluations of four mite predators (Acari: Phytosiidae) released for suppression of spider mite infesting protected crop of sweet pepper (*Capsicum annuum* L.). *Afr. J. Agric. Res.* 2011a, 6(15): 3509–3514.

Sarwar, M.; Shad, N. A.; Batool. R. Integrated management of vectored viral diseases of plants. In: *Applied Plant Virology: Advances, Detection, and Antiviral Strategies*, Awasthi, L.P. (Ed.), Elsevier Inc., London. 2020, pp. 707–724.

Sarwar, M.; Xuenong, X.; Kongming, W. Effects of different flours on the biology of the pry *Tyrophagus putrescentiae* (Schrank) (Acarina: Acaridae) and the predator *Neoseiulus pseudolongispinosus* (Xin, Liang and Ke) (Acari: Phytoseiidae). *International Journal of Acarology.* 2010, 36(5): 363–369.

Sarwar M.; Xuenong, X.; Kongming, W. Suitability of webworm *Loxostege sticticalis* L. (Lepidoptera: Crambidae) eggs for consumption by immature and adults of the predatory mite *Neoseiulus pseudolongispinosus* (Xin, Liang and Ke) (Acarina: Phytoseiidae). *Spanish Journal of Agricultural Research.* 2012, 10(3): 786–793.

Sarwar, M.; Xuenong, X.; Wang, E.; Kongming, W. The potential of four mite species (Acari: Phytoseiidae) as predators of sucking pests on protected cucumber (*Cucumis sativus* L.) crop. *Afr. J. Agric. Res.* 2011b, 6(1): 73–78.

Seeman, O.; Beard, J. Identification of exotic pest and Australian native and naturalised species of Tetranychus (Acari: Tetranychidae). *Zootaxa.* 2011, 2961: 1–72.

Sivira, A.; Sanabria, M.; Valera, N.; Vásquez, C. Toxicity of ethanolic extracts from *Lippia origanoides* and *Gliricidia sepium* to *Tetranychus cinnabarinus* (Boisduval) (Acari: Tetranychidae). *Neotrop. Entomol.* 2011, 40(3): 375–379.

Soto, A.; Oliveira, H.; Pallini, A. Integración de control biológico y de productos alternativos contra *Tetranychus urticae* (Acari: Tetranychidae. *Rev. U.D.C.A. Actualidad Divulgación Científica. Enero-Junio.* 2011, 14 (1): 23–29.

Soto, A.; Venzon, M.; Oliveira, R.; Oliveira, H.; Pallini, A. 2010, Alternative control of *Tetranychus evansi* Baker & Pritchard (Acari: Tetranychidae) on tomato plants grown in greenhouses. *Neotrop. Entomol.* 2010, 39(4): 638–644.

Tomczyk, A. *Physiological and Chemical Responses of Different Host Plants to Infestation by Spider Mites (Acarina: Tetranychidae).* Warsaw Agricultural University Press, Warsaw, Poland.1989, 112 pp.

Tripathi, A.; Prajapati, V.; Aggarwal, K. K.; Khanuja, S.P.S.; Kumar, S. Repellency and toxicity of oil from *Artemisia annua*to certain stored-product beetles. *J. Econ. Entomol.* 2000, 93(1): 43–47.

Vasquez, C.; Balza, D.; Jiménez, M. A.; Colmenárez, Y.; Rios, Y. Use of plant extracts as an alternative control method against phytophagous mites in South America. *Current Topics in Phytochemistry.* 2016, 13: 35–41.

Section XX

Biotechnological Trends in Insect Pest Control Strategy

50

Biotechnological Trends in Insect Pests Control Strategy

Noreen Akram, Muhammad Sarwar, and Hafiz Sanaullah Babar

CONTENTS

50.1 Introduction

There are more than 10,000 species of insects that damage food crops and less than 10% of pests are considered to be major pests (Dhaliwal et al., 2007). Human, crops, and farm animals are directly or indirectly afflicted by insects attack (Atwal and Dhaliwal, 2015). Agricultural pest populations are decreased by a novel approach of infertile insect methods and they rivaled with wild insects (Sarwar, 2017a,d). The Mediterranean fruit fly ("Medfly") *Ceratitis capitata* (Wiedemann), attacks 250 host plants and causes yield loss (Sarwar, 2016a; 2018). A number of vegetables and oilseed crops have been affected by *Plutella xylostella* (Yakob et al., 2008). An estimated annual loss of US$ 4–5 billion is faced worldwide, through a combination of yield losses and costs of management (Zalucki et al., 2012; Furlong et al., 2013).

50.2 Genetics-Based Methods for IMP

50.2.1 Alternatives to Neonicotinoid Insecticides for Pest Control Case Studies in Agriculture and Forestry

In agricultural pests management, neonicotinoid insecticides are extensively used for the control of insects. This provides the impetus for exploring alternatives to neonicotinoid insecticides for controlling insect pests. (Jeschke et al., 2011). A continued analysis into alternatives is secured, but equally pressing is emphases for info transfer and training for farmers and managers to encourage the adoption of IPM strategies (Sarwar, 2016b; 2019).

50.2.2 Toward the Elements of Successful Insect RNAi

RNA obstructions (RNAi) have changed insect pest science investigations, since they empower the analyst to smother a gene of interest and subsequently introduce a phenotype to the insect. Professionals are increasingly mindful that RNAi in insects can be erratic; efficacy varies over insect taxa, among their qualities, mode of conveyance, and even between different research facilities (Belles, 2010; Terenius et al., 2011). Significantly, there's no single protocol for a proper RNAi experiment, partially because the effectivity of RNAi methods varies among insect teams (Chintapalli et al., 2007).

50.2.3 RNA Interference: Applications and Advances in Insect Toxicology and Insect Pests Management

RNA impedance (RNAi) is a succession explicit post-transcriptional quality quieting procedure evoked by double-stranded

RNA (dsRNA) that happens generally among plants, creatures, and microorganisms. We center around ongoing applications and advances of RNAi in the field of insect toxicology and host-prompted quality quieting as it is identified with insects across the board (Price and Gatehouse, 2008).

50.3 Role of Nanotechnology in Agriculture with Special Reference to Management of Insect Pests

Nanotechnology provides experienced and economical alternatives for the management of insect pests in agriculture while not harming the environment (Biyela et al., 2004; Levy, 2002).

50.4 Consequences of Resistance Evolution in a Cas9-Based Sex-Conversion Suppression Gene Drive for IMP

The utilization of CRISPR-Cas9 frameworks as a homing-based gene drive apparatus to modify the genotype of insect populations has hypothetically and for all intents and purposes been demonstrated to be plausible. This has recently brought up unique considerations about pests and disease vectors control. Efforts concentrate principally on disease vector mosquitoes such as Anopheles (Hammond et al., 2016; Unckless et al., 2017).

Consequently, the exceptionally adjustable nature of CRISPR/Cas9 permits basic structure of HCGDs to drive any wanted characteristic, even those subsequent in sterility, into wild populations as long as the cost of this phenotype does not go beyond a specific boundary (Deredec et al., 2008). An ongoing report (Hammond et al., 2016) distinguishes a lot of genes whose knockout brought about female-explicit sterility in Anopheles. As anticipated by scientific population hereditary gene models, the spread of female-explicit sterility characteristics in a population utilizing HCGDs ought to in the end bring about a population breakdown, as well as neighborhood or worldwide end of the objective species (Esvelt et al., 2014).

In *Aedes aegypti*, for example, a type of drive element known as a "Killer-Y chromosome" is ready to break the X chromosome during spermatogenesis, and in this manner all posterity of mosquitoes conveying such a chromosome will be male (Sarwar, 2014b). The bending of the sex proportion utilizing an X chromosome-explicit CRISPR/Cas9 framework has likewise been demonstrated to be effective in *An. gambiae* (Sarwar, 2014a).

The blend of germline homing at the tra locus (which results in the spread of the drive allele) and substantial focusing of the wild-type tra allele (which results in sex change) is expected to empower our proposed concealment gene drive system to be compelling. To accomplish this, various kinds of advertisers or blends thereof could be utilized (Esvelt et al., 2014; Galizi et al. (2014;).

A germline-explicit advertiser could be joined with an early zygotic advertiser from a cellularization quality for high and universal blastoderm articulation. Such cellularization advertisers have just been effectively applied for transgenic approaches in *C. capitata* (Schetelig et al., 2009). It is imperative to note that these early cellularization qualities are most certainly not communicated in the primordial germ cell (PGC) cores (Martinho et al., 2004), which are along these lines not presented to NHEJ-based change in the early developing life (Champer et al., 2017).

In *D. melanogaster*, one of these cellularization genes, Sry-α, is in actuality communicated both in a physically restricted way in the blastoderm and in the PGCs at later formative stages, and its advertiser in this way may be adequate for both germline homing and substantial sex transformation. Since Pol II-dependent interpretation is effectively stifled in the PGCs (Timinszky et al., 2008), a universal cell cycle-explicit advertiser, for example, the DNA pol-α180 promoter (Hirose et al., 1993), could bring about uniform focusing everything being equal during improvement with the exception of beginning time PGCs. The fatherly as it is transmission of our proposed gene drive system will help defeat the issue of DNA cleavage at early embryonic stages at the point when HDR is unlikely (Champer et al., 2016). In this way, the presentation of a germline-explicit advertiser, for example, the Rcd 1rpromoter, which has recently been appeared to bring about effective homing-based quality drive in *D. melanogaster* (Sarwar, 2020), may by itself be adequate to drive Cas9 articulation for the two purposes *D. melanogaster* XX males are constantly sterile, the substantial sex change forces a solid wellness cost on the XX people conveying the drive allele, which blocks the spread of the drive allele in the population (Esvelt et al., 2014), rendering *D. melanogaster* a safe model framework to examine this concealment quality drive procedure at the subatomic level. In any case, our outcomes show that low discharge frequencies (< 1%) are normal if there should arise an unplanned discharge, the drive allele moves toward becoming dispensed with from the population at beginning periods with no noteworthy impact on the wild populace. Thusly, it is naive to expect that such a drive framework is naturally restricted in *D. melanogaster* and along these lines meets the suggestions for gene drive tests (Lunshof, 2015, Akbari et al., 2015).

50.5 Recombinant Entomopathogenic Agents: A Review of Biotechnological Approaches to Pest Insects Control

Contrasting entomopathogens and concoction bug sprays is normally done dependent on viability and cost (Lacey et al., 2001); the extreme utilization of synthetic bug sprays is a risk to human wellbeing and the biological environment (Azizoglu et al., 2011). This could make it conceivable to create biobug spray items that target nearby bugs. Moreover, progresses in biotechnology may deliver increasingly strong and cost-effective items (Azizoglu et al., 2015).

50.5.1 Recombinant Entomopathogenic Infections

Baculoviruses have stood out due to their entomopathogenicity (Hails, 2001). Recombinant baculoviruses are being built to expand the executing rate of the specialists and diminish the resistance of the host. For these purposes, the genes have been controlled utilizing additions and erasures.

50.6 Gene Insertion

The principal transmitted insecticidal quality encoding a peptide neurotoxin is BeIT, which has been acquired from the lesser Asian scorpion, *Buthus* (now Mesobuthus) *eupeus* (Carbonell et al., 1988). At the point when moved to *B. mori* nucleopolyhedrovirus (BmNPV), the AaIT neurotoxin gene in the recombinant baculovirus, BmAaIT, prompted a 40% improvement in controlling *B. mori*. Essentially, genetically built *Autographa californica* different nucleopolyhedrovirus (AcMNPV) with AaIT is 12–half more powerful than non-recombinant strains against *Helicoverpa armigera, H. virescens, Spodoptera exigua*, and *Trichoplusia ni* (Rajput et al., 2003; Sarwar, 2012; 2013; 2017c; Sarwar et al., 2009; 2011). Other insecticidal peptides, for example, TxP, the tox21Agene item (vermin), Dol m(hornet), μ-Aga, TalTX and DTX (creepy crawly), and As and Sh (ocean anemone) have been assessed for recombinant baculovirus development, and their movement has likewise been resolved against bugs (Kroemer et al., 2015).

50.7 Toxins

Neurotoxin qualities from scorpions, parasites, bugs, ocean anemones, and ants have been embedded into baculoviruses and surveyed against pest insects. Communicated poisons from genetically designed baculoviruses are 10–60% more viable than those of wild-type infections dependent on resistance suspension, loss of motion, and decrease in larval lifespan (Inceoglu hormone, helicokinin II) (Raina et al., 2007).

50.8 Enzymes

Adolescent hormone esterase and chitinolytic and proteolytic chemicals have additionally been utilized to improve the viability of baculoviruses. A recombinant strain of AcMNPV conveying the *Manduca sexta* chitinase quality is designed. The kill rate of this recombinant strain is increased by over-communicating the chitinase-delivering quality in the *Spodoptera frugiperda* larvae (Gopalakrishnan et al., 1995; Sarwar, 2017b).

A constitutive advertiser, glyceraldehyde-3-phosphate dehydrogenase (gpd) from *Aspergillus nidulans* is additionally used to drive articulation; also, the bug's endurance time and resistance are diminished by 25 and 40%, respectively, contrasted with the wild-type contagious strain. Overarticulation of chitinase CHIT1 from *Beauveria bassiana* improved its destructiveness by 23% (Fang et al., 2005). Overexpression of chitinase 2 (Chi2) from *M. anisopliae* is more effective on the cotton stainer bug, *Dysdercus perivianu*. Essentially, recombinant *Trichoderma konigii* expressing chitinase-40 moved from *M. anisopliae* caused insecticidal action in the Asian corn borer, *Ostrinia furnacalis*, and the silkworm, *B. mori*. Overarticulation of the corrosive trehalase, ATM1, quickened the development of *M. acridumin* grasshopper hemocoel, decreasing the conidia number for half mortality (LC 50) by 8.3-overlay contrasted with the wild-type strain (Peng et al., 2015).

50.9 Gene Deletion

Gene removing has additionally been assessed to build adequacy and advance the baculovirus murder speed. Removing the ecdysteroid UDP-glucosyltransferase (egt) quality from the AcMNPV genome expanded the executing speed and diminished encouraging by roughly 40% in *T. ni* and *S. frugiperda* hatchlings contrasted with the wild-type AcMNPV (O'Reilly and Miller, 1991).

50.10 Recombinant Entomopathogenic Fungi

Recombinant entomopathogenic organisms can be built to encode an insect poison. Another ecologically benevolent option in contrast to concoction bug sprays is the utilization of parasitic entomopathogens (Figure 50.1) in biocontrol programs for horticultural vermin and infection vectors. Hereditary designing is a productive apparatus for improving the adequacy of myco-bug sprays by improving their destructiveness and resilience to ecological pressure (Zhao et al., 2016). Hereditary designing methodologies center essentially on the poisons, hormones, compounds, and other physiological controllers.

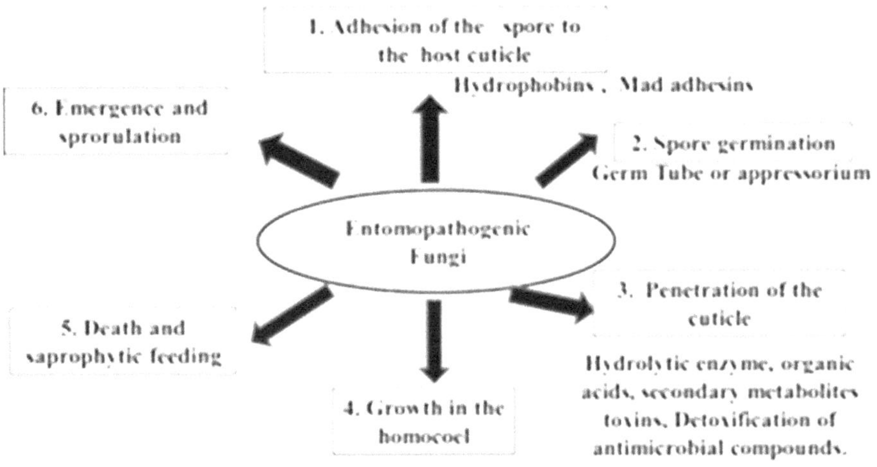

FIGURE 50.1 Mode of action of fungi.

Poison qualities from various arthropods, for example scorpions, have likewise been assessed for developing recombinant parasitic operators of insects control. The scorpion poison gene, AaIT1, that squares sodium diverted in insects is acquired from *A. australis* and embedded into *Metarhiziuma nisopliae* with safe related flagging pathway inhibitors expanded the insecticidal action against *G. mellonella* and the green peach aphid, *Myzus persicae*. Trypsin tweaking of static components is likewise assessed to design recombinant *B. bassiana*. The *Isaria fumosorosea* strain (IfB01-TLR7) is built to express the dsRNA focusing on the immunity related gene, TLR7, in the whitefly, *Bemisia tabaci*. TLR7 is effectively knocked down after whitefly sprites are contaminated with the recombinant strain, IfB01-TLR7 (Chen et al., 2015)

50.11 Recombinant Entomopathogenic Bacteria

Bacterial entomopathogens (Figure 50.2) have been utilized to control agrarian insect nuisances and oversee therapeutically significant sickness vectors, for example, nematodes. Recombinant DNA strategies have been utilized to improve bacterial bug spray adequacy by particularly expanding insecticidal protein amalgamation and empowering new endotoxin mixes (Federici et al., 2003).

50.11.1 Bacterium *Bacillus thuringiensis*

Bacterium *Bacillus thuringiensis* (Bt) has been disengaged in 1901 by Japanese researcher, Shigetane Ishiwatari named this *Bacillus sotto* because it caused sotto malady in silkworms. After ten years, Berliner segregated similar microbes in the Thuringia locale of Germany from *Ephestia kuehniella* and named it *Bacillus thuringiensis* (Siegel, 2000).

Bacterium Bt can be secluded from environmental situations, for example, soil, the rhizosphere, new water, plants, bugs, and items put away in distribution centers (Hendriksen and Hansen, 2002; Raymond et al., 2010; Yilmaz et al., 2011, 2017). The Bt life cycle comprises two phases: vegetative cell division and the sporulation stage. Since the precious stone (Cry) proteins created by Bt during sporulation have explicit pesticidal movement (Figure 50.3), they are generally utilized around the world in organic irritation control applications. Cry proteins are dangerous to

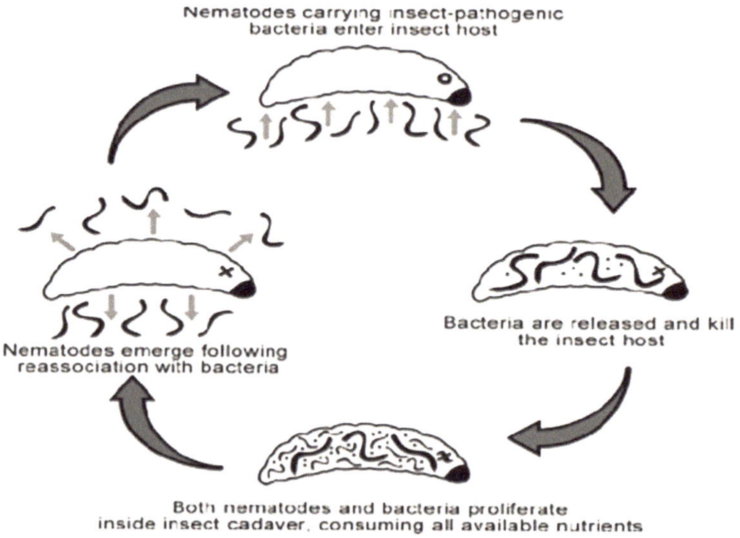

FIGURE 50.2 Mode of action of bacteria.

FIGURE 50.3 Diagram of Bt gene action.

specific types of insects from the orders Lepidoptera, Diptera, Coleoptera, Hymenoptera, and Hemiptera. They are likewise harmful to invertebrate creatures, for example nematodes and insects (Sarwar, 2020).

50.11.2 *Bacillus thuringiensis* Chitinases

Numerous chitinase qualities from various strains of Bt kenyea, pakistani, colmeneri, canadiensis, entomocidus, kurstaki have been added to the gene bank. Among these, 58% are cloned in China, 18% cloned in Mexico, and the others cloned in other countries, for example the United States, Egypt, Turkey, Thailand, India, Pakistan, Tunisia, and Korea. Most Bt chitinases have 93.8–99.6% similarity with each other. Bt chitinases have a secluded structure comprising of a reactant space and a potential chitin restricting area (ChBD).

Models 2 and 3 are options that diminish synthetic utilization. Shrimp is a conventional food in fish cafés; nonetheless, the unpalatable waste from shrimp, counting the head and exoskeletons, relates to 56–60% of these creatures by volume, and conists of chitin, protein, and mineral waste. The synergistic movement of a chitinase and protease arrangement combined by Bt are used to corrupt this squander. Debased shrimp waste is utilized in pharmaceuticals, medicines, the food industry, water treatment, color extraction, development media; animal feed, decrease of soil nematode populations, and single-cell protein creation (Rojas-Avelizapa et al., 1999). In addition, Bt chitinases have appeared to increase the insecticidal action of Cry poisons by processing the peritrophic film. Overarticulation of Bt homolog chitinases for biotechnological purposes is cultivated by a hereditary strategy for increasing Cry protein articulation (Park et al., 1998). Furthermore, Bt chitinases likewise show inhibitory action against phytopathogenic parasitic species, for example Fusarium species and *Sclerotium rolfsii*. Numerous chitinase (chi) qualities have been acquired from various microscopic organisms and chi genes have been designed into Bt to increase insecticidal action. *Serratia marcescens* chitinase genes (chiB and chiC) are designed into Bt, and an improved insecticidal action recorded utilizing a recombinant Bt strain (Ozgen et al., 2013). Comparable outcomes have likewise been reported by other scientists.

50.12 Pesticide-Induced Stress in Arthropod Pests

In spite of the high costs of utilization and the overall drive toward sustainable farming development, pesticide use is expanding (Ghimire et al., 2013; Schreinemachers et al., 2012). Administrative organizations center generally around transient endpoints when choosing to enlist mixes, long haul impacts may happen. Over time, long-term impacts may happen, and even changing mortality rates in arthropod insects may not be the essential endpoint to consider (Casida et al., 2013; Della et al., 2014), a point again and again dismissed by the scholarly world and governments alike.

50.13 Conclusion and Future Prospects

Modern techniques in biotechnology and genetic engineering are extremely helpful in manipulating the desired traits in entomopathogens, which can further improve their bioactivity. The requirement for new developments to manage rising farming pests and infections has never been more prominent, and this is a fascinating time for the science and research of genetic control of pests. A portion of the GM advancements portrayed in the present chapter are now being demonstrated in the field. The wave of atomic techniques is being applied to disease-transmitting mosquitoes and this is starting to reach over to horticulturally significant species. Consideration is being given to administrative perspectives to empower the protected and proper usage of these organic, gene-based procedures. Altogether, these improvements advance the possibilities for huge rural and financial advantages in plant protection, and control of vectors.

REFERENCES

Akbari, O. S.; Bellen, H. J. Safeguarding gene drive experiments in the laboratory. *Science*. 2015, 349, 927–929.

Atwal, A. S; Dhaliwal, G. S. *Agricultural Pests of South Asia and Their Management*, Kalyani Publishers, New Delhi. 2015.

Azizoglu, U.; Yilmaz, S.; Ayvaz, A.; Karaborklü, S.; Akbulut, M. Characterization of local *Bacillus thuringiensis* isolates and their toxicity to *Ephestia kuehniella* (Zeller) and Plodia interpunctella (Hübner) larvae. *Egypt. Biol. Pest Control*. 2011, 21, 143–150.

Azizoglu, U.; Yilmaz, S.; Ayvaz, A.; Karaborklü, S. Effects of *Bacillus thuringiensis* subsp *kurstaki* HD1 spore-crystal mixture on the adults of egg parasitoid *Trichogramma evanescens* (Hymenoptera: Trichogrammatidae). *Biotechnol. Biotechnol. Equip*. 2015, 29, 653–658.

Belles, X. Beyond Drosophila: RNAi in vivo and functional genomics in insects. *Ann. Rev. Entomol*. 2010, 55, 111–128.

Biyela, P. T.; Lin, L.; Bezuidenhout, C. C. The role of aquatic ecosystems as reservoirs of antibiotic resistant bacteria and antibiotic resistance genes. *Water Sci. Technol*. 2004, 50(1), 45–50.

Carbonell, L. F.; Hodge, M. R.; Tomalski, M. D.; Miller, L. K. Synthesis of a gene coding for an insect-specific scorpion neurotoxin and attempts to express it using baculovirus vectors. *Gene*. 1988, 73, 409–418.

Casida, J. E.; Durkin, K. A. Neuro active insecticides: Targets, selectivity, resistance, and secondary effects. *Annu. Rev. Entomol*. 2013, 58, 99–117.

Champer, J.; Buchman, A.; Akbari, O. S. Cheating evolution: Engineering gene drives to manipulate the fate of wild populations. *Nat. Rev. Genet*. 2016, 17, 146–159.

Champer, J.; Riona, R.; Suh, Y. O. Novel CRISPR/Cas9 gene drive constructs reveal insights into mechanisms of resistance allele formation and drive efficiency in genetically diverse populations. *PLoS Genet*. 2017, 13, e1006796.

Chen, X.; Li, L.; Hu, Q.; Zhang, B.; Wu, W.; Jin, F.; Jiang, J. Expression of dsRNA in recombinant *Isaria fumosorosea* strain targets the TLR7 gene in *Bemisia tabaci*. *BMC Biotech*. 2015, 15, 1–8.

Chintapalli, V. R.; Wang, J.; Dow, J. A. T. Using fly atlas to identify better *Drosophila melanogaster* models of human disease. *Nat. Genet*. 2007, 39, 715–720.

Della, T. M. C.; Gandra, L. C.; Guedes, R. N. C. Managing leaf-cutting ants: Peculiarities, trends and challenges. *Pest Manage. Sci*. 2014, 70, 14–23.

Deredec, A.; Burt, A.; Godfray, H. C. J. The population genetics of using homing endonuclease genes in vector and pest management. *Genetics*. 2008, 179, 2013–2026.

Dhaliwal, G. S.; Dhawan, A. K.; Singh, R. Biodiversity and ecological agriculture: Issues and perceptives. *Indian J. Ecol.* 2007, 34(2), 100–108.

Esvelt, K. M.; Smidler, A. L.; Catteruccia, F.; Church, G. M. Concerning RNA-guided gene drives for the alteration of wild populations. *eLife*. 2014, 3, 3401.

Fang, W. G.; Leng, B.; Xiao, Y.H.; Jin, K.; Ma, J. C.; Fan, Y. H.; Feng, J.; Yang, X. Y.; Zhang, Y. J.; Pei, Y. Cloning of *Beauveria bassiana chitinase* gene *Bbchit 1* and its application to improve fungal strain virulence. *Appl. Environ. Microbiol.* 2005, 71(1), 363–370.

Federici, B. A.; Park, H. W.; Bideshi, D. K.; Wirth, M. C.; Johnson, J. J. Recombinant bacteria for mosquito control. *J. Exp. Biol.* 2003, 206, 3877–3885.

Furlong, M. J.; Wright, D.J.; Dosdall, L. M. Diamondback moth ecology and management: Problems, progress, and prospects. *Ann. Rev. Entomol.* 2013, 58, 517–541.

Galizi, R.; Lindsey, A.; Miriam, M. A synthetic sex ratio distortion system for the control of the human malaria mosquito. *Nat. Commun.* 2014, 5, 3977.

Ghimire, N.; Woodward, R. T. Under- and over-use of pesticides: An international analysis. *Ecol. Econ.* 2013, 89, 73–81.

Gopalakrishnan, K.; Muthukrishnan, S.; Kramer, K. J. Baculovirus-mediated expression of a *Manduca sexta* chitinase gene: Properties of the recombinant protein. *Insect Biochem. Mol. Biol.* 1995, 25, 255–265.

Hails, R. S. Natural and genetically modified baculoviruses: Environmentally friendly pest control or an ecological threat? *Outlook Agric.* 2001, 30, 171–178.

Hammond, A.; Roberto, G.; Kyros, K. A CRISPR-Cas9 gene drive system targeting female reproduction in the malaria mosquito vector *Anopheles gambiae. Nat. Biotechnol.* 2016, 34, 78–83.

Hendriksen, N. B.; Hansen, B. M. Long-term survival and germination of *Bacillus thuringiensis* var. *kurstaki* in a field trial. *Can. J. Microbiol.* 2002, 48, 256–261.

Hirose, F.; Yamaguchi, M.; Handa, H.; Inomata, Y.; Matsukage, A. Novel 8-base pair sequence (Drosophila DNA replication-related element) and specific binding factor involved in the expression of Drosophila genes for DNA polymerase alpha and proliferating cell nuclear antigen. *J. Biol. Chem.* 1993, 268, 2092–2099.

Jeschke, P.; Nauen, R.; Schindler, M.; Elbert, A. Overview of the status and global strategy for neonicotinoids. *J. Agric. Food Chem.* 2011, 59, 2897–2908.

Kroemer, J. A.; Bonning, B. C.; Harrison, R. L. Expression, delivery and function of insecticidal proteins expressed by recombinant baculoviruses. *Viruses*. 2015, 7, 422–455.

Lacey, L. A.; Frutos, R.; Kaya, H. K.; Vail, P. Insect pathogens as biological control agents: Do they have a future? *Biol. Control.* 2001, 21, 230–248.

Levy, S. B. Factors impacting on the problem of antibiotic resistance. *J. Anti. Chem.* 2002, 49, 25–30.

Lunshof, J. Regulate gene editing in wild animals. *Nature.* 2015, 521, 127.

Martinho, R. G.; Kunwar, P. S.; Casanova, J.; Lehmann, R. A noncoding RNA is required for the repression of RNA pol II-dependent transcription in primordial germ cells. *Curr. Biol.* 2004, 4, 159–165.

O'Reilly, D. R; Miller, L. K. Improvement of a baculovirus pesticide by deletion of the egt gene. *Biotechnology.* 1991, 9, 1086–1089.

Ozgen, A.; Sezen, K.; Demir, I.; Demirbag, Z.; Nalcacioglu, R. Molecular characterization of chitinase genes from a local isolate of *Serratia marcescens* and their contribution to the insecticidal activity of *Bacillus thuringiensis* strains. *Curr. Microbiol.* 2013, 67, 499–504.

Park, H. W.; Ge, B.; Bauer, L. S.; Federici, B. A. Optimization of Cry3A yields in *Bacillus thuringiensis* by use of sporulation-dependent promoters in combination with the STAB-SD mRNA sequence. *Appl. Environ. Microbiol.* 1998, 64, 3932–3938.

Peng, G.; Jin, K.; Liu, Y.; Xia, Y. Enhancing the utilization of host trehalose by fungal trehalase improves the virulence of fungal insecticide. *Appl. Microbiol. Biotechnol.* 2015, 99, 8611–8618.

Price, D. R.; Gatehouse, J. A. RNAi-mediated crop protection against insects. *Trends Biotechnol.* 2008, 26, 393–400.

Raina, A. K.; Vakharia, V. N.; Leclerc, R. F.; Blackburn, M. B. Engineering a recombinant baculovirus with a peptide hormone gene and its effect on the corn earworm, *Helicoverpa zea. Biopestic. Int.* 2007, 3, 43–52.

Rajput, A. A.; Sarwar, M.; Ahmad, N.; Siddiqui, Q. H.; Tofique, M. Evaluation for resistance in some local and exotic chickpea genotypes against *Helicoverpa armigera* (Hubner). *Pak. J. Biolo. Sci.* 2003, 6 (18): 1612–1615.

Raymond, B.; Johnston, P. R.; Nielsen-LeRoux, C. Lereclus, D.; Crickmore, N. *Bacillus thuringiensis*: An impotent pathogen. *Trends Microbiol.* 2010, 18, 189–194.

Rojas-Avelizapa, L. I.; Cruz-Camarillo, R.; Guerrero, M. I.; Rodríguez-Vázquez, R.; Ibarra, J. E. Selection and characterization of proteo-chitinolytic strain of *Bacillus thuringiensis* able to grow in shrimp waste media. *World J. Microbiol. Biotechnol.* 1999, 15, 299–308.

Sarwar, M. Area-wide integrated management of fruit flies (Diptera: Tephritidae) pest in vegetables cultivation. *J. Biol. Environ. Eng.* 2016a, 1 (2): 10–16.

Sarwar, M. An area-wide integrated management of fruit fly (Diptera: Tephritidae) pests in fruits production. *Int. J. Plant Sci. Ecol.* 2018, 4 (1): 1–7.

Sarwar, M. Biological parameters of pink bollworm *Pectinophora gossypiella* (Saunders) (Lepidoptera: Gelechiidae): A looming threat for cotton and its eradication opportunity. *Int. J. Res. Agric. For.* 2017a, 4 (7): 25–36.

Sarwar, M. Bollworms *Earias vittella* (Fabricius) and *Earias insulana* (Lepidoptera: Noctuidae) impairment in cotton and integrated crop management. *Scholars J. Res. Agric. Biol.* 2016b, 1 (1): 1–7.

Sarwar, M. Characteristics of armyworm *Spodoptera litura* (Fabricius) (Lepidoptera: Noctuidae) occurrence and control on cotton. *Scholars J Res Agric Biol.* 2017b, 2 (1): 33–38.

Sarwar, M. Competency of natural and synthetic chemicals in controlling gram pod borer, *Helicoverpa armigera* (Hubner) on Chickpea Crop. *Int. J. Agric. Sci.* 2012, 2 (4): 132–135.

Sarwar, M. Cotton Bollworm *Helicoverpa armigera* (Hubner, 1809) (Lepidoptera: Noctuidae) and development of integrated pest management platform. *Scholars J. Res. Agric. Biol.* 2017c, 2 (1): 63–70.

Sarwar, M. Defeating malaria with preventative treatment of disease and deterrent measures against anopheline vectors (Diptera: Culicidae). *J. Pharmacol. Toxicol. Stud.* 2014a, 2 (4): 1–6.

Sarwar, M. Exploration on resource of resistance in chickpea (*Cicer arietinum* L.) genotypes to gram pod borer *Helicoverpa armigera* (Hubner) (Lepidoptera). *Afr. J. Agric. Res.* 2013, 8 (26): 3431–3435.

Sarwar, M. Multiple integrated pest management (IPM) and area-wide management (AWM) approaches for insect pests of chickpea *Cicer arietinum* (L.) walp crop. In: *Handbook of Chickpeas: Nutritional Value, Health Benefits and Management.* A. T. Lund and N. D. Schultz (Eds.). Nova Science Publishers, Inc., New York, 2019; pp. 237–263.

Sarwar, M. Pink bollworm *Pectinophora gossypiella* (Saunders) [Lepidoptera: Gelechiidae] and practices of its integrated management in cotton. *Int. J. Plant Sci. Ecol.* 2017d, 3 (1): 1–6.

Sarwar, M. Proposals for the control of principal dengue fever virus transmitter *Aedes aegypti* (Linnaeus) mosquito (Diptera: Culicidae). *J. Ecol. Environ. Sci.* 2014b, 2 (2): 24–28.

Sarwar M. Typical flies: Natural history, lifestyle and diversity of Diptera, In: *Typical Flies: Natural History, Lifestyle and Diversity of Diptera*, M. Sarwar (Ed.), InTech, London, UK. 2020, p. 50.

Sarwar, M.; Ahmad, N.; Tofique, M. Host plant resistance relationships in chickpea (*Cicer arietinum* L.) against gram pod borer (*Helicoverpa armigera*) (Hubner). *Pak. J. Bot.* 2009, 41 (6): 3047–3052.

Sarwar, M. Experimental Induction of Insect Growth Regulators in Controls of Insect Vectors as well as Crops and Stored Products Pests. *Specialty Journal of Agricultural Sciences.* 2020, 6 (1): 32–41.

Sarwar, M.; Ahmad, N.; Tofique, M. Identification of susceptible and tolerant gram (*Cicer arietinum* L.) genotypes against gram pod borer (*Helicoverpa armigera*) (Hubner). *Pak. J. Bot.* 2011, 43 (2): 1265–1270.

Schetelig, M.; Caceres, C.; Zacharopoulou, A.; Franz, G.; Wimmer, E. A. Conditional embryonic lethality to improve the sterile insect technique in Ceratitiscapitata (Diptera: Tephritidae). *BMC Biol.* 2009, 7, 4.

Schreinemachers, P.; Tipraqsa, P. Agriculture pesticides and land use intensification in high, middle and low income countries. *Food Policy.* 2012, 37, 616–26.

Siegel, J. P. Bacteria. In: Lacey, L. L.; Kaya, H. K. (Eds.) *Field Manual of Techniques in Invertebrate Pathology.* Kluwer Scientific Publishers, Dordrecht. 2000, pp. 209–230.

Terenius, O.; Papanicolaou, A.; Garbutt, J. S.; Eleftherianos, I.; Huvenne, H.; Kanginakudru, S.; Albrechtsen, M.; An, C.; Aymeric, J. L.; Barthel, A. RNA interference in Lepidoptera: An overview of successful and unsuccessful studies and implications for experimental design. *J. Insect Physiol.* 2011, 57, 231–245.

Timinszky, G.; Bortfeld, M.; Ladurner, A. G. Repression of RNA polymerase II transcription by a Drosophila oligopeptide. *PLoS One.* 2008, 3, e2506.

Unckless, R. L.; Clark, A. G.; Messer, P. W. Evolution of resistance against CRISPR/Cas9 gene drive. *Genetics.* 2017, 205, 827–841.

Yakob, L.; Alphey, L.; Bonsall, M. B. *Aedes aegypti* control: The concomitant role of competition, space and transgenic technologies. *J. Appl. Ecol.* 2008, 45, 1258–1265.

Yilmaz, S.; Azizoglu, U.; Ayvaz, A.; Temizgül, R.; Atcıyurt, Z. B.; Karabörklü, S. Cloning and expression of cry2Aa from native *Bacillus thuringiensis* strain SY49–1 and its insecticidal activity against *Culex pipiens* (Diptera: Culicidae). *Microb. Pathog.* 2017, 105, 81–85.

Yilmaz, S.; Demirezen, N.; Azizoglu, U.; Karabörklü, S.; Ayvaz, A.; Akbulut, M.; Tekin, S. Identification of some cry1 genes and toxicity determination in *Bacillus thuringiensis* isolates obtained from mills and warehouses in Turkey. *Egypt J. Biol. Pest Control.* 2011, 21, 189–195.

Zalucki, M. P.; Shabbir, A.; Silva, R.; Adamson, D.; Shu-Sheng, L.; Furlong, M. J. Estimating the economic cost of one of the world's major insect pests, *Plutella xylostella* (Lepidoptera: Plutellidae): Just how long is a piece of string? *J. Econ. Entomol.* 2012, 105, 1115–1129.

Zhao, H.; Lovett, B.; Fang, W. Genetically engineering entomopathogenic fungi. *Adv. Genet.* 2016, 94, 137–163.

Section XXI

Challenges in the Popularization of Biopesticides in Organic Farming

51

Challenges to Popularizing Biopesticides in Organic Farming

Riccardo Bugiani and Massimo Bariselli

CONTENTS

51.1 Introduction: Factors Involved in the Development and Growth of the Biopesticide Market

Agriculture in the third millennium is rapidly changing and in the near future it will face some very important challenges that can be summarized as follows: (1) providing sustainable production; (2) increase profitability and production; (3) reduce the level of residues in food; (4) reduce the environmental impact; and (5) making new defense technologies available to farmers.

Organic and integrated production in Europe is primarily driven by the demands of consumers who prefer natural products and free of pesticide residues. To meet consumer demands, European distribution chains quickly adapted to bring fruit and vegetables produced according to organic production to the market. Consequently, the European Union has also started to finance integrated production but, even more, organic production in all agrifood contexts.

However, organic agriculture, as well as integrated or conventional agriculture, is substantially influenced, often in a negative way, by two constantly increasing phenomena, "global warming" and the spread on a global scale of invasive alien species (IAS).

51.1.1 Climate Change

In the last 50–100 years, the planet has undergone a record global warming of about 0.8°C. Such a high value in the last million years had already been reached five more times in the past, but such changes had occurred in much longer times (in the order of 40,000–100,000 years). This abrupt warming is affecting and will significantly affect ecosystems making them less and less stable. Climate warming will also influence crops, making them more susceptible to pest and disease attacks. The crops most affected by climate change are perennial crops. The high level of relative humidity in some areas of the Po Valley caused considerable damage in the last two years caused by *Stemphylium vesicarium* agent of the brown spot of European pear and *Erwinia amylovora* on pome fruits. At the same time, bacterial and fungal infections on peaches caused by *Xanthomonas campestris* pv *pruni, Monilia fructicola* and *Phomopsis amygdali* were favored by higher temperatures and by wetter springs and autumns. Arable crops have also seen massive pest attacks. Potato cultivation is suffering all over Europe from attacks caused by wireworms (*Agriotes* spp.) While maize, especially in areas without irrigation, is strongly affected by water stress and aflatoxin producer *Aspergillus* spp. The latter, together with Fusarium head blight on wheat, poses serious risks of grain contamination by mycotoxins making the production not marketable. On top of all that, the reduction of the active ingredients available to control such pests have also to be considered along with the more likely occurrence of pest populations resistant to some agrochemicals.

51.1.2 Invasive Alien Species (IAS)

The continuous spread of new invasive species represents a significant economic problem for the damage that such species may cause to agriculture. The impact of IAS on the environment is often devastating. In Europe, southern countries are those most exposed to the accidental introduction of exotic organisms, and favorable climatic conditions may allow their acclimation. In recent years, the Mediterranean countries have seen the epidemic explosion of bacterial kiwi canker caused by *Pseudomonas syringae* pv *actinidiae*, *Xylella fastidiosa* on the olive tree, and marmorated stink bug (*Halyomorpha halys*) on many fruit crops in northern Italy. The renewed virulence with which some old pests occur and new pests accidentally introduced and spread uncontrollably thanks to more favorable climatic conditions, are putting at risk some crops and the profitability of farmers especially if, as in organic agriculture, they are without plant protection products capable of controlling them adequately.

51.2 The Biopesticide Market

Biopesticides, more commonly called BCA (Biocontrol Agents), in 2013 represent a 3.5% share of the world plant health market with a total value of $1.5 billion. Although Europe currently represents around 25% of the market, the tendency is nevertheless for rapid growth in the sector which, according to forecasts, should, in the coming years, cover a market share ranging between 15% and 23% (Olson, 2015; Cavoski et al., 2011). With these growth prospects, the promising sector of biopesticides is also of considerable interest for traditional agrochemical companies producing chemicals whose numbers, following the European revision of plant protection products, are continuously contracting. This will certainly give a further boost to the sector both for their investment and research capacity and for their experience in the formulation and authorization of new products.

51.3 Regulatory Aspects

The crop protection in Europe is experiencing a period of profound and important changes, primarily due to two the regulations issued by the European Union The first European act on plant protection products (PPPs) was the EU Regulation no.1107/2009, which defined the rules for the authorization of plant protection products and their marketing, providing for a rigorous assessment of plant protection products regarding the usefulness of active substances, efficacy, and potential risks. The other European provision is EU Directive No. 128/09 on the "sustainable use of plant protection products," which was implemented by each member state issuing a National Action Plan (NAP), which defines the actions to be taken to comply with the guidelines on the rationalization and sustainable use of plant protection products. The EU Reg.1107/09 in force today aims to harmonize the processes and requirements at the European level for the registration of new plant protection products, introduces the concept of zonal evaluation for the formulated products, and establishes "cut-off criteria" that classify some active substances as "candidates for substitution" on the basis of their toxicological properties (Damalas et al., 2018). The entire process of developing a new crop protection product necessarily produces a huge amount of data and studies. Obviously, this is a very expensive process that is becoming a strong obstacle for the registration of new plant protection products, as well as for biopesticides that still occupy a market "niche." At present, in Europe there is not a precise European regulatory framework that facilitates the authorization and commercialization of biopesticides. In fact, except macro-organisms, all biocontrol agents are currently subject to the same rules that are applied to traditional agrochemicals according to the methods defined by the EU Reg. 1107/09 currently in force. This is an articulated and complex regulatory framework that makes research aimed at developing biopesticides very expensive and time-consuming.

On the other hand, research on biocontrol agents is having a strong impetus from the need to find effective and sustainable alternatives to traditional chemical pesticide, both with a view to their use in the organic agriculture, which has always lacked solutions for effective crop protection products, and in conventional and integrated agriculture where their use could have the advantage, from one hand, of reducing the load of chemical residues on food, and on the other hand, reducing the risk of the onset of pest populations resistant to certain agrochemicals. In this new perspective the biopesticides will assume greater importance. The global biopesticide market yet appears fairly static; some innovations from both Europe and the United States are being authorized, but the time for obtaining authorizations do not differ from those necessary for traditional agrochemicals and are very long. The biopesticide sector is of growing importance in Europe; however, legislation continues to be an element that hinders the development of this new sector. Therefore, a simplification of the approval procedures for "low impact substances" is needed. With this respect, an EPPO guideline has been prepared dedicated to products considered to be "low risk" (EPPO, 2017).

The need to find effective biopesticides in Europe represents an essential factor particularly for the development of the organic agriculture sector today when European Union has limited copper application to 28 kg/ha in seven years for the control of bacterial and fungal diseases. Copper is considered the only effective active ingredient to control bacterial diseases which, due to climate change, are the crop diseases with the greatest risk of increase.

51.4 Biopesticides Authorized in Europe and Research

Thanks to a better toxicological profile, shorter preharvest times, shorter re-entry period, and reduction of residues on food, biopesticides in pest control strategies can have several technical advantages for the farmer both from the point of view of environmental safety and production, compared to conventional agrochemicals. In addition, the use of biopesticides may allow effective antiresistance strategies to be implemented by alternating them with traditional chemicals with different mode of action, therefore reducing the risk of pest populations resistant to chemicals.

51.4.1 Biofungicides

Forty years after the first attempts, biological control of plant diseases is still in its infancy and only a few biocontrol agents (BCA) are available on the European market. It is therefore interesting to review the progress and failures in biological control research and their implementation to avoid repeating the same mistakes made in the past. European project ENDURE reports that successes were much more for experiments performed under controlled conditions (in-vitro or in-planta) than for field tests. This is a very important point which points out the difficulties that are being encountered in transferring in the field the results of research carried out in the laboratory (Alabouvette et al., 2006).

The low impact biocontrol agents can be divided into two groups: natural microbiological agents, plant protection products based on one or more microorganisms (BCAs), and botanicals (active ingredients extracted from plants that are subsequently formulated in commercial products).

The mode of action of BCAs are complex and include:

- Antibiosis: inhibition of spore germination and or growth of the pathogen
- Exclusion: competition and compartmentalization of phytopathogenic fungi by microbial biofungicides
- Mycoparasitism of pathogenic fungi by microbial biofungicides
- Induction of the plant's endogenous defense mechanisms

Many microbial biofungicides can colonize and grow easily on treated soils or, more rarely, the phylloplane, providing various sources of different long-term microbial and biochemical modes of action. Genus *Bacillus* includes several strains that have long been commercially available as biofungicides. Some *Bacillus* strains such as *B. subtilis*, *B. amyloliquefaciens*, and *B. licheniformis*, and *B. pumilus* among others, produce antimicrobial metabolites such as some lipopeptides, which show biocidal or surfactant activity and a wide spectrum of action. Thanks to their lipoprotein structure, lipopeptides are capable to penetrate lipid membranes such as those surrounding the cell cytoplasm, affecting their integrity and permeability. Lipopetides such as Iturine and Fengicine show a strong antifungal activity for one rather broad spectrum of phytopathogenic fungi, with the Fengicine, which seem to have a greater efficacy for the control of Oomycetes such as *Pythium* and *Phytophthora*. *Bacillus subtilis* var. *amyloliquefaciens* strain FZB24 is active against soilborne diseases (*Rhizoctonia*, *Fusarium*, *Sclerotinia*, *Phytophthora*, and *Pythium*) and foliar diseases (grape downy mildew, powdery mildew, late blight and early blight of potato, as well as bacterial diseases), on vegetable crops, in fruit orchards and ornamental crops. *Bacillus subtilis* strain QST 713 acts as a stimulator of the plant's natural defenses and improves the physiology of the plant itself. At present, the microorganism is authorized on various fruit crops and grapevine to control some fungal and bacterial diseases. Microbial biofungicides can compete with pathogens for nutrients and infection sites on leaf surfaces and in the soil or for producing antimicrobial metabolites such as some lipopeptides, which show biocidal or surfactant activity. Some of them are capable to metabolize the radical exudates that stimulate the pathogen's germination, therefore reducing its potential ability to infect the plant. In addition to *Trichoderma*-based products, which have already become a reality in the European market (against Grapevine Esca disease and many soilborne pathogens) (Woo et al., 2014), others have recently been introduced such as *Ampelomyces quisqualis*, against *Uncinula necator*, and *Pythium olygandrum* M1 strain, against grapevine gray mold and collar rot and several vegetable and ornamental crops (Kohl et al., 2019; O'Brien, 2015).

Plants are capable to react to the attacks of plant pathogens and to promote the production of endogenous defense mechanisms to contain them. Chito-oligosaccharides and oligo-galacturonides of natural origin are authorized as elicitor for the control of grapevine powdery mildew. Laminarines, natural oligosaccharides extracted from the brown alga *Laminaria digitate*, are the active ingredient eliciting biochemical and morphological processes in the plant to contain grapevine and strawberry powdery mildew, lettuce downy mildew, pome fruit scab and lenticel rot, and bacterial necrosis of the buds. Cerevisane, an inert fraction of the yeast *Saccharomices cerevisiae*, is the active ingredient eliciting the production of phytoalexins and hydrogen peroxide directly involved in the control of *Botrytis cynerea*, *Plasmopara viticola*, *Erysiphe necator*, *Phytophthora infestans*, and *Bremia lactucae*. At present, the formulation is authorized on grapevine, lettuce, solanaceous and cucurbitaceous.

In Europe, the authorized plant essential oils (botanicals) include tea-trees essential oil, sweet orange essential oil, eugenol, thymol, and geraniol. Essential oil of sweet orange has been authorized for the control of grapevine powdery mildew and downy mildew, and powdery mildew of several vegetable and ornamental crops. The formulation is effective by direct contact, dehydrating and dissolving the mycelium and spores of powdery mildew pathogens. It also has insecticidal and repellent activity against several pests. The mixture of eugenol, thymol, and geraniol has been authorized for the control of grapevine gray mold. Tea tree essential oil is authorized in Europe against several fungal diseases. Group of "botanicals," highly effective but very volatile compounds, in the future, could be successfully developed if suitable formulation could increase their persistence, because their application methodology can be considered very similar to chemicals.

51.4.2 Insects and Mites

Most of the studies (93.5%) focused on biological control of insects and only four studies focused on biological control of mites (Alabouvette et al., 2006). The antagonists used in biocontrol mainly referred to microbial bioinsecticides, including *Bacillus thuringiensis*, fungi, and entomopathogenic nematodes. Currently, *B. thuringiensis* is the leader of the BCA market, which in Europe currently represents 25% of the market with important growth perspective. Although BCAs generally have less efficacy than chemical products, a potential market is estimated for these products which, in 2020, could be worth from $6 to 8 billion compared with the market of traditional agrochemicals, which is reducing its growth rate. Schematically, bioinsecticides can be divided into three categories: microbial insecticides, bioderived insecticides, and bioprotectors incorporated in plants. In the first category, bacteria, viruses, and entomopathogenic nematodes and fungi can be included. The entomopathogenic fungi acting as pathogenic parasites for insects (and for other invertebrates such as phytophagous nematodes) belong to the Ascomycetes (such as *Beauveria bassiana*) and *Zygomycetes* divisions. Among the entomopathogenic bacteria, are various strains of *Bacillus thuringiensis* (BT), which in Europe have a long and consolidated history of use for the control of some species of pests, particularly Lepidoptera. "Bioderived insecticides" are compounds commonly found in nature that are often produced by plants to defend themselves against pests, but they may also include microbial extracts capable of controlling certain species of insects (e.g., Spinosad), or substances capable of inducing mechanisms of resistance in the plants themselves. In this wide category we can also include the pheromones (sexual or aggregate) used in pest monitoring or mass trapping of harmful insects. A "chitosan hydrochloride" derived by mushroom or crustaceous chitin, is capable of inducing systemic acquired resistance (SAR) against certain species of insects and has currently been included in the list of basic substances as plant elicitor. Plant extracts such as garlic are capable to act as insect repellent, antifeeding but they can also have a direct effect on harmful insects. The third group includes bioinsecticides based on the inclusion of a portion of foreign DNA in the genetic material of the crop (GMO), which are currently not allowed by some European member states' legislation.

51.4.3 Bioherbicides

Weeding management has become a fixed practice on all crops, whether extensive, horticultural, or fruit, and today one-third of sold agrochemicals are herbicides. Weed management, is currently, one of the major obstacles to the affirmation of organic agriculture. The traditional alternative to weeding is the manual or mechanical removal of weeds. It is certainly a safe method for the environment and for human health but also time-consuming and more expensive. Alternatively, other agronomic practices limiting weeds such as mulching or flame weeding can be carried out but in both cases they are significantly more expensive and less practical methods. Unfortunately, the repeated use of chemical herbicides has created some problems such as the development of resistant weed populations and the increase of residues in food and water, which could have a much lesser impact if bioherbicides were available. For these reasons, research is moving towards the formulation of bioherbicides that may be applied as traditional herbicides, without having the same environmental impact. Although in the world of agrochemicals, the bioherbicides sector is currently the most lacking in immediate solutions. the characteristics that the herbicide of the near future should have are efficacy, low toxicity for the operator and the environment, poor residuality, specific or target-oriented. Bioherbicides on which research is working fall into two main categories: (1) microbial bioherbicides; and (2) phytotoxin herbicides. The first group includes mycoherbicides that may have a different mode of action, ranging from enzymes degrading the plant cell wall, and phytotoxic molecules such as diterpene, and growth-regulating hormones in plants such as indol-acetic acid (IAA).

The second group includes molecules having the same mode of action used by plants and microorganisms to exclude other species of invasive plants in their immediate vicinity, producing one or more substances (allelochemicals) that negatively influence their growth and survival. Phytotoxin of this kind are sorgoleone (a substance extracted from the roots of *Sorghum bicolor*), inhibiting protein D1 of photosystem II; leptospermone, an allelochemical herbicide produced by a blueberry (Owens et al., 2013). *Calistemon citrinus*, a separate category, is represented by the essential oils of many plants such as those extracted by *Thymus vulgaris* and *Satureia hortensis*. At present, in Europe, only pelargonic acid, herbicides with non-selective leaf absorption only acting by contact, destroying the leaf cuticle, is authorized.

51.5 Conclusion and Future Prospects

In the future we will have to get used to another concept of pest control, trying to move from the simple use of a biocontrol agent with a generally partial effect replacing a traditional agrochemical, to a global strategy that combines different techniques leading to safe and effective and competitive crop protection. BCAs generally have a lower efficacy compared to a traditional agrochemical. Therefore, they cannot completely replace them. However, they can nevertheless support the agrochemicals integrating their activity and alternate with them, particularly in presence of medium to low epidemic pressure. Therefore, their efficacy must be assessed in the context of a global and integrated pest control strategy and not only for their own performance. These difficulties could however be offset by the benefit of image and eco-sustainability that the use of these products can bring. Despite the progress made in understanding the modalities of action of BCA, the possibilities of biological control of plant diseases in Europe are still very limited. To meet the European regulatory requirements, biopesticides should have the following characteristics: efficacy; low toxicity for the operator and the environment; low residuality, target-oriented (Dhakal et al., 2019).

51.5.1 Advantages

Biopesticides in pest control strategies may have various technical advantages for the farmer both from the point of view of greater environmental safety and production, due to a better toxicological profile, shorter preharvest time and shorter reentry times, and for the reduction of residues on food production. In addition, the use of biopesticides may allow effective antiresistance strategies to be implemented by alternating them with chemicals having different modes of action, therefore reducing the risk of populations resistant to chemicals.

51.5.2 Disadvantages

Compared to traditional agrochemicals, disadvantages in using biopesticides are a lower efficacy and lower user-friendliness. Each formulation has its own characteristics which must be known in order to obtain the best efficacy. A wider biological and epidemiological knowledge of the pest or disease that farmers want to control, is needed by the farmer. In fact, BCAs if used in the wrong epidemic phase of the disease, may have no efficacy. Another disadvantage of some BCAs relies on the stability of the formulation which, if not properly stored can affect the effectiveness of the treatment. Finally, biopesticide should have good compatibility with agrochemicals. This is a factor that must be carefully considered in defining an integrated pest and disease control strategy (Figure 51.1).

To increase the use of biopesticide there are some gaps that should be overcome (Alabouvette et al., 2006):

I. lack of information on biocontrol against pests and diseases of some of the most important crops, particularly fruit orchards
II. lack of information on biocontrol against various pests of economically important crops such as rusts and downy mildews

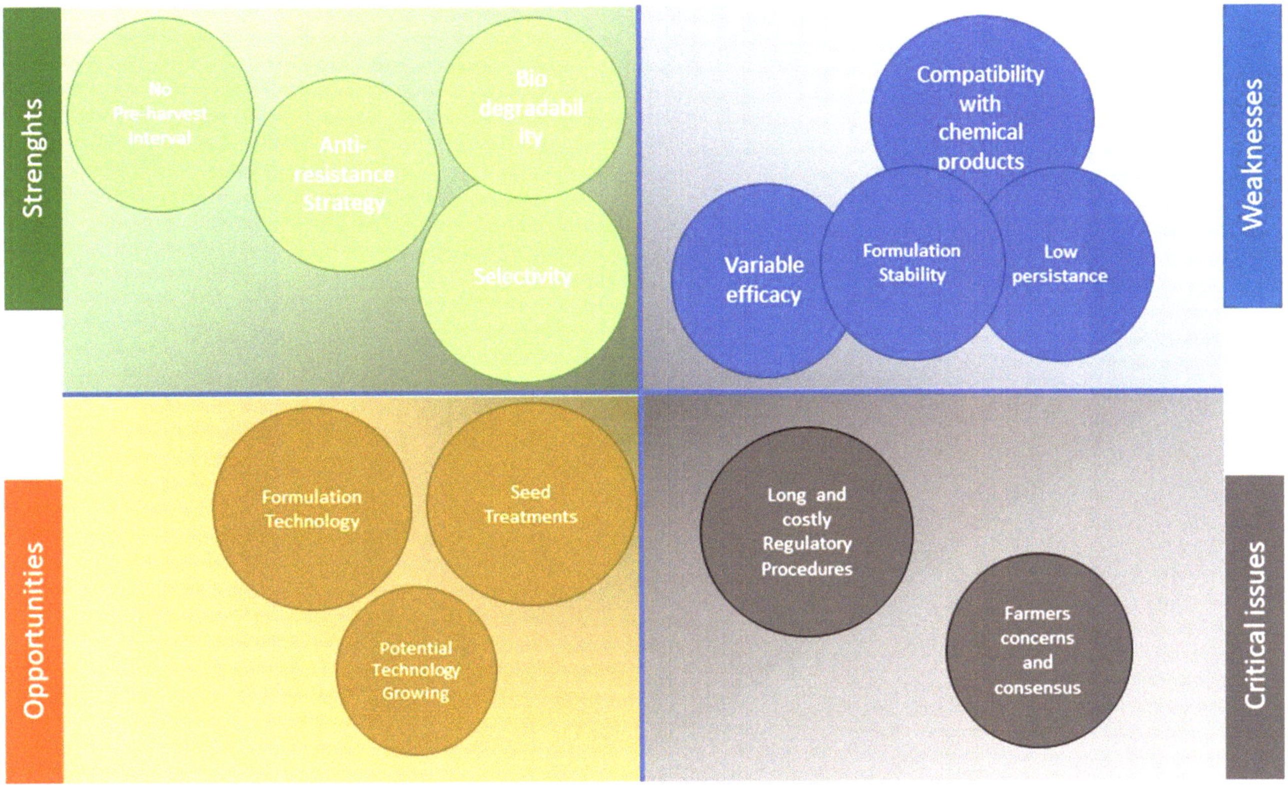

FIGURE 51.1 Strengths, weaknesses, critical issues, and opportunities for the development and use of biocontrol agents in pest and disease control strategies of the crops.

III. still limited knowledge on specific mode of action of some biopesticides. The knowledge available at the molecular level only focused on some biocontrol agents (e.g., *Trichoderma* spp. and *Pseudomonas* spp.)
IV. still very limited information on secondary metabolites produced by biocontrol agents
V. lack of understanding for the generally low field efficacy of compounds that induce resistance
VI. lack of knowledge on the variability of plant pathogens susceptibility to the action of BCA and on the possible consequences for efficacy in the field and its long-term duration

REFERENCES

Alabouvette C., Olivain C., Steinberg C. (2006). Biological control of plant diseases: The European situation. *European Journal of Plant Pathology* 114:329–341.

Cavoski I., Caboni P., Miano T. (2011). Natural pesticides and future perspectives. In *Pesticides in the Modern World – Pesticides Use and Management.* Ed. Margarita Stoytcheva. 169–190 published by InTech, Janeza Trdine 9, 51000 Rijeka, Croatia.

Damalas C. A., Koutroubas S. D. (2018). Current status and recent developments in biopesticide use. *Agriculture* 8: 13, doi:10.3390/agriculture8010013.

Dhakal R., Singh D. N. (2019). Biopesticides: A key to sustainable agriculture. International *Journal of Pure and Applied Biosciences.* 7 (3): 391–396.

EPPO PP 1/296 (1). (2017). Principles of efficacy evaluation for low-risk plant protection products. *Bulletin OEPP/EPPO Bulletin* 47 (3): 297–304.

Köhl J., Kolnaar R., Ravensberg W. J. (2019). Mode of action of microbial biological control agents against plant diseases: Relevance beyond efficacy. *Frontiers in Plant Sciences* 10: 845. www.frontiersin.org.

O'Brien P. A. (2017). Biological control of plant diseases. *Australasian Plant Pathology.* 46 (4): 1–12.

Olson S. (2015). An Analysis of the biopesticide market now and where it is going. *Outlooks on Pest Management* 26 (5): 203–206.

Owens D. K., Nanayakkara N. P. D., Dayan F. E. (2013). Inplant a mechanism of action of leptospermone: Impact of its physico-chemical properties on uptake, translocation, and metabolism. *Journal of Chemical Ecology* 39: 262–270.

Woo S. L., Ruocco M., Vinale F., Nigro M., Marra R., Lombardi N., Pascale A., Lanzuise S., Manganiello G., Lorito M. 2014. Trichoderma-based products and their widespread use in agriculture. *The Open Mycology Journal* 8 (Suppl-1, M4): 71–126.

52

Constraints and Challenges in Popularizing Biopesticides

Surendra K. Dara

CONTENTS

52.1 Introduction

Biopesticides refer to formulations of biological origin to control insects, mites, plant pathogens, plant-parasitic nematodes, weeds, and other pests. These could be based on microorganisms such as bacteria, fungi, nematodes, viruses, and yeasts; fermentation products of microbes or other biological molecules that do not contain live microorganisms; or plant extracts and essential oils (Table 52.1). Organically approved pesticides of non-biological origin are not included in this category. The demand for organic acreage, increased awareness of sustainable food production, and development of resistance to chemical pesticides are some of the reasons responsible for the increase in the biopesticide market around the world, both in organic and conventional production systems. Among various biologicals, the global market value of biopesticides was estimated to be $3-4 billion in 2018 and continues to increase (Marrone et al. 2019). However, in general, biopesticide use is still not a mainstream choice. This chapter will discuss the major constraints and challenges in popularizing biopesticides.

52.2 Farmers' Perception and Cost of Biopesticides

Compared to synthetic chemical pesticides, biopesticides are generally perceived to be slow, less effective, and inconsistent in their control. Lack of quality control to ensure efficacy, far less advanced technology to produce stable and efficacious formulations, lack of proper knowledge and guidelines in using biopesticides, and limited availability of field efficacy data to build confidence were some of the major limitations for widespread use of biopesticides in earlier years. While stricter regulatory and quality control procedures improved the quality and modern formulation technologies improved the shelf life and stability of the biopesticide formulations, the perception of low efficacy continues to be a major challenge. The knowledge of various biopesticide active ingredients, their modes of action or infection, and the best time and strategy for their use are also lacking at the farm level. Compared to synthetic chemical pesticides, biopesticides are usually expensive as mass production of certain active ingredients on a commercial scale is a complex process. Many farmers find the higher cost of biopesticides as a major deterrent to being their first choice or at least a part of the integrated pest management (IPM) program. Biopesticide development was limited to small companies catering to certain niche markets until very recently. With the recent increase in the demand for sustainable pest management, major pesticide companies have also been investing in biopesticide development, with several new products arriving on the market in recent years. Further improvement in production technology to optimize the cost, increased field research to generate efficacy data, and effective outreach to educate the farming community about biopesticide use strategies are critical in addressing these constraints. Studies that show the environmental cost of using synthetic chemicals or long-term economic benefits of using biopesticides would also help change farmers' perception and encourage them to use biopesticides.

52.3 Typical Choices and Associated Risks

Biopesticides containing active ingredients with modes of actions similar or close to those of synthetic chemical pesticides are generally more popular than others. This is due to their efficacy and speed of control, as typically seen with chemical pesticides. Although other biopesticides, especially those based on live organisms, can be very effective in pest suppression with minimal or no non-target effect, options with a rapid knockdown effect are generally preferred. For example, pyrethrins from the flowers of *Chrysanthemum cinerariaefolium* and spinosad derived from the bacterium *Saccharopolispora spinosa*

TABLE 52.1

Examples of Biopesticides and Their Primary Targets (Some Are Used for Controlling More Than One Group of Target Pests)

Group	Source	Target Pest
Bacteria	*Bacillus thuringiensis* subsp. *aizawai*	Insects
	B. thuringiensis subsp. *israelensis*	Insects
	B. thuringiensis subsp. *kurstaki*	Insects
	B. thuringiensis subsp. *tenebrionis*	Insects
	Paenibacillus popilliae	Insects
	*Chromobacterium subtsugae**	Insects and mites
	*Saccharopolyspora spinosa**	Insects and mites
	*Burkholderia rinojensis**	Insects, mites, and plant-parasitic nematodes
	Bacillus amyloliquefaciens	Plant pathogens
	Bacillus subtilis	Plant pathogens
	Streptomyces lydicus	Plant pathogens
	Pseudomonas fluorescens	Plant pathogens and weeds
	Bacillus megaterium	Weeds
	Bacillus firmus	Plant-parasitic nematodes
	Burkholderia cepacian	Plant-parasitic nematodes
	Pasteuria penetrans	Plant-parasitic nematodes
Fungi	*Beauveria bassiana*	Insects and mites
	Hirsutella thompsonii	Insects and mites
	Isaria fumosorosea	Insects and mites
	Lecanicillium lecanii	Insects and mites
	Lecanicillium longisporum	Insects and mites
	Metarhizium acridum	Insects
	Metarhizium anisopliae	Insects and mites
	Metarhizium brunneum	Insects and mites
	Myrothecium verrucaria	Plant-parasitic nematodes
	Paecilomyces lilacinus	Plant-parasitic nematodes
	Pochonia chlamydosporia	Plant-parasitic nematodes
	Coniothyrium minitans	Plant pathogens
	Trichoderma asperellum	Plant pathogens
	Trichoderma gamsii	Plant pathogens
	Trichoderma harzianum	Plant pathogens
	Trichoderma viridae	Plant pathogens
	Trichoderma koningiopsis	Weeds
	Colletotrichum gloeosporioides f. sp. *aeschynomene*	Weeds
Nematodes	*Heterorhabditis bacteriophora*	Insects
	Steinernema carpocapsae	Insects
	Steinernema feltiae	Insects
	Heterorhabditis heliothidis	Insects
	Heterorhabditis bacteriophora	
Viruses	Granuloviruses (GV):	Insects
	Cydiapomonella GV	Codling moth
	Pieris rapae GV	Imported cabbageworm
	Plutella xylostella GV	Diamondback moth
	Nucelopolyhedroviruses (NPV):	Insects
	Helicoverpazea NPV	Corn earworm
	Spodoptera exigua NPV	Beet armyworm
	Lymantria dispar NPV	Gypsy moth
	Tobacco mild green mosaic tobamovirus	Weeds
Yeasts	Cerevisane from *Saccahromyces cerevisae*	Plant pathogenic fungi
Spider	Spider venom peptide GS-omega/kappa-Hxtx-Hv1a*	Insects and mites
Plants	Azadirachtin from *Azadirachta indica*	Insects, mites, plant pathogens, and plant-parasitic nematodes
	Pyrethrins from *Chrysanthemum cinerariaefolium*	Insects and mites
	Clove oil from *Syzygium aromaticum*	Insects, mites, and weeds
	Peppermint oil from *Mentha x piperita*	Insects and mites
	Rosemary oil from *Salvia rosmarinus*	Insects and mites
	Cinnamaldehyde from *Cinnamomum verum*	Insects, mites, and plant pathogenic fungi
	Extract of *Reynoutria sachalinensis*	Plant pathogenic fungi
	Thymol from *Thymus vulgaris*	Plant pathogenic fungi

*Products with non-viable microorganisms.

Sources: Hedke et al. 2001; Lacey 2017; Dara et al. 2018; Abd-Wlgawad and Askary 2018; and product labels.

are among the most popular biopesticides and are the bases for their synthetic counterparts, pyrethroids and spinetoram, respectively. They have activity comparable to their synthetic analogs. Among microorganisms, *Bacillus thuringiensis*-based formulations are the most popular, especially against lepidopteran pests. The high efficacy of *B. thuringiensis* is due to a toxic crystalline protein complex rather than the infection process alone. Entomopathogenic fungi and nematodes cause insect mortality primarily due to infection, which can take a few days, so they are relatively less popular than toxin-based biopesticides. While some biopesticides seem to be more effective than others, similar to synthetic chemicals, their excessive use can lead to insect resistance. There are several examples of insect resistance to pyrethrins and their synthetic analogs, pyrethroids. Resistance to pyrethrins and other pesticides has been reported for several decades in pests of stored products, such as the wheat weevil, *Sitophilus granaries* (Lloyd 1969), the Indian meal moth, *Plodia interpunctella*, and the almond moth, *Cadra cautella* (Zettle et al. 1973), the red flour beetle, *Tribolium castaneum* (Lloyd and Ruczkowski 1980), and the pests of field crops such as the beet armyworm, *Spodoptera exigua* (Wu et al. 1995), the tarnished plant bug (*Lygus lineolaris*), (Snodgrass 1996), and the corn rootworm, *Diabrotica virgifera virgifera* (Souza 2019). Similarly, resistance of the diamondback moth, *Plutella xylostella*, to spinosad, abamectic, and *B. thuringiensis* (Sayyed et al. 2004); the western flower thrips, *Frankliniella occidentalis* (Herron and James 2005) and *S. exigua* (Moulton et al. 2000) to spinosad; of multiple insect species to *B. thuringiensis* applications and transgenic crops (Tabashnik 1994), viruses, and other biopesticides (Siegwart et al. 2015; Dara 2017) have been well documented. When there is a limited number of biopesticides that work like synthetic chemicals and their overuse leads to resistance issues, the resulting lack of control can have a negative impact on popularizing biopesticides. It is important to use pesticide risk mitigation strategies for biopesticides in organic and conventional agriculture.

52.4 Impact on Non-Target Organisms

While pyrethrins are very effective, they are not selective and harm natural enemy populations. Azadirachtin from *Azadirachta indica* is a multipurpose pesticide, with fungicidal, nematicidal, and insecticidal activity, as well as antifeedant, repellency, and insect growth regulation properties. Both pyrethrins and azadirachtin negatively impact natural enemies. Nikolova et al. (2015) reported that azadirachtin and pyrethrum reduced the number of predatory thrips, *Aeolothrips intermedius*, by 17.2 and 26.8%, respectively, and the number of predatory lady beetle *Coccinella septempunctata* by 11.2 and 16.4%, respectively in spring forage pea (*Pisum sativum*). While the authors claimed these reductions are not significantly harmful, the reported reductions in predator numbers are still important. Pyrethrins were highly toxic to *Anagyrus pseudococci*, a parasitoid of the vine mealybug (*Planococcus ficus*), resulting in 100% mortality compared to spirotetramat, which had less than 20% mortality, similar to untreated control (Veylit 2013). Medina et al. (2004) found that the fecundity of the predatory *Chrysoperla carnea* was reduced from azadirachtin, but it was reversible. Spollen and Isaman (1996) evaluated the impact of azadirachtin on predatory mites *Phytoseiulus persimilis* and *Amblyseius cucumeris* and the aphid predator, *Aphidoletes aphidimyza*. While some differences were not significant in the fecundity and adult mortality of predatory mites, fewer adults were found on treated surfaces. Larval mortality and adult emergence of *A. aphidimyza* were also significantly impacted by azadirachtin. Similarly, spinosad was as toxic as acephate, acetamiprid, imidacloprid, and some other chemicals to the acarophagous thrips, *Scolothrips takahashii*, and the lesser mite destroyer, *Stethorus japonicus*, while chemicals such as bifenazate, buprofezin, fluacrypyrim, and pymetrozine were tolerated by the predators (Mori and Gotoh 2009). In general, spinosad is more toxic to parasitoids than predators and 78–86% of laboratory and field-type studies showed that spinosad is moderately harmful to harmful towards parasitoids (Williams 2003). Spinosad and other biopesticides can also be toxic to pollinators (Tomé et al. 2015; Challa et al. 2019), which can be a limiting factor for their use in crops that require pollinators. While biopesticides are generally perceived to be safer to natural enemies and the environment, these examples demonstrate that they can be equally or sometimes more harmful to natural enemies than synthetic chemicals suggesting their judicious use.

52.5 Biopesticide Storage, Handling, and Compatibility

Certain biopesticides, especially those containing live organisms, require special storage and handling, which can be an additional burden when compared to synthetic pesticides. While several new formulations eliminate the need for refrigeration or storage below certain temperatures, there are certain products, such as those based on entomopathogenic fungi and nematodes, that still require special arrangements to maintain the viability of infective stages. Similarly, the shelf life of some biopesticides is very short, leading to economic losses or ineffective results. Additionally, mixing multiple agricultural inputs such as foliar nutrients, fungicides, acaricides, and insecticides to save on the application costs is a common practice in many cropping systems, and certain biopesticides may not be compatible with other tank mix materials. Furthermore, application of fungicides for disease control when entomopathogenic fungi are used for controlling arthropod pest, or application of certain biopesticides when biocontrol agents are used can impact pest management efficacy. These are some practical problems of biopesticide use, but there are also solutions for these issues. Compatibility of biopesticides with other tank mix materials or biocontrol agents is available either on pesticide labels and supplementary materials or in extension publications. Understanding the compatible combinations and avoiding incompatible ones will help increase effective biopesticide use. For example, the entomopathogenic fungus *Beauveria bassiana* is incompatible with captan and thiram, but compatible with several other fungicides (Dara et al. 2014). The fungus *Microdochium dimerum* used to control the gray mold fungus *Botrytis cinerea*, the giant knotweed extract (*Reynoutria sachalinensis*) used against powdery mildew by *Oidium neolycopersici*, and the entomopathogenic fungus *Lecanicillium muscarium* used against the silverleaf whitefly (*Bemisia tabaci*) were all compatible and did not affect the efficacy of each other when used together in greenhouse tomato (Bardin et al. 2008). In some greenhouse and open-field cropping systems, pollinators are used to deliver entomopathogenic fungi or fungus-based mycofungicides with minimal or no negative impact on pollinator health in a concept known as entomo-vectoring or bee-vectoring (Mommaerts and Smagghe 2011; Smagghe et al. 2012; Smagghe et al. 2013).

52.6 Conclusions and Prospects

The cost and efficacy of biopesticides and the knowledge for their effective use in IPM are the major challenges in their widespread use. Recent advances in biopesticide development, marketing, and knowledge of their role in IPM have made a significant impact in addressing some of the constraints and challenges associated with popularizing biopesticides. However, there is a dearth of biopesticide field efficacy data compared to chemical pesticides, which is necessary to build farmer confidence. This is partially due to limited expertise and funding for biopesticide research. For the long-term promotion of biopesticide use, there are three main recommendations: (1) invest in training students to increase the number of next-generation researchers and educators, (2) invest in research and development of effective and affordable biopesticides, and (3) improve the knowledge of farmers on biopesticide use through effective outreach methods.

REFERENCES

Abd-Elgawad, M. M. M. and T. H. Askary. 2018. Fungal and bacterial nematicides in integrated nematode management strategies. *Egyptian Journal of Biological Pest Control*. 28: 74. https://doi.org/10.1186/s41938-018-0080-x.

Bardin, M., J. Fargues, and P. C. Nicot. 2008. Compatibility between biopesticides used to control grey mould, powdery mildew and whitefly on tomato. *Biological Control*. 46: 476–483. https://doi.org/10.1016/j.biocontrol.2008.05.012.

Challa, G. K., D. M. Firake, and G. T. Behere. 2019. Bio-pesticide applications may impair the pollination services and survival of foragers of honey bee, *Apiscerana* Fabricius in oilseed brassica. *Environmental Pollution* 249: 598–609. https://doi.org/10.1016/j.envpol.2019.03.048.

Dara, S. K. 2017. Insect resistance to biopesticides. *UCANR E-Journal of Entomology and Biologicals*. http://www.ucanr.org/blogs/blogcore/postdetail.cfm?postnum=25819 (accessed 2 April 2020).

Dara, S. K., T. A. Goble, and D. I. Shapiro-Ilan. 2018. Leveraging the ecology of invertebrate pathogens in microbial control. In *Ecology of Invertebrate Diseases*, eds. A. E. Hajek and D. I. Shapiro-Ilan, 469–493. Oxford: Wiley.

Dara, S. S. R., S. S. Dara, A. Sahoo, H. Bellam, and S. K. Dara. 2014. Can entomopathogenic fungus *Beauveria bassiana* be used for pest management when fungicides are used for disease management? *UCANR E-Journal of Entomology and Biologicals* http://www.ucanr.org/blogs/blogcore/postdetail.cfm?postnum=15671 (accessed 2 April 2020).

Hedke, K., P. Lüth, and A. von Tiedemann. 2001. Contans-first biocontrol agent against *Sclerotinia sclerotiorum* in oilseed rape. In Proceedings of the 10th International Rapeseed Congress, pp. 26–29.

Herron, G. A. and T. M. James. 2005. Monitoring insecticide resistance in Australian *Frankliniella occidentalis* Pergande (Thysanoptera: Thripidae) detects fipronil and spinosad resistance. *Australian Journal of Entomology* 44: 299–303. https://doi.org/10.1111/j.1440-6055.2005.00478.x.

Lacey, L. A. 2017. *Microbial Control of Insect and Mite Pests: From Theory to Practice*. New York: Academic Press.

Lloyd, C. J. 1969. Studies on the cross-tolerance to DDT-related compounds of a pyrethrin-resistant strain of *Sitophilus granarius* (L.) (Coleoptera, Crculionidae). *Journal of Stored Products Research* 5: 337–356. https://doi.org/10.1016/0022-474X(69)90007-1.

Lloyd, C. J. and G. E. Ruczkowski. 1980. The cross-resistance to pyrethrins and eight synthetic pyrethroids, of an organophosphorus-resistant strain of the rust-red flour beetle *Tribolium castaneum* (Herbst). *Pesticide Science* 11: 331–340. https://doi.org/10.1002/ps.2780110307.

Marrone, P. 2019. Pesticidal natural products - status and future potential. Pest Management Science 75: 2325-2340. https://doi.org/10.1002/ps.5433

Medina, P., F. Budia, P. Del Estal, and E. Viñuela. 2004. Influence of azadirachtin, a botanical insecticide, on *Chrysoperla carnea* (Stephens) reproduction: Toxicity and ultrastructural approach. *Journal of Economic Entomology* 97: 43–50. https://doi.org/10.1093/jee/97.1.43.

Mommaerts, V. and G. Smagghe. 2011. Entomovectoring in plant protection. *Arthropod-Plant Interactions* 5: 81–95. https://doi.org/10.1007/s11829-011-9123-x.

Mori, K. and T. Gotoh. 2009. Effects of pesticides on the spider mite predators, *Scolothrips takahashii* (Thysanoptera: Thripidae) and *Stethorus japonicus* (Coleoptera: Coccinellidae). *International Journal of Acarology* 27: 299–302. https://doi.org/10.1080/01647950108684270.

Moulton, J. K., D. A. Pepper, and T. J. Dennehy. 2000. Beet armyworm (*Spodoptera exigua*) resistance to spinosad. *Pesticide Management Science* 56: 842–848. https://doi.org/10.1002/1526-4998(200010)56:10%3C842::AID-PS212%3E3.0.CO;2-H.

Nikolova, I., N. Georgieva, and N. Tahsin. 2015. Toxicity of neem and pyrethrum products applied alone and in combination with different organic products to some predators and their population density. *Romanian Agricultural Research* 32: 291–301.

Sayyed, A. H., D. Omar, and D. J. Wright. 2004. Genetics of spinosad resistance in a multi-resistant field-selected population of *Plutellaxylostella*. *Pest Management Science* 60: 827–832. https://doi.org/10.1002/ps.869.

Siegwart, M., B. Graillot, C. B. Lopez, S. Besse, M. Bardin, P. C. Nicot, and M. Lopez-Ferber. 2015. Resistance to bio-insecticides or how to enhance their sustainability: A review. *Frontiers in Plant Science* 6: 381. https://doi.org/10.3389/fpls.2015.00381.

Smagghe, G., L. De Meyer, I. Meeus, and V. Mommaerts. 2013. Safety and acquisition potential of *Metarhizium anisopliae* in entomovectoring with bumble bees, *Bombus terrestris*. *Journal of Economic Entomology* 106: 277–282. https://doi.org/10.1603/EC12332.

Smagghe, G., V. Mommaerts, H. Hokkanen, and I. Menzler-Hokkanen. 2012. Multitrophic interactions: The entomovector technology. In *Arthropod-Plant Interactions. Progress in Biological Control*, eds. G. Smagghe and I. Diaz, 127–157. Dordrecht: Springer. https://doi.org/10.1007/978-94-007-3873-7_5.

Snodgrass, G. L. 1996. Insecticide resistance in field populations of the tarnished plant bug (Heteroptera: Miridae) in cotton in the Mississippi Delta. *Journal of Economic Entomology* 89: 783–790. https://doi.org/10.1093/jee/89.4.783.

Souza, D. S. O. 2019. Characterization of pyrethroid resistance in the western corn rootworm *Diabrotica virgifera virgifera* LeConte. PhD Dissertation, The University of Nebraska-Lincoln, pp. 303. https://digitalcommons.unl.edu/cgi/viewcontent.cgi?article=1059&context=entomologydiss (accessed 2 April 2020).

Spollen, K. M. and M. B. Isman. 1996. Acute and sublethal effects of a neem insecticide on the commercial biological control agents *Phytoseiulus persimilis* and *Amblyseiuscucumeris* (Acari: Phytoseiidae) and *Aphidoletesaphidimyza* (Diptera: Cecidomyiidae). *Journal of Economic Entomology* 89: 1379–1386. https://doi.org/10.1093/jee/89.6.1379.

Tabashnik, B. E. 1994. Evolution of resistance to *Bacillus thuringiensis*. *Annual Review of Entomology* 39: 47–79. https://doi.org/10.1146/annurev.en.39.010194.000403.

Tomé, H. V., W. F. Barbosa, G. F. Martins, R. N. C. Guedes. 2015. Spinosad in the native stingless bee *Melipona quadrifasciata*: Regrettable non-target toxicity of a bioinsecticide. *Chemosphere* 124: 103–109. https://doi.org/10.1016/j.chemosphere.2014.11.038.

Veylit, L. 2013. Testing the toxicity of organic and IPM compatible insecticides on *Anagyrus pseudococci*, a parasitoid of the vine mealybug, *Planococcus ficus*, over multiple generations. *Pesticides and Parasitoids*. https://nature.berkeley.edu/classes/es196/projects/2013final/VeylitL_2013.pdf. (accessed 1 April 2020).

Williams, T., J. Valle, and E. Viñuela. 2003. Is the naturally derived insecticide Spinosad® compatible with insect natural enemies? *Biocontrol Science and Technology* 13: 459–475. https://doi.org/10.1080/0958315031000140956.

Wu, S. C., Y. Z. Gu, Z. L. Sheng, and H. Z. Zheng. 1995. Monitoring of insecticides resistance and chemical control in beet army moth. *Acta Phytophylacica Sinica* 22: 95–96.

Zettle, J. L., L. L. McDonald, L. M. Redlinger, and R. D. Jones. 1973. *Plodia interpunctella* and *Cadracautella* resistance in strains to malathion and synergized pyrethrins. *Journal of Economic Entomology* 66: 1049–1050. https://doi.org/10.1093/jee/66.5.1049.

Section XXII

Certification Process and Standards of Organic Farming

53

Certification Process and Standards of Organic Farming

Prakruthi M., Mahesh M.S., and Shekhara Naik R.

CONTENTS

53.1 Introduction

Organic agriculture is a management system which nurtures and strengthens the agro-ecosystem health, including biodiversity, biological cycles and soil biological activity, which is accomplished (Kalainith 2013) by using on farm agronomic, biological and mechanical methods and largely excluding the use of synthetic inputs (such as fertilizers, pesticides, hormones, feed additives etc.) (FAO 1999).

Globally, 0.9% of the land is certified agricultural land, out of which 31 million hectares of land is farmed with wild organic collections (berries in Finland, honey in Zambia, and medicinal plants in India). There are nearly 1.9 million organic producers in 164 countries (e.g., India 600,000, Uganda 190,000, Mexico 170,000) (Nadia 2014).

53.2 Organic Certification

Organic certification system is a quality assurance initiative, aspired to ensure quality, prevent fraud, and promote commerce, based on a set of standards and principles. The system was developed in the early 1970s and by the 1980s there were organic certification bodies in most OECD countries. Today, there are 70 countries that have a domestic certification organization, and a dozen internationally active organizations that render their organic certification services effectively all over the world (TOS 2005).

The "organicness" of a product cannot be proved by looking at the harvested product or by testing it. Rather, it is ascertained through documentation and inspection of the whole production process (Rundgren 2008). It is a process for producers of organic food and other organic plant products. The standards required for organic production varies from country to country and generally entail a set of production standards for growing, storage, processing, packaging, and shipping that include:

- Avoidance of synthetic chemical inputs and genetically modified organisms.
- Farmland that has been free from chemicals for a number of years (often, three or more) shall be utilized for cultivation.
- Record-keeping practices, which must include written documents of the whole production process.
- Separation of organic and non-organic products.
- Regular on-site inspections.

53.3 Standards and Certification Systems

In the world arena, the FAO/WHO Codex Alimentarius Commission (CAC) has set off guidelines for production, labeling, and marketing of organically produced foods, to guide the producers and to protect the consumers against deception and fraud. These guidelines have been agreed upon by all member states of the CAC and are regularly updated. The CAC approved a plant protection guideline in June 1999, and that of animal production in July 2001.

Codex Alimentarius Commission guidelines include accepted management principles for the production of plants, livestock, bees, and their products; for handling, storage, processing, packaging and transportation of products, and a list of permitted substances used in the production and processing of organic foods (Dittrich 2012). These particular guidelines of the criteria for the permitted substances and the process by which the inspection is carried out and the certification is held are regularly reviewed within CAC along with other specifications. The CAC organic guidelines are significant in the frame of the WTO/TBT agreement and officially have the status of recommendations but carry a certain legal value as the official international standard. The government of the developing countries should always be encouraged in developing national standards in order to follow and ensure compatibility with the CAC guidelines.

The International Basic Standards for Organic Production and Processing guidelines are created by the International Federation of Organic Agricultural Movements (IFOAM) for the private sector. IFOAM basic standards—together with the IFOAM accreditation criteria—constitute the IFOAM norms, which provide a framework for certification bodies and standard-setting organizations worldwide to develop their own certification standards. They reflect the current scenario of organic production and processing methods and include a list of substances permitted in the production.

53.3.1 International Federation of Organic Agricultural Movements (IFOAM)

The IFOAM standards and the IFOAM accreditation requirement (IAR) are used by the international organic accreditation service (IOAS) in the IFOAM accreditation process for organic certification bodies. The IOAS assesses the standards (used by the certifier) against the IFOAM standard, as well as the certification body's performance against the IFOAM accreditation requirements (IFOAM 2014).

Certification bodies must implement the requirements of the IFOAM standard relevant to the certified farming or processing operations in order to become IFOAM Accredited Certification Bodies (ACBs). The IFOAM standards may also be used by non-accredited certification and standard-setting organizations as a way to outsource their standard-setting activity to IFOAM-organics international. In addition, governments and other standard setters may use the IFOAM standard as a reference to develop their own regulation or standard (IFOAM 2014).

53.3.2 United States Department of Agriculture (USDA Organic)

Organic certification authenticates that a farm or handling facility located anywhere in the world complies with the USDA organic regulations and permits the producer to sell, label, and represent the products as organic. Producers can use the word "organic" or the USDA organic seal on food, feed, or fiber products provided that the specific standards which are described by the regulating body are satisfied. The USDA National Organic Program (NOP) administers these regulations, with substantial input from its citizen advisory board and the public (USDA 2012).

NOP, part of USDA's Agricultural Marketing Service (AMS), has regulatory oversight responsibilities over the USDA organic standards and the accreditation of organic certifying agents. The NOP also has the authority to take appropriate legal action to enforce the organic standards and thus protect the integrity of the USDA organic standards, from farm to market, around the world (USDA-NOP 2016).

53.3.3 Asia Region Organic Standards (AROS)

The Asia Region Organic Standards describes the requirements for organic production. It covers plant (including mushrooms) production, collection of wild products, and also the processing and labeling of products obtained from these activities. This standard provides a mechanism to define the expectations for organic products. When these processes comply with the guidelines AROS, it accredits the producers to label their products as organic. However, the standard does not cover procedures for verification, such as inspection or certification of products.

The main purpose of formulating the AROS is to facilitate equivalence between organic standards and certification programs within and beyond the Asia region. It was also anticipated that AROS could be adapted to serve as the national standard for some of the individual countries in the region (AROS 2012). The Global Organic Market Access (GOMA) Working Group for Cooperation on Organic Labelling and Trade for Asia agreed on the vision and objectives for AROS. The standard can be adapted as a national standard, particularly by countries that have not yet developed and promulgated their own standard.

53.3.3.1 Objectives

Objectives are: to facilitate access of organic products in both intra- and inter-regional markets of the Asian countries, promote regional cooperation, enhance private and public sector dialogues, and enhance exchange of information. A long-term goal is to foster global recognition of regional standards.

53.3.4 National Program for Organic Production (NPOP)

The National Programme for Organic Production (NPOP) provides the Standards for the organic production, systems, criteria, and procedure for accreditation of certification bodies, the National (India Organic) Logo, and the regulations governing its use. The standards and procedures have been formulated in harmony with other International Standards regulating import and export of organic products (APEDA 2014). The NPOP also proposes to provide an institutional mechanism for the implementation of the National Standards for Organic Production (NSOP). The aims of the NPOP include the following:

i. To encourage the development of organic farming and organic processing.
ii. To facilitate certification of organic products in conformity with the NSOP as well as with the importing countries organic standards as per equivalence agreement between the two countries or as per importing country requirements.
iii. To evaluate the certification program for organic agriculture and products as per the approved criteria.
iv. To accredit certification programs of certification bodies seeking accreditation under this programme.

53.4 General Requirements for Organic Food Processing and Handling

The processing and handling of organic products should be optimized to maintain the quality and integrity of the product. The producer or the operator must develop an organic production and handling plan, which must include the following:

- Description of practices and procedures to be employed.
- List of each substance/inputs used during production, storage, and handling, indicating its composition, source, locations where it will be used shall be mentioned.
- Description of the monitoring practices and procedures that are followed and maintained to verify the plan should be effectively implemented.
- Description of the record-keeping system implemented to comply with the requirements of respective governing bodies (NOP, NPOP, IFOAM, AROS, etc.).
- Description of the management practices and separation measures established to prevent interweaving of organic and non-organic products during each and every step of processing and handling.
- Sources and chances of pollution should be identified and controlled to avoid contamination.
- Each and every product shall be adequately identified throughout the process.
- Certification program shall regulate the means the measures to be allowed.
- Regular operations of decontamination, cleaning including disinfection should be done in the organic food processing and storage area.

53.4.1 Pest Control

- A plan for pest prevention and control should be established.
- Pests should be avoided by GMPs.
- Employment of pest regulating agents must thus be regarded as the last call.
- Physical barriers, sound, ultra-sound, light and UV-light, traps (pheromone traps and static bait traps), temperature control, controlled atmosphere, and diatomaceous earth are the recommended pest control treatments (NPOP 2014).
- For pest management and control the following measures shall be used in order of priority:
  i. Preventive methods such as disruption, elimination of habitat, and access to facilities.
  ii. Mechanical, physical, and biological methods.
  iii. Pesticides prescribed by the respective national standards.
  iv. Other substances used in traps.
- Irradiation processing is prohibited.
- There shall never be direct or indirect contact between organic products and the substances prohibited.
- Carcinogenic pesticides and disinfectants are not permitted.

53.4.2 Ingredients

- The ingredients used in processing should be 100% organic, except where an organic ingredient is not available in sufficient quality or quantity, non-organic ingredients may be used to a minimum extent only if in case of essential technological demands or for particular nutritional purposes. If that non-organic raw material is not genetically engineered. The accredited certification body may approve the use of non-organic raw materials subject to periodic re-evaluation (NPOP 2014).
- The same ingredient within a single product shall not be derived from both organic and non-organic origin.
- Preparations of micro-organisms and enzymes commonly used in food processing may be used, until and unless it is genetically engineered micro-organisms and their products. For the production of enzymes and other microbiological products, the medium shall be composed of organic ingredients.
- The use of minerals (including trace elements), vitamins, and similar isolated ingredients shall not be permitted. The certification program may grant exceptions

where use is legally required or where severe dietary reasons or nutritional deficiency can be demonstrated.

- Ethylene gas is permitted for ripening.

53.4.3 Processing Methods

- Processing methods shall be based on physical and biological processes.
- The essential qualities of an organic ingredient shall be maintained throughout steps employed in the processing methods and shall be chosen to limit the number and quantity of additives and processing aids.
- The processing methods such as mechanical and physical, biological, smoking, and extraction precipitation filtration are approved to be employed in the production and processing of organic food.
- Extraction shall be either with water, ethanol, plant and animal oils, vinegar, carbon dioxide, nitrogen, or carboxylic acids. These shall be of food grade, appropriate for the purpose (NPOP 2014).
- Filtration substances shall not be made with the permeate substances that may negatively affect the safety of the product.
- Irradiation is not allowed.

53.5 The General Certification Process

To acquire organic certification for a product, a producer should go through and follow the process.

Producer or handler adopts organic practics; submits application and fees to certifying agent

↓

Certifying agents reviews applications to verify that practices comply with the regulations of the respective country

↓

Inspector conducts an on-site inspection of the applicant's operatiion.

↓

Certifying agent reviews the application and the inspector's repot to determine if the applicant complies with organic standard regulations of the respective country

↓

Certifying agent issues organic certificate

Procedure for General certification process
Source: USDA 2012. NOP, agricultural marketing services.

53.6 Packaging of Organic Products

Once the organic produce is harvested, the products need to be protected for the safety and suitability of the product for consumption. Because of the organicness of the product, there are specifications on packaging materials as well.

53.6.1 Packaging Requirements

i. Biodegradable, recyclable, reusable systems and eco-friendly packaging materials shall be used wherever possible.
ii. The packaging material used shall not contaminate food. Certain additives for use in manufacturing of packaging films for packaging of organic foodstuffs are allowed for restricted use (listed in Table 53.1).
iii. The packages shall be closed in such a manner that substitution of the content cannot be achieved without manipulation or damage of the seal.
iv. The accredited certification body shall approve the packaging material for use.

When the standards of the requirements of the organic agriculture of the respective certification board are fulfilled, the products can be sold as "produce of organic agriculture" labeled on the package. A certain set of requirements need to be met during labeling the organic produce.

53.7 Labeling Requirements of Organic Products

i. Labeling shall convey unambiguous and accurate information on the organic status of the product.
ii. The label for conversion products shall be clearly distinguishable from the label for organic products by mentioning the year of conversion.
iii. The name and address of the person or company legally responsible for the production or processing of the product shall be mentioned on the label.
iv. Every element of additives and processing aids shall be proclaimed. Product labels should mention the processing procedures, which influence the product properties in any way, not immediately but eventually.
v. Additional product information shall be provided on request.
vi. Ingredients or products extracted from wild production shall be mentioned as such.

53.7.1 Labeling Requirements for Processed Products

- Single-ingredient products may be labeled as "organic" when all standard requirements have been met.
- Multi-ingredient products where not all ingredients, including additives, are of organic origin may be labeled in the following way (raw material weight):
  i. If a minimum of 95% of the ingredients used in the product are of certified organic origin, the product may be labeled "certified organic" and should carry the logo of the certification program.

TABLE 53.1

Permitted and Restricted Packaging Materials

S. No.	Products	Limitations
1.	4,4′-Bis(2-benzoxazolyl)stilbene	Restricted
2.	9,9-Bis(methoxymethyl)fluorine	Restricted
3.	Carbonic acid, copper salt	
4.	Diethyleneglycol	Restricted
5.	2-(4,6-Diphenyl-1,3,5-triazin-2-yl)-5-[(hexyl)oxy]-phenol	
6.	Ethylenediaminetetraacetic acid, copper salt	Restricted
7.	2-(2′-Hydroxy-3′,5′-di-tert-butylphenyl)-5-chlorobenzotriazole	
8.	2-Methyl-4-isothiazolin-3-one	Restricted
9.	Phosphoric acid, trichlorocthylester	
10.	Polyesters of 1,2 propanediol and/or 1,3-and 1,-4 butanediol and/or polypropyleneglycol with adipic acid, also end-capped with acetic acid or fatty acids C10-C18 or n-octanol and/or n-decanol	Restricted
11.	1,1,1-Trimethylolpropane	
12.	3-Hydroxybutanoic acid 3-hydroxypentanoic acid, copolymer	Restricted

ii. When less than 95% and greater than 70% of the ingredients are of certified organic origin then, the product may not be called "organic"; instead the word "organic" may be applied on the principal display in statements like "made with organic ingredients" provided there is a clear statement of the proportion of the organic ingredients. An indication that the product is covered by the certification program should be used, close to the indication of the proportion of organic ingredients (NPOP 2014).

iii. If less than 70% of the ingredients are of certified organic origin, such a product may not be called "organic," because the indication that an ingredient is organic may appear in the ingredients list.

- Added water and salt shall not be included in the percentage calculations of organic ingredients. The use of iodized salt for aquaculture products shall be referred on the labels.
- The raw materials of a multi-ingredient product shall be labeled on the package of the product according to their weight percentage and shall also be mention that the ingredient is of certified organic origin or is a non-organic ingredient. All the additives used in the processing shall be listed with their full name.
- If herbs and/or spices constitute less than 2% of the total weight of the product, they may be listed as "spices" or "herbs" without stating the percentage.
- Organic products shall not be labeled as GE (genetic engineering) or GM (genetic modification) free in order to avoid potentially misleading claims about the end product. Any reference to genetic engineering on product labels shall be limited to the production method (NPOP 2014).
- The label of a certified organic product must delineate the name and logo of the accredited certification body, accreditation number.
- Before the labels are used, the accredited certification body shall verify the labeling requirement and approve the labels of their certified operators.

53.8 Conclusion and Future Prospects

Each and every step of the entire process of organic farming, i.e., from the procurement of seed till the sale of the final product should comply with the guidelines and standards given by the certifying bodies (IFOAM, USDA, APEDA, etc.) of the respective country in order to sell or label the product as "organic." Thus, fair practices that comply with the set guidelines and standards certifying bodies of the respective countries are required in each and every step of the production process in order to name the product as "organic."

REFERENCES

APEDA. 2014. National programme for organic production, Department of commerce. Ministry of Commerce and Industry. New Delhi. https://apeda.gov.in/apedawebsite/organic/organic_contents/Chapter_1.pdf (Accessed on November 29, 2019).

AROS. 2012. Global Organic Market Access. Food and Agriculture Organization of the United Nations, Rome.

Dittrich, P. 2012. Organic agriculture. In *Rural Development, Food Security and Nutrition*. European commission, Brussels, Belgium.

FAO (Food and Agriculture Organization of the United Nations). 2001. *Guidelines for the Production, Processing, Labelling and Marketing of Organically Produced Foods* (Gl 32 – 1999, Rev. 1 – 2001). FAO, Rome.

IFOAM-Organics International. Annual Report 2014. The IFOAM (International Federation of Organic Agriculture Movements (IFOAM - Organics International)) norms for organic production and processing.

Kalainith. 2013. Organic farming, agropedia. http://agropedia.iitk.ac.in/content/organic-farming-6 (Accessed on January 30, 2020).

National Programme for Organic Production. 2014. APEDA. Department of Commerce. Ministry of Commerce and Industry. New Delhi. https://apeda.gov.in/apedawebsite/organic/organic_contents/Appendix_5_Processing_and_Handling.pdf (Accessed on November 29, 2019).

Rundgren, G. 2008. *Best Practices for Organic Policy What Developing Country Can Do to Promote the Organic Agriculture Sector*. United Nations, New York.

Scialabba, N.E.H. 2014. Paper presented in the Intersessional Meeting of the Intergovernmental Group on Tea, Rome.

TOS (The Organic Standard). 2005. The organic certification directory. In the organic standard. http://www.organicstandard.com.

USDA (United States Department of Agriculture). 2012. Organic certification. http://www.fieldsourcefoods.com/about/organic-certification.

USDA-NOP (United States Department of Agriculture-National Organic Program). 2016. The national organic program. http://www.ams.usda.gov/nop (Accessed on November 29, 2019).

Section XXIII

Marketing and Export Potential of Organic Products

54

Marketing and Export Potential of Organic Products

Meena Thakur and Enespa

CONTENTS

54.1 Introduction

The increased use of synthetic substances met with fierce criticism and gave birth to multiple organic farming movements that are gaining gradual momentum across the world (Ronald and Adamchak, 2017). Organic agriculture (OA) is defined by the IFOAM as "a production system that sustains the health of soils, ecosystems, and people". It emphasizes the use of management practices in preference to the use of off-farm inputs, considering that regional conditions require locally adapted systems (Mgbenka et al., 2015; Nandwani and Nwosisi, 2016). This is accomplished by using, where possible, agronomic, biological, and mechanical methods, as opposed to using synthetic materials, to fulfill any specific function within the system (Kavitha, 2016). Organic agriculture combines tradition, innovation, and science to benefit the shared environment and promote fair relationships and good quality of life for all involved (Shetty et al., 2013; Eneizan, 2017).

54.1.1 Principles of Organic Agriculture

The main principle of organic agriculture is to work as much as possible within a closed system, and draw upon local resources to:

- Maintain the long-term fertility of soils.
- Avoid all forms of pollution that may result from agricultural techniques.
- Produce foodstuffs of high quality and sufficient quantity.
- Reduce the use of fossil energy in agricultural practices to a minimum.
- Make it possible for agricultural producers to earn a living through their work and develop their potential as human beings.

Growing awareness of health and environmental issues in agriculture has demanded production of organic food which is emerging as an attractive source of rural income generation (Poudel et al., 2012; Rana and Paul, 2017). While trends of rising consumer demand for organics are becoming discernible, sustainability in production of crops has become the prime concern in agriculture development (Barański et al., 2014). The categorization of a product as organic implies two main things:

- First, it is free from toxic persistent pesticides, synthetic fertilizers, growth hormones, and antibiotics or genetically modified organisms (GMOs).
- Second, stringent organic cultivation standards are followed, with respect to impact on soil, water, and air.

54.1.2 Key Drivers for Organic Agriculture

Organic agriculture systems and products are not always certified and are referred to as "non-certified organic agriculture or products" (Seufert et al., 2012). This excludes agriculture systems that do not use synthetic inputs by default, i.e., systems that lack soil building practices and degrade land (Jayne et al., 2014; Gomiero, 2016). Three different driving forces can be identified for organic agriculture:

- *Consumer or market-driven organic agriculture*: Products are clearly identified through certification and labeling. Consumers take a conscious decision on how their food is produced, processed, handled, and marketed (Sillani and Nassivera, 2015; Grunert and Aachmann, 2016). The consumer therefore has a strong influence over organic production.
- *Service-driven organic agriculture*: In countries and regions such as in the European Union (EU), subsidies for organic agriculture are available to generate environmental goods and services, such as reducing groundwater pollution or creating a more biologically diverse landscape (Reganold and Wachter, 2016).
- *Farmer-driven organic agriculture*: Some farmers believe that conventional agriculture is unsustainable and have developed alternative modes of production to improve their family health, farm economies, and/or self-reliance (Trauger and Passidomo, 2012). In many developing countries, organic agriculture is adopted as a method to improve household food security or to achieve a reduction of input costs (Panneerselvam et al., 2012). Produce is not necessarily sold on the market or is sold without a price distinction as it is not certified. In developed countries, small farmers are increasingly developing direct channels to deliver non-certified organic produce to consumers (Méndez et al., 2010; Barham and Weber, 2012). In the United States of America (USA), farmers marketing small quantities of organic products are formally exempt from certification.

54.1.3 Benefits of Organic Agriculture

The consumers are increasingly becoming aware of the food safety issues and environmental issues because of their increased concern about health, the environment's health, and its global implications (Roitner-Schobesberger et al., 2008; Pouta et al., 2010). Organic food now has become an available alternative for an increasing number of consumers worried about the presence of chemical residue and the environmental consequences of such activities (Pearson et al., 2011; Rana and Paul, 2017). Organic farming also helps stabilize and even increase farmers' income due to growing market demand and support from the government. The benefits of organic products are many as summarized below:

1. *Health benefits*: Organic agriculture regulates how food is grown and processed. In addition to meeting the health and safety requirements of conventional food, organic food must also meet the additional safety standards of organic farming such as tougher regulations on manure use (Reganold and Wachter, 2016; Seufert et al., 2017). Studies have shown that organic food contains a lower concentration of pesticides as compared to conventionally grown food (Barański et al., 2014). Hence, there is a good reason to believe that organic farming reduces health risks (Azadi and Ho, 2010). The health benefits of organic products include:
   - *Toxin and GMO-free*: Organic products are the most heavily regulated food products and come with a guarantee that no toxic persistent pesticides, synthetic fertilizers, or GMOs are used in their production (Gunasekaran and Murugan, 2020). Additionally, no antibiotics or growth hormones are given to livestock.
   - *Safer to consume*: Recent studies have found that organic fruits, vegetables, and grains have fewer nitrates and cadmium and fewer pesticide residues than non-organic crops, making them safer to consume (Matt et al., 2011; Reganold and Wachter, 2016).
2. *Environmental benefits*: Organic cultivation promotes conservation of biodiversity, and enhances ecological functions and ecosystem services via ecofriendly practices (Baudron and Giller, 2014). The stringent organic certification standards enable several measures in mitigating climate change (Creutzig et al., 2015).
3. *Economic benefits*: Certified organic foods fetch higher prices as most of the organic produce is directly procured by companies thereby excluding the middlemen (Halder and Pati, 2011). The retail price of organic food is high and eventually, organic production reduces cost also. This provides farmers a good return on an organized supply chain system.

54.1.4 Global Organic Market

The market and the area under organic cultivation has grown remarkably since the 1990swith the growing demand of organic produce globally. The market for organic products (CAGR 16%) is growing faster than the global markets (CAGR 10%) for conventional products (Xie et al., 2013). Organic farming is practiced with varying levels of success in 178 countries (Thøgersen, 2010). However, the North American and European Union regions (as single markets) generate the bulk of global sales. The global sales increased to \$89.7 billion in 2016 from \$7.9 billion in 2000 (Kushwah et al., 2019). Country-wise, the top consumers of organic products are the United States (\$43.1 billion), followed by Germany (\$10.5 billion) and France (\$7.5 billion). The increase in demand has led to a considerable increase in the area subject to organic management techniques globally, surging from 11 million ha in 1999 to 69.8 million ha in 2017 (Willer and Lernoud, 2019). The wild harvest and other non-agriculture organic collection area also increased to 39.9 million ha in 2016 from 4.1 million ha in 1999. The three countries with the largest area under organic cultivation are Australia (27.1 million ha), followed by Argentina (3.0 million ha) and China (2.3 million ha). The three countries with the largest wild harvest area for organic products are Finland (11.6 million ha), followed by Zambia (6.7 million ha) and India (4.2 million ha) (Rao and Rao, 2019).

The transition from a "production-driven supply chain" to a "demand-driven value chain" has been phenomenal, especially in the last five years with the organic market growing significantly (Adikenavar and Monica, 2018). In the organic food and beverages segment alone, India has witnessed a market growth of INR40,000 million in 2016–2017 from INR675 crore in

2009–2010, with an annual growth rate of 25% (Bandhu, 2019). While, the growth rate for conventional food processing industries is 10–15% (Miyasaka et al., 2020), 97% of the companies surveyed by ICRIER in 2017 reported 10%–40% higher profit margins on organic products (Mukherjee et al., 2017; Mehdi et al., 2019). This differential growth rate is observed in multiple market segments, including food and beverages, textiles, health and wellness, and beauty and personal care, among others. The rapid growth of the organic market can be attributed to various factors. The increasing emphasis on good health, proliferation of consumption-related ailments, an increased awareness regarding the health benefits of organic products among consumers, enhanced income levels and standard of living, together with government initiatives aimed at promoting organic products are key drivers of this exponential market expansion.

54.2 Organic Market in India

The organic markets in India are largely spread across the food and beverages, health and wellness, beauty and personal care, and textile industries. The highest growth is observed in the organic food segment, followed by textiles, beauty, and personal care (Mitra, 2017). The current Indian domestic market is estimated at INR40,000 million, which is likely to increase by INR100,000 million—INR120,000 million by 2020 with a similar increase in exports (Boland and Singh, 2019). Organic packaged food and beverages is an emerging niche market in India and its primary consumers are in urban areas with fast forward life. The total market size for organic packaged food in India in 2016 was INR533 million, growing at 17% over 2015, and is expected to reach INR871 million by 2021 (Ueasangkomsate and Santiteerakul, 2016).

In addition to the growing domestic market, India is the second largest exporter of organic products in Asia after China (Rogers, 2011). India's exports of organic products have increased by 17% between 2015–2016 and 2016–2017. The increasing export market coupled with the government's support is making organic cultivation in India highly successful. Indian organic food exports were estimated at $299 million during 2015–2016 with total volume of 263,688 MT. The major export destinations were the United States, European Union, Canada, and New Zealand. It is assumed that most of the remaining quantity is sold in local markets. In other words, the sale of the remainder is uncategorized. Oilseeds comprised half of India's overall organic food export, followed by processed food products at 25% (Centric et al., 2015). The growth in the export of organic products was shown in Figure 54.1.

54.2.1 Key Growth Drivers of Organic Cultivation and Markets in India

There are several growth drivers in the organic sector in India; some of the key points include:

- *Increasing health awareness*: In recent years, there has been an increase in health concerns among Indian consumers. They have started giving importance to the nutrient content and the quality of the food they eat, thereby resulting in an increase in the consumption of organic food (Kearney, 2010).
- *Rising disposable income*: Rising income and education levels are leading to an increase in the expenditure of people on health care products. The per capita income of Indians at current prices increased by 10.4% to INR103,007 in 2016–2017, as against INR93,293 during 2015–2016. Average household incomes are expected to triple in the next two decades, making India the world's third-largest consumer economy by 2025, up from the current seventh position (Patidar and Roy, 2017).
- *Urbanization and working population*: Increase in urban population implies growth in per capita purchasing power as well as a change in lifestyle and food habits. Urbanization is growing rapidly and 45% of the Indian population will be residing in urban areas by 2030. About 70% of the consumption growth in India in the next 15 years is expected to come from the working population (people aged 15–59 years). Urban

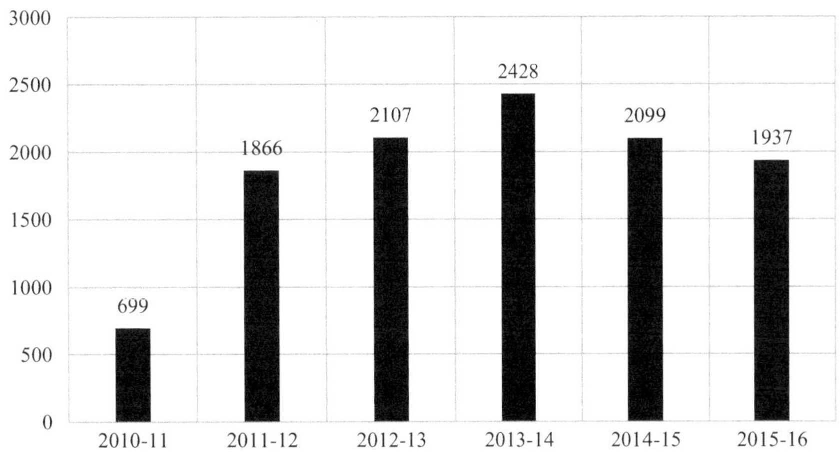

FIGURE 54.1 Growth in the export of organic products. Source: APEDA Annual Reports (14, 15, 16, 17, 18).

consumers are open to paying increased prices for organic food products (Kennedy et al., 2011).

- *Increasing support from the Government of India*: The government of India is promoting organic farming and the consumption of organic food in the country. Financial assistance is provided to farmers who are adopting organic farming under various central sector schemes such as the National Mission for Sustainable Agriculture (NMSA), Mission for Integrated Development of Horticulture (MIDH), National Food Security Mission (NFSM), and Rashtriya Krishi Vikas Yojana (RKVY). In the union budget of 2016, the government of India proposed to allocate 500,000 ha in the country under organic farming and develop value chains in the northeastern region (NER) (Grover and Singh, 2020).
- *Technological developments*: Innovative technological solutions using artificial intelligence, imaging, and renewable energy among others are being developed by private companies for the organic food industry. Ecozen Solutions, a Pune-based startup, has developed solar-powered cold storage for fresh farm produce, which can operate in remote areas with irregular power supply. Hyperspectral cameras are being promoted that use infrared rays to gauge the quality and shelf life of the product. Developments like these on a larger scale may help tackle various challenges faced by the organic sector. Finally, through IT and digital interventions, the traceability challenge of organic products has been resolved (Davis, 2020).

54.3 Export Potential and Market Opportunities

The growth in the organic sector in terms of area, number of producers, and market size grew significantly from 2016 to 2017. The organic agriculture area grew to 69.8 million ha representing a growth of 20 percent or 11.7 million ha over 2016, whereas organic producers grew to 2.9 million ha, which is 5 percent more than in 2016 (Arbenz et al., 2017). As per FiBL survey (2019), India continues to be the country with the highest number of producers (835,000), followed by Uganda (210,352) and Mexico (210,000) in 2016. Currently, the organic exports from India are in the range of Rs. 3450 crores (2017–2018) (Peng, 2019). The market research company Ecovia Intelligence estimates that the global market for organic food reached $97 billion in 2017 (approx. 90 billion euros).

Currently, the percentage of organic processed products export from India is only 5.5% of the total organic food export. Only a limited range of products (mango pulp, puree, by-products of oil crops, soya meal, cakes, and ready-to-eat products) and single processed products such as sugar, tea, edible oils, coffee, and essential oils are exported from India. India can look at exports of a whole range of value-added fruits and vegetables, IQF fruits and vegetables, ready-to-eat products, pickles, soups and sauces, dairy products, processed livestock, aquaculture products, textile products, etc. Some of the ideas shared by stakeholders for promotion of value-added organic exports (Singh, 2018) include:

54.3.1 Marketing and Branding

The organic exporters have suggested that the gestation period for establishing an organic product in the retail market is long and the cost of launching is very high. Support for product registration, support for buying shelf space in the organic retail chains, etc., have been floated by the exporters (Ortega and Tschirley, 2017).

54.3.2 Developing Uniform Quality and Packaging Standards

There is increasing demand for value-added, ready-to-eat, and ethnic food across the globe. In the era of globalization, quality-conscious consumers across the world are becoming aware of the health benefits of what they consume. Offering a one-stop solution and providing a range of ethnic products ranging from healthy and organic to processed and convenient food can boost the organic market. This can create an excellent opportunity for entrepreneurs to leverage the traditional food heritage and invest in innovation in taste, packaging, fusion food, and shelf life of food (Roy, 2020).

54.3.3 Organic Products in North East

Although quality organic products are available in northeastern regions in India, export of these products is not very great. Development of big cooperatives especially in the eastern and northeastern regions can be a game-changer for northeastern states in initiating and boosting the export of organic produce. It will create discipline in procurement and well-structured distribution network for the products from the region. Such cooperatives can undertake grading/sorting/testing of products, which can be further sold or processed (Jackson and Viehoff, 2016).

54.4 Way Forward

A domestic market segment including mass organic products, mid-segment organic products, and premium organic products and e-commerce platforms are opportunities that can be explored. The government is optimistic about organic cultivation and farmers are encouraged through various schemes. However, there is a need to have an organic policy and organic cell to monitor organic cultivation and schemes, and to take necessary steps to discourage the use of fertilizers and pesticide, coupled with the promotion of biofertilizers and biopesticides to decrease the cost of cultivation. The FSSAI notification lays emphasis on appropriate labeling, certification, and district display of organic products. This will enable the creation of further credibility among consumers. Apart from organic food and beverage markets, health and wellness, beauty and personal care market opportunities have grown immensely and good examples of these are companies like Forest Essentials and Himalaya Herbal emerging as successful brands.

REFERENCES

Adikenavar, RN; Monica, M; *A Study on Consumer Buying Behaviour and Marketing Mix at Grami Super Foods*, 2018, Acharya Institute of Technology, Bangalore.

Arbenz, M; Gould, D; Stopes, C; ORGANIC 3.0: The vision of the global organic movement and the need for scientific support. *Organic Agric*, 2017, *7*(3), 199–207.

Azadi, H; Ho, P; Genetically modified and organic crops in developing countries: A review of options for food security. *Biotechnol Adv*, 2010, *28*(1), 160–168.

Bandhu, R; Governments calls it as Investment, T; Yearly Archive, 2019.

Barański, M; Średnicka-Tober, D; Volakakis, N; Seal, C; Sanderson, R; Stewart, GB; Gromadzka-Ostrowska, J; Higher antioxidant and lower cadmium concentrations and lower incidence of pesticide residues in organically grown crops: A systematic literature review and meta-analyses. *British J Nutr*, 2014, *112*(5), 794–811.

Barham, BL; Weber, JG; The economic sustainability of certified coffee: Recent evidence from Mexico and Peru. *World Dev*, 2012, *40*(6), 1269–1279.

Baudron, F; Giller, KE; Agriculture and nature: Trouble and strife? *Biol Conserv*, 2014, *170*,232–245.

Boland, M; Singh, H; *Milk Proteins: From Expression to Food*, 2019, Academic Press, Cambridge, MA.

Centric, C; Login, ES; User, ASI; General info. *Policy*, 2015, *38*, 2020.

Creutzig, F; Ravindranath, NH; Berndes, G; Bolwig, S; Bright, R; Cherubini, F; Fargione, J; Bioenergy and climate change mitigation: An assessment. *GCB Bioenergy*, 2015,*7*(5), 916–944.

Davis, G; Sustaining Queensland's agricultural sector: Challenges and opportunities from the bioeconomy and the circular economy. In *Sustaining Our Environ for Better Future*, 2020 (pp. 117–144), Springer, Singapore.

Eneizan, BM; Critical obstacles to adopt the organic farming in Jordan: From marketing perspective. *Eur J Bus Manage*, 2017, *9*(13), 38–43.

Gomiero, T; Soil degradation, land scarcity and food security: Reviewing a complex challenge. *Sustainability*, 2016, *8*(3), 281.

Grover, A; Singh, RB; Health policy, programmes and initiatives. In *Urban Health and Wellbeing*, 2020 (pp. 251–266), Springer, Singapore.

Grunert, KG; Aachmann, K; Consumer reactions to the use of EU quality labels on food products: A review of the literature. *Food Control*, 2016, *59*, 178–187.

Gunasekaran, V; Murugan, S; Challenges and opportunities of organic products market. *Stud Indian Place Names*, 2020, *40*(18), 2626–2640.

Halder, P; Pati, S; A need for paradigm shift to improve supply chain management of fruits & vegetables in India. *Asian J Agric Rural Dev*, 2011, *1*(393-2016-23908), 1–20.

Jackson, P; Viehoff, V; Reframing convenience food. *Appetite*, 2016, *98*, 1–11.

Jayne, TS; Chamberlin, J; Headey, DD; Land pressures, the evolution of farming systems, and development strategies in Africa: A synthesis. *Food Policy*, 2014, *48*, 1–17.

Kavitha, MS; A study on customers' attitude towards selected agricultural organic food product with special reference to Coimbatore city. *Int J Res Social Sci*, 2016, *6*(9), 1–7.

Kearney, J; Food consumption trends and drivers. *Philos Trans R Soc B: Biol Sci*, 2010, *365*(1554), 2793–2807.

Kennedy, L; Robbins, G; Scott, D; Sutherland, C; Denis, E; Andrade, J; Bon, B; *The Politics of Large-Scale Economic and Infrastructure Projects in Fast-Growing Cities of the South*, Literature Review, 3, 2011.

Kushwah, S; Dhir, A; Sagar, M; Understanding consumer resistance to the consumption of organic food. A study of ethical consumption, purchasing, and choice behaviour. *Food Qual Preference*, 2019, *77*, 1–14.

Luik, A; Pehme, S; Matt, D; Peetsmann, E; Rembialkowska, E; Quality of organic vs. conventional food and effects on health report, 2011.

Matt, D; Rembialkowska, E; Luik, A; Peetsmann, E; Pehme, S; Quality of Organic vs. Conventional Food and Effects on Health Report. Williams, Ingrid Helvi (Ed.) (2011). *Estonian University of Life Sciences*, Tartu, Estonia. 106pp.

Mehdi, A; Tomar, P; Chaudhry, D; Joshi, P; *Streamlining Food Safety Compliance Ecosystem in India: Facilitating Ease of Doing Business in India's Food Sector.* Report- Indian Council for Research on International Economic Relations. AF PRESS No. 35, Sector 7, IMT Manesar, Gurugram - 122050. Haryana (India). 54pp. http://hdl.handle.net/11540/10515, 2019.

Méndez, VE; Bacon, CM; Olson, M; Petchers, S; Herrador, D; Carranza, C; Mendoza, A; Effects of fair trade and organic certifications on small-scale coffee farmer households in Central America and Mexico. *Renewable Agric Food Syst*, 2010, *25*(3), 236–251.

Mgbenka, RN; Onwubuya, EA; Ezeano, CI; Organic farming in Nigeria: Need for popularization and policy. *World J Agric Sci*, 2015, *11*(6), 346–355.

Mitra, SK; *Politics in India: Structure, Process and Policy*, 2017, Routledge, London.

Miyasaka, T; Kulkarni, A; Kim, GM; Öz, S; Jena, AK; Perovskite solar cells: Can we go organic-free, lead-free, and dopant-free? *Adv Energy Mater*, 2020, *10*(13), 1902500.

Mukherjee, A; Dutta, S; Mendiratta, S; Kapoor, A; Goyal, TM; Promoting organic food products and exports: Status, issues and way forward. Working Paper, No. 342, 2017.

Nandwani, D; Nwosisi, S; Global trends in organic agriculture. In *Organic Farming for SustainableAgriculture*, 2016 (pp. 1–35), Springer, Cham.

Ortega, DL; Tschirley, DL; Demand for food safety in emerging and developing countries: A research agenda for Asia and Sub-Saharan Africa, *Journal of Agribusiness in Developing and Emerging Economies*, 2017, *7*(1), 21–34.

Panneerselvam, P; Halberg, N; Lockie, S; Consequences of organic agriculture for smallholder farmers' livelihood and food security. In *Organic Agriculture for Sustainable Livelihoods*, 2012 (pp. 37–60), Routledge, London.

Patidar, G; Roy, R; Contribution of smart phones in India's ecommerce growth. *Asian J Res Mark*, 2017, *6*(5), 67–76.

Pearson, D; Henryks, J; Jones, H; Organic food: What we know (and do not know) about consumers. *Renewable Agric Food Syst*, 2011, *26*(2), 171–177.

Peng, M; The growing market of organic foods: Impact on the US and global economy. In *Safety and Practice for Organic Food*, 2019 (pp. 3–22), Academic Press, Cambridge, MA.

Poudel, KL; Yamamoto, N; Johnson, TG; Comparing technical efficiency of organic and conventional coffee farms in Nepal using data envelopment analysis (DEA) approach, In Conference on International Association of Agricultural Economists (IAAE), August 18–24, 2012, Foz do Iguaçu, Brazil,2012, *1007*, 2016-79397.

Pouta, E; Heikkilä, J; Forsman-Hugg, S; Isoniemi, M; Mäkelä, J; Consumer choice of broiler meat: The effects of country of origin and production methods. *Food Qual Preference*, 2010, *21*(5), 539–546.

Rana, J; Paul, J; Consumer behavior and purchase intention for organic food: A review and research agenda. *J Retail Consum Serv*, 2017, *38*, 157–165.

Rao, HS; Rao, DS; Production and marketing of coffee in high altitude zone of Visakhapatnam. In National Symposium on the Choice of Farming Models for Income Security: Towards Equity and Sustainability, 2019, 170.

Reganold, JP; Wachter, JM; Organic agriculture in the twenty-first century. *Nat Plants*, 2016, *2*(2), 1–8.

Rogers, H; Shale gas- the unfolding story. *Oxford Rev Econ Policy*, 2011, *27*(1), 117–143.

Roitner-Schobesberger, B; Darnhofer, I; Somsook, S; Vogl, CR; Consumer perceptions of organic foods in Bangkok, Thailand. *Food Policy*, 2008, *33*(2), 112–121.

Ronald, PC; Adamchak, RW; *Tomorrow's Table: Organic Farming, Genetics, and the Future of Food*, 2017, Oxford University Press, Oxford, UK.

Roy, MP; Improving nutrition in India: Encouraging informed food choices. *BMJ*, 2020, 368.

Seufert, V; Ramankutty, N; Foley, JA; Comparing the yields of organic and conventional agriculture. *Nature*, 2012, *485*(7397), 229–232.

Seufert, V; Ramankutty, N; Mayerhofer, T; What is this thing called organic?—How organic farming is codified in regulations. *Food Policy*, 2017, *68*, 10–20.

Shetty, PK; Hiremath, MB; Murugan, M; Status of organic forming in agro ecosystems in India. *Indian J Sci Technol*, 2013, *6*(8), 5083–5088.

Sillani, S; Nassivera, F; Consumer behavior in choice of minimally processed vegetables and implications for marketing strategies. *Trends Food Sci Technol*, 2015, *46*(2), 339–345.

Singh, C; India since demonetisation. IIM Bangalore Research Paper, 567, 2018.

Thøgersen, J; Country differences in sustainable consumption: The case of organic food. *J Macromarketing*, 2010, *30*(2), 171–185.

Trauger, A; Passidomo, C; Towards a post-capitalist-politics of food: cultivating subjects of community economies. *ACME: Int E-J Crit Geographies*, 2012, *11*(2), 282–303.

Ueasangkomsate, P; Santiteerakul, S; A study of consumers' attitudes and intention to buy organic foods for sustainability. *Procedia Environ Sci*, 2016, *34*, 423–430.

Wani, SP; Jakkula, VS; Singh, D; Doubling farmers' income: KISAN–MITrA. In Proceedings of National Workshop on Doubling Farmers' Income through Scaling up: KISAN–MITrA (Knowledge-based Integrated Sustainable Agriculture, Network–Mission India for Transforming Agriculture), 2017.

Willer, H; Lernoud, J; The world of organic agriculture—Statistics and emerging trends. In FiBl and IFOAM-Organics International, Frick. Switzerland, 2019, 346.

Xie, B; Qin, J; Yang, H; Wang, X; Wang, YH; Li, TY; Organic aquaculture in China: A review from a global perspective. *Aquaculture*, 2013, *414*, 243–253.

Index

J

K

L

M

R

S